W9-CDL-495

Essential # UNIVERSITY PHYSICS

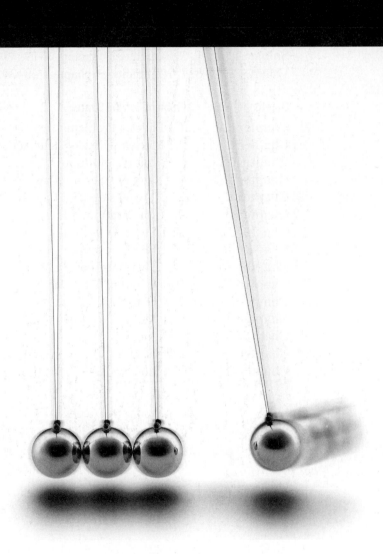

Problem Solving Strategies

Volume 1 (pp. 1–326) includes chapters 1–19.
Volume 2 (pp. 327–727) includes chapters 20–39.

Essential UNIVERSITY PHYSICS

FIRST EDITION

Volume 1 **Richard Wolfson**
Chapters 1–19 *Middlebury College*

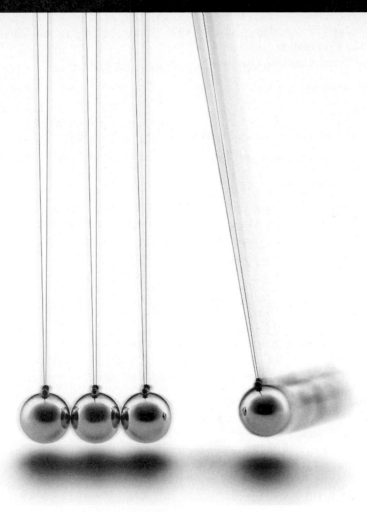

PEARSON
Addison
Wesley

San Francisco Boston New York
Cape Town Hong Kong London Madrid Mexico City
Montreal Munich Paris Singapore Tokyo Toronto

Editor-in-Chief: Adam Black, Ph.D.
Development Editor: Brad Patterson
Project Manager: Laura Kenney
Project Editor: Martha Steele
Managing Producer: Claire Masson
Director of Marketing: Christy Lawrence
Managing Editor: Corinne Benson
Production Supervisor: Nancy Tabor
Production Service: Elm Street Publishing Services, Inc.
Composition: Preparé, Inc.
Text Design: Elm Street Publishing Services, Inc.
Cover Design: Hespenheide Design
Illustrations: Rolin Graphics
Photo Research: Cypress Integrated Systems
Director, Image Resource Center: Melinda Patelli
Rights and Permissions Manager: Zina Arabia
Cover Printer: Phoenix Color Corporation
Printer and Binder: Von Hoffmann Graphics Owensville

Cover Image: Tyler Boley, Workbook

Photo Credits: See page C-1.

ISBN 0-8053-3829-2 Volume 1 without MasteringPhysics

ISBN 0-8053-3978-7 Volume 1 with MasteringPhysics

ISBN 0-8053-3981-7 Volume 1 Professional copy

Copyright © 2007 Pearson Education, Inc., publishing as Addison-Wesley 1301 Sansome St.,
San Francisco, CA 94111. All rights reserved. Manufactured in the United States of America.
This publication is protected by Copyright and permission should be obtained from the
publisher prior to any prohibited reproduction, storage in a retrieval system, or transmission
in any form or by any means, electronic, mechanical, photocopying, recording, or likewise.
To obtain permission(s) to use material from this work, please submit a written request to
Pearson Education, Inc., Permissions Department, 1900 E. Lake Ave., Glenview, IL 60025.
For information regarding permissions, call 847/486/2635.

Many of the designations used by manufacturers and sellers to distinguish their products are
claimed as trademarks. Where those designations appear in this book, and the publisher was
aware of a trademark claim, the designations have been printed in initial caps or all caps.

Library of Congress Cataloging-in-Publication Data
Wolfson, Richard.
    Essential university physics
        p. cm.
    Includes index.
    ISBN 0-8053-9212-2
        1. Physics—Textbooks.   I. Title.

QC21.3.W65 2007
530—dc22   2007040286

    6 7 8 9 10 — 0928 — 09

www.aw.com/physics

Brief Contents

About the Author

Richard Wolfson

Richard Wolfson is the Benjamin F. Wissler Professor of Physics at Middlebury College, where he has taught since 1976. He did undergraduate work at MIT and Swarthmore College, and he holds an M.S. degree from the University of Michigan and Ph.D. from Dartmouth. His ongoing research on the Sun's corona and climate change has taken him to sabbaticals at the National Center for Atmospheric Research in Boulder, Colorado; St. Andrews University in Scotland; and Stanford University.

Rich is a committed and passionate teacher. This is reflected in his many publications for students and the general public, including the video series *Einstein's Relativity and the Quantum Revolution: Modern Physics for Nonscientists* (The Teaching Company, 1999) and *Physics in Your Life* (The Teaching Company, 2004); books *Nuclear Choices: A Citizen's Guide to Nuclear Technology* (MIT Press, 1993), *Simply Einstein: Relativity Demystified* (W. W. Norton, 2003), and *Energy, Environment, and Climate* (W. W. Norton, 2007); and articles for *Scientific American* and the *World Book Encyclopedia*.

Outside of his research and teaching, Rich enjoys hiking, canoeing, gardening, cooking, and watercolor painting.

Introductory physics texts have grown ever larger, more massive, more encyclopedic, more colorful, and more expensive. *Essential University Physics* bucks that trend—without compromising coverage, pedagogy, or quality. The text benefits from the author's three decades of teaching introductory physics, seeing firsthand the difficulties and misconceptions that students face as well as the "Got It!" moments when big ideas become clear. It also builds of the author's honing multiple editions of a previous calculus-based textbook and on feedback from hundreds of instructors and students.

Goals of This Book

Physics is the fundamental science, at once fascinating, challenging, and subtle—and yet simple in a way that reflects the few basic principles that govern the physical universe. My goal is to bring this sense of physics alive for students in a range of academic disciplines who need a solid calculus-based physics course—whether they're engineers, physics majors, premeds, biologists, chemists, geologists, mathematicians, computer scientists, or other majors. My own courses are populated by just such a variety of students, and among my greatest joys as a teacher is having students who took a course only because it was required say afterwards that they really enjoyed their exposure to the ideas of physics. More specifically, my goals include:

- Helping students build the analytical and quantitative skills and confidence needed to apply physics in problem-solving for science and engineering.

- Adressing key misconceptions and helping students build a stronger conceptual understanding.

- Helping students see the relevance and excitement of the physics they're studying with wide-ranging and contemporary applications in science, technology, and everyday life.

- Helping students develop an appreciation of the physical universe at its most fundamental level.

- Engaging students with an informal, conversational writing style that balances precision with approachability.

Pedagogical Innovations

This book is *concise*, but it's also *progressive* in its embrace of proven techniques from physics education research and *strategic* in its approach to learning physics. Chapter 1 introduces the IDEA framework for problem solving, and every one of the book's subsequent **worked examples** employs this framework. IDEA—an acronym for Identify, Develop, Evaluate, Assess—is not a "cookbook" method for students to apply mindlessly, but rather a tool for organizing students' thinking and discouraging equation hunting. It begins with an interpretation of the problem and an identification of the key physics concepts involved; develops a plan for reaching the solution; carries out the mathematical evaluation; and assesses the solution to see that it makes sense, to compare the example with others, and to mine additional insights into physics. In nearly all of the text's worked examples, the Develop phase includes making a drawing, and most of these use a hand-drawn style to encourage students to make their own drawings—a step that research suggests they often skip. IDEA provides a common approach to all physics problem solving, an approach that emphasizes the conceptual unity of physics and helps break the

- Complete edition Volumes 1–2 (shrinkwrapped) (ISBN 0-8053-9212-2): Chapters 1–39
- Volume 1 (ISBN 0-8053-3829-2): Chapters 1–19
- Volume 2 (ISBN 0-8053-3838-1): Chapters 20–39
- Complete edition Volumes 1–2 (shrinkwrapped) with MasteringPhysics (ISBN 0-8053-0296-4): Chapters 1–39
- Volume 1 with MasteringPhysics™ (ISBN 0-8053-3978-7): Chapters 1–19
- Volume 2 with MasteringPhysics™ (ISBN 0-8053-4004-1): Chapters 20–39

typical student view of physics as a hodgepodge of equations and unrelated ideas. In addition to IDEA-based worked examples, other pedagogical features include:

- **Problem-Solving Strategy boxes** that follow the IDEA framework to provide detailed guidance for specific classes of physics problems, such as Newton's second law, conservation of energy, thermal-energy balance, Gauss's law, or multiloop circuits.

- **Tactics boxes** that reinforce specific essential skills such as differentiation, setting up integrals, vector products, drawing free-body diagrams, simplifying series and parallel circuits, or ray tracing.

- **Got It? boxes** that provide quick checks for students to test their conceptual understanding. Many of these use a multiple-choice or quantitative ranking format to probe student misconceptions and facilitate their use with classroom-response systems.

- **Tips** that provide helpful problem-solving hints or warn against common pitfalls and misconceptions.

- **Chapter openers** that include a forward-looking **To Learn** list of new concepts and skills in the chapter ahead, and a backward-looking **To Know** list of important ideas on which the chapter builds. Both lists reference specific chapter sections by number.

- **Applications**, self-contained presentations typically shorter than half a page, provide interesting and contemporary instances of physics in the real world, such as: bicycle stability; flywheel energy storage; laser vision correction; ultracapacitors; wind energy; magnetic resonance imaging; global climate change; combined-cycle power generation; circuit models of the cell membrane; CD, DVD, and "Blu-Ray" technologies; and radiocarbon dating.

- **For Thought and Discussion** questions at the end of each chapter designed for peer learning or for self-study to enhance students' conceptual understanding of physics.

- **Annotated figures** that adopt the research-based approach of including simple "instructor's voice" commentary to help students read and interpret pictorial and graphical information.

- **End-of-chapter** problems that begin with simpler exercises keyed to individual chapter sections and ramp up to more challenging and often multistep problems that synthesize chapter material. At the end are context-rich problems focusing on real-world situations.

- **Chapter summaries** that combine text, art, and equations to provide a synthesized overview of each chapter. Each summary is hierarchical, beginning with the chapter's "big picture" ideas, then focusing on key concepts and equations, and ending with a list of "Cases and Uses"—specific instances or applications of the physics presented in the chapter.

Organization

This contemporary book is *concise*, *strategic*, and *progressive*, but it's *traditional* in its organization. Following the introductory Chapter 1, the book is divided into six parts. Part One (Chapters 2–12) develops the basic concepts of mechanics, including Newton's laws and conservation principles as applied to single particles and multiparticle systems. Part Two (Chapters 13–15) extends mechanics to oscillations, waves, and fluids. Part Three (Chapters 16–19) covers thermodynamics. Part Four (Chapters 20–29) deals with electricity and magnetism. Part Five (Chapters 30–32) treats optics, first in the geometrical optics approximation and then including wave phenomena. Part Six (Chapters 33–39) introduces relativity and quantum physics. Each part begins with a brief description of its coverage, and ends with a conceptual summary and a challenge problem that synthesizes ideas from several chapters.

Essential University Physics is available in two paperback volumes, so students can purchase only what they need—making the low-cost aspect of this text even more attractive. Volume 1 includes Parts One, Two, and Three, mechanics through thermodynamics. Volume 2 contains Parts Four, and Five, and Six, electricity and magnetism along with optics and modern physics.

Instructor Supplements

The **Instructor Solutions Manual, Chapters 1–39** (ISBN 0-8053-0342-1), written by Edw. S. Ginsberg, University of Massachusetts, with contributing author Sen-Ben Liao, University of California, provides complete and detailed solutions to all end-of-chapter problems. Solutions to odd-numbered problems follow the Interpret/Develop/Evaluate/Assess problem-solving framework used in the textbook.

The cross-platform **Instructor Resource CD-ROMs** (ISBN 0-8053-4008-4) consist of the **Simulation and Image Presentation CD-ROM** and the **Instructor Supplement CD-ROM**. The *Simulation and Image Presentation CD-ROM* provides all line figures and tables from the textbook in JPEG format. In addition, all the key equations, problem-solving strategies, and chapter summaries are provided in multiple editable Word formats. In-class weekly multiple-choice questions for use with various Classroom Response Systems (CRS) are also provided, based on the Got It? questions in the text. The *Instructor Supplement CD-ROM* provides the Instructor Solutions Manual in convenient editable multiple Word formats and as PDFs.

MasteringPhysics™ (www.masteringphysics.com) is the most advanced, educationally effective, and widely used physics homework and tutorial system in the world. It provides instructors with a library of extensively pretested end-of-chapter problems and rich, Socratic tutorials that incorporate a wide variety of answer types, wrong-answer feedback, and adaptive help (comprising hints or simpler subproblems upon request). MasteringPhysics™ allows instructors to quickly build wide-ranging homework assignments of just the right difficulty and duration and provides them with efficient tools to analyze class trends—or the work of any student—in unprecedented detail and to compare the results either with the national average or with the performance of previous classes.

The **Transparency Acetates** (ISBN 0-8053-0404-5) provide more than 200 key figures from *Essential University Physics,* 1st edition.

The **Printed Test Bank** (ISBN 0-8053-0404-5) provides more than 2000 multiple-choice, true-false, and conceptual questions.

The **Computerized Test Bank** (ISBN 0-8053-0401-0) provides all of the questions from the Printed Test Bank on a cross-platform CD-ROM. More than half of the questions can be assigned with randomized numerical values.

Student Supplements

Student Solutions Manuals, Chapters 1–19 (ISBN 0-8053-4009-2) and **Chapters 20–39** (ISBN 0-8053-0405-3), written by Edw. S. Ginsberg, University of Massachusetts, with contributing author Sen-Ben Liao, University of California, provide detailed, step-by-step solutions to all of the odd-numbered and selected other end-of-chapter problems from the textbook. These solutions consistently follow the same Interpret/Develop/Evaluate/Assess problem-solving framework used in the textbook, reinforcing good problem-solving techniques.

MasteringPhysics™ (www.masteringphysics.com) is the most advanced, widely used, and educationally proven physics tutorial system in the world. It is the result of eight years of detailed studies of how real students work physics problems and of precisely where they need help. Studies show that students who use MasteringPhysics™ significantly improve their scores on final exams and conceptual tests such as the Force Concept Inventory. MasteringPhysics™ achieves this improvement by providing students with instantaneous feedback specific to their wrong answers, simpler subproblems upon request when they get stuck, and partial credit for their method(s) used. This individualized, 24/7 tutor system is recommended by 9 out of 10 students to their peers as the most effective and time-efficient way to study.

Acknowledgments

A project of this magnitude isn't the work of its author alone. First and foremost among those I thank for their contributions are the now several thousand students I've taught in calculus-based introductory physics courses at Middlebury College. Over the years your questions have taught me how to convey physics ideas in many different ways appropriate to your diverse learning styles. You've helped identify the "sticking points" that challenge introductory physics students, and you've showed me ways to help you avoid and "unlearn" the misconceptions that many students bring to introductory physics.

Thanks also go to my Middlebury faculty colleagues and to numerous instructors and students from around the world who have contributed valuable suggestions that were incorporated in the revisions of my earlier introductory physics text, *Physics for Scientists and Engineers* (Wolfson and Pasachoff, third edition: Addison-Wesley, 1999). I've heard you, and you'll find still more of your suggestions implemented in *Essential University Physics*.

Experienced physics instructors thoroughly reviewed every chapter of this book, and reviewers' comments resulted in substantive changes—and sometimes in major rewrites—to the first drafts of the manuscript. We list all these reviewers below. But first, special thanks are due to six individuals who made exceptional contributions to the quality and in some cases the very existence of this book. First is Professor Jay Pasachoff of Williams College, whose willingness more than two decades ago to take a chance on an inexperienced coauthor has made writing introductory physics a large part of my professional career. Dr. Adam Black, physics editor and Ph.D. physicist at Addison-Wesley, had the vision to see promise in a new introductory text that would respond to the rising chorus of complaints about massive, encyclopedic, and expensive physics texts. Brad Patterson, the book's development editor, brought his graduate-level knowledge of physics to a role that made him a real collaborator and the closest this book has to a coauthor. Brad is responsible for many of the book's innovative features, and it was a pleasure to work with him. Mark Hollabaugh of Normandale Community College is responsible for most of the new context-rich problems. We've gone to great lengths to make this book as error-free as possible, and much of the credit for that happy situation goes to Linda Barton of the Rochester Institute of Technology. Not only did she check the numbers for every worked example, but she also read the entire book in page proof with a professionally critical eye, and she's responsible for many improvements to both text and art made even at that late stage. Very special thanks are due to Edw. Ginsberg, of the University of Massachusetts, with whom I've collaborated since the first edition of *Physics for Scientists and Engineers*. Ed developed new calculus-based problems for this book and, most important, provided valuable suggestions for the entire book when it was in the final manuscript phase. Ed Ginsberg's stamp is on every chapter, and readers will be better served because of his contributions.

Finally, I thank Martha Steele, Nancy Tabor, and Laura Kenney at Addison-Wesley, and Heather Johnson of Elm Street Publishing Services, for their highly professional efforts in shepherding this book through its vigorous production schedule. And, as always, I thank my family, my colleagues, and my students for the patience they showed during the intensive process of writing this book.

Reviewers

John R. Albright, *Purdue University–Calumet*
Rama Bansil, *Boston University*
Richard Barber, *Santa Clara University*
Linda S. Barton, *Rochester Institute of Technology*

Rasheed Bashirov, *Albertson College of Idaho*
Chris Berven, *University of Idaho*
David Bixler, *Angelo State University*
Ben Bromley, *University of Utah*
Charles Burkhardt, *St. Louis Community College*
Susan Cable, *Central Florida Community College*
George T. Carlson, Jr., *West Virginia Institute of Technology–West Virginia University*
Catherine Check, *Rock Valley College*
Norbert Chencinski, *College of Staten Island*
Carl Covatto, *Arizona State University*
David Donnelly, *Texas State University–San Marcos*
David G. Ellis, *University of Toledo*
Tim Farris, *Volunteer State Community College*
Paula Fekete, *Hunter College of The City University of New York*
Idan Ginsburg, *Harvard University*
James Goff, *Pima Community College*
Mark Hollabaugh, *Normandale Community College*
Rex W. Joyner, *Indiana Institute of Technology*
Nikos Kalogeropoulos, *Borough of Manhattan Community College–The City University of New York*
Kevin T. Kilty, *Laramie County Community College*
Duane Larson, *Bevill State Community College*
Kenneth W. McLaughlin, *Loras College*
Tom Marvin, *Southern Oregon University*
Perry S. Mason, *Lubbock Christian University*
Mark Masters, *Indiana University–Purdue University Fort Wayne*
Jonathan Mitschele, *Saint Joseph's College*
Gregor Novak, *United States Air Force Academy*
Richard Olenick, *University of Dallas*
Robert Philbin, *Trinidad State Junior College*
Russell Poch, *Howard Community College*
Steven Pollock, *Colorado University–Boulder*
James Rabchuk, *Western Illinois University*
George Schmiedeshoff, *Occidental College*
Natalia Semushkina, *Shippensburg University of Pennsylvania*
Anwar Shiekh, *Dine College*
David Slimmer, *Lander University*
Chris Sorensen, *Kansas State University*
Ronald G. Tabak, *Youngstown State University*
Gajendra Tulsian, *Daytona Beach Community College*
Henry Weigel, *Arapahoe Community College*
Arthur W. Wiggins, *Oakland Community College*
Fredy Zypman, *Yeshiva University*

Preface to the Student

Welcome to physics! Maybe you're taking introductory physics because you're majoring in a field of science or engineering that requires a semester or two of physics. Maybe you're premed, and you know that medical schools are increasingly interested in seeing calculus-based physics on your transcript. Perhaps you're really gung-ho and plan to major in physics. Or maybe you want to study physics further as a minor associated with related fields like math or chemistry or to complement a discipline like economics, environmental studies, or even music. Perhaps you had a great high-school physics course, and you're eager to continue. Maybe high-school physics was an academic disaster for you, and you're approaching this course with trepidation. Or perhaps this is your first experience with physics. Whatever your reason for taking introductory physics, welcome!

And whatever your reason, my goals for you are similar: I'd like to help you develop an understanding and appreciation of the physical universe at a deep and fundamental level; I'd like you to become aware of the broad range of natural and technological phenomena that physics can explain; and I'd like to help you strengthen your analytic and quantitative problem-solving skills. Even if you're studying physics only because it's a requirement, I want to help you engage the subject and come away with an appreciation for this fundamental science and its wide applicability. One of my greatest joys as a physics teacher is having students tell me after the course that they had taken it only because it was required, but found they really enjoyed their exposure to the ideas of physics.

Physics is fundamental. To understand physics is to understand how the world works, both in everyday life and on scales of time and space so small and so large as to defy intuition. For that reason I hope you'll find physics fascinating. But you'll also find it challenging. Learning physics will challenge you with the need for precise thinking and language; with subtle interpretations of even commonplace phenomena; and with the need for skillful application of mathematics. But there's also a simplicity to physics, a simplicity that results because there are in physics only a very few really basic principles to learn. Those succinct principles encompass a universe of natural phenomena and technological applications.

I've been teaching introductory physics for decades, and this book distills everything my students have taught me about the many different ways to approach physics; about the subtle misconceptions students often bring to physics; about the ideas and types of problems that present the greatest challenges; and about ways to make physics engaging, exciting, and relevant to your life and interests.

I have some specific advice for you that grows out of my long experience teaching introductory physics. Keeping this advice in mind will make physics easier (but not necessarily easy!), more interesting, and, I hope, more fun:

- *Read* each chapter thoroughly and carefully before you attempt to work any problem assignments. I've written this text with an informal, conversational style to make it engaging. It's not a reference work to be left alone until you need some specific piece of information; rather, it's an unfolding "story" of physics—its big ideas and their applications in quantitative problem solving. You may think physics is hard because it's mathematical, but in my long experience I've found that failure to *read* thoroughly is the biggest single reason for difficulties in introductory physics.

- *Look for the big ideas*. Physics isn't a hodge-podge of different phenomena, laws, and equations to memorize. Rather, it's a few big ideas from which flow myriad applications, examples, and special cases. In particular, don't think of physics as a jumble of equations that you choose among when solving a problem. Rather, identify those few big ideas and the equations that represent them, and try to see how seemingly distinct examples and special cases relate to the big ideas.

■ *When working problems*, *re-read* the appropriate sections of the text, paying particular attention to the worked examples. Follow the IDEA strategy described in Chapter 1 and used in every subsequent worked example. Don't skimp on the final Assess step. Always ask: Does this answer make sense? How can I understand my answer in relation to the big principles of physics? How was this problem like others I've worked, or like examples in the text?

■ *Don't confuse physics with math*. Mathematics is a tool, not an end in itself. Equations in physics aren't abstract math, but statements about the physical world. Be sure you understand each equation for what it says about physics, not just as an equality between mathematical terms.

■ *Work with others*. Getting together informally in a room with a blackboard is a great way to explore physics, to clarify your ideas and help others clarify theirs, and to learn from your peers. I urge you to discuss physics problems together with your classmates, to contemplate together the "For Thought and Discussion" questions at the end of each chapter, and to engage one another in lively dialog as you grow your understanding of physics, the fundamental science.

Detailed Contents

PART FIVE

Optics   529

1 Doing Physics

■ Which realms of physics are involved in the workings of your DVD player?

▶ **To Learn**
By the end of this chapter you should be able to
■ Describe the different realms of physics and their applications in both natural and technological systems (1.1).
■ Explain the SI unit system and convert units (1.2).
■ Express and manipulate numbers using scientific notation (1.3).
■ Explain the importance of accuracy and significant figures and handle them in calculations (1.3).
■ Make quick order-of-magnitude estimates (1.3).
■ Describe the strategic steps used in solving physics problems (1.4).

◀ **To Know**
■ Physics is a quantitative science, and we'll begin right off using algebra. Soon we'll add trigonometry and later calculus (which you might be studying now, concurrently with physics). However, you don't need to have taken physics to get a full understanding from this book.

You slip a DVD into your player and settle in to watch a favorite movie. The DVD spins, and a precisely focused laser beam "reads" its content. Electronic circuitry processes the information, sending it to your video display and to loudspeakers that turn electrical signals into sound waves. Every step of the way, the principles of physics are behind the delivery of the movie from DVD to you.

1.1 Realms of Physics

That DVD player is a metaphor for all of **physics**—the science that describes the fundamental workings of physical reality. Physics explains natural phenomena ranging from the behavior of atoms and molecules to thunderstorms and rainbows and on to the evolution of stars, galaxies, and the universe itself. Technological applications of physics are the basis for everything from microelectronics to medical imaging to cars, airplanes, and space flight.

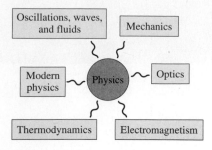

FIGURE 1.1 Realms of physics.

At its most fundamental, physics provides a nearly unified description of all physical phenomena. However, it's convenient to divide physics into several distinct realms (Fig. 1.1). Your DVD player encompasses essentially all of those realms. **Mechanics**, the branch of physics that deals with motion, describes the spinning disc. Mechanics also explains the motion of a car, the orbits of the planets, and the stability of a skyscraper. Part 1 of this book deals with the basic ideas of mechanics.

Those sound waves coming from your loudspeakers represent **wave motion**. Other examples include the ocean waves that pound Earth's coastlines, the wave of standing spectators that sweeps through a football stadium, and the undulations of Earth's crust that spread the energy of an earthquake. Part 2 of this book covers wave motion and other phenomena involving the motion of fluids like air and water.

When you burn your own CD or DVD, the high temperature produced by an intensely focused laser beam alters the material properties of a writable CD or DVD, thus storing audio, video, or computer information. That's an example of **thermodynamics**—the study of heat and its effects on matter. Thermodynamics also describes the delicate balance of energy-transfer processes that keeps our planet at a habitable temperature and puts serious constraints on our ability to meet the burgeoning energy demands of modern society. Part 3 comprises four chapters on thermodynamics.

What spins your DVD? An electric motor, a device that converts electrical energy to the energy of motion. Electric motors are ubiquitous and essential in modern society, running everything from subway trains to washing machines to your computer's hard drive. Conversely, electric generators convert the energy of motion to electricity, providing virtually all of our electrical energy. Motors and generators are two applications of **electromagnetism** in modern technology. Others include computers, audiovisual electronics, microwave ovens, digital watches, and even the humble light bulb; without these electromagnetic technologies our lives would be very different. Equally electromagnetic are all the wireless technologies that enable modern communications, from satellite TV to cell phones to wireless computer networks, mice, and keyboards. And even light itself is an electromagnetic phenomenon. Part 4 presents the principles of electromagnetism and their many applications.

The precise focusing of laser light in your DVD player allows hours of video to fit on a small plastic disc. The details and limitations of that focusing are governed by the principles of **optics**, the study of light and its behavior. Applications of optics range from simple magnifiers to contact lenses to sophisticated instruments such as microscopes, telescopes, and spectrometers. Optical fibers carry your e-mail, web pages, and music downloads over the global Internet. Natural optical systems include your eye and the raindrops that deflect sunlight to form rainbows. Part 5 of the book explores optical principles and their applications.

That laser light in your DVD player is an example of an electromagnetic wave, but an atomic-level look at the light's interaction with matter reveals particle-like "bundles" of electromagnetic energy. This is the realm of **quantum physics**, which deals with the often counterintuitive behavior of matter and energy at the atomic level. Quantum phenomena also explain how that DVD laser works and, more profoundly, the structure of atoms and the periodic arrangement of the elements that is the basis of all chemistry. Quantum physics is one of the two great developments of **modern physics**. The other is Einstein's **theory of relativity**. Relativity and quantum physics arose during the 20th century, and together they've radically altered our commonsense notions of time, space, and causality. Part 6 of the book surveys the ideas of modern physics.

1.2 Measurements and Units

"A long way" means different things to a sedentary person, a marathon runner, a pilot, and an astronaut. We need to quantify our measurements. Today, all of science uses the **metric system**, in which the fundamental quantities length, mass, and time are measured in meters, kilograms, and seconds, respectively. The modern version of the metric system is **SI**, for Système International d'Unités (International System of Units), which incorporates scientifically precise definitions of the fundamental quantities.

Length

The **meter** was first defined as one ten-millionth of the distance from the equator to the north pole. In 1889 a standard meter was fabricated to replace the Earth-based unit, and in 1960 that gave way to a standard based on the wavelength of light. Such an **operational definition**, a measurement standard based on a laboratory procedure, has the advantage that scientists anywhere can reproduce the standard meter. More recently, even this wavelength standard became insufficient. By the 1970s, the speed of light had become one of the most accurately determined quantities. As a result, in 1983 the meter was given a new operational definition:

> The meter is the length of the path traveled by light in vacuum during a time interval of 1/299,792,458 of a second.

This definition of the meter also means that the speed of light is now a defined quantity; its value is exactly 299,792,458 m/s.

Time

The **second** used to be defined by Earth's rotation until it was recognized that Earth's rotation is slowing slightly. The second was then redefined as a specific fraction of the year 1900. In 1967 the second was again redefined, in terms of the vibrations of a specific atom, to give our current definition:

> The second is the duration of 9,192,631,770 periods of the radiation corresponding to the transition between the two hyperfine levels of the ground state of the cesium-133 atom.

The device that implements this definition—which will seem a lot less obscure once you've studied some atomic physics—is called an atomic clock. Like the definition of the meter, the atomic-clock definition of the second is reproducible in any laboratory with the appropriate equipment.

Mass

The standard of mass is now the least satisfactory. Unlike the operational definitions of length and time, which are based on procedures that can be repeated by scientists anywhere, the unit of mass is defined in terms of a particular object—the international prototype **kilogram** kept at the International Bureau of Weights and Measures at Sèvres, France.

The prototype kilogram is made of a special platinum-iridium alloy that is very hard, not subject to corrosion, and very dense. Nevertheless, it could conceivably change, and in any event comparison with such a standard is less convenient than an operational definition that can be checked in a laboratory. So scientists are working on techniques based on counting the number of silicon atoms in a given volume, to scale up from the mass of a single atom to a new definition of the kilogram.

Other SI Units

SI includes seven independent base units: In addition to the three we've just defined, there are the ampere (A) for electric current, the kelvin (K) for temperature, the mole (mol) for the amount of a substance, and the candela (cd) for luminosity. Two supplementary units are used to measure angle: the radian (rad) for ordinary angles (Fig. 1.2) and the steradian (sr) for solid angles. Units for all other quantities are derived from these base units.

You could specify the length of a bacterium (e.g., 0.00001 m) or the distance to the next city (e.g., 58,000 m) in meters, but the results are unwieldy—too small in the first case and too large in the latter. So we use prefixes to indicate multiples of the SI base units. For example, the prefix k (for "kilo") means 1000; 1 km is 1000 m, and the distance to the next

The angle θ in radians is defined as the ratio of the subtended arc length s to the radius r: $\theta = \frac{s}{r}$.

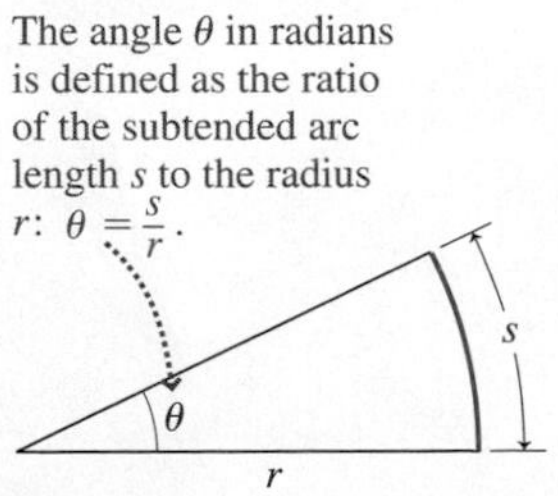

FIGURE 1.2 The radian is the SI unit of angle.

TABLE 1.1 SI Prefixes

Prefix	Symbol	Power
yotta	Y	10^{24}
zetta	Z	10^{21}
exa	E	10^{18}
peta	P	10^{15}
tera	T	10^{12}
giga	G	10^{9}
mega	M	10^{6}
kilo	k	10^{3}
hecto	h	10^{2}
deca	da	10^{1}
—	—	10^{0}
deci	d	10^{-1}
centi	c	10^{-2}
milli	m	10^{-3}
micro	μ	10^{-6}
nano	n	10^{-9}
pico	p	10^{-12}
femto	f	10^{-15}
atto	a	10^{-18}
zepto	z	10^{-21}
yocto	y	10^{-24}

city is more easily expressed as 58 km. Similarly, the prefix μ (the lowercase Greek "mu") means "micro," or 10^{-6}. So our bacterium is 10 μm long. The SI prefixes are listed in Table 1.1, which is repeated inside the front cover. We'll use the prefixes routinely in examples and problems.

When two units are used together, a hyphen appears between them—for example, newton-meter. Each unit has a symbol, such as m for meter or N for newton (the SI unit of force). Symbols are ordinarily lowercase, but those named after people are uppercase. Thus "newton" is written with a small "n" but its symbol is a capital N. The exception is the unit of volume, the liter, defined as the volume of a cube one-tenth of a meter on each side. Since the lowercase "l" is easily confused with the number one, the symbol for liter is a capital L. When two units are multiplied, their symbols are separated by a centered dot: N·m for newton-meter. Division of units is expressed by using the slash (/) or by writing the symbol for the denominator unit raised to the -1 power. Thus the SI unit of speed is the meter per second, written m/s or m·s^{-1}.

Other Unit Systems

The inches, feet, yards, miles, and pounds of the so-called English system still dominate measurement in the United States. Other non-SI units such as the hour for time are often mixed with English or SI units, as when speed limits are given in miles per hour or kilometers per hour. In some specialized fields of physics there are good reasons for using non-SI units. We'll discuss these as the need arises, and we will occasionally use non-SI units in examples and problems. We'll also often find it convenient to use degrees rather than radians in measuring angles. The vast majority of examples and problems in this book, however, use strictly SI units.

Changing Units

Sometimes we need to change from one unit system to another—for example, from English to SI units. Appendix C contains extensive tables for converting among unit systems; you should familiarize yourself with this and the other appendices and refer to them often.

For example, Appendix C shows that 1 ft = 0.3048 m. Since 1 ft and 0.3048 m represent the same physical distance, multiplying any distance by their ratio will change the units but not the actual physical distance. Thus the 1815-ft CN Tower in Toronto—the world's tallest free-standing structure—is

$$(1815 \text{ ft})\left(\frac{0.3048 \text{ m}}{1 \text{ ft}}\right) = 553.2 \text{ m}$$

high (Fig. 1.3). If we had wanted to convert meters to feet, we would have used the ratio 1 ft/0.3048 m.

Often you'll need to change several units in the same expression. Keeping track of the units through a chain of multiplications helps prevent you from carelessly inverting any of the conversion factors. A numerical answer cannot be correct unless it has the right units!

553 m
1815 ft

FIGURE 1.3 Toronto's CN tower is the world's tallest free-standing structure.

APPLICATION **Units Matter: A Bad Day on Mars**

In September 1999 the Mars Climate Orbiter was destroyed when the spacecraft passed through Mars's atmosphere and experienced stresses and heating it was not designed to tolerate. Why did this $125-million craft enter the Martian atmosphere when it was supposed to remain in the vacuum of space as shown in the art? Although a host of navigation-related errors contributed, NASA identified the "root cause" as a failure to convert the English units one team used to specify rocket thrust to the SI units another team expected. Units matter!

EXAMPLE 1.1 **Changing Units: Speed Limits**

Express a 65 mi/h speed limit in meters per second.

EVALUATE In Appendix C, we find that $1 \text{ mi} = 1609 \text{ m}$, so we can multiply miles by the ratio 1609 m/mi to get meters. Similarly, we use the conversion factor 3600 s/h to convert hours to seconds. So we have

$$65 \text{ mi/h} = \left(\frac{65 \text{ mi}}{\text{h}}\right)\left(\frac{1609 \text{ m}}{\text{mi}}\right)\left(\frac{1 \text{ h}}{3600 \text{ s}}\right) = 29 \text{ m/s} \quad \blacksquare$$

1.3 Working with Numbers

Scientific Notation

The range of measured quantities in the universe is enormous; lengths alone go from about 1/1,000,000,000,000,000 m for the radius of a proton to 1,000,000,000,000,000,000,000 m for the size of a galaxy; our telescopes see 100,000 times farther still. Therefore we frequently express numbers in **scientific notation**, in which a number of reasonable size is multiplied by some power of 10. For example, 4,185 is 4.185×10^3 and 0.00012 is 1.2×10^{-4}. Table 1.2 suggests the range of measurements we work with for the fundamental quantities of length, time, and mass. Take a minute (about 10^2 heartbeats, or 3×10^{-8} of a typical human lifespan) to peruse this table along with Fig. 1.4.

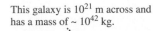

This galaxy is 10^{21} m across and has a mass of $\sim 10^{42}$ kg.

TABLE 1.2 Distances, Times, and Masses (rounded to one significant figure)

Radius of observable universe	1×10^{26} m
Earth's radius	6×10^6 m
Tallest mountain	9×10^3 m
Height of person	2 m
Diameter of red blood cell	1×10^{-5} m
Size of proton	1×10^{-15} m
Age of universe	4×10^{17} s
Earth's orbital period (1 year)	3×10^7 s
Human heartbeat	1 s
Wave period, microwave oven	5×10^{-10} s
Time for light to cross a proton	3×10^{-24} s
Mass of Milky Way galaxy	1×10^{42} kg
Mass of mountain	1×10^{18} kg
Mass of human	70 kg
Mass of red blood cell	1×10^{-13} kg
Mass of uranium atom	4×10^{-25} kg
Mass of electron	1×10^{-30} kg

Scientific calculators handle numbers in scientific notation. But straightforward rules allow you to manipulate scientific notation if you don't have such a calculator handy.

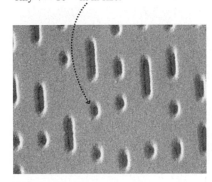

Your movie is stored on a DVD in "pits" only 4×10^{-7} m in size.

FIGURE 1.4 Large and small.

TACTICS 1.1 **Using Scientific Notation**

Addition/Subtraction To add (or subtract) numbers in scientific notation, first give them the same exponent and then add (or subtract):

$$3.75 \times 10^6 + 5.2 \times 10^5 = 3.75 \times 10^6 + 0.52 \times 10^6 = 4.27 \times 10^6$$

Multiplication/Division To multiply (or divide) numbers in scientific notation, multiply (or divide) the digits and add (or subtract) the exponents:

$$(3.0 \times 10^8 \text{ m/s})(2.1 \times 10^{-10} \text{ s}) = (3.0)(2.1) \times 10^{8 + (-10)} \text{ m} = 6.3 \times 10^{-2} \text{ m}$$

Powers/Roots To raise numbers in scientific notation to any power, raise the digits to the given power and multiply the exponent by the power:

$$\sqrt{(3.61 \times 10^4)^3} = \sqrt{3.61^3 \times 10^{(3)(4)}} = (47.04 \times 10^{12})^{1/2}$$
$$= \sqrt{47.04} \times 10^{(12)(1/2)} = 6.86 \times 10^6$$

EXAMPLE 1.2 Scientific Notation: Tsunami Warnings

Earthquake-generated tsunamis are so devastating because the entire ocean, from surface to bottom, participates in the wave motion. For such waves, the wave speed is given by $v = \sqrt{gh}$, where $g = 9.8$ m/s^2 is the acceleration due to gravity, h is the depth in meters, and v is in meters per second. Determine a tsunami's speed in open ocean with a depth of 3.0 km.

EVALUATE That 3-km depth is 3×10^3 m, so we have

$$v = \sqrt{gh} = [(9.8 \text{ m/s}^2)(3.0 \times 10^3 \text{ m})]^{1/2} = (29.4 \times 10^3 \text{ m}^2/\text{s}^2)^{1/2}$$
$$= (2.94 \times 10^4 \text{ m}^2/\text{s}^2)^{1/2} = \sqrt{2.94} \times 10^2 \text{ m/s} = 1.7 \times 10^2 \text{ m/s}$$

where we wrote 29.4×10^3 m^2/s^2 as 2.94×10^4 m^2/s^2 in the second line in order to calculate the square root more easily. Now, converting the speed to km/h, we find

$$1.7 \times 10^2 \text{ m/s} = \left(\frac{1.7 \times 10^2 \text{ m}}{\text{s}}\right)\left(\frac{1 \text{ km}}{1.0 \times 10^3 \text{ m}}\right)\left(\frac{3.6 \times 10^3 \text{ s}}{\text{h}}\right)$$
$$= 6.1 \times 10^2 \text{ km/h}$$

This speed—about 600 km/h—shows why even distant coastlines have little time to prepare for the arrival of a tsunami. ∎

Accuracy and Significant Figures

How accurate is that 1.7×10^2 m/s we calculated in Example 1.2? The two **significant figures** in this number imply that the value is closer to 1.7 than to 1.6 or 1.8. The fewer significant figures, the less accurately we can claim to know a given quantity.

In Example 1.2 we were, in fact, given two significant figures for both quantities. The mere act of calculating can't add accuracy, so we rounded our answer to two significant figures as well. Calculators and computers often give us numbers with many figures, but most of those are usually meaningless.

What's Earth's circumference? It's $2\pi R_E$, of course. And π is approximately 3.14159. ... But if you only know Earth's radius as 6.37×10^6 m, knowing π to more significant figures doesn't mean you can claim to know the diameter any more accurately. This example suggests a rule for handling calculations involving numbers with different accuracies:

> In multiplication and division, the answer should have the same number of significant figures as the least accurate of the quantities entering the calculation.

You're engineering an access ramp to a bridge whose main span is 1.248 km long. The ramp will be 65.4 m long. What will be the overall length of the structure? A simple calculation gives 1.248 km + 0.0654 km = 1.3134 km. How should you round this? You know the bridge length to ±0.001 km, so even the addition of an amount this small is significant. Thus your answer should have three digits to the right of the decimal point, giving 1.313 km as the appropriate answer. Thus:

> In addition and subtraction, the answer should have the same number of digits to the right of the decimal point as the term in the sum or difference that has the smallest number of digits to the right of the decimal point.

In subtraction, this rule can quickly lead to loss of accuracy, as Example 1.3 illustrates.

EXAMPLE 1.3 Significant Figures: Nuclear Fuel

A uranium fuel rod is 3.241 m long before it is inserted in a nuclear reactor. After insertion, heat from the nuclear reaction has increased its length to 3.249 m. What is the increase in its length?

EVALUATE Subtraction gives 3.249 m − 3.241 m = 0.008 m or 8 mm. Should this be 8 mm or 8.000 mm? Just 8 mm. Subtraction affected only the last digit of the four-significant-figure lengths, leaving only one significant figure in the answer. ∎

✓ **TIP** Intermediate Results

Although it's important that your final answer reflect the accuracy of the numbers that went into it, any intermediate results should have at least one extra significant figure. Otherwise rounding of intermediate results could alter your answer.

GOT IT? 1.1 Rank the numbers according to (a) their size and (b) the number of significant figures. Some may be of equal rank. 0.0008, 3.14×10^7, 2.998×10^{-9}, 55×10^6, 0.041×10^9

Estimation

Some problems in physics and engineering call for precise numerical answers. We need to know exactly how long to fire a rocket to put a space probe on course toward a distant planet, or exactly what size to cut the tiny quartz crystal whose vibrations set the pulse of a digital watch. But for many other purposes, we need only a rough idea of the size of a given physical effect. And we can often make a rough estimate to check whether the results of more difficult calculations make sense.

EXAMPLE 1.4 **Estimation: Counting Brain Cells**

Estimate the mass of your brain and the number of cells it contains.

EVALUATE My head is about 6 in. or 15 cm wide, but there's a lot of skull bone in there, so maybe my brain is about 10 cm or 0.1 m across. I don't know its exact shape, but for estimating, I'll just take it to be a cube. Then its volume is $(10 \text{ cm})^3 = 1000 \text{ cm}^3$, or 10^{-3} m^3. I'm mostly water, and water has a density of 1 gram per cubic centimeter (1 g/cm^3), so my 1000-cm^3 brain has a mass of about 1 kg.

How big is a brain cell? I don't know, but Table 1.1 lists the diameter of a red blood cell as about 10^{-5} m. If brain cells are roughly the same size, then each cell has a volume of approximately $(10^{-5} \text{ m})^3 = 10^{-15} \text{ m}^3$. Then the number of cells in my 10^{-3}-m^3 brain is roughly

$$N = \frac{10^{-3} \text{ m}^3/\text{brain}}{10^{-15} \text{ m}^3/\text{cell}} = 10^{12} \text{ cells/brain}$$

Crude though they are, these estimates aren't bad. The average adult brain's mass is about 1.3 kg, and it contains at least 10^{11} cells. ∎

1.4 Strategies for Learning Physics

You can learn *about* physics, and you can learn to *do* physics. This book is for science and engineering students, so it emphasizes both. Learning about physics will help you appreciate the role of this fundamental science in explaining both natural and technological phenomena. Learning to do physics will make you adept at solving quantitative problems—finding precise answers to questions about how the natural world works and about how we forge the technologies at the heart of modern society.

Physics: Challenge and Simplicity

Physics problems can be challenging, calling for clever insight and mathematical agility. That challenge is what gives physics both intellectual interest and a reputation as a difficult subject. But underlying all of physics is only a handful of basic principles. Because physics is so fundamental, it's also inherently simple. There are only a few basic ideas to learn; if you really understand those, you can apply them in a wide variety of situations. This book is written in a spirit that emphasizes the underlying simplicity of physics by reminding you how the many examples, applications, and problems are manifestations of the same few basic principles. If you approach physics as a hodgepodge of unrelated laws and equations, you'll miss the point and make things difficult. But if you look for the basic principles, for connections among seemingly unrelated phenomena and problems, then you'll discover the underlying simplicity that reflects the scope and power of physics—the fundamental science.

Problem Solving: An IDEA Strategy

Solving a quantitative physics problem always starts with basic principles or concepts and ends with a precise answer expressed as either a numerical quantity or an algebraic expression. Whatever the principle, whatever the realm of physics, and whatever the specific situation, the path from principle to answer follows four simple steps—steps that make up a comprehensive strategy for approaching all problems in physics. Their acronym, IDEA, will help you remember these steps, and they'll be reinforced as we apply them over and over again in worked examples throughout the book. We'll generally write all four steps out separately, although in the examples of this chapter we've cut right to the EVALUATE phase. And in some chapters we'll introduce versions of this strategy tailored to specific material.

The IDEA strategy is not a "cookbook" formula for working physics problems. Rather, it's a tool for organizing your thoughts, clarifying your conceptual understanding, developing and executing plans for solving problems, and assessing your answers. Here's the big IDEA:

PROBLEM SOLVING STRATEGY 1.1 **Physics Problems**

INTERPRET You can't begin a problem unless you're sure what it's about. So the first step is to *interpret* the problem to be sure you know what it's asking. Then *identify* the applicable concepts and principles—Newton's laws of motion, the conservation of energy principle, the first law of thermodynamics, Gauss's law for electricity, and so forth. Also *identify* the players in the situation—the object whose motion you're asked to describe, the forces acting on an object, the thermodynamic system whose heat flow you're going to determine, the charges that produce an electric field, the components in an electric circuit whose power consumption you're after, the light rays that will help you locate an image, and so on.

DEVELOP The second step is to *develop* a plan for solving the problem. As part of this step, it's always helpful and often essential to *draw* a diagram showing the essential aspects of the situation. Your drawing should indicate objects, forces, and other physical entities. Labeling masses, positions, forces, velocities, heat flows, electric or magnetic fields, and other quantities will be a big help. Next, *determine* the relevant mathematical formulas—namely, those that contain the quantities you're given in the problem as well as the unknown(s) you're solving for. Don't just grab equations—rather, think about how each reflects the underlying concepts and principles that you've identified as applying to this problem. The plan you develop might include calculating intermediate quantities, finding values in a table or in one of this text's several appendices, or even solving a preliminary problem whose answer you need to get your final result.

EVALUATE Physics problems have numerical or symbolic answers, and you need to *evaluate* your answer. In this step you *execute* your plan, going in sequence through the steps you've outlined. Here's where your math skills come in. Use algebra, trig, or calculus, as needed, to solve your equations. It's a good idea to keep all numerical quantities, whether known or not, in symbolic form as you work through the solution of your problem. At the end you can plug in numbers and work the arithmetic to *evaluate* the numerical answer, if the problem calls for one.

ASSESS Don't be satisfied with your answer until you *ask* whether it makes sense! Are the units correct? Do the numbers sound reasonable? Does the algebraic form of your answer work in obvious special cases, like perhaps "turning off" gravity or making an object's mass zero or infinite? Checking special cases not only helps you decide whether your answer makes sense but can also give you insights into the underlying physics. In worked examples, we'll often use this step to enhance your knowledge of physics by relating the example to other applications of physics.

Don't memorize the IDEA problem-solving strategy. Instead, grow to understand it as you see it applied in examples and as you apply it yourself in working end-of-chapter problems. This book has a number of additional features and supplements, discussed in the Preface, to help you grow your problem-solving skills.

Big Picture

Physics is the fundamental science. It's convenient to consider several realms of physics, which together describe all that's known about physical reality:

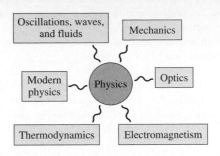

Key Concepts and Equations

Numbers describing physical quantities must have units. The SI unit system comprises seven fundamental units:

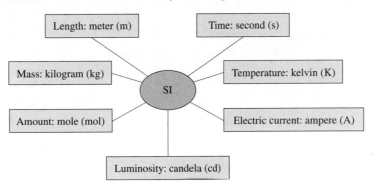

In addition, physics uses geometric measures of angle.

Numbers are often written with prefixes or in scientific notation to express powers of 10. Accuracy is shown by the number of significant figures:

$$\text{Earth's radius} = 6.37 \times 10^6 \, \text{m} = 6.37 \, \text{Mm}$$

Power of 10

Three significant figures

SI prefix for "$\times 10^6$"

Cases and Uses

The IDEA strategy for solving physics problems follows four steps: Interpret, Develop, Evaluate, and Assess.

For Thought and Discussion

1. Explain why measurement standards based on laboratory procedures are preferable to those based on specific objects such as the international prototype kilogram.
2. Which measurement standards are now defined operationally? Which are not?
3. When a computer that carries seven significant figures adds 1.000000 and 2.5×10^{-15}, what answer does it display? Why?
4. Why does Earth's rotation not provide a suitable standard of time?
5. To raise a power of 10 to another power, you multiply the exponent by the power. Explain why this works.
6. A scientist and a creationist are arguing about the age of the Earth. What facts might the scientist use in estimating this age?
7. How would you determine the length of a curved line?
8. Write $1/\sqrt{x}$ as x to some power.
9. Emissions of carbon dioxide from fossil-fuel combustion are often expressed in gigatonnes per year, where 1 tonne = 1000 kg. But sometimes CO_2 emissions are given in petagrams per year. How are the two units related?

Exercises and Problems

Exercises

Section 1.2 Measurements and Units

10. The power output of a typical large power plant is 1000 megawatts (MW). Express this result in (a) W, (b) kW, and (c) GW.
11. The diameter of a hydrogen atom is about 0.1 nm, and the diameter of a proton is about 1 fm. How much bigger is a hydrogen atom than a proton?
12. Use the definition of the meter to determine how far light travels in 1 ns.
13. How long, in nanoseconds, is the period of the cesium-133 radiation used to define the second?
14. Lake Baikal in Siberia holds the world's largest quantity of fresh water, about 14 Eg. How many kilograms is that?
15. A hydrogen atom is about 0.1 nm in diameter. How many hydrogen atoms lined up side by side would make a line 1 cm long?
16. How long a piece of wire would you need to form a circular arc subtending an angle of 1.4 rad, if the radius of the arc is 8.1 cm?
17. Making a turn, a jetliner flies 2.1 km on a circular path of radius 3.4 km. Through what angle does it turn?
18. A car is moving at 35.0 mi/h. Express its speed in (a) m/s and (b) ft/s.
19. I have enough postage for a 1-oz letter but only a metric scale. What's the maximum mass for my letter, in grams?
20. A year is very nearly $\pi \times 10^7$ s. By what percentage is this figure in error?
21. How many cubic centimeters (cm^3) are in a cubic meter (m^3)?
22. By what percentage do the 1500-m and 1-mile races differ?
23. A gallon of paint covers 350 ft^2. What is its coverage in m^2/L?
24. Superhighways in Canada have speed limits of 100 km/h. How does this compare with the 65 mi/h speed limit common in the United States?
25. One m/s is how many km/h?
26. A 3.0-lb box of grass seed will seed 2100 ft^2 of lawn. Express this coverage in m^2/kg.
27. A radian is how many degrees?

Section 1.3 Working with Numbers

28. Add 3.6×10^5 m and 2.1×10^3 km.
29. Divide 4.2×10^3 m/s by 0.57 ms, and express your answer in m/s^2.
30. Add 5.1×10^{-2} cm and 6.8×10^3 μm, and multiply the result by 1.8×10^4 N (1 N is the SI unit of force).
31. What is the cube root of 6.4×10^{19}? Do this without a calculator.
32. Add 1.46 m and 2.3 cm.
33. A 3.6-cm-long radio antenna is added to the front of an airplane 41 m long. What is the overall length?
34. Repeat Exercise 33 given that the airplane's length is 41.05 m.

Problems

35. To see that it is important to carry more digits in intermediate calculations, determine $(\sqrt{3})^3$ to three significant figures in two ways: (a) Find $\sqrt{3}$ and round to three significant figures, then cube and again round; and (b) find $\sqrt{3}$ to four significant figures, then cube and round to three significant figures.
36. Paper is made from wood pulp. Estimate the number of trees that must be cut down to make one day's run of a big city's daily newspaper. Assume no recycling.
37. The average dairy cow produces about 10^4 kg of milk per year. Estimate the number of dairy cows needed to keep the United States supplied with milk.
38. How many Earths would fit inside the Sun?
39. The average American uses electrical energy at the rate of about 1.5 kilowatts (kW). Solar energy reaches Earth's surface at an average rate of about 300 watts on every square meter. What fraction of the United States' land area would have to be covered with solar cells to provide all of our electrical energy? Assume the cells are 20% efficient at converting sunlight to electricity.
40. (a) Estimate the volume of water going over Niagara Falls each second. (b) The falls provides the outlet for Lake Erie; if the falls were shut off, estimate how long it would take Lake Erie to rise 1 m.
41. Estimate the number of air molecules in your dormitory room.
42. A human hair is about 100 μm across. Estimate the number of hairs in a typical braid.
43. The density of bubble gum is about 1 g/cm^3. You blow an 8-g wad of gum into a bubble 10 cm in diameter. What is the thickness of the bubble? *Hint:* Think about spreading the bubble into a flat sheet. The surface area of a sphere is $4\pi r^2$.
44. The Moon barely covers the Sun at a solar eclipse. Given that the Moon is 4×10^5 km from Earth and that the Sun is 1.5×10^8 km from Earth, determine how much bigger the Sun's diameter is than the Moon's. If the Moon's radius is 1800 km, how big is the Sun?
45. The semiconductor chip at the heart of a personal computer is a square 4 mm on a side and contains 10^9 electronic components. (a) If each component is a square, what is the distance across each component? (b) If a calculation requires that electrical impulses traverse 10^4 elements on the chip, each a million times, how many such calculations can the computer perform each

second? The maximum speed of an electrical impulse is close to the speed of light, 3×10^8 m/s.

46. Estimate the number of (a) atoms and (b) cells in your body.

47. When we write the number 3.6 as typical of a number with two significant figures, we're saying that the actual value is closer to 3.6 than to 3.5 or 3.7; that is, the actual value lies between 3.55 and 3.65. Show that the percent uncertainty implied by two-significant-figure accuracy varies with the value of the number, being the lowest for numbers beginning with 9 and the highest for numbers beginning with 1. In particular, what is the percent uncertainty implied by the numbers (a) 1.1, (b) 5.0, and (c) 9.9?

48. Continental drift occurs at about the rate at which your fingernails grow. Estimate the age of the Atlantic Ocean, assuming the eastern and western hemispheres have been drifting apart.

49. In the 1908 London Olympic Games, the originally intended marathon distance of 26 miles was extended by 385 yards so that the end was in front of the royal reviewing stand. This distance subsequently became standard. What is the marathon distance in kilometers, to the nearest meter?

50. Express the following with appropriate units and significant figures: (a) 1.0 m plus 1 mm, (b) 1.0 m times 1 mm, (c) 1.0 m minus 999 mm, (d) 1.0 m divided by 999 mm.

51. Estimation problems were a favorite of the famous physicist Enrico Fermi. Here is one such problem attributed to him: What is the number of piano tuners in Chicago? Explain how you would estimate this number.

52. You are the owner of a small manufacturing company and wish to install some new computer-aided manufacturing equipment. A sales representative tells you the computer his company sells contains "super chips" that no one else has. He claims each chip contains more than 10 billion electronic components. Each chip measures 5.0 mm by 5.0 mm, and you know that each component is 90 nm on a side. Is the sales rep correct?

53. Café Milagro in Costa Rica sells coffee via the Internet. A 0.5-kg bag of coffee costs $8.95, excluding shipping. How much does this coffee cost per pound? If you order six 0.5-kg bags, the shipping costs $6.66. How much does a bag cost when you include the shipping?

54. Suppose you drive your old classic car to an auto show in Canada. Your speedometer shows only miles per hour (mph). All the Canadian speed limits are in km/h (kph)! What speed limit in mph corresponds to these Canadian speed limits: 40 kph, 50 kph, 80 kph, and 100 kph?

55. While in Canada, you go to a grocery store and purchase some deli meats and cheese for sandwiches. You normally request one-half pound of each in the United States. About how many kilograms of meat and cheese should you ask the clerk to slice?

Answers to Chapter Questions

Answer to Chapter Opening Question

All of them!

Answers to GOT IT? Question

1.1 (a) 2.998×10^{-9}, 0.0008, 3.14×10^7, 0.041×10^9, 55×10^6

(b) 0.0008, 0.041×10^9 and 55×10^6 (with two significant figures each), 3.14×10^7, 2.998×10^{-9}

Mechanics

■ Hikers check their position using signals from GPS satellites.

A wilderness hiker uses the Global Positioning System to follow her chosen route. A farmer plows a field with centimeter-scale precision, guided by GPS and saving precious fuel as a result. One scientist uses GPS to track endangered elephants, another to study the accelerated flow of glaciers as Earth's climate warms. Our deep understanding of motion is what lets us use a constellation of satellites, 20,000 km up and moving faster than 10,000 km/hour, to find positions on Earth so precisely.

Motion occurs at all scales, from the intricate dance of molecules at the heart of life's cellular mechanics, to the everyday motion of cars, baseballs, and our own bodies, to the trajectories of GPS and TV satellites and of spacecraft exploring the distant planets, to the stately motions of the celestial bodies themselves and the overall expansion of the universe. The study of motion is called **mechanics**. The 11 chapters of Part 1 introduce the physics of motion, first for individual bodies and then for complicated systems whose constituent parts move relative to one another.

We explore motion here from the viewpoint of Newtonian mechanics, which applies accurately in all cases except the subatomic realm and when relative speeds approach that of light. The Newtonian mechanics of Part 1 provides the groundwork for much of the material in subsequent parts, until, in the book's final chapters, we extend mechanics into the subatomic and high-speed realms.

2 Motion in a Straight Line

► **To Learn**

By the end of this chapter you should be able to

- Explain the meanings of and relationships among displacement, velocity, and acceleration in one-dimensional motion (2.1–2.3).

- Calculate average and instantaneous velocities and accelerations (2.1–2.3).

- Solve problems involving motion under constant acceleration, including the acceleration of gravity at Earth's surface (2.4, 2.5).

◄ **To Know**

- From Chapter 1, you should be comfortable working with SI units, and you should be able to convert from other unit systems (1.2).

- You should be able to express quantitative answers with the correct number of significant figures (1.3).

- You should be ready to apply the four problem-solving steps of the IDEA strategy (1.4).

■ The server tosses the tennis ball straight up and hits it on its way down. Right at its peak height, the ball has zero velocity, but what's its acceleration?

Electrons swarming around atomic nuclei, cars speeding along a highway, blood coursing through your arteries, galaxies rushing apart in the expanding universe—all these are examples of matter in motion. The study of motion without regard to its cause is called **kinematics** (from the Greek "kinema," or motion, as in motion pictures). This chapter deals with the simplest case: a single object moving in a straight line. Later, we generalize to motion in more dimensions and with more complicated objects. But the basic concepts and mathematical techniques we develop here remain fundamental even in those situations.

2.1 Average Motion

You drive 15 minutes to a pizza place 10 km away, grab your pizza, and return home in another 15 minutes. You've traveled a total distance of 20 km, and the trip took half an hour, so your **average speed**—distance divided by time—was 40 kilometers per hour. To describe your motion more precisely, we introduce the quantity x that gives your position at

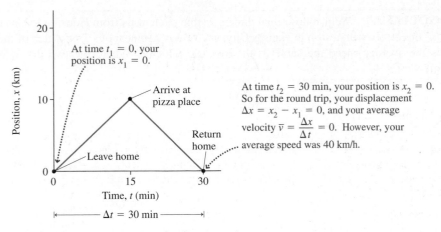

FIGURE 2.1 Position versus time for the pizza trip.

any time t. We then define **displacement**, Δx, as the net change in position: $\Delta x = x_2 - x_1$, where x_1 and x_2 are your starting and ending positions, respectively. Your **average velocity**, $\bar{v}$, is displacement divided by the time interval:

$$\bar{v} = \frac{\Delta x}{\Delta t} \quad \text{(average velocity)} \tag{2.1}$$

where $\Delta t = t_2 - t_1$ is the interval between your ending and starting times. The bar in $\bar{v}$ indicates an average quantity (and is read "v bar"). The symbol Δ (capital Greek delta) stands for "the change in." For the round trip to the pizza place, your overall displacement was zero and therefore your average velocity was also zero—even though your average *speed* was not (Fig. 2.1).

Directions and Coordinate Systems

It matters whether you go north or south, east or west. Displacement therefore includes not only *how far* but also in what *direction*. For motion in a straight line, we can describe both properties by taking position coordinates x to be positive going in one direction from some origin, and negative in the other. This gives us a one-dimensional **coordinate system**. The choice of coordinate system—both of origin and of which direction is positive—is entirely up to you. The coordinate system isn't physically real; it's just a convenience we create to help in the mathematical description of motion.

Figure 2.2 shows some midwestern cities that lie on a north-south line. We've established a coordinate system with northward direction positive and origin at Kansas City. Arrows show displacements from Houston to Des Moines and from International Falls to Des Moines; the former is approximately $+1300$ km, and the latter is approximately -750 km, with the minus sign indicating a southward direction. Suppose the Houston-to-Des Moines trip takes 2.6 hours by plane; then the average velocity is $(1300 \text{ km})/(2.6 \text{ h}) = 500$ km/h. If the International Falls-to-Des Moines trip takes 10 h by car, then the average velocity is $(-750 \text{ km})/(10 \text{ h}) = -75$ km/h; again, the minus sign indicates southward.

In calculating average velocity, all that matters is the overall displacement. Maybe that trip from Houston to Des Moines was a nonstop flight going 500 km/h. Or maybe it involved a faster plane that stopped for a while in Kansas City. Maybe the plane even went first to Minneapolis and then backtracked to Des Moines. No matter: The displacement remains 1300 km and, as long as the total time is 2.6 h, the average velocity remains 500 km/h.

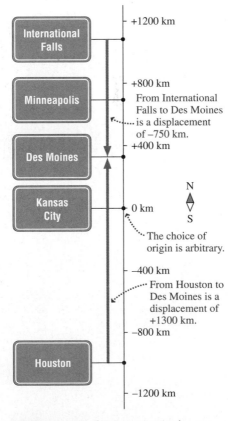

FIGURE 2.2 Describing motion in the central United States.

GOT IT? 2.1 We just described three possible plane trips from Houston to Des Moines: (a) direct; (b) with a stop in Kansas City, and (c) via Minneapolis. For which of these trips is the average speed the same as the average velocity? Where the two differ, which is greater?

EXAMPLE 2.1 **Speed and Velocity: Flying with a Connection**

To get a cheap flight from Houston to Kansas City—a distance of 1000 km—you have to connect in Minneapolis, 700 km north of Kansas City. The flight to Minneapolis takes 2.2 h, then you have a 30-min layover, and then a 1.3-h flight to Kansas City. What are your average velocity and your average speed on this trip?

INTERPRET We interpret this as a one-dimensional kinematics problem involving the distinction between velocity and speed, and we identify three distinct travel segments: the two flights and the layover. We identify the key concepts as speed and velocity; their distinction is clear from our pizza example.

DEVELOP Figure 2.2 is our drawing. We determine that Equation 2.1, $\bar{v} = \Delta x / \Delta t$, will give the average velocity, and that the average speed is the total distance divided by the total time. We develop our plan: Find the displacement and the total time, and use those values to get the average velocity; then find the total distance traveled and use that along with the total time to get the average speed.

EVALUATE You start in Houston and end up in Kansas City, for a displacement of 1000 km—regardless of how far you actually traveled. The total time for the three segments is $\Delta t = 2.2\ \text{h} + 0.50\ \text{h} + 1.3\ \text{h} = 4.0\ \text{h}$. Then the average velocity, from Equation 2.1, is

$$\bar{v} = \frac{\Delta x}{\Delta t} = \frac{1000\ \text{km}}{4.0\ \text{h}} = 250\ \text{km/h}$$

However, that Minneapolis connection means you've gone an extra $2 \times 700\ \text{km}$, for a total distance of 2400 km in 4 hours. Thus your average speed is $(2400\ \text{km})/(4.0\ \text{h}) = 600\ \text{km/h}$, more than twice your average velocity.

ASSESS: Make sense? Average velocity depends only on the net displacement between the starting and ending points. Average speed takes into account the actual distance you travel—which can be a lot longer on a circuitous trip like this one. So it's entirely reasonable that the average speed should be greater. ∎

2.2 Instantaneous Velocity

Geologists determine the velocity of a lava flow by dropping a stick into the lava and timing how long it takes the stick to go a known distance (Fig. 2.3a). Dividing the distance by the time then gives the average velocity. But did the lava flow faster at the beginning of the interval? Or did it speed up and slow down again? Clearly, velocity can change over time. To understand motion in all its detail, we need to know the velocity at each instant.

Geologists could explore that detail with a series of observations taken over smaller intervals of time and distance (Fig. 2.3b). As the size of the intervals shrinks, a more detailed picture of the motion emerges. In the limit of very small intervals, we're measuring the velocity at a single instant. This is the **instantaneous velocity**, or simply the **velocity.** The magnitude of the instantaneous velocity is the **instantaneous speed**.

You might object that it's impossible to achieve that limit of an arbitrarily small time interval. With observational measurements that's true, but calculus lets us go there. Figure 2.4a is a plot of position versus time for the stick in the lava flow shown in Fig. 2.3. Where the curve is steep, the position changes rapidly with time—so the velocity is greater. Where the curve is flatter, the velocity is lower. Study the clocks in Fig. 2.3b and you'll see that the stick starts out moving rapidly, then slows, and then speeds up a bit at the end. The curve in Fig. 2.4a reflects this behavior.

Suppose we want the instantaneous velocity at the time marked t_1 in Fig. 2.4a. We can approximate this quantity by measuring the displacement Δx over the interval Δt between t_1 and some later time t_2; the ratio $\Delta x / \Delta t$ is then the average velocity over this interval. Note that this ratio is the slope of a line drawn through points on the curve that mark the ends of the interval.

Figure 2.4b shows what happens as we make the time interval Δt arbitrarily small: Eventually, the line between the two points becomes indistinguishable from the tangent line to the curve. That tangent line has the same slope as the curve right at the point we're interested in, and therefore it defines the instantaneous velocity at that point. We write this

The average velocity as the stick
goes from A to B is $\bar{v} = \Delta x/\Delta t$.

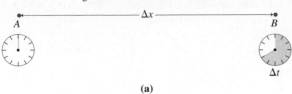

(a)

Using shorter distance intervals gives details
about how the velocity changes.

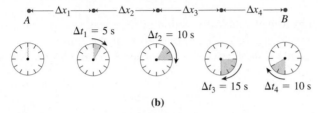

(b)

FIGURE 2.3 Determining the velocity of a lava flow.

mathematically by saying that the instantaneous velocity is the limit, as the time interval Δt becomes arbitrarily close to zero, of the ratio of displacement Δx to Δt:

$$v = \lim_{\Delta t \to 0} \frac{\Delta x}{\Delta t} \tag{2.2a}$$

You can imagine making the interval Δt as close to zero as you like, getting ever better approximations to the instantaneous velocity. Given a graph of position versus time, an easy approach is to "eyeball" the tangent line to the graph at a point you're interested in; its slope is the instantaneous velocity (Fig. 2.5).

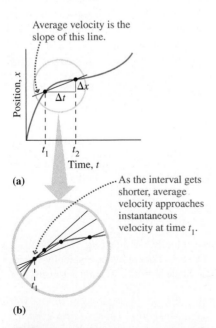

(a)

(b)

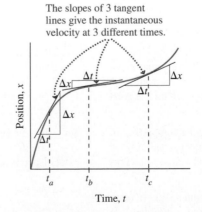

FIGURE 2.4 Position-versus-time graph for the motion in Fig. 2.3.

FIGURE 2.5 The instantaneous velocity is the slope of the tangent line.

GOT IT? 2.2 The figure shows position-versus-time graphs for four objects. Which object is moving with constant speed? Which reverses direction? Which starts slowly and then speeds up?

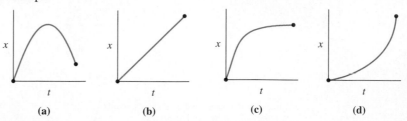

(a) (b) (c) (d)

Given position as a mathematical function of time, calculus provides a quick way to find instantaneous velocity. In calculus, the result of the limiting process described in Equation 2.2a is called the **derivative** of x with respect to t and is given the special symbol dx/dt:

$$\frac{dx}{dt} = \lim_{\Delta t \to 0} \frac{\Delta x}{\Delta t}$$

The quantities dx and dt are called **infinitesimals**; they represent vanishingly small quantities that result from the limiting process. We can then write Equation 2.2a as

$$v = \frac{dx}{dt} \quad \text{(instantaneous velocity)} \tag{2.2b}$$

Given position x as a function of time t, calculus shows how to find the velocity $v = dx/dt$. Consult Tactics 2.1 if you haven't yet seen derivatives in your calculus class, or if you need a refresher.

TACTICS 2.1 **Taking Derivatives**

You don't have to go through an elaborate limiting process every time you want to find an instantaneous velocity. That's because calculus provides formulas for the derivatives of common functions. For example, any function of the form $x = bt^n$, where b and n are constants, has derivative

$$\frac{dx}{dt} = bnt^{n-1} \tag{2.3}$$

Appendix A lists derivatives of other common functions.

EXAMPLE 2.2 **Instantaneous Velocity: A Space Shuttle Ascends**

The altitude of a space shuttle in the first half-minute of its ascent is given by $x = bt^2$, where the constant b is 2.90 m/s^2. Find a general expression for the shuttle's velocity as a function of time and from it the instantaneous velocity at $t = 20$ s. Also find an expression for the average velocity, and compare your two velocity expressions.

INTERPRET We interpret this as a problem involving the comparison of two distinct but related concepts: instantaneous velocity and average velocity. We identify the space shuttle as the object whose velocities we're interested in.

DEVELOP Equation 2.2b, $v = dx/dt$, gives the instantaneous velocity and Equation 2.1, $\bar{v} = \Delta x/\Delta t$, gives the average velocity. Our plan is to use Equation 2.3, $dx/dt = bnt^{n-1}$, to evaluate the derivative that gives the instantaneous velocity. Then we can use Equation 2.1 for the average velocity but first we'll need to determine the displacement from the equation we're given for the shuttle's position.

EVALUATE Applying Equation 2.2b with position given by $x = bt^2$ and using Equation 2.3 to evaluate the derivative, we have

$$v = \frac{dx}{dt} = \frac{d(bt^2)}{dt} = 2bt$$

for the instantaneous velocity. Evaluating at $t = 20$ s with $b = 2.90$ m/s^2 gives $v = 116$ m/s. For the average velocity we need the total displacement at 20 s. Since $x = bt^2$, Equation 2.1 gives

$$\bar{v} = \frac{\Delta x}{\Delta t} = \frac{bt^2}{t} = bt$$

where we've used $x = bt^2$ for Δx and t for Δt because both position and time are taken to be zero at liftoff. Comparison with our earlier result shows that the average velocity from liftoff to any particular time is exactly half the instantaneous velocity at that time. *cont'd.*

ASSESS Make sense? Yes. The shuttle's speed is always increasing, so its velocity at the end of any time interval is greater than the average velocity over that interval. The fact that the average velocity is exactly half the instantaneous velocity results from the quadratic (t^2) dependence of position on time.

✓**TIP** Language
Language often holds clues to the meaning of physical concepts. In this example we speak of the *instantaneous* velocity *at* a particular time. That wording should remind you of the limiting process that focuses on a single instant. In contrast, we speak of the *average* velocity *over* a time interval, since averaging explicitly involves a range of times.

2.3 Acceleration

When velocity changes, as in Example 2.2, an object is said to undergo **acceleration**. Quantitatively, we define acceleration as the rate of change of velocity, just as we defined velocity as the rate of change of position. The **average acceleration** over a time interval Δt is

$$\bar{a} = \frac{\Delta v}{\Delta t} \quad \text{(average acceleration)} \tag{2.4}$$

where Δv is the change in velocity and the bar on $\bar{a}$ indicates that this is an average value. Just as we defined instantaneous velocity through a limiting procedure, we define **instantaneous acceleration** as

$$a = \lim_{\Delta t \to 0} \frac{\Delta v}{\Delta t} = \frac{dv}{dt} \quad \text{(instantaneous acceleration)} \tag{2.5}$$

As we did with velocity, we also use the term *acceleration* alone to mean instantaneous acceleration.

In one-dimensional motion, acceleration is either in the direction of the velocity or opposite it. In the former case the accelerating object speeds up, whereas in the latter it slows (Fig. 2.6). Although slowing is sometimes called *deceleration*, it's simpler to use *acceleration* to describe the time rate of change of velocity no matter what's happening. With two-dimensional motion, we'll find much richer relationships between the directions of velocity and acceleration.

Since acceleration is the rate of change of velocity, its units are (distance per time) per time, or distance/time^2. In SI, that's m/s^2. Sometimes acceleration is given in mixed units; for example, a car going from 0 to 60 mi/h in 10 s has an average acceleration of 6 mi/h/s.

Position, Velocity, and Acceleration

Figure 2.7 shows graphs of position, velocity, and acceleration for an object undergoing one-dimensional motion. In Fig. 2.7a the rise and fall of the position-versus-time curve show that the object first moves away from the origin, turns around, and then reaches the origin again at $t = 4$ s. It then continues moving into the region $x < 0$. Velocity, shown in Fig. 2.7b, is the slope of the position-versus-time curve in Fig. 2.7a. Note that the velocity is great where the curve in Fig. 2.7a is steep—that is, where the position is changing most rapidly. At the peak of the position curve, the object is momentarily at rest as it turns around, so there the position curve is flat and the velocity is zero. Once the object turns around, at about 2.7 s, it is heading in the negative x direction and so its velocity is negative.

Just as velocity is the slope of the position-versus-time curve, acceleration is the slope of the velocity-versus-time curve. Initially that slope is positive—velocity is increasing—but eventually it peaks at the point of maximum velocity and zero acceleration and then it decreases. That velocity decrease corresponds to a negative acceleration, as shown clearly in the region of Fig. 2.7c beyond about 1.3 s.

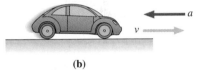

FIGURE 2.6 Acceleration and velocity.

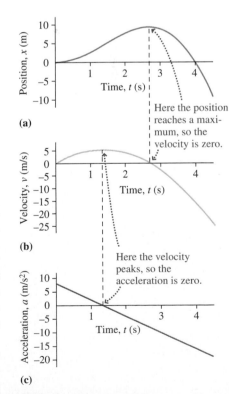

FIGURE 2.7 (a) Position, (b) velocity, and (c) acceleration versus time.

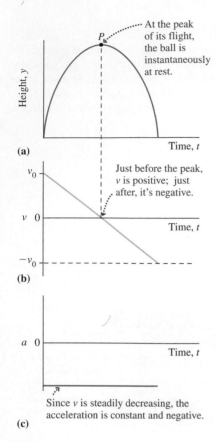

At the peak of its flight, the ball is instantaneously at rest.

(a)

Just before the peak, v is positive; just after, it's negative.

(b)

Since v is steadily decreasing, the acceleration is constant and negative.

(c)

FIGURE 2.8 (a) Position, (b) velocity, and (c) acceleration versus time for a ball thrown straight up.

In Fig. 2.7 notice that the *value* of the velocity is unrelated to the *value* of the position. The velocity is the *slope* of the position curve—and the slope depends on how the position is *changing*, not on its actual value. Similarly, the value of the acceleration depends only on the rate of *change* of velocity, not on the velocity itself. In particular, Figs. 2.7b and c show that we can have zero velocity and still be accelerating!

How can something be accelerating if it's not moving? Consider a ball thrown straight up (Fig. 2.8). Right at the peak of its flight, its velocity is instantaneously zero. But just before the peak it was moving upward, and just after it will be moving downward. No matter how small a time interval you consider about the peak, the velocity is always changing. Therefore the ball is accelerating, even right at the instant its velocity is zero.

Acceleration is the rate of change of velocity, and velocity is the rate of change of position. That makes acceleration the rate of change of the rate of change of position. This sounds less awkward mathematically: We say that acceleration is the **second derivative** of the position with respect to time. Symbolically, we write the second derivative as d^2x/dt^2; this is just a symbol and doesn't mean that anything is actually squared. Then the relationship among acceleration, velocity, and position can be written

$$a = \frac{dv}{dt} = \frac{d}{dt}\left(\frac{dx}{dt}\right) = \frac{d^2x}{dt^2} \qquad (2.6)$$

2.4 Constant Acceleration

The description of motion has an especially simple form when acceleration is constant. Suppose an object starts at time $t = 0$ with some initial velocity v_0 and constant acceleration a. Later, at some time t, it has velocity v. Because the acceleration doesn't change, its average and instantaneous values are identical, so we can write

$$a = \bar{a} = \frac{\Delta v}{\Delta t} = \frac{v - v_0}{t - 0}$$

or, rearranging,

$$v = v_0 + at \qquad \text{(for constant acceleration only)} \qquad (2.7)$$

This equation says that the velocity changes from its initial value by an amount that is the product of acceleration and time.

✓**TIP** Know Your Limits

Many equations we develop are special cases of more general laws, and they're limited to special circumstances. Equation 2.7 is a case in point: It applies *only when acceleration is constant.*

Having determined velocity as a function of time, we now consider position. With constant acceleration, the velocity increases at a steady rate—and thus the average velocity over an interval is the average of the velocities at the beginning and the end of that interval. So we can write

$$\bar{v} = \frac{1}{2}(v_0 + v) \qquad (2.8)$$

for the average velocity over the interval from $t = 0$ to some later time when the velocity is v. We can also write the average velocity as the change in position divided by the time interval. Suppose that at time 0 our object was at some position x_0. Then its average velocity over a time interval from 0 to time t is

$$\bar{v} = \frac{\Delta x}{\Delta t} = \frac{x - x_0}{t - 0}$$

where x is the object's position at time t. Equating this expression for $\bar{v}$ with the expression in Equation 2.8 gives

$$x = x_0 + \bar{v}t = x_0 + \tfrac{1}{2}(v_0 + v)t \qquad (2.9)$$

But we already found the instantaneous velocity v that appears in this expression; it's given by Equation 2.7. Substituting and simplifying then give the position as a function of time:

$$x = x_0 + v_0 t + \tfrac{1}{2}at^2 \quad \text{(for constant acceleration only)} \qquad (2.10)$$

Does Equation 2.10 make sense? With no acceleration $(a=0)$, the position would increase linearly with time, at a rate given by the initial velocity v_0. With constant acceleration, the additional term $\tfrac{1}{2}at^2$ describes the effect of the ever-changing velocity; time is squared because the longer the object travels, the faster it moves, so the more distance it covers in a given time. Figure 2.9 shows the meaning of the terms in Equation 2.10.

How much runway do I need to land a jetliner, given the touchdown speed and a constant acceleration? Sometimes we're interested in questions like this, where we want to relate position, velocity, and acceleration directly without explicit mention of time. We can solve Equation 2.7 for time:

$$t = \frac{v - v_0}{a}$$

Substituting this expression for t in Equation 2.9, we can write

$$x - x_0 = \tfrac{1}{2}\frac{(v_0 + v)(v - v_0)}{a}$$

or, since $(a+b)(a-b) = a^2 - b^2$,

$$v^2 = v_0^2 + 2a(x - x_0) \qquad (2.11)$$

Equations 2.7, 2.9, 2.10, and 2.11 link all possible combinations of position, velocity, and acceleration for motion with constant acceleration. We summarize them in Table 2.1, and remind you that they apply *only* in the case of constant acceleration.

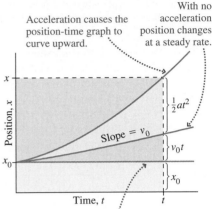

Acceleration causes the position-time graph to curve upward.

With no acceleration position changes at a steady rate.

Slope = v_0

$\tfrac{1}{2}at^2$

$v_0 t$

x_0

With $v = 0$ and $a = 0$, position doesn't change.

FIGURE 2.9 Meaning of the terms in Equation 2.10.

TABLE 2.1 Equations of Motion for Constant Acceleration

Equation	Contains	Number
$v = v_0 + at$	v, a, t; no x	2.7
$x = x_0 + \tfrac{1}{2}(v_0 + v)t$	x, v, t; no a	2.9
$x = x_0 + v_0 t + \tfrac{1}{2}at^2$	x, a, t; no v	2.10
$v^2 = v_0^2 + 2a(x - x_0)$	x, v, a; no t	2.11

Using the Equations of Motion

The equations in Table 2.1 fully describe motion under constant acceleration. Don't regard them as a set of separate laws, but instead recognize them as complementary descriptions of a single underlying phenomenon—one-dimensional motion with *constant acceleration*. Having several equations provides convenient starting points for approaching problems. Don't memorize these equations, but grow familiar with them as you work problems. We now offer a strategy for solving problems about one-dimensional motion with *constant acceleration* using these equations.

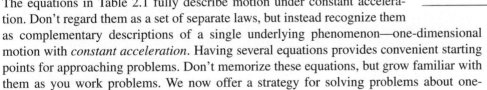

PROBLEM SOLVING STRATEGY 2.1 **Motion with Constant Acceleration**

INTERPRET Interpret the problem to be sure it asks about motion with *constant acceleration*. Next, identify the object(s) whose motion you're interested in.

DEVELOP Draw a diagram with appropriate labels, and choose a coordinate system. For instance, sketch the initial and final physical situations, or draw a position-versus-time graph. Then determine which equations of motion from Table 2.1 contain the quantities you're given and will be easiest to solve for the unknown(s).

EVALUATE Solve the equations in symbolic form and then evaluate numerical quantities.

ASSESS Does your answer make sense? Are the units correct? Do the numbers sound reasonable? What happens in special cases—for example, when a distance, velocity, acceleration, or time becomes very large or very small?

The next two examples are typical of problems involving constant acceleration. Example 2.3 is a straightforward application of the equations we've just derived to a single object. Example 2.4 involves two objects, in which case we need to write equations describing the motions of both objects.

EXAMPLE 2.3 Motion with Constant Acceleration: Landing a Jetliner

A jetliner touches down at 270 km/h. The plane then decelerates (i.e., undergoes acceleration directed opposite its velocity) at 4.5 m/s². What's the minimum runway length on which this aircraft can land?

INTERPRET We interpret this as a problem involving one-dimensional motion with constant acceleration and identify the airplane as the object of interest.

DEVELOP We determine that Equation 2.11, $v^2 = v_0^2 + 2a(x - x_0)$, relates distance, velocity, and acceleration, so our plan is to solve that equation for the minimum runway length. We want the airplane to come to a stop, so the final velocity v is 0, and v_0 is the initial touchdown velocity. If x_0 is the touchdown point, then the quantity $x - x_0$ is the distance we're interested in; we'll call this Δx.

EVALUATE Setting $v = 0$ and solving Equation 2.11 then give

$$\Delta x = \frac{-v_0^2}{2a} = \frac{-[(270 \text{ km/h})(1000 \text{ m/km})/(3600 \text{ s/h})]^2}{(2)(-4.5 \text{ m/s}^2)} = 625 \text{ m}$$

Note that we used a negative value for the acceleration because the plane's acceleration is directed opposite its velocity—which we chose as the positive x direction. We also converted the speed to m/s for compatibility with the SI units given for acceleration.

ASSESS Make sense? That 625 m is just over one-third of a mile, which seems a bit short. However, this is an absolute minimum with no margin of safety. For full-size jetliners, the standard for minimum landing runway length is about 5000 feet or 1.5 km.

✓**TIP** Be Careful with Mixed Units

Frequently problems are stated in units other than SI. Although it's possible to work consistently in other units, when in doubt, convert to SI. In this problem, the acceleration is originally in SI units but the velocity is not—a sure indication of the need for conversion.

EXAMPLE 2.4 Motion with Two Objects: Speed Trap!

A speeding motorist zooms through a 50 km/h zone at 75 km/h (that's 21 m/s) without noticing a stationary police car. The police officer heads after the speeder, accelerating at 2.5 m/s². When the officer catches up to the speeder, how far down the road are they, and how fast is the police car going?

INTERPRET We interpret this as *two* problems involving one-dimensional motion with constant acceleration. We identify the objects in question as the speeding car and the police car. Their motions are related because we're interested in the point where the two coincide.

DEVELOP It's helpful to draw a sketch showing qualitatively the position-versus-time graphs for the two cars. Since the speeding car moves with constant speed, its graph is a straight line. The police car is accelerating from rest, so its graph starts flat and gets increasingly steeper. Our sketch in Fig. 2.10 shows clearly the point we're interested in, when the two cars coincide for the second time. Equation 2.10,

$x = x_0 + v_0 t + \frac{1}{2}at^2$, gives position versus time with constant acceleration. Our plan is (1) to write versions of this equation specialized to each car, (2) to equate the resulting position expressions to find the time when the cars coincide, and (3) to find the corresponding position and the police car's velocity. For the latter we'll use Equation 2.7, $v = v_0 + at$.

EVALUATE Let's take the origin to be the point where the speeder passes the police car and $t = 0$ to be the corresponding time, as marked in Fig. 2.10. Then $x_0 = 0$ in Equation 2.10 for both cars, while the speeder has no acceleration and the police car has no initial velocity. Thus our two versions of Equation 2.10 are

$$x_s = v_{s0}t \text{ (speeder)} \quad \text{and} \quad x_p = \tfrac{1}{2}a_p t^2 \quad \text{(police car)}$$

Equating x_s and x_p tells when the speeder and the police car are at the same place, so we write $v_{s0}t = \frac{1}{2}a_p t^2$. This equation is satisfied when $t = 0$ or $t = 2v_{s0}/a_p$. Why two answers? We asked for *any* times when the two cars are in the same place. That includes the initial encounter at $t = 0$ as well as the later time $t = 2v_{s0}/a_p$ when the police car catches the speeder; both points are shown on our sketch. *Where* does this occur? We can evaluate using $t = 2v_{s0}/a_p$ in the speeder's equation:

$$x_s = v_{s0}t = v_{s0}\frac{2v_{s0}}{a_p} = \frac{2v_{s0}^2}{a_p} = \frac{(2)(21 \text{ m/s})^2}{2.5 \text{ m/s}^2} = 350 \text{ m}$$

Equation 2.7 then gives the police car's speed at this time:

$$v_p = a_p t = a_p \frac{2v_{s0}}{a_p} = 2v_{s0} = 150 \text{ km/h}$$

ASSESS Make sense? As Fig. 2.10 shows, the police car starts from rest and undergoes constant acceleration, so it has to be going faster at the point where the two cars meet. In fact, it's going twice as fast—again, as in Example 2.2, that's because the police car's position depends quadratically on time. That quadratic dependence also tells us that the police car's position-versus-time graph in Fig. 2.10 is a parabola. ■

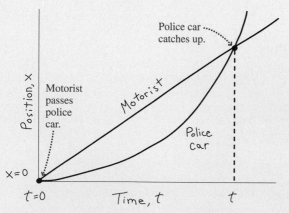

FIGURE 2.10 Our sketch of position versus time for the cars in Example 2.4.

GOT IT? 2.3 The police car in Example 2.4 starts with zero velocity and is going at twice the car's velocity when it catches up to the car. So at some intermediate instant it must be going at the same velocity as the car. Is that instant (a) halfway between the times when the two cars coincide, (b) closer to the time when both cars are at the stoplight, or (c) closer to the time when the police car catches the speeder?

2.5 The Acceleration of Gravity

Drop an object, and it falls at an increasing rate, accelerating because of gravity (Fig. 2.11). The acceleration is constant for objects falling near Earth's surface, and furthermore it has the same value for all objects. This value, the **acceleration of gravity**, is designated g and is approximately 9.8 m/s^2 near Earth's surface.

The acceleration of gravity applies strictly only in **free fall**—motion under the influence of gravity alone. Air resistance, in particular, may dramatically alter the motion, giving the false impression that gravity acts differently on lighter and heavier objects. As early as the year 1600, Galileo is reputed to have shown that all objects have the same acceleration by dropping objects off the Leaning Tower of Pisa. More recently, astronauts verified that a feather and a hammer fall with the same acceleration on the airless Moon—although that acceleration is less than on Earth.

We emphasize again that the values of acceleration and velocity are independent. An object in free fall has a downward acceleration of 9.8 m/s^2 regardless of whether it's moving downward (in which case it's gaining speed) or upward (in which case it's slowing).

Although g is approximately constant near Earth's surface, it varies slightly with latitude and even local geology. The variation with altitude becomes substantial over distances of tens to hundreds of kilometers. But nearer Earth's surface it's a good approximation to take g as strictly constant. Then an object in free fall undergoes constant acceleration, and the equations of Table 2.1 apply. In working gravitational problems, we usually replace x with y to designate the vertical direction. If we make the arbitrary but common choice that the upward direction is positive, then the acceleration a becomes $-g$ because the acceleration is downward.

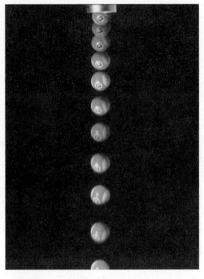

FIGURE 2.11 Strobe photo of a falling ball. Successive images are farther apart, showing that the ball is accelerating.

EXAMPLE 2.5 **Constant Acceleration Due to Gravity: Cliff Diving**

A diver drops from a 10-m-high cliff. At what speed does he enter the water, and how long is he in the air?

INTERPRET This is a case of constant acceleration due to gravity, and the diver is the object of interest. The diver drops a known distance starting from rest, and we want to know the speed and time when he hits the water.

DEVELOP Figure 2.12 is a sketch showing what the diver's position versus time should look like. We've incorporated what we know: the initial position 10 m above the water, the start from rest, and the downward acceleration that results in a parabolic position-versus-time curve. Given the dive height, Equation 2.11, $v^2 = v_0^2 + 2a(x - x_0)$, determines the speed v. Since the diver starts from rest, $v_0 = 0$ and the equation becomes $v^2 = -2g(y - y_0)$. So our plan is first to solve for the speed at the water; then use Equation 2.7, $v = v_0 + at$, to get the time.

EVALUATE Our sketch shows that we've chosen $y = 0$ at the water, so $y_0 = 10 \text{ m}$ and Equation 2.11 gives

$$|v| = \sqrt{-2g(y - y_0)} = \sqrt{(-2)(9.8 \text{ m/s}^2)(0 \text{ m} - 10 \text{ m})}$$
$$= 14 \text{ m/s}$$

This is the magnitude of the velocity, hence the absolute value sign; the actual value is $v = -14 \text{ m/s}$, with the minus sign indicating downward

motion. Knowing the initial and final velocities, we use Equation 2.7 to find how long the dive takes. Solving that equation for t gives

$$t = \frac{v_0 - v}{g} = \frac{0 \text{ m/s} - (-14 \text{ m/s})}{9.8 \text{ m/s}^2} = 1.4 \text{ s}$$

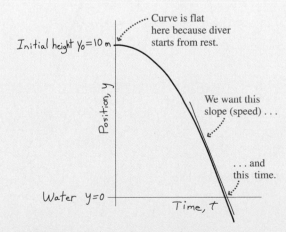

FIGURE 2.12 Our sketch for Example 2.5.

cont'd.

Note the careful attention to signs here; we wrote v with its negative sign and used $a = -g$ in Equation 2.7 because we defined downward to be the negative direction in our coordinate system.

ASSESS Make sense? Our expression for v gives a higher speed with a greater acceleration or a greater distance $y - y_0$—both as expected. Our approach here isn't the only one possible; we could also have found the time by solving Equation 2.10 and then evaluating the speed using Equation 2.7. ∎

In Example 2.5 the diver was moving downward, and the downward gravitational acceleration steadily increased his speed. But the acceleration of gravity is downward regardless of an object's motion. Throw a ball straight up, for example, and it's accelerating *downward* even while moving *upward*. Since velocity and acceleration are in opposite directions, the ball slows until it reaches its peak.

EXAMPLE 2.6 **Constant Acceleration Due to Gravity: Tossing a Ball**

You toss a ball straight up at 7.3 m/s; it leaves your hand at 1.5 m above the floor. Find when it hits the floor, the maximum height it reaches, and its speed when it passes your hand on the way down.

INTERPRET We have constant acceleration due to gravity, and here the object of interest is the ball. We want to find time, height, and speed.

DEVELOP The ball starts by going up, eventually comes to a stop, and then heads downward. Figure 2.13 is a sketch of the height versus time that we expect, showing what we know and the three quantities we're after. Equation 2.10, $x = x_0 + v_0 t + \frac{1}{2} a t^2$, determines position (here, height y) as a function of time, so our plan is to use that equation to find the time the ball hits the floor. Then we can use Equation 2.11, $v^2 = v_0^2 + 2a(x - x_0)$, to find the height at which $v = 0$—that is, the peak height. Finally, Equation 2.11 will also give us the speed at any height, letting us answer the question about the speed when the ball passes the height of 1.5 m on its way down.

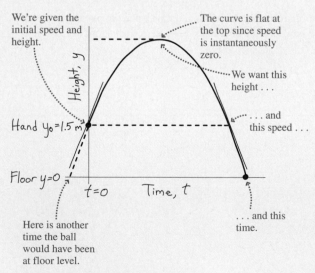

We're given the initial speed and height.

The curve is flat at the top since speed is instantaneously zero.

We want this height . . .

. . . and this speed . . .

Hand $y_0 = 1.5$ m

Floor $y = 0$

$t = 0$ Time, t

Here is another time the ball would have been at floor level.

. . . and this time.

FIGURE 2.13 Our sketch for Example 2.6.

EVALUATE Our sketch shows that we've taken $y = 0$ at the floor, so when the ball is at the floor, Equation 2.10 becomes $0 = y_0 + v_0 t - \frac{1}{2} g t^2$, which we can solve for t using the quadratic formula (Appendix A; $t = \left(v_0 \pm \sqrt{v_0^2 + 2 y_0 g} \right) / g$). Here v_0 is the initial velocity, 7.3 m/s; it's positive because the motion is initially upward. The initial position is the hand height so $y_0 = 1.5$ m, and g of course is 9.8 m/s² (we accounted for the downward acceleration by putting $a = -g$ in Equation 2.10). Putting in these numbers gives $t = 1.7$ s or -0.18 s; the answer we want is 1.7 s. At the peak of its flight, the ball's velocity is instantaneously zero because it's moving neither up nor down. So we set $v^2 = 0$ in Equation 2.11 to get $0 = v_0^2 - 2g(y - y_0)$. Solving for y then gives the peak height:

$$y = y_0 + \frac{v_0^2}{2g} = 1.5 \text{ m} + \frac{(7.3 \text{ m/s})^2}{(2)(9.8 \text{ m/s}^2)} = 4.2 \text{ m}$$

To find the speed when the ball reaches 1.5 m on the way down, we set $y = y_0$ in Equation 2.11. The result is $v^2 = v_0^2$, so $v = \pm v_0$ or ± 7.3 m/s. Once again, there are two answers. The equation has given us *all* the velocities the ball has at 1.5 m—including the initial upward velocity and the later downward velocity. We've shown here that an upward-thrown object returns to its initial height with the same speed it had initially.

ASSESS Make sense? With no air resistance to sap the ball of its energy, it seems reasonable that the ball comes back down with the same speed—a fact we'll explore further when we introduce energy conservation in Chapter 7. But why are there two answers for the time and velocity? Equation 2.10 doesn't "know" about your hand or the floor; it "assumes" the ball has always been undergoing downward acceleration g. We asked of Equation 2.10 when the ball would be at $y = 0$. The second answer, 1.7 s, was the one we wanted. But if the ball had always been in free fall, it would also have been on the floor 0.18 s earlier, heading upward. That's the meaning of the other answer, -0.18 s, as we've indicated on our sketch. Similarly, Equation 2.11 gave us all the velocities the ball had at a height of 1.5 m, including both the initial upward velocity and the later downward velocity. ∎

✓**TIP** Multiple Answers

Frequently the mathematics of a problem gives more than one answer. Think about what each answer means before discarding it! Sometimes an answer isn't consistent with the physical assumptions of the problem, but other times all answers are meaningful even if they aren't all what you're looking for.

GOT IT? 2.4 Standing on a roof, you simultaneously throw one ball straight up and drop another from rest. Which hits the ground first? Which hits the ground moving faster?

APPLICATION   **Keeping Time**

The NIST-F1 atomic clock, shown here with the developers, sets the U.S. standard of time. The clock gets its remarkable accuracy by monitoring a super-cold clump of freely falling cesium atoms for what is, in this context, a long time period of about 1 second. The atom clump is put in free fall by a more sophisticated version of the ball toss in Example 2.6. In the NIST-F1 clock, laser beams gently "toss" the ball of atoms upward with a speed that gives it an up-and-down travel time of about 1 second (see Problem 66). For this reason NIST-F1 is called an **atomic fountain clock**. In the photo you can see the clock's towerlike structure that accommodates this atomic fountain.

Big Picture

The big ideas here are those of **kinematics**—the study of motion without regard to its cause. **Position, velocity,** and **acceleration** are the quantities that characterize motion:

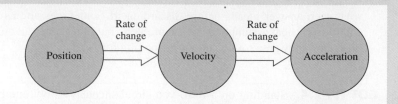

Key Concepts and Equations

Average velocity and acceleration involve changes in position and velocity, respectively, occurring over a time interval Δt:

$$\bar{v} = \frac{\Delta x}{\Delta t}$$

$$\bar{a} = \frac{\Delta v}{\Delta t}$$

Here Δx is the **displacement**, or change in position, and Δv is the change in velocity.

Instantaneous values are the limits of infinitesimally small time intervals and are given by calculus as the time derivatives of position and velocity:

$$v = \frac{dx}{dt}$$

$$a = \frac{dv}{dt}$$

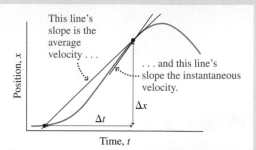

This line's slope is the average velocity . . .

. . . and this line's slope the instantaneous velocity.

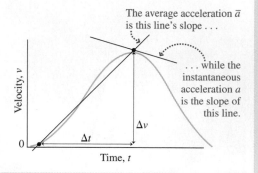

The average acceleration $\bar{a}$ is this line's slope . . .

. . . while the instantaneous acceleration a is the slope of this line.

Cases and Uses

Constant acceleration is a special case that yields simple equations describing one-dimensional motion:

$$v = v_0 + at$$

$$x = x_0 + v_0 t + \tfrac{1}{2}at^2$$

$$v^2 = v_0^2 + 2a(x - x_0)$$

These equations apply only in the case of constant acceleration.

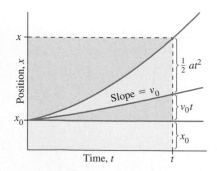

An important example is the acceleration of gravity, essentially constant near Earth's surface, with a magnitude of approximately 9.8 m/s².

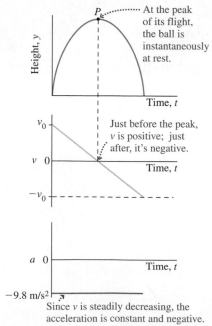

At the peak of its flight, the ball is instantaneously at rest.

Just before the peak, v is positive; just after, it's negative.

Since v is steadily decreasing, the acceleration is constant and negative.

For Thought and Discussion

1. Under what conditions are average and instantaneous velocity equal?
2. You're driving straight at a steady 80 km/h but stop a while for a picnic lunch. How does the stop affect your average velocity?
3. Does a speedometer measure speed or velocity?
4. You check your odometer at the beginning of a day's driving and again at the end. Under what conditions would the difference between the two readings represent your displacement?
5. Consider two possible definitions of average speed: (a) average speed is the average of the values of the instantaneous speed over a time interval; and (b) average speed is the magnitude of the average velocity. Are these definitions equivalent? Give examples to demonstrate your conclusion.
6. Is it possible to be at position $x = 0$ and still be moving?
7. Is it possible to have zero velocity and still be accelerating?
8. If you know the initial velocity v_0 and the initial and final heights y_0 and y, you can use Equation 2.10 to solve for the time t when the object will be at height y. But the equation is quadratic in t, so you'll get two answers. Physically, why is this?
9. Starting from rest, an object undergoes an acceleration given by $a = bt$, where t is time and b is a constant. Can you use the expression bt for a in Equation 2.10 to predict the object's position as a function of time? Why or why not?
10. In which of the velocity-versus-time graphs shown in Fig. 2.14 would the average velocity over the interval shown equal the average of the velocities at the ends of the interval?

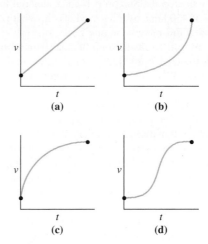

FIGURE 2.14 For Thought and Discussion 10

11. If you travel in a straight line at 50 km/h for 1 h and at 100 km/h for another hour, is your average velocity 75 km/h? If not, is it more or less?
12. If you travel in a straight line at 50 km/h for 50 km and then at 100 km/h for another 50 km, is your average velocity 75 km/h? If not, is it more or less?

Exercises and Problems

Exercises

Section 2.1 Average Motion

13. In 2005 Asafa Powell of Jamaica set a world record in the 100-m dash, with a time of 9.77 s. What was his average speed?
14. In 2004 Mizuki Noguchi of Japan won the Women's Olympic Marathon, completing the 26-mi, 385-yd course in 2 h, 26 min, 20 s. What was Noguchi's average speed, in meters per second?
15. Starting from home, you bicycle 24 km north in 2.5 h and then turn around and pedal straight home in 1.5 h. What are your (a) displacement at the end of the first 2.5 h, (b) average velocity over the first 2.5 h, (c) average velocity for the homeward leg of the trip, (d) displacement for the entire trip, and (e) average velocity for the entire trip?
16. On January 14, 2005, the Huygens probe from the Cassini Orbiter landed on Saturn's moon Titan, approximately 1.2 billion km from Earth. How long did it take Huygens's radio signals, traveling at the speed of light, to reach Earth?
17. Australian Peter Robertson won the 2005 triathlon world championship, completing the 1500-m swim, 40-km bicycle ride, and 10-km run in 1 h, 49 min, 31 s. What was Robertson's average speed?
18. Taking Earth's orbit to be a circle of radius 1.5×10^8 km, determine the speed of Earth's orbital motion in (a) meters per second and (b) miles per second.
19. What is the conversion factor from meters per second to miles per hour?
20. If the average American driver goes 5000 mi each year on interstate highways, how much less time did the average driver spend on interstate highways each year as a result of the 1995 increase in the speed limit from 55 mi/h to 65 mi/h?

Section 2.2 Instantaneous Velocity

21. On a single graph, plot distance versus time for the two trips from Houston to Des Moines described on page 15. For each trip, identify graphically the average velocity and, for each segment of the trip, the instantaneous velocity.
22. For the motion plotted in Fig. 2.15, estimate (a) the greatest velocity in the positive x direction, (b) the greatest velocity in the negative x direction, (c) any times when the object is instantaneously at rest, and (d) the average velocity over the interval shown.

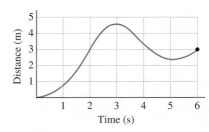

FIGURE 2.15 Exercise 22

23. A model rocket is launched straight upward. Its altitude y as a function of time is given by $y = bt - ct^2$, where $b = 82$ m/s, $c = 4.9$ m/s^2, t is the time in seconds, and y is in meters. (a) Use differentiation to find a general expression for the rocket's velocity as a function of time. (b) When is the velocity zero?

Section 2.3 Acceleration

24. A giant eruption on the Sun propels solar material from rest to a final speed of 450 km/s over a period of 1 h. What is the average acceleration of this material in m/s²?

25. Starting from rest, a subway train first accelerates to 25 m/s and then begins to brake. Forty-eight seconds after starting, it is moving at 17 m/s. What is its average acceleration in this 48-s interval?

26. A space shuttle's main engines cut off 8.5 min after launch, at which time the shuttle's speed is 7.6 km/s. What is the shuttle's average acceleration during this interval?

27. An egg drops from a second-story window, taking 1.12 s to fall and reaching a speed of 11.0 m/s just before hitting the ground. On contact with the ground, the egg stops completely in 0.131 s. Calculate the average magnitudes of its acceleration while falling and while stopping.

28. An airplane's takeoff speed is 320 km/h. If its average acceleration is 2.9 m/s², how long is it on the runway after starting its takeoff roll?

29. ThrustSSC, the world's first supersonic car, accelerates from rest to 1000 km/h in 16 s. What is its acceleration in m/s²?

Section 2.4 Constant Acceleration

30. You're driving at 70 km/h when you accelerate with constant acceleration to pass another car. Six seconds later, you're doing 80 km/h. How far did you go in this time?

31. Differentiate both sides of Equation 2.10, and show that you get Equation 2.7.

32. Electrons that "paint" the picture in a TV tube undergo constant acceleration over a distance of 3.8 cm. If they reach a final speed of 1.2×10^7 m/s, what are (a) the electrons' acceleration and (b) the time spent accelerating?

33. A rocket rises with constant acceleration to an altitude of 85 km, at which point its speed is 2.8 km/s. (a) What is its acceleration? (b) How long does the ascent take?

34. Starting from rest, a car accelerates at a constant rate, reaching 88 km/h in 12 s. (a) What is its acceleration? (b) How far does it go in this time?

35. A car moving initially at 50 mi/h begins decelerating at a constant rate 100 ft short of a stoplight. If the car comes to a full stop just at the light, what is the magnitude of its deceleration?

36. In an X-ray tube, electrons are accelerated to a velocity of 10^8 m/s and then slammed into a tungsten target. The electrons undergo rapid deceleration, producing X rays. If the stopping time for an electron is on the order of 10^{-19} s, approximately how far does an electron move while decelerating? Assume constant deceleration.

37. The Barringer meteor crater in northern Arizona is 180 m deep and 1.2 km in diameter. The fragments of the meteor lie just below the bottom of the crater. If these fragments decelerated at a constant rate of 4×10^5 m/s² as they ploughed through the Earth in forming the crater, what was the speed of the meteor's impact at Earth's surface?

38. A gazelle accelerates from rest at 4.1 m/s² over a distance of 60 m to outrun a predator. What is its final speed?

Section 2.5 The Acceleration of Gravity

39. You drop a rock into a deep well and 4.4 s later hear the splash. How far down is the water? Neglect the travel time of the sound.

40. Your friend is sitting 6.5 m above you in a tree branch. How fast should you throw an apple so that it just reaches her?

41. A model rocket leaves the ground, heading straight up at 49 m/s. (a) What is its maximum altitude? What are its speed and altitude at (b) 1 s, (c) 4 s, and (d) 7 s?

42. A foul ball leaves the bat going straight upward at 23 m/s. (a) How high does it rise? (b) How long is it in the air? Neglect the distance between the bat and the ground.

43. A Frisbee is lodged in a tree branch 6.5 m above the ground. A rock thrown from below must be going at least 3 m/s to dislodge the Frisbee. How fast must such a rock be thrown upward if it leaves the thrower's hand 1.3 m above the ground?

44. Space pirates kidnap an earthling and hold him imprisoned on one of the planets of the solar system. With nothing else to do, the prisoner amuses himself by dropping his watch from eye level (170 cm) to the floor. He observes that the watch takes 0.95 s to fall. On what planet is he being held? *Hint:* Consult Appendix E.

Problems

45. You allow yourself 40 min to drive 25 mi to the airport, but you're caught in heavy traffic and average only 20 mi/h for the first 15 min. What must your average speed be on the rest of the trip if you are to get to the airport on time?

46. A fast base runner can get from first to second base in 3.4 s. If he leaves first base as the pitcher throws a 90 mi/h fastball the 61-ft distance to the catcher, and if the catcher takes 0.45 s to catch and rethrow the ball, how fast does the catcher have to throw the ball to second base to make an out? Home plate to second base is the diagonal of a square 90 ft on a side.

47. You drive the 4600 km from coast to coast of the United States at 65 mi/h (105 km/h), stopping an average of 30 min for rest and refueling after every 2 h of driving. (a) What is your average velocity for the entire trip? (b) How long does the trip take?

48. I can run 90 m/s, 20% faster than my kid brother. How much head start should I give him in order to have a tie race over 100 m?

49. A jetliner leaves San Francisco for New York, 4600 km away. With a strong tailwind, its speed is 1100 km/h. At the same time, a second jet leaves New York for San Francisco. Flying into the wind, it makes only 700 km/h. When and where do the two planes pass each other?

50. The position of an object as a function of time is given by $x = bt + ct^3$, where $b = 1.50$ m/s and $c = 0.640$ m/s³. To study the limiting process leading to the definition of instantaneous velocity, calculate the average velocity of the object over time intervals from (a) 1.00 s to 3.00 s, (b) 1.50 s to 2.50 s, and (c) 1.95 s to 2.05 s. (d) Find the instantaneous velocity as a function of time by differentiating, and compare its value at 2 s with your average velocities.

51. The position of an object as a function of time is given by $x = bt^4$, where b is a constant. Find an expression for the instantaneous velocity as a function of time, and show that the average velocity over the interval from $t = 0$ to any time t is one-fourth of the instantaneous velocity at t.

52. In a drag race, the position of a car as a function of time is given by $x = bt^2$, with $b = 2.000$ m/s². In an attempt to determine the car's velocity midway down a 400-m track, two observers stand at the 180-m and 220-m marks and note the time when the car passes them. (a) What value do the two observers compute for the car's velocity? Give your answer to four significant figures. (b) By what percentage does this observed value differ from the actual instantaneous value at $x = 200$ m?

53. The position of an object is given by $x = bt^3$, where x is in meters, t is in seconds, and the constant b is 1.5 m/s^3. Determine (a) the instantaneous velocity and (b) the instantaneous acceleration at the end of 2.5 s. Find (c) the average velocity and (d) the average acceleration during the first 2.5 s.

54. If you square Equation 2.7, you'll have an expression for v^2. Equation 2.11 also gives an expression for v^2. Equate the two expressions for v^2, and show that the resulting equation reduces to Equation 2.10.

55. On packed snow, the use of computerized antilock brakes can reduce the stopping distance for a car by 55%. By what percentage is the stopping *time* reduced?

56. A particle leaves its initial position x_0 at time $t = 0$, moving in the positive x direction with speed v_0 but undergoing acceleration of magnitude a in the negative x direction. Find expressions for (a) the time when it returns to the position x_0 and (b) its speed when it passes that point.

57. A hockey puck moving at 32 m/s slams through a wall of snow 35 cm thick. It emerges moving at 18 m/s. (a) How much time does it spend in the snow? (b) How thick a wall of snow would be needed to stop the puck entirely?

58. Amtrak's 20th-Century Limited is en route from Chicago to New York at 110 km/h when the engineer spots a cow on the track. The train brakes to a halt in 1.2 min, stopping just in front of the cow. (a) What is the magnitude of the train's (constant) acceleration? (b) What is the direction of the acceleration? (c) How far was the train from the cow when the engineer first applied the brakes?

59. A jetliner touches down at 220 km/h, reverses its engines to provide braking, and comes to a halt 29 s later. What is the shortest runway on which this aircraft can land, assuming constant deceleration starting at touchdown?

60. A motorist suddenly notices a stalled car and slams on the brakes, decelerating at the rate of 6.3 m/s^2. Unfortunately this isn't good enough, and a collision ensues. From the damage sustained, police estimate that the car was moving at 18 km/h at the time of the collision. They also measure skid marks 34 m long. (a) How fast was the motorist going when the brakes were first applied? (b) How much time elapsed from the initial braking to the collision?

61. A racing car undergoing constant acceleration covers 140 m in 3.6 s. (a) If it's moving at 53 m/s at the end of this interval, what was its speed at the beginning of the interval? (b) How far did it travel *from rest* to the end of the 140-m distance?

62. The maximum deceleration of a car on a dry road is about 8 m/s^2. If two cars are moving head-on toward each other at 88 km/h (55 mi/h), and their drivers apply their brakes when they are 85 m apart, will they collide? If so, at what relative speed? If not, how far apart will they be when they stop? On the same graph, plot distance versus time for both cars.

63. After 35 minutes of running, at the 9-km point in a 10-km race, you find yourself 100 m behind the leader and moving at the same speed. What should your acceleration be if you are to catch up by the finish line? Assume that the leader maintains a constant speed throughout the entire race.

64. You're speeding at 85 km/h when you notice that you're only 10 m behind the car in front of you, which is moving at the legal speed limit of 60 km/h. You slam on your brakes, and your car decelerates at 4.2 m/s^2. Assuming the car in front of you continues at constant speed, will you collide? If so, at what relative speed? If not, what will be the distance between the cars at their closest approach?

65. The Mars rover Spirit landed in 2004 to explore the Martian surface. Its landing was cushioned by airbags, and the rover bounced some 15 m vertically after its first impact. Assuming no loss of speed at contact with the Martian surface, what was Spirit's impact speed?

66. Calculate the speed with which cesium atoms must be "tossed" in the NIST-F1 atomic clock so that their up-and-down travel time is 1.0 s. See Application on page 25.

67. A falling object travels one-fourth of its total distance in the last second of its fall. From what height was it dropped?

68. The defenders of a castle throw rocks down on their attackers from a 15-m-high wall. If the rocks are thrown with an initial speed of 10 m/s, how much faster are they moving when they hit the ground than if they were simply dropped?

69. Two divers jump from a 3.00-m platform. One jumps upward at 1.80 m/s, and the second steps off the platform as the first passes it on the way down. (a) What are their speeds as they hit the water? (b) Which hits the water first and by how much?

70. A balloon is rising at 10 m/s when its passenger throws a ball straight up at 12 m/s. How much later does the passenger catch the ball?

71. Landing on the Moon, a spacecraft fires its retrorockets and comes to a complete stop just 12 m above the lunar surface. It then drops freely to the surface. How long does it take to fall, and what is its impact speed? *Hint*: Consult Appendix E.

72. Launched from the ground, a rocket accelerates vertically upward at 4.6 m/s^2. It passes through a band of clouds 5.3 km thick, extending upward from an altitude of 1.9 km. How long is it in the clouds?

73. A subway train is traveling at 80 km/h when it approaches a slower train 50 m ahead traveling in the same direction at 25 km/h. If the faster train begins decelerating at 2.1 m/s^2 while the slower train continues at constant speed, how soon and at what relative speed will they collide?

74. You toss a book into your dorm room, just clearing a windowsill 4.2 m above the ground. (a) If the book leaves your hand 1.5 m above the ground, how fast must it be going to clear the sill? (b) How long after it leaves your hand will it hit the floor, 0.87 m below the windowsill?

75. Consider an object traversing a distance L, part of the way at speed v_1 and the rest of the way at speed v_2. Find expressions for the average speeds (a) when the object moves at each of the two speeds for half the total *time* and (b) when the object moves at each of the two speeds for half the *distance*.

76. The position of a particle as a function of time is given by $x = x_0 \sin \omega t$, where x_0 and ω are constants. (a) Take derivatives to find expressions for the velocity and acceleration. (b) What are the maximum values of velocity and acceleration? *Hint:* Consult the table of derivatives in Appendix A.

77. Ice skaters, ballet dancers, and basketball players executing vertical leaps often give the illusion of "hanging" almost motionless near the top of the leap. To see why this is, consider a leap that takes an athlete up a vertical distance h. Of the total time spent in the air, what fraction is spent in the upper half (i.e., at $y > \frac{1}{2}h$)?

78. A student is staring idly out her dormitory window when she sees a water balloon fall past. If the balloon takes 0.22 s to cross the 130-cm-high window, from what height above the top of the window was it dropped?

79. A police radar has an effective range of 1.0 km, and a motorist's radar detector has a range of 1.9 km. The motorist is going 110 km/h in a 70 km/h zone when the radar detector beeps. At what rate must the motorist decelerate to avoid a speeding ticket?

80. For the trip in Exercise 15, find the average speed for the entire trip. Show that this speed is equal to the time-weighted average of the speeds for the individual trip segments.

81. An object starts moving in a straight line from position x_0, at time $t = 0$, with velocity v_0. Its acceleration is given by $a = a_0 + bt$, where a_0 and b are constants. Find expressions for (a) the instantaneous velocity and (b) the position, as functions of time.

82. You are keeping pace with another runner, at 6 minutes per mile, but are 10 m behind her when she is 100 m from the finish line. What constant acceleration do you need to catch her at the finish if she maintains a constant speed? At what pace will you be running at the finish? Express acceleration in m/s^2 and pace in minutes per mile. (Pace is the reciprocal of speed.)

83. For the ball tossed in Example 2.6, (a) find its velocity just before it hits the floor. Suppose you had tossed a second ball straight down at 7.3 m/s (from the same place 1.5 m above the floor). (b) What would its velocity be just before it hits the floor? (c) When would the second ball hit the floor? (Interpret any multiple answers.)

84. Undaunted when you threw him out of your office on a previous visit (see Problem 52 in Chapter 1), a computer sales representative now tries to sell you on the speed of his computer. Your manufacturing process requires that the computer be located such that the travel time between the computer and the machinery is 10 μs or less. The sales rep claims it will take data 8 μs to travel between the computer and the machine if the distance is 2 ft. Is he still trying to fool you?

85. Your roommate is an aspiring novelist. Because you are taking physics your roommate asks your opinion on a matter of physics. It seems the central character in the novel is kept awake at night by a leaky faucet. The sink is 19.6 cm below the leaky faucet. At the instant one drop leaves the faucet, another strikes the sink below and two more drops are in between on the way down. How many drops per second are keeping the annoyed protagonist awake?

86. You and your physics major roommate become involved in a sinister plot to drop water balloons on students entering your dorm. Your room is 64 ft above the sidewalk. You plan to place an X on the sidewalk to mark the spot a student must be when the balloon is dropped. The student will walk the distance from the X to the place the balloon hits in the time the balloon falls. After observing several students, you conclude most students walk at about 2 m/s when coming into the dorm. How far from the impact point do you place the X?

87. Estimate the time it would take you to travel from Los Angeles to New York (about 2800 miles) if you walk at 3 mi/h, ride a bike at 25 km/h, or drive a car at an average speed of 55 mi/h.

Answers to Chapter Questions

Answer to Chapter Opening Question

Although the ball's velocity is zero at the top of its motion, its acceleration is -9.8 m/s^2, as it is throughout the toss.

Answers to GOT IT? Questions

2.1 (a) and (b); average speed is greater for (c).

2.2 (b) moves with constant speed; (a) reverses; (d) speeds up.

2.3 Halfway between the times. Because its acceleration is constant, the police car's speed increases by equal amounts in equal times. So it gets from 0 to half its final velocity—which is twice the car's velocity—in half the total time.

2.4 The dropped ball hits first; the thrown ball hits moving faster.

3 Motion in Two and Three Dimensions

■ At what angle should this penguin leave the water to maximize the range of its jump?

What's the speed of an orbiting satellite? How should I leap to win the long-jump competition? How do I engineer a curve in the road for safe driving? These and many other questions involve motion in more than one dimension. In this chapter we extend the ideas of one-dimensional motion to these more complex—and more interesting—situations.

3.1 Vectors

We've seen that quantities describing motion have direction as well as magnitude. In Chapter 2, a simple plus or minus sign took care of direction. But now, in two or three dimensions, we need a way to account for all possible directions. We do this with mathematical quantities called **vectors**, which express both magnitude and direction. Vectors stand in contrast to **scalars**, which are quantities that have no direction.

Position and Displacement

The simplest vector quantity is position. Given an origin, we can characterize any position in space by drawing an arrow from the origin to that position. That arrow is a pictorial representation of a **position vector**, which we call $\vec{r}$. The arrow over the r indicates that this

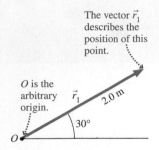

The vector $\vec{r}_1$ describes the position of this point.

O is the arbitrary origin.

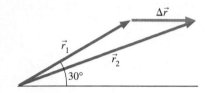

FIGURE 3.1 A position vector $\vec{r}_1$.

FIGURE 3.2 Vectors $\vec{r}_1$ and $\Delta\vec{r}$ sum to $\vec{r}_2$.

is a vector quantity, and it's crucial to include the arrow whenever you're dealing with vectors. Figure 3.1 shows a position vector in a two-dimensional coordinate system; this vector describes a point a distance of 2 m from the origin, in a direction 30° from the horizontal axis.

Suppose you walk from the origin straight to the point described by the vector $\vec{r}_1$ in Fig. 3.1, and then you turn right and walk another 1 m. Figure 3.2 shows how you can tell where you end up. Draw a second vector whose length represents 1 m and that points to the right; we'll call this vector $\Delta\vec{r}$ because it's a **displacement vector**, representing a *change* in position. Put the tail of $\Delta\vec{r}$ at the head of the vector $\vec{r}_1$; then the head of $\Delta\vec{r}$ shows your ending position. The result is the same as if you had walked straight from the origin to this position. So the new position is described by a third vector $\vec{r}_2$, as indicated in Fig. 3.2. What we've just described is the process of **vector addition**. To add two vectors, put the second vector's tail at the head of the first; the sum is then the vector that extends from the tail of the first vector to the head of the second, as does $\vec{r}_2$ in Fig. 3.2.

Remember that a vector has both magnitude and direction—but because that's all the information it contains, it doesn't matter where a vector starts. So you're free to move a vector around anywhere as needed to form vector sums. Figure 3.3 shows some examples of vector addition and also shows that vector addition obeys simple rules you know for regular arithmetic.

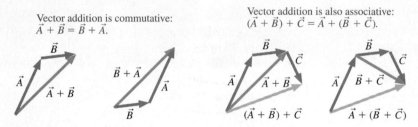

FIGURE 3.3 Vector addition is commutative and associative.

Multiplication

You and I both jog in the same direction, but you go twice as far. Your displacement vector, $\vec{B}$, is then twice as long as my displacement vector, $\vec{A}$; mathematically, we write $\vec{B} = 2\vec{A}$. That's what it means to multiply a vector by a scalar; simply rescale the magnitude of the vector by that scalar. If the scalar is negative, then the vector direction reverses—and that provides a way to subtract vectors. In Fig. 3.2, for example, you can see that $\vec{r}_1 = \vec{r}_2 + (-1)\Delta\vec{r}$, or simply $\vec{r}_1 = \vec{r}_2 - \Delta\vec{r}$. Later, we'll see ways to multiply two vectors, but for now the only multiplication we consider is a vector multiplied by a scalar.

Vector Components

You can always add vectors by the graphical method shown in Fig. 3.2, or you can use geometric relationships like the laws of sines and cosines to accomplish the same thing algebraically. In both these approaches, you specify a vector by giving its magnitude and its direction. But often it's more convenient instead to describe vectors in terms of their **components** in a given coordinate system.

A **coordinate system** is a framework for describing positions in space. It's a mathematical construct, and you're free to choose whatever coordinate system you want. You've already seen **Cartesian** or **rectangular coordinate systems**, in which a pair of numbers (x, y) represents each point in a plane. You could also think of each point as representing the head of a position vector, in which case the numbers x and y are the vector components. The components tell how much of the vector is in the x direction and how much is in the y direction. Not all vectors represent actual positions in space; for example, there are velocity, acceleration, and force vectors. The lengths of these vectors represent the magnitudes of the corresponding physical quantities, and we can still talk about components in a chosen coordinate system. For an arbitrary vector quantity $\vec{A}$, we designate the components A_x and A_y (Fig. 3.4). Note that the components themselves aren't vectors but scalars.

In two dimensions it takes two quantities to specify a vector—either its magnitude and direction or its components. How are the two related? By the Pythagorean theorem and the definitions of the trig functions, as shown in Fig. 3.4. Summarizing, we have

$$A = \sqrt{A_x^2 + A_y^2} \quad \text{and} \quad \tan\theta = \frac{A_y}{A_x} \qquad (\text{vector magnitude and direction}) \quad (3.1)$$

Note that a vector's symbol without the arrow above it stands for the magnitude of the vector.

Going the other way, we have

$$A_x = A\cos\theta \quad \text{and} \quad A_y = A\sin\theta \qquad (\text{vector components}) \quad (3.2)$$

If a vector $\vec{A}$ has zero magnitude, we write $\vec{A} = \vec{0}$, where the vector arrow on the zero indicates that both components must be zero.

Unit Vectors

It's cumbersome to say "a vector of magnitude 2 m at 30° to the x axis" or, equivalently, "a vector whose x and y components are 1.73 m and 1.0 m, respectively." We can express this more succinctly using the **unit vectors** $\hat{i}$ and $\hat{j}$ (read as "i hat"). These unit vectors have magnitude 1, no units, and point along the x and y axes, respectively. In three dimensions we add a third unit vector, $\hat{k}$, along the z axis. Any vector in the x direction can be written as some number—perhaps with units, such as meters or meters per second—times the unit vector $\hat{i}$, and analogously in the y direction using $\hat{j}$. That means any vector in a plane can be written as a sum involving the two unit vectors: $\vec{A} = A_x\hat{i} + A_y\hat{j}$ (Fig. 3.5a). Similarly, any vector in space can be written with the three unit vectors (Fig. 3.5b).

The unit vectors convey only directional information; the numbers that multiply them tell about size and units. Together they provide compact mathematical representations of all vectors, including their units. The displacement vector $\vec{r}_1$ in Fig. 3.1, for example, is $\vec{r}_1 = 1.7\hat{i} + 1.0\hat{j}$ m.

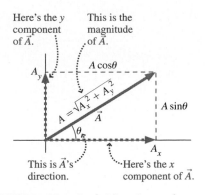

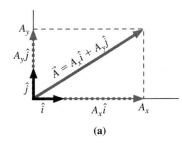

FIGURE 3.4 Magnitude/direction and component representations of vector $\vec{A}$.

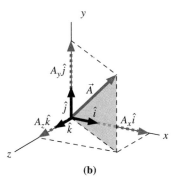

FIGURE 3.5 Vectors in (a) a plane and (b) space, expressed using unit vectors.

EXAMPLE 3.1 **Unit Vectors: Taking a Drive**

You drive to a city 160 km from home, going 35° north of east. Express your new position in unit vector notation, using an east-west/north-south coordinate system.

INTERPRET We interpret this as a problem about writing a vector in unit vector notation, given its magnitude and direction.

DEVELOP Unit vector notation multiplies a vector's x and y components by the unit vectors $\hat{i}$ and $\hat{j}$ and sums the results, so we draw a sketch showing those components (Fig. 3.6). Our plan is to solve for the two components, multiply by the unit vectors, and then add. Equations 3.2 determine the components.

EVALUATE We have $x = r\cos\theta = (160 \text{ km})(\cos 35°) = 131 \text{ km}$ and $y = r\sin\theta = (160 \text{ km})(\sin 35°) = 92 \text{ km}$. Then the position of the city is

$$\vec{r} = 131\hat{i} + 92\hat{j} \text{ km}$$

ASSESS Make sense? Figure 3.6 suggests that the x component should be longer than the y component, as our answer indicates. Our sketch

shows the component values and the final answer. Note that we treat $131\hat{i} + 92\hat{j}$ as a single vector quantity, labeling it at the end with the appropriate unit, km. ∎

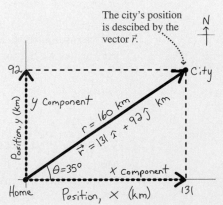

FIGURE 3.6 Our sketch for Example 3.1.

Vector Arithmetic with Unit Vectors

Vector addition is simple with unit vectors; we just add the corresponding components. If $\vec{A} = A_x\hat{i} + A_y\hat{j}$ and $\vec{B} = B_x\hat{i} + B_y\hat{j}$, for example, then their sum is

$$\vec{A} + \vec{B} = (A_x\hat{i} + A_y\hat{j}) + (B_x\hat{i} + B_y\hat{j}) = (A_x + B_x)\hat{i} + (A_y + B_y)\hat{j}$$

Subtraction and multiplication by a scalar are similarly straightforward.

GOT IT? 3.1 Which vector describes a displacement of 10 units in a direction $30°$ below the positive x axis? (a) $10\hat{i} - 10\hat{j}$, (b) $5.0\hat{i} - 8.6\hat{j}$, (c) $8.6\hat{i} - 5.0\hat{j}$, (d) $10(\hat{i} + \hat{j})$

3.2 Velocity and Acceleration Vectors

We defined velocity in one dimension as the rate of change of position. In two or three dimensions it's the same thing, except now the change in position—displacement—is a vector. So we write

$$\overline{\vec{v}} = \frac{\Delta\vec{r}}{\Delta t} \quad \text{(average velocity vector)} \tag{3.3}$$

for the average velocity, in analogy with Equation 2.1. Here division by Δt simply means multiplying by $1/\Delta t$. As before, instantaneous velocity is given by a limiting process:

$$\vec{v} = \lim_{\Delta t \to 0} \frac{\Delta\vec{r}}{\Delta t} = \frac{d\vec{r}}{dt} \quad \text{(instantaneous velocity vector)} \tag{3.4}$$

Again, that derivative $d\vec{r}/dt$ is shorthand for the result of the limiting process, taking ever smaller time intervals Δt and the corresponding displacements $\Delta\vec{r}$. Another way to look at Equation 3.4 is in terms of components. If $\vec{r} = x\hat{i} + y\hat{j}$, then we can write

$$\vec{v} = \frac{d\vec{r}}{dt} = \frac{dx}{dt}\hat{i} + \frac{dy}{dt}\hat{j} = v_x\hat{i} + v_y\hat{j}$$

where the velocity components v_x and v_y are the derivatives of the corresponding position components.

Acceleration is the rate of change of velocity, so we write

$$\overline{\vec{a}} = \frac{\Delta\vec{v}}{\Delta t} \quad \text{(average acceleration vector)} \tag{3.5}$$

for the average acceleration and

$$\vec{a} = \lim_{\Delta t \to 0} \frac{\Delta\vec{v}}{\Delta t} = \frac{d\vec{v}}{dt} \quad \text{(instantaneous acceleration vector)} \tag{3.6}$$

for the instantaneous acceleration. We can also express instantaneous acceleration in components, as we did for velocity:

$$\vec{a} = \frac{d\vec{v}}{dt} = \frac{dv_x}{dt}\hat{i} + \frac{dv_y}{dt}\hat{j} = a_x\hat{i} + a_y\hat{j}$$

Velocity and Acceleration in Two Dimensions

Motion in a straight line may or may not involve acceleration, but motion on curved paths in two or three dimensions is *always* accelerated motion. Why? Because moving in multiple dimensions means *changing direction*—and *any* change in velocity, including direction, involves acceleration. Get used to thinking of acceleration as meaning more than "speeding up" or "slowing down." It can equally well mean "changing direction," whether or not speed is also changing. Whether acceleration results in a speed change, a direction change, or both depends on the relative orientation of the velocity and acceleration vectors.

Suppose you're driving down a straight road at speed v_0 when you step on the gas to give a constant acceleration $\vec{a}$ for a time Δt. Equation 3.5 shows that the change in your velocity is $\Delta\vec{v} = \vec{a}\Delta t$. In this case the acceleration is in the same direction as your velocity and, as Fig. 3.7a shows, the result is an increase in the magnitude of your velocity; that is, you speed up. Step on the brake, and your acceleration is opposite your velocity, and you slow down (Fig. 3.7b).

✓ **TIP** Vectors Tell It All

Are you thinking there should be a minus sign in Fig. 3.7b because the speed is decreasing? Nope: Vectors have both magnitude and direction, and the vector addition $\vec{v} = \vec{v}_0 + \vec{a}\Delta t$ tells it all. In Fig. 3.7b, $\Delta\vec{v}$ points to the left, and that takes care of the "subtraction."

In two dimensions acceleration and velocity do not have to be colinear but can be at any angle. In general, acceleration then changes both the magnitude and the direction of the velocity (Fig. 3.8). Particularly interesting is the case when $\vec{a}$ is perpendicular to $\vec{v}$; then only the direction of motion changes. If acceleration is constant—in both magnitude and direction—then the two vectors won't stay perpendicular once the direction of $\vec{v}$ starts to change, and the magnitude will change, too. But in the special case where acceleration changes direction so it's always perpendicular to velocity, then it's strictly true that only the direction of motion changes. Figure 3.9 illustrates this point, which we'll soon explore quantitatively.

GOT IT? 3.2 An object is accelerating downward. Which, if any, of the following must be true? (a) The object cannot be moving upward. (b) The object cannot be moving in a straight line. (c) The object is moving directly downward. (d) If the object's motion is instantaneously horizontal, it can't continue to be so.

3.3 Relative Motion

You stroll down the aisle of a plane, walking toward the front at a leisurely 4 km/h. Meanwhile the plane is moving relative to the ground at 1000 km/h. Obviously, then, you're moving at 1004 km/h relative to the ground. As this example suggests, velocity is meaningful only when we know the answer to the question, Velocity relative to what? That "what" is called a **frame of reference**. Often we know an object's velocity relative to one frame of reference—for example, your velocity relative to the plane—and we want to know its velocity relative to some other reference frame—in this case the ground. In this one-dimensional case, we can simply add the two velocities. If you had been walking toward the back of the plane, then the two velocities would have opposite signs and you would be going at 996 km/h relative to the ground.

The same idea works in two dimensions, but here we need to recognize that velocity is a vector. Suppose that airplane is flying with velocity $\vec{v}\,'$ relative to the air. If a wind is blowing, then the air is moving with some velocity $\vec{V}$ relative to the ground. The plane's velocity $\vec{v}$ relative to the ground is the vector sum of its velocity relative to the air and the air's velocity relative to the ground:

$$\vec{v} = \vec{v}\,' + \vec{V} \qquad \text{(relative velocity)} \qquad (3.7)$$

Here we use lowercase letters for the velocities of an object relative to two different reference frames; we distinguish the two with the prime on one of the velocities. The capital $\vec{V}$

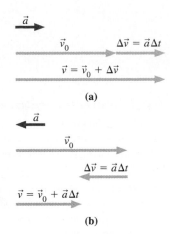

FIGURE 3.7 When $\vec{v}$ and $\vec{a}$ are colinear, only the speed changes.

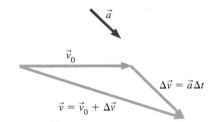

FIGURE 3.8 In general, acceleration changes both the magnitude and the direction of velocity.

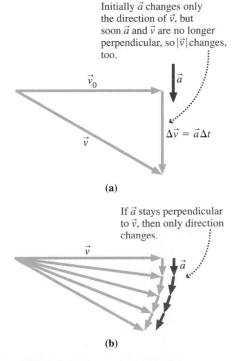

FIGURE 3.9 Acceleration that is always perpendicular to velocity changes only the direction.

is the relative velocity between the two frames. In general, Equation 3.7 lets us use the velocity of an object in one reference frame to find its velocity relative to another frame—provided we know that relative velocity $\vec{V}$. Example 3.2 illustrates the application of this idea to aircraft navigation.

EXAMPLE 3.2 **Relative Velocity: Navigating a Jetliner**

A jetliner flies at 960 km/h relative to the air. It's going from Houston to Omaha, 1290 km northward. At cruising altitude a wind is blowing eastward at 190 km/h. In what direction should the plane fly? How long will the trip take?

INTERPRET This is a problem involving relative velocities. We identify the given information: the plane's speed, but not its direction, in the reference frame of the air; the plane's direction, but not its speed, in the reference frame of the ground; and the wind velocity, both speed and direction.

DEVELOP Equation 3.7, $\vec{v} = \vec{v}\,' + \vec{V}$, applies, and we identify $\vec{v}$ as the plane's velocity relative to the ground, $\vec{v}\,'$ as its velocity relative to the air, and $\vec{V}$ as the wind velocity. Equation 3.7 shows that $\vec{v}\,'$ and $\vec{V}$ add vectorially to give $\vec{v}$; that, with the given information, helps us draw the situation (Fig. 3.10). Measuring the angle of $\vec{v}\,'$ and the length of $\vec{v}$ in the diagram would then give the answers. However, we'll work the problem algebraically using vector components. Since the plane is flying northward and the wind is blowing eastward, a suitable coordinate system has x axis eastward and y axis northward. Our plan is to work out the vector components in these coordinates and then apply Equation 3.7.

EVALUATE Using Equations 3.2 for the vector components, we can express the three vectors as

$$\vec{v}\,' = v'\cos\theta\,\hat{\imath} + v'\sin\theta\,\hat{\jmath}, \quad \vec{V} = V\hat{\imath}, \quad \text{and} \quad \vec{v} = v\hat{\jmath}$$

Here we know the magnitude v' of the velocity $\vec{v}\,'$, but we don't know the angle θ. We know the magnitude V of the wind velocity $\vec{V}$, and we also know its direction—toward the east. So $\vec{V}$ has only an x component. Meanwhile we want the velocity $\vec{v}$ relative to the ground to be purely northward, so it has only a y component—although we don't know its magnitude v. We're now ready to put the three velocities into Equation 3.7. Since two vectors are equal only if all their components are equal, we can express the vector Equation 3.7 as two separate scalar equations for the x and y components:

x component: $\quad v'\cos\theta + V = 0$

y component: $\quad v'\sin\theta + 0 = v$

The rest is math, evaluating the unknowns θ and v. Solving the x equation gives

$$\theta = \cos^{-1}\left(-\frac{V}{v'}\right) = \cos^{-1}\left(-\frac{190\text{ km/h}}{960\text{ km/h}}\right) = 101.4°$$

This angle is measured from the x axis (eastward; see Fig. 3.10), so it amounts to a flight path 11° west of north. We can then evaluate v from the y equation:

$$v = v'\sin\theta = (960\text{ km/h})(\sin 101.4°) = 941\text{ km/h}$$

That's the plane's speed relative to the ground. Going 1290 km will then take $(1290\text{ km})/(941\text{ km/h}) = 1.4$ h.

ASSESS Make sense? The plane's heading of 11° west of north seems reasonable compensation for an eastward wind blowing at 190 km/h, given the plane's airspeed of 960 km/h. If there were no wind, the trip would take 1 h, 20 min (1290 km divided by 960 km/h), so our time of 1 h, 24 min with the wind makes sense. ∎

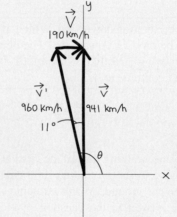

FIGURE 3.10 Our vector diagram fo Example 3.2.

3.4 Constant Acceleration

When acceleration is constant, the individual components of the acceleration vector are themselves constant. Furthermore, the component of acceleration in one direction has no effect on the motion in a perpendicular direction (Fig. 3.11). Then with constant acceleration, the separate components of the motion must obey the constant-acceleration formulas we developed in Chapter 2 for one-dimensional motion. Using vector notation, we can then generalize Equations 2.7 and 2.10 to read

$$\vec{v} = \vec{v}_0 + \vec{a}t \qquad \text{(for constant acceleration only)} \tag{3.8}$$

$$\vec{r} = \vec{r}_0 + \vec{v}_0 t + \tfrac{1}{2}\vec{a}t^2 \qquad \text{(for constant acceleration only)} \tag{3.9}$$

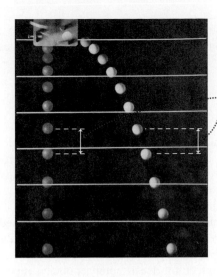

Vertical spacing is
the same, showing
that vertical and
horizontal motion
are independent.

FIGURE 3.11 Two marbles, one
dropped and the other projected
horizontally.

where $\vec{r}$ is the position vector. In two dimensions, each of these vector equations represents a pair of scalar equations describing constant acceleration in two mutually perpendicular directions. Equation 3.9, for example, contains the pair $x = x_0 + v_{x0}t + \frac{1}{2}a_x t^2$ and $y = y_0 + v_{y0}t + \frac{1}{2}a_y t^2$. (Remember that the components of the displacement vector $\vec{r}$ are just the coordinates x and y.) In three dimensions there would be a third equation for the z component of the motion. Starting with these vector forms of the equations of motion, you can apply Problem-Solving Strategy 2.1 to problems in two or three dimensions.

EXAMPLE 3.3   Acceleration in Two Dimensions: Windsurfing

You're windsurfing at 7.3 m/s when a wind gust hits, accelerating your sailboard at 0.82 m/s² at a 60° angle to your original direction. If the gust lasts 8.7 s, what is the board's net displacement during this time?

INTERPRET This is a problem involving constant acceleration in two dimensions. The key concept is that motion in two perpendicular directions is independent, so we can treat the problem as involving two separate one-dimensional motions.

DEVELOP Equation 3.9, $\vec{r} = \vec{r}_0 + \vec{v}_0 t + \frac{1}{2}\vec{a}t^2$, will give the board's displacement. We need a coordinate system, so we take the x axis along the board's initial motion, with the origin at the point where the wind gust first hits. Our plan is to find the components of the acceleration vector and then apply the two components of Equation 3.9 to get the components of the displacement. In Fig. 3.12 we draw the acceleration vector to determine its components.

EVALUATE With the x direction along the initial velocity, $\vec{v}_0 = 7.3\hat{\imath}$ m/s. As Fig. 3.12 shows, the acceleration is $\vec{a} = 0.41\hat{\imath} + 0.71\hat{\jmath}$ m/s². Our choice of origin gives $x_0 = y_0 = 0$, so the two components of Equation 3.9 are

$$x = v_{x0}t + \tfrac{1}{2}a_x t^2 = 79.0 \text{ m}$$
$$y = \tfrac{1}{2}a_y t^2 = 26.9 \text{ m}$$

where we used the appropriate components of $\vec{a}$ and where $t = 8.7$ s. The new position vector is then $\vec{r} = x\hat{\imath} + y\hat{\jmath} = 79.0\hat{\imath} + 26.9\hat{\jmath}$ m, giving a net displacement of $r = \sqrt{x^2 + y^2} = 84$ m.

ASSESS Make sense? Figure 3.13 shows how the acceleration deflects the sailboard from its original path and also increases its speed somewhat. Since the acceleration makes a fairly large angle with the initial velocity, the change in direction is the greater effect. ∎

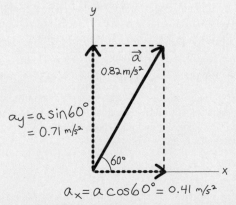

FIGURE 3.12 Our sketch of the sailboard's acceleration components.

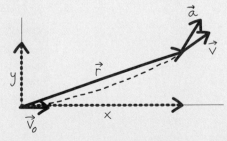

FIGURE 3.13 Our sketch of the displacement $\vec{r}$ and curved path (dashed) for the sailboard.

GOT IT? 3.3 An object is moving initially in the $+x$ direction. Which of the following accelerations, all acting for the same time interval, will cause the greatest change in its speed? In its direction? (a) $10\hat{\imath}$ m/s^2, (b) $10\hat{\jmath}$ m/s^2, (c) $10\hat{\imath} + 5\hat{\jmath}$ m/s^2, (d) $2\hat{\imath} - 8\hat{\jmath}$ m/s^2

3.5 Projectile Motion

A **projectile** is an object that's launched into the air and then moves predominantly under the influence of gravity. Examples are numerous; baseballs, jets of water, fireworks, missiles, ejecta from volcanoes, drops of ink in an ink-jet printer, and leaping dolphins are all projectiles.

To treat projectile motion, we make two simplifying assumptions: (1) We neglect any variation in the direction or magnitude of the gravitational acceleration, and (2) we neglect air resistance. The first assumption is equivalent to neglecting the curvature of the Earth and is valid for projectiles whose displacements are small compared with Earth's radius. Air resistance has a more variable effect; for dense, compact objects it is often negligible, but for objects whose ratio of surface area to mass is large—like ping-pong balls and parachutes—air resistance may dramatically alter the motion.

To describe projectile motion, it's convenient to choose a coordinate system with the y axis vertically upward and the x axis horizontal. With the only acceleration provided by gravity, $a_x = 0$ and $a_y = -g$, so the components of Equations 3.8 and 3.9 become

$$v_x = v_{x0} \qquad\qquad (3.10)$$

$$v_y = v_{y0} - gt \qquad (3.11)$$
(for constant gravitational

$$x = x_0 + v_{x0}t \qquad (3.12)$$
acceleration)

$$y = y_0 + v_{y0}t - \tfrac{1}{2}gt^2 \qquad (3.13)$$

Here we take g to be a positive number, and account for the downward direction using minus signs. Equations 3.10–3.13 tell us mathematically what Fig. 3.14 tells us physically: Projectile motion can be broken down into two perpendicular and independent components—horizontal motion with constant velocity and vertical motion with constant acceleration.

PROBLEM SOLVING STRATEGY 3.1 **Projectile Motion**

INTERPRET Make sure that you have a problem involving the constant acceleration of gravity near Earth's surface, and that the motion involves both horizontal and vertical components. Identify the object or objects in question and whatever initial or final positions and velocities are given. Know what quantities you're being asked to find.

DEVELOP Establish a horizontal/vertical coordinate system, and write the separate components of the equations of motion (Equations 3.10–3.13). The equations for different components will be linked by a common variable—namely, the time. Draw a sketch showing the initial motion and a rough trajectory.

EVALUATE Solve your individual equations simultaneously for the unknowns of the problem.

ASSESS Check that your answer makes sense. Consider special cases, like purely vertical or horizontal initial velocities. Because the equations of motion are quadratic in time, you may have two answers. One answer may be the one you want, but you gain more insight into physics if you consider the meaning of the second answer, too.

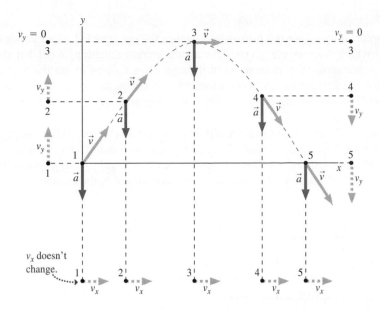

FIGURE 3.14 Velocity and acceleration at five points on a projectile's path. Also shown are horizontal and vertical components.

EXAMPLE 3.4 **Finding the Horizontal Distance: Washout!**

A raging flood has washed away a section of highway, creating a gash 1.7 m deep. A car moving at 31 m/s goes straight over the edge. How far from the edge of the washout does it land?

INTERPRET This is a problem involving projectile motion, and it asks for the horizontal distance the car moves after it leaves the road. We're given the car's initial speed and direction (horizontal) and the distance it falls.

DEVELOP Fig. 3.15a shows the situation, and we've sketched the essentials in Fig. 3.15b. Since there's no horizontal acceleration, Equation 3.12, $x = x_0 + v_{x0}t$, would determine the unknown distance if we knew the time. But horizontal and vertical motions are independent, so we can find the time until the car hits the ground from the vertical motion alone, as determined by Equation 3.13, $y = y_0 + v_{y0}t - \frac{1}{2}gt^2$. So our plan is to get the time from Equation 3.13 and then use that time in

Equation 3.12 to get the horizontal distance. If we choose the origin as the bottom of the washout, then $y_0 = 1.7$ m. Then we want the time when $y = 0$.

EVALUATE Solving Equation 3.13 for t gives

$$t = \sqrt{\frac{2y_0}{g}} = \sqrt{\frac{(2)(1.7 \text{ m})}{(9.8 \text{ m/s}^2)}} = 0.589 \text{ s}$$

During this time the car continues to move horizontally at $v_{x0} = 31$ m/s, so Equation 3.12 gives $x = v_{x0}t = (31 \text{ m/s})(0.589 \text{ s}) = 18$ m.

ASSESS Make sense? About half a second to drop 1.7 m or about 6 ft seems reasonable, and at 31 m/s an object will go somewhat farther than 15 m in this time. ∎

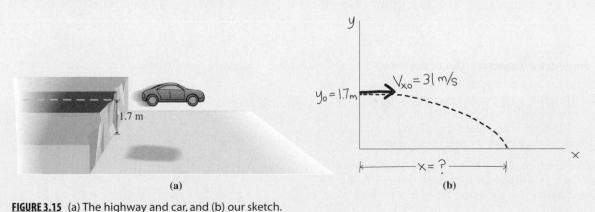

(a)   (b)

FIGURE 3.15 (a) The highway and car, and (b) our sketch.

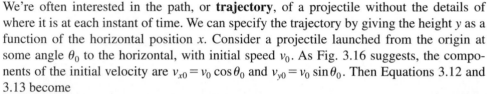

✔ **TIP** Handling Multistep Problems

Example 3.4 asked for the horizontal distance the car traveled. For that we needed the time—which we weren't given. This is a common situation in all but the simplest physics problems. You need to work through several steps to get the answer—in this case solving first for the unknown time and then for the distance. In essence, we solved two problems in Example 3.4: the first involving vertical motion and the second horizontal motion.

Instead of calculating t numerically in Example 3.4, we could have kept $t = \sqrt{2y_0/g}$ in symbolic form until the end. That would avoid roundoff error and having to keep track of numerical digits and units. And you can often gain more physical insight from an answer that's expressed symbolically before you put in the numbers.

Projectile Trajectories

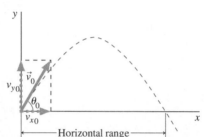

FIGURE 3.16 Parabolic trajectory of a projectile.

We're often interested in the path, or **trajectory**, of a projectile without the details of where it is at each instant of time. We can specify the trajectory by giving the height y as a function of the horizontal position x. Consider a projectile launched from the origin at some angle θ_0 to the horizontal, with initial speed v_0. As Fig. 3.16 suggests, the components of the initial velocity are $v_{x0} = v_0 \cos\theta_0$ and $v_{y0} = v_0 \sin\theta_0$. Then Equations 3.12 and 3.13 become

$$x = v_0 \cos\theta_0 t \quad \text{and} \quad y = v_0 \sin\theta_0 t - \tfrac{1}{2}gt^2$$

Solving the x equation for the time t gives

$$t = \frac{x}{v_0 \cos\theta_0}$$

Using this result in the y equation, we have

$$y = v_0 \sin\theta_0\left(\frac{x}{v_0 \cos\theta_0}\right) - \tfrac{1}{2}g\left(\frac{x}{v_0 \cos\theta_0}\right)^2$$

or

$$y = x \tan\theta_0 - \frac{g}{2v_0^2 \cos^2\theta_0}x^2 \quad (\text{projectile trajectory}) \tag{3.14}$$

Equation 3.14 gives a mathematical description of the projectile's trajectory. Since y is a quadratic function of x, the trajectory is a parabola. However, there's no new physics here, since Equation 3.14 follows directly from the constant-acceleration Equations 3.12 and 3.13.

EXAMPLE 3.5 **Finding the Trajectory: Out of the Hole**

A construction worker stands in a 2.6-m-deep hole, 3.1 m from the edge of the hole. He tosses a hammer to a companion outside the hole. If the hammer leaves his hand 1.0 m above the bottom of the hole at an angle of 35°, what's the minimum speed it needs to clear the edge of the hole? How far from the edge of the hole does it land?

INTERPRET We're concerned about *where* an object is but not *when*, so we interpret this as a problem about the trajectory—specifically, the minimum-speed trajectory that just grazes the edge of the hole.

DEVELOP We draw the situation in Fig. 3.17. Equation 3.14 determines the trajectory, so our plan is to find the speed that makes the trajectory pass just over the edge of the hole at $x = 3.1$ m, $y = 1.6$ m, where Fig. 3.17 shows that we've chosen a coordinate system with its origin at the worker's hand.

We want v_0 so that the hammer will just clear the point $x = 3.1$ m, $y = 1.6$ m.

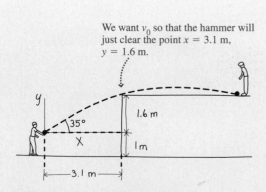

FIGURE 3.17 Our sketch for Example 3.5.

cont'd.

EVALUATE To find the minimum speed we solve Equation 3.14 for v_0, using the coordinates of the hole's edge for x and y:

$$v_0 = \sqrt{\frac{gx^2}{2\cos^2\theta_0\,(x\tan\theta_0 - y)}} = 11 \text{ m/s}$$

To find where the hammer lands, we need to know the horizontal position when $y = 1.6$ m. Rearranging Equation 3.14 into the standard form for a quadratic equation gives $(g/2v_0^2\cos^2\theta_0)x^2 - (\tan\theta_0)x + y = 0$. Applying

the quadratic formula (Appendix A) gives $x = 3.1$ m and $x = 8.7$ m; the second value is the one we want. That 8.7 m is the distance from our origin at the worker's hand, and amounts to 8.7 m − 3.1 m = 5.6 m from the hole's edge.

ASSESS Make sense? The other answer to the quadratic, $x = 3.1$ m, is a clue that we did the problem correctly. That 3.1 m is the distance to the edge of the hole. The fact that we get this position when we ask for a vertical height of 1.6 m confirms that the trajectory does indeed just clear the edge of the hole. ∎

The Range of a Projectile

How far will a soccer ball go if I kick it at 12 m/s at 50° to the horizontal? If I can throw a rock at 15 m/s, can I get it across a 30-m-wide pond? How far off vertical can a rocket's trajectory be and still land within 50 km of its launch point? As in these examples, we're frequently interested in the **horizontal range** of a projectile—that is, how far it moves horizontally over level ground.

For a projectile launched on level ground, we can determine when the projectile will return to the ground by setting $y = 0$ in Equation 3.14:

$$0 = x\tan\theta_0 - \frac{g}{2v_0^2\cos^2\theta_0}x^2 = x\left(\tan\theta_0 - \frac{gx}{2v_0^2\cos^2\theta_0}\right)$$

There are two solutions: $x = 0$, corresponding to the launch point, and

$$x = \frac{2v_0^2}{g}\cos^2\theta_0\tan\theta_0 = \frac{2v_0^2}{g}\sin\theta_0\cos\theta_0$$

But $\sin 2\theta_0 = 2\sin\theta_0\cos\theta_0$, so this becomes

$$x = \frac{v_0^2}{g}\sin 2\theta_0 \qquad \text{(horizontal range)} \qquad (3.15)$$

✓**TIP** Know Your Limits

We emphasize that Equation 3.15 gives the *horizontal* range—the distance a projectile travels horizontally before returning *to its starting height*. From the way it was derived—setting $y = 0$—you can see that it does *not* give the horizontal distance when the projectile returns to a different height (Fig. 3.18).

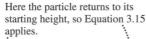
Here the particle returns to its starting height, so Equation 3.15 applies.

(a)

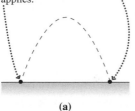
Here the particle lands at a different height, so Equation 3.15 doesn't apply.

(b)

FIGURE 3.18 Equation 3.15 applies in (a) but not in (b).

| APPLICATION | **Line Drives, Pop Flies, and Hang Time** |

Although air resistance significantly influences baseball trajectories, to a first approximation a baseball behaves like a projectile. Because horizontal and vertical motions are independent, the time a baseball spends in the air—the "hang time"—depends only on how high the ball goes. A line drive—a nearly horizontal trajectory that rises only slightly above where it was hit—is therefore in the air for only a short time, making it difficult for an opposing player to judge its trajectory, get to the ball, and make a successful catch. A pop fly, even if hit with the same speed, spends far more time in the air and is therefore easier to catch (see photo). If you worked Problem 77 in Chapter 2, you learned further that a projectile spends 71% of its time in the upper half of its trajectory, which is what makes a fly ball seem to "hang" so long before it comes speeding down.

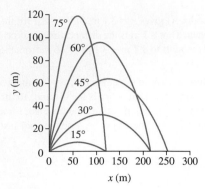

FIGURE 3.19 Trajectories for a projectile launched at 50 m/s.

Does Equation 3.15 make sense? When $\theta_0 = 0$, the range is zero, and a projectile launched horizontally immediately hits the ground. When $\theta_0 = 90°$, $\sin 2\theta_0 = 0$, and again the range is zero. Here the projectile is launched vertically upward, so it returns to the same point. When is the range a maximum? The largest value of $\sin 2\theta_0$ is 1, which occurs when $\theta_0 = 45°$. Figure 3.19 shows the trajectories of projectiles launched at different angles with the same initial speed. At angles smaller than 45° the projectile has greater horizontal velocity, but it doesn't get as high and therefore isn't in the air as long; the net effect is a shorter range. At angles larger than 45°, the projectile rises higher and is in the air longer, but its horizontal velocity is lower and the net effect is again a shorter range. In fact, as you can prove in Problem 70, the range is the same for angles equally spaced on either side of 45°.

✓ **TIP** Know the Fundamentals

Equations 3.14 for the trajectory and 3.15 for the range of a projectile are useful, but they're hardly fundamental equations of physics. Both follow directly from the equations for constant acceleration. If you think that specialized results like Equations 3.14 and 3.15 are on an equal footing with more fundamental equations and principles, then you're seeing physics as a hodgepodge of equations and missing the big picture of a science with a few underlying principles from which all else follows.

EXAMPLE 3.6   **Projectile Range: Probing the Atmosphere**

A rocket is to probe the atmosphere. After a short engine firing, it reaches 4.6 km/s. If the rocket must land within 50 km of its launch site, what's the maximum allowable deviation from a vertical trajectory?

INTERPRET Although we're asked about the launch angle, the 50-km criterion is a clue that we can interpret this as a problem about the horizontal range. That "short engine firing" means we can neglect the distance over which the rocket fires and consider it a projectile that leaves the ground at $v_0 = 4.6$ km/s.

DEVELOP Equation 3.15, $x = (v_0^2/g) \sin 2\theta_0$, determines the horizontal range, so our plan is to solve that equation for θ_0 with range $x = 50$ km.

EVALUATE We have $\sin 2\theta_0 = gx/v_0^2 = 0.0232$. There are two solutions, corresponding to $2\theta_0 = 1.33°$ and $2\theta_0 = 180° - 1.33°$. The second is the one we want, giving a launch angle $\theta_0 = 90° - 0.67°$. Therefore the launch angle must be within 0.67° of vertical.

ASSESS Make sense? At 4.6 km/s, this rocket goes quite high, so with even a small deviation from vertical it will land far from its launch point. Again we've got two solutions. The one we rejected is like the low trajectories of Fig. 3.19; although it gives a 50-km range, it isn't going to get our rocket high into the atmosphere.   ∎

3.6 Uniform Circular Motion

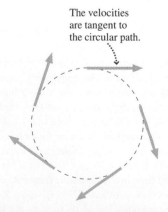

The velocities are tangent to the circular path.

FIGURE 3.20 Velocity vectors in circular motion are tangent to the circular path.

An important case of accelerated motion in two dimensions is **uniform circular motion**—the motion of an object describing a circular path at constant speed. Although the speed is constant, the motion is accelerated because the *direction* of the velocity is changing.

Examples of uniform circular motion are numerous. Many spacecraft are in circular orbits, and the orbits of the planets are approximately circular. The Earth's daily rotation carries you around in uniform circular motion. Pieces of rotating machinery describe uniform circular motion, and you're temporarily in circular motion as you drive around a curve. Even electrons undergo circular motion when they encounter magnetic fields.

Here we derive an important relationship among the acceleration, speed, and radius of uniform circular motion. Figure 3.20 shows several velocity vectors for an object moving with speed v around a circle of radius r. Note that the velocity vectors are tangent to the circle, indicating the instantaneous direction of motion. In Fig. 3.21a we focus on two nearby points described by position vectors $\vec{r}_1$ and $\vec{r}_2$, showing also the velocity vectors $\vec{v}_1$ and $\vec{v}_2$. Figure 3.21b shows the displacement $\Delta\vec{r}$ that represents the difference between $\vec{r}_1$ and $\vec{r}_2$; similarly, Fig. 3.21c shows the velocity difference $\Delta\vec{v}$.

Because $\vec{v}_1$ is perpendicular to $\vec{r}_1$, and $\vec{v}_2$ is perpendicular to $\vec{r}_2$, the angles θ shown in all three parts of Fig. 3.21 are the same. Therefore the triangles in Fig. 3.21b and c are similar, and we can write

$$\frac{\Delta v}{v} = \frac{\Delta r}{r}$$

Now suppose the angle θ is small, corresponding to a short time interval Δt for motion from position $\vec{r}_1$ to $\vec{r}_2$. Then the length of the vector $\Delta \vec{r}$ is approximately the length of the circular arc joining the endpoints of the position vectors, as suggested in Fig. 3.21b. The length of this arc is the distance the object travels in the time Δt, or $v\,\Delta t$, so $\Delta r \simeq v\,\Delta t$. Then the relation between similar triangles becomes

$$\frac{\Delta v}{v} \simeq \frac{v\,\Delta t}{r}$$

Rearranging this equation gives an approximate expression for the magnitude of the average acceleration:

$$\bar{a} = \frac{\Delta v}{\Delta t} \simeq \frac{v^2}{r}$$

Taking the limit $\Delta t \rightarrow 0$ gives the instantaneous acceleration; in this limit the angle θ approaches 0, the circular arc and $\Delta \vec{r}$ become indistinguishable, and the relation $\Delta r \simeq v\,\Delta t$ becomes exact. So we have

$$a = \frac{v^2}{r} \quad \text{(uniform circular motion)} \tag{3.16}$$

for the magnitude of the instantaneous acceleration of an object moving in a circle of radius r at constant speed v. What about the direction? As Fig. 3.21c suggests, $\Delta \vec{v}$ is very nearly perpendicular to both velocity vectors; in the limit $\Delta t \rightarrow 0$, $\Delta \vec{v}$ and the acceleration $\Delta \vec{v}/\Delta t$ become exactly perpendicular to the velocity. The direction of the acceleration vector is therefore toward the center of the circle.

Clearly, our geometric argument would work for any point on the circle, so we conclude that the acceleration has constant magnitude v^2/r and always points toward the center of the circle. Isaac Newton coined the term *centripetal* to describe this center-pointing acceleration. However, we'll use that term sparingly because we want to emphasize that centripetal acceleration is fundamentally no different from any other acceleration: It's simply a vector describing the rate of change of velocity.

Does Equation 3.16 make sense? Yes. An increase in speed v means the time Δt for a given change in direction of the velocity becomes shorter. Not only that, but the associated change $\Delta \vec{v}$ in velocity is larger. These two effects combine to give an acceleration that depends on the *square* of the speed. On the other hand, an increase in the radius with a fixed speed increases the time Δt associated with a given change in velocity, so the acceleration is inversely proportional to the radius.

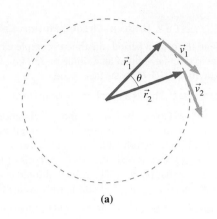

(a)

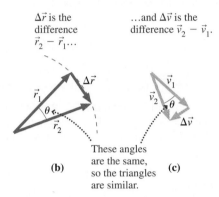

$\Delta \vec{r}$ is the difference $\vec{r}_2 - \vec{r}_1$...

...and $\Delta \vec{v}$ is the difference $\vec{v}_2 - \vec{v}_1$.

(b)

(c)

These angles are the same, so the triangles are similar.

FIGURE 3.21 Position and velocity vectors for two nearby points on the circular path.

✓ **TIP** Circular Motion and Constant Acceleration

The direction toward the center changes as an object moves around a circular path, so the acceleration vector is *not constant*, even though its magnitude is. Uniform circular motion is *not* motion with constant acceleration, and our constant-acceleration equations *do not apply*. In fact, we know that constant acceleration in two dimensions implies a parabolic trajectory, not a circle.

EXAMPLE 3.7 **Uniform Circular Motion: Calculating a Space Shuttle Orbit**

Find the orbital period (the time to complete one orbit) of a space shuttle in circular orbit at an altitude of 250 km, where the acceleration of gravity is 93% of its surface value.

INTERPRET This is a problem about uniform circular motion.

DEVELOP Given the radius and acceleration, we could use Equation 3.16, $a = v^2/r$, to determine the orbital speed. But we're given the altitude, not the orbital radius, and we want the period, not the speed. So our plan is to write the speed in terms of the period and use the result in Equation 3.16. The orbital altitude is the distance from Earth's surface, so we'll need to add Earth's radius to get the orbital radius r.

EVALUATE The speed v is the orbital circumference, $2\pi r$, divided by the period T. Using this in Equation 3.16 gives

$$a = \frac{v^2}{r} = \frac{(2\pi r/T)^2}{r} = \frac{4\pi^2 r}{T^2}$$

Appendix E lists Earth's radius as $R_E = 6.37$ Mm, giving an orbital radius $r = R_E + 250$ km $= 6.62$ Mm. Solving our acceleration expression for the period then gives $T = \sqrt{4\pi^2 r/a} = 5355$ s $= 89$ min, where we used $a = 0.93g$.

ASSESS Make sense? You've probably heard that astronauts orbit Earth in about an hour and a half, experiencing multiple sunrises and sunsets in a 24-hour day. Our answer of 89 min is certainly consistent with that. There's no choice here; for a given orbital radius, Earth's size and mass determine the period. Because astronauts' orbits are limited to a few hundred kilometers, a distance small compared with R_E, variations in g and T are minimal. Any such "low Earth orbit" has a period of approximately 90 min. At higher altitudes, the decline in the strength of gravity becomes more noticeable and periods lengthen; the Moon, for example, orbits in 27 days. We'll discuss orbits more in Chapter 8. ∎

EXAMPLE 3.8 **Uniform Circular Motion: Engineering a Road**

An engineer is designing a flat, horizontal road for an 80 km/h speed limit (that's 22.2 m/s). If the maximum acceleration of a vehicle on this road is 1.5 m/s², what's the minimum safe radius for curves in the road?

INTERPRET Even though a curve is only a portion of a circle, we can still interpret this problem as involving uniform circular motion.

DEVELOP Equation 3.16, $a = v^2/r$, determines the acceleration given the speed and radius. Here we have the acceleration and speed, so our plan is to solve for the radius.

EVALUATE Using the given numbers, we have $r = v^2/a = (22.2$ m/s$)^2/1.5$ m/s² $= 329$ m.

ASSESS Make sense? A speed of 80 km/h is pretty fast, so we need a wide curve to keep the required acceleration below its design value. If the curve is sharper, vehicles may slide off the road. We'll see more clearly in subsequent chapters how vehicles manage to negotiate high-speed curves. ∎

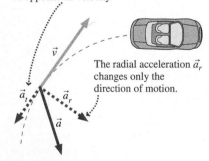

The car is slowing, so its tangential acceleration $\vec{a}_t$ is opposite its velocity.

The radial acceleration $\vec{a}_r$ changes only the direction of motion.

FIGURE 3.22 Acceleration of a car that slows as it rounds a curve.

Nonuniform Circular Motion

What if an object moves in a circular path but its speed changes? Then it has components of acceleration both perpendicular and parallel to its velocity. The former, the **radial acceleration** a_r, is what changes the direction to keep the object in circular motion. Its magnitude is still v^2/r, with v now the instantaneous speed. The parallel component of acceleration, also called **tangential acceleration** a_t because it's tangent to the circle, changes the speed but not the direction. Its magnitude is therefore the rate of change of speed, or dv/dt. Figure 3.22 shows these two acceleration components for a car rounding a curve.

Finally, what if the radius of a curved path changes? At any point on a curve we can define a **radius of curvature**. Then the radial acceleration is still v^2/r, and it can vary if either v or r changes along the curve. The tangential acceleration is still tangent to the curve, and it still describes the rate of change of speed. So it's straightforward to generalize the ideas of uniform circular motion to cases where the motion is nonuniform either because the speed changes, or because the radius changes, or both.

GOT IT? 3.4 The figure shows velocity vectors for four points on a noncircular path. Rank order the centripetal accelerations at these points given $v_1 = v_4$ and $v_2 = v_3$.

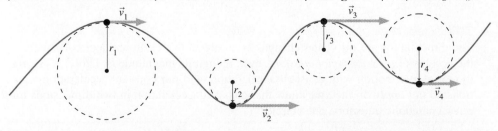

Big Picture

Quantities characterizing motion in two and three dimensions have both **magnitude** and **direction** and are described by **vectors**. Position, velocity, and acceleration are all vector quantities, related as they are in one dimension:

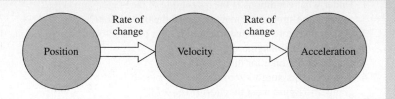

These vector quantities need not have the same direction. In particular, acceleration that's perpendicular to velocity changes the direction but not the magnitude of the velocity. Acceleration that's colinear changes only the magnitude of the velocity. In general, both change.

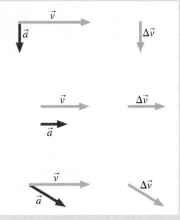

Components of motion in two perpendicular directions are independent. This reduces problems in two and three dimensions to sets of one-dimensional problems that can be solved with the methods of Chapter 2.

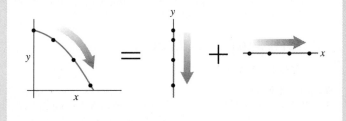

Key Concepts and Equations

Vectors can be characterized by magnitude and direction or by components. In two dimensions these representations are related by

$$A = \sqrt{A_x^2 + A_y^2} \quad \text{and} \quad \theta = \tan^{-1}\frac{A_y}{A_x}$$

$$A_x = A\cos\theta \quad \text{and} \quad A_y = A\sin\theta$$

A compact way to express vectors involves unit vectors that have magnitude 1, no units, and point along the coordinate axes:

$$\vec{A} = A_x\hat{i} + A_y\hat{j}$$

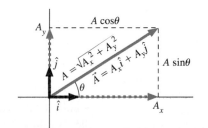

Velocity is the rate of change of the position vector $\vec{r}$:

$$\vec{v} = \frac{d\vec{r}}{dt}$$

Acceleration is the rate of change of velocity:

$$\vec{a} = \frac{d\vec{v}}{dt}$$

Cases and Uses

When acceleration is constant, motion is described by vector equations that generalize the one-dimensional equations of Chapter 2:

$$\vec{v} = \vec{v}_0 + \vec{a}t \qquad \vec{r} = \vec{r}_0 + \vec{v}_0 t + \tfrac{1}{2}\vec{a}t^2$$

An important case of constant-acceleration motion in two dimensions is **projectile motion** under the influence of gravity.

$$y = x\tan\theta_0 - \frac{g}{2v_0^2\cos^2\theta_0}x^2$$

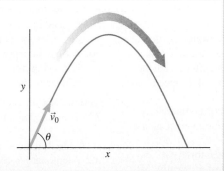

In **uniform circular motion** the magnitudes of velocity and acceleration remain constant, but their directions continually change. For an object moving in a circular path of radius r, the magnitudes of $\vec{a}$ and $\vec{v}$ are related by $a = v^2/r$.

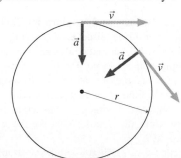

For Thought and Discussion

1. Under what conditions is the magnitude of the vector sum $\vec{A} + \vec{B}$ equal to the sum of the magnitudes of the two vectors?
2. Can two vectors of equal magnitude sum to zero? Can two vectors of unequal magnitude sum to zero?
3. Repeat Question 2 for three vectors.
4. Three vectors sum to zero. If they are placed head to tail, what geometric figure must they form? Explain.
5. Is it meaningful to talk about vectors without mentioning coordinate systems or components?
6. Can an object have a southward acceleration while moving northward? A westward acceleration while moving northward?
7. Is there any way to negotiate a curved path without accelerating?
8. A satellite glides through space at a steady 29,000 km/h in a circular orbit around Earth. Is the satellite accelerating? If so, in what direction? If not, why not?
9. You're moving northward when you accelerate briefly toward the east. Is your subsequent motion strictly eastward? Explain.
10. In what sense is Equation 3.8 really two (or three) equations?
11. Is the speed of a projectile constant throughout its parabolic trajectory?
12. What is the vertical component of a projectile's velocity at the peak of its trajectory?
13. Is there any point on a projectile's trajectory where the velocity and acceleration are perpendicular?
14. Projectiles launched at 30° and 60° have the same range. Does this mean they stay in the air the same amount of time?
15. A friend who's not taking physics insists that you can't be accelerating when you drive around a curve since the speedometer reading remains steady. Refute this argument.
16. How is it possible for an object to be moving in one direction but accelerating in another?

Exercises and Problems

Exercises

Section 3.1 Vectors

17. You walk west 220 m and then north 150 m. What are the magnitude and direction of your displacement vector?
18. An ion in a mass spectrometer (a device that sorts atomic-size particles) follows a semicircular path of radius 15.2 cm. What are (a) the distance it travels and (b) the magnitude of its displacement?
19. A migrating whale follows the west coast of Mexico and North America toward its summer home in Alaska. It first travels 360 km due northwest to just off the coast of northern California and then turns due north and travels 400 km toward its destination. Determine graphically the magnitude and direction of its displacement vector.
20. A city's streets are laid out with its north-south blocks twice as long as its east-west blocks. You walk 8 blocks east and 3 blocks north. Determine (a) the total distance you've walked and (b) the magnitude of your displacement vector. Express answers in units of east-west blocks.
21. Vector $\vec{A}$ has magnitude 3.0 m and points to the right; vector $\vec{B}$ has magnitude 4.0 m and points vertically upward. Find the magnitude and direction of a vector $\vec{C}$ such that $\vec{A} + \vec{B} + \vec{C} = \vec{0}$.
22. Vector $\vec{V}$ represents a displacement of 120 km at 29° counterclockwise from the x axis. Write $\vec{V}$ in unit vector notation.
23. Find the magnitude of the vector $34\hat{i} + 13\hat{j}$ m and determine the angle it makes with the x axis.
24. (a) What is the magnitude of $\hat{i} + \hat{j}$? (b) What angle does it make with the x axis?

Section 3.2 Velocity and Acceleration Vectors

25. An asteroid is discovered heading straight toward Earth at 15 km/s. An international team manages to attach a giant rocket engine to the asteroid. The rocket fires for 10 min, after which the asteroid is moving at 28° to its original path at a speed of 19 km/s. Find its average acceleration.
26. An object is moving at 18 m/s at an angle of 220° counterclockwise from the x axis. What are the x and y components of its velocity?
27. A car drives north at 40 mi/h for 10 min and then turns east and goes 5.0 mi at 60 mi/h. Finally, it goes southwest at 30 mi/h for 6.0 min. Draw a vector diagram and determine (a) the car's displacement and (b) its average velocity for this trip.

28. The minute hand of a clock is 5.5 cm long. What is the average velocity vector for the tip of the hand during the interval from the hour to 20 minutes past the hour, expressed in a coordinate system with the y axis toward noon and the x axis toward 3:00?
29. A car, initially going eastward, rounds a 90° curve and ends up heading southward. If the speedometer reading remains constant, what is the direction of the car's average acceleration vector?
30. What are (a) the average velocity and (b) the average acceleration of the tip of the 2.4-cm-long hour hand of a clock in the interval from 12 PM to 6 PM? Express the answers in unit vector notation, with the x axis pointing toward 3 PM and the y axis toward 12 PM.
31. A skater is gliding along the ice at 2.4 m/s, when she undergoes an acceleration of magnitude 1.1 m/s² for 3.0 s. At the end of that time she is moving at 5.7 m/s. What must be the angle between the acceleration vector and the initial velocity vector?
32. An object is moving in the x direction at 1.3 m/s when it is subjected to an acceleration given by $\vec{a} = 0.52\hat{j}$ m/s². What is its velocity vector after 4.4 s of acceleration?

Section 3.3 Relative Motion

33. A jetliner with an airspeed of 1000 km/h sets out on a 1500-km flight due south. To maintain a southward direction, however, the plane must be pointed 15° west of south. If the flight takes 100 min, what is the wind velocity?
34. You wish to row straight across a 63-m-wide river. You can row at a steady 1.3 m/s relative to the water, and the river flows at 0.57 m/s. (a) In what direction should you head? (b) How long will it take you to cross the river?
35. An airplane with airspeed of 370 km/h flies perpendicularly across the jet stream. To achieve this flight, the plane must be pointed into the jet stream at an angle of 32° from the perpendicular direction of its flight. What is the speed of the jet stream?
36. A flock of geese is attempting to migrate due south, but the wind is blowing from the west at 5.1 m/s. If the birds can fly at 7.5 m/s relative to the air, in what direction should they head?

Section 3.4 Constant Acceleration

37. The position of an object as a function of time is given by $\vec{r} = (3.2t + 1.8t^2)\hat{i} + (1.7t - 2.4t^2)\hat{j}$ m, where t is the time in seconds. What are the magnitude and direction of the acceleration?
38. A sailboard is sailing at 6.5 m/s when a gust of wind hits, causing it to accelerate at 0.48 m/s² at a 35° angle to its original direction of motion. If the acceleration lasts 6.3 s, what is the board's net displacement during the wind gust?

Section 3.5 Projectile Motion

39. You toss an apple horizontally at 8.7 m/s from a height of 2.6 m. Simultaneously, you drop a peach from the same height. How long does each take to reach the ground?

40. A carpenter tosses a shingle off an 8.8-m-high roof, giving it an initially horizontal velocity of 11 m/s. (a) How long does it take the shingle to reach the ground? (b) How far does it move horizontally in this time?

41. An arrow fired horizontally at 41 m/s travels 23 m horizontally before it hits the ground. From what height was it fired?

42. Ink droplets in an ink-jet printer are ejected horizontally at 12 m/s and travel a horizontal distance of 1.0 mm to the paper. How far do they fall in this interval?

43. Protons in a particle accelerator drop 1.2 μm over the 1.7-km length of the accelerator. What is their approximate average speed?

44. If you can hit a golf ball 180 m on Earth, how far can you hit it on the Moon? (Your answer is an underestimate because the distance on Earth is restricted by air resistance as well as by a larger g.)

Section 3.6 Uniform Circular Motion

45. How fast would a car have to round a turn with a radius of 75 m for its acceleration to be numerically equal to that of gravity?

46. Estimate the acceleration of the Moon, which completes a nearly circular orbit of 385,000 km radius in 27 days.

47. Global Positioning System satellites are at altitudes of approximately 20,000 km, where the acceleration of gravity is only 5.8% of its surface value. To the nearest hour, what's the orbital period of the GPS satellites?

Problems

48. Two vectors $\vec{A}$ and $\vec{B}$ have the same magnitude A and are at right angles. Find the magnitude of the vectors (a) $\vec{A} + 2\vec{B}$ and (b) $3\vec{A} - \vec{B}$.

49. Vector $\vec{A}$ has magnitude 1.0 m and points at 35° clockwise from the x axis. Vector $\vec{B}$ has magnitude 1.8 m. What angle should $\vec{B}$ make with the x axis so that $\vec{A} + \vec{B}$ is purely vertical?

50. Let $\vec{A} = 15\hat{\imath} - 40\hat{\jmath}$ and $\vec{B} = 31\hat{\jmath} + 18\hat{k}$. Find a vector $\vec{C}$ such that $\vec{A} + \vec{B} + \vec{C} = \vec{0}$.

51. A biologist studying the motion of bacteria notes a bacterium at position $\vec{r}_1 = 2.2\hat{\imath} + 3.7\hat{\jmath} - 1.2\hat{k}$ μm $(1\,\mu\text{m} = 10^{-6}\text{ m})$. After 6.2 s the bacterium is at $\vec{r}_2 = 4.6\hat{\imath} + 1.9\hat{k}$ μm. What is its average velocity? Express the answer in unit vector notation, and calculate the magnitude.

52. An object's position as a function of time is given by $\vec{r} = 12t\hat{\imath} + (15t - 5.0t^2)\hat{\jmath}$ m, where t is time in seconds. (a) What is the object's position at $t = 2.0$ s? (b) What is its average velocity in the interval from $t = 0$ to $t = 2.0$ s? (c) What is its instantaneous velocity at $t = 2.0$ s?

53. Attempting to stop on a slippery road, a car moving at 80 km/h skids across the road at a 30° angle to its initial motion, coming to a stop in 3.9 s. Determine the average acceleration in m/s², using a coordinate system with the x axis in the direction of the car's original motion and the y axis toward the side of the road to which the car skids.

54. An object undergoes acceleration of $2.3\hat{\imath} + 3.6\hat{\jmath}$ m/s² over a 10-s interval. At the end of this time, its velocity is $33\hat{\imath} + 15\hat{\jmath}$ m/s. (a) What was its velocity at the beginning of the 10-s interval? (b) By how much did its speed change? (c) By how much did its direction change? (d) Show that the speed change is *not* given by the magnitude of the acceleration multiplied by the time. Why not?

55. The sweep-second hand of a clock is 3.1 cm long. What are the magnitude of (a) the average velocity and (b) the average acceleration of the hand's tip over a 5.0-s interval? (c) What is the angle between the average velocity and acceleration vectors?

56. A ferryboat sails between two towns directly opposite each other on a river. If the boat sails at 15 km/h relative to the water, and if the current flows at 6.3 km/h, at what angle should the boat head?

57. The sum, $\vec{A} + \vec{B}$, of two vectors is perpendicular to the difference, $\vec{A} - \vec{B}$. How do the magnitudes of the two vectors compare?

58. Write an expression for a unit vector that lies at 45° between the positive x and y axes.

59. An object is moving initially in the x direction at 4.5 m/s, when an acceleration is applied in the y direction for a period of 18 s. If the object moves equal distances in the x and y directions during this time, what is the magnitude of its acceleration?

60. A particle leaves the origin with initial velocity $\vec{v}_0 = 11\hat{\imath} + 14\hat{\jmath}$ m/s. It undergoes a constant acceleration given by $\vec{a} = -1.2\hat{\imath} + 0.26\hat{\jmath}$ m/s². (a) When does the particle cross the y axis? (b) What is its y coordinate at the time? (c) How fast is it moving, and in what direction, at that time?

61. A kid fires water horizontally from a squirt gun held 1.6 m above the ground. It hits another kid 2.1 m away square in the back, at a point 0.93 m above the ground. What was the initial speed of the water?

62. In a chase scene, a movie stuntman is supposed to run right off the flat roof of one city building and land on another roof 1.9 m lower. If the gap between the buildings is 4.5 m wide, how fast must he run?

63. Standing on the ground 3.0 m from the wall of a building, you want to throw a package from your 1.5-m shoulder level to someone in a second-floor window 4.2 m above the ground. At what speed and angle should you throw the package so it just barely reaches the window?

64. Derive a general formula for the horizontal distance covered by a projectile launched horizontally at speed v_0 from height h.

65. Compare the travel times for the projectiles launched at 30° and 60° in Fig. 3.19, both of which have the same starting and ending points.

66. You toss a chocolate bar to your hiking companion located 8.6 m up a 39° slope, as shown in Fig. 3.23. Determine the initial velocity vector so that the chocolate bar will reach your friend moving horizontally.

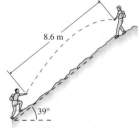

FIGURE 3.23 Problem 66

67. Prove that a projectile launched on level ground reaches its maximum height midway along its trajectory.

68. A projectile launched at an angle θ to the horizontal reaches a maximum height h. Show that its horizontal range is $4h/\tan\theta$.

69. A motorcyclist driving in a 60 km/h zone hits a stopped car on a level road. The cyclist is thrown from his bike and lands 39 m down the road. Was the cyclist speeding? To answer, find the minimum speed he could have been going just before the accident.

70. Show that, for a given initial speed, the horizontal range of a projectile is the same for launch angles $45° + \alpha$ and $45° - \alpha$, where α is between 0° and 45°.

71. A basketball player is 15 ft horizontally from the center of the basket, which is 10 ft off the ground. At what angle should the player aim the ball if it is thrown from a height of 8.2 ft with a speed of 26 ft/s?

72. A jet is diving vertically downward at 1200 km/h. If the pilot can withstand a maximum acceleration of 5g (i.e., 5 times Earth's gravitational acceleration) before losing consciousness, at what height must the plane start a quarter turn to pull out of the dive? Assume the speed remains constant.

73. An alpine rescue team is using a slingshot to send an emergency medical packet to climbers stranded on a ledge, as shown in Fig. 3.24. What should be the launch speed from the slingshot?

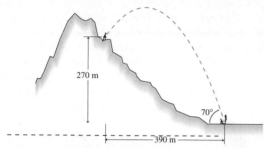

FIGURE 3.24 Problem 73

74. If you can throw a stone straight up to a height of 16 m, how far could you throw it horizontally over level ground? Assume the same throwing speed and optimum launch angle.

75. In a conversion from military to peacetime use, a missile with a maximum horizontal range of 180 km is being adapted for studying the upper atmosphere. What is the maximum altitude it can achieve if launched vertically?

76. I can kick a soccer ball 28 m on level ground, giving it an initial velocity at 40° to the horizontal. At the same initial speed and angle to the horizontal, what horizontal distance can I kick the ball on a 15° upward slope?

77. A diver leaves a 3-m board on a trajectory that takes her 2.5 m above the board and then into the water a horizontal distance of 2.8 m from the end of the board. At what speed and angle did she leave the board?

78. In your calculus class, you may have learned that you can find the maximum or minimum of a function by differentiating and setting the result to zero. Do this for Equation 3.15, differentiating with respect to θ, and thus verify that the maximum range occurs for $\theta = 45°$.

79. A well-engineered ski jump is less dangerous than it looks because skiers hit the ground with very small velocity components perpendicular to the ground. Skiers leave the Olympic ski jump in Lake Placid, New York, at 28 m/s at an angle of 9.5° below the horizontal. Their landing zone is a horizontal distance of 55 m from the end of the jump. The ground at that point is contoured so skiers' trajectories make an angle of only 3.0° with the ground on landing, as suggested in Fig. 3.25. What is the slope of the ground in the landing zone?

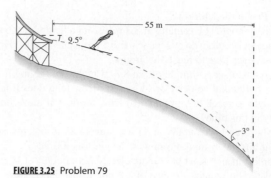

FIGURE 3.25 Problem 79

80. Differentiate the trajectory Equation 3.14 to find its slope, $\tan \theta = dy/dx$, and show that the slope is in the direction of the projectile's velocity, as given by Equations 3.10 and 3.11.

81. A projectile has horizontal range R and maximum height h. Find an expression for its initial speed.

82. (a) Show that the position of a particle on a circle of radius R with center at the origin $(x=0, y=0)$ is given in unit vectors by $\vec{r} = R(\cos\theta\hat{i} + \sin\theta\hat{j})$, where θ is the angle of the position vector with the x axis. (b) If the particle moves with constant speed v and period T, starting on the x axis at $t=0$, find an expression for θ in terms of the time and T. (c) Differentiate the position vector twice with respect to time to find the acceleration, and show that it is the centripetal acceleration, whose magnitude is given by Equation 3.16 and whose direction is toward the center of the circle. (Note that the unit vectors $\hat{i}$ and $\hat{j}$ are constants, independent of time.)

83. A friend is writing a science fiction screenplay about an asteroid on a collision course with Earth. Despite your admonitions about other films with the same plot, she asks you to calculate some numbers so her scenario will be correct. Astronauts will attach a rocket engine to the asteroid in an attempt to divert it. The asteroid is moving at 21 km/s. The rocket will provide an acceleration of 0.035 km/s² at a right angle to the original motion. This will change the direction of the asteroid's motion by 22.6° and displace it 5.36×10^3 km, enough to miss Earth and save civilization. Your friend asks you whether a rocket firing lasting 4 min will be enough to give these results. How do you advise her?

84. An archery coach at your college wishes to know the approximate speed with which arrows are fired. She tells you the archer fires an arrow horizontally from a height of 1.54 m. The arrow hits the ground 23 m downrange from the archer. Ignoring air resistance, what speed do you report to the coach?

85. Using your textbook as a guide to projectile motion, search the Internet for information on the range of baseballs and golf balls. What factors affect the range? The theoretical best launch angle for maximum range is 45°, but in reality this is not the case. What angle gives the greatest range for a baseball or a golf ball? Why is the potential range of a baseball greater in Denver than in Minneapolis?

Answers to Chapter Questions

Answer to Chapter Opening Question

Assuming negligible air resistance, the penguin should leave the water at a 45° angle.

Answers to GOT IT? Questions

3.1 (c).

3.2 (d) only.

3.3 (c) gives the greatest change in speed; (b) gives the greatest change in direction.

3.4 $a_2 > a_3 > a_4 > a_1$.

4 Force and Motion

■ What forces govern the motion of the sailboard?

An interplanetary spacecraft moves effortlessly, yet its engines shut down years ago. Why does it keep moving? A baseball heads toward the batter. The batter swings, and suddenly the ball is heading toward left field. Why did its motion change?

Questions about the "why" of motion are the subject of **dynamics**. Here we develop the basic laws that answer those questions. Isaac Newton first stated these laws more than 300 years ago, yet they remain a vital part of physics and engineering today, helping us guide spacecraft to distant planets, develop better cars, and manipulate the components of individual cells.

4.1 The Wrong Question

We began this chapter with two questions: one about why a spacecraft *moved* and the other about why a baseball's motion *changed*. For nearly 2000 years following the work of Aristotle (384–322 BCE), the first question—Why do things move?—was the crucial one.

▶ **To Learn**
By the end of this chapter you should be able to

■ Explain the concept of force and its role in causing *change* in motion (4.1).

■ Describe the fundamental forces of physics (4.3).

■ State Newton's three laws of motion (4.2, 4.6).

■ Explain the force of gravity and the distinction between mass and weight (4.4).

■ Apply Newton's laws to one-dimensional motion (4.5, 4.6).

◀ **To Know**

■ Newton's laws relate force and acceleration. Therefore you need a solid understanding of acceleration, here based on the one-dimensional analysis of Chapter 2 (2.3).

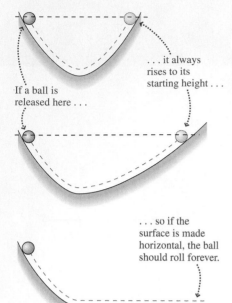

FIGURE 4.1 Galileo considered balls rolling on inclines and concluded that a ball on a horizontal surface should roll forever.

And the answer seemed obvious: It took a force—a push or a pull—to keep something moving. This idea makes sense: Stop exerting yourself when jogging, and you stop moving; take your foot off the gas pedal, and your car soon stops. Everyday experience seems to suggest that Aristotle was right, and most of us carry in our heads the Aristotelian idea that motion requires a cause—something that pushes or pulls on a moving object to keep it going.

Actually, What keeps things moving? is the wrong question. In the early 1600s, Galileo Galilei did experiments that convinced him that a moving object has an intrinsic "quantity of motion" and needs no push to keep it moving (Fig. 4.1). Instead of answering What keeps things moving?, Galileo declared that the question needs no answer. In so doing, he set the stage for centuries of progress in physics, beginning with the achievements of Issac Newton and culminating in the work of Albert Einstein.

The Right Question

Our first question—Why does the spacecraft keep moving?—is the wrong question. So what's the right question? It's the second one, about why the baseball's motion *changed*. The study of dynamics isn't about what causes motion itself; it's about what causes *changes* in motion. Changes include starting and stopping, speeding up and slowing down, and changing direction. Any *change* in motion begs an explanation, but motion itself does not. Get used to this important idea and you'll have a much easier time with physics. But if you remain a "closet Aristotelian," secretly looking for causes of motion, you'll find it difficult to understand and apply the simple laws that actually govern motion.

Galileo identified the right question about motion. But it was Isaac Newton who formulated the quantitative laws describing how motion changes. We use those laws today for everything from designing antilock braking systems, to building skyscrapers, to guiding spacecraft.

4.2 Newton's First and Second Laws

What caused the baseball's motion to change? Obviously, it was the bat pushing the ball. We use the term **force** to describe a push or a pull. And the essence of dynamics is simply this:

> Force causes change in motion.

We'll soon quantify this notion, writing equations and solving numerical problems. But the essential point is in the simple sentence above. If you want to change an object's motion, you need to apply a force. If you see an object's motion change, you know there's a force acting. Contrary to Aristotle, and probably to your own intuitive sense as well, it does *not* take a force to keep something in unchanging motion; force is needed *only* to *change* an object's motion.

The Net Force

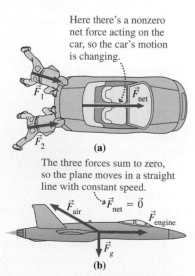

FIGURE 4.2 The net force determines the change in an object's motion.

You can push a ball left or right, up or down. Your car's tires can push the car forward or backward, or make it round a curve. Force has direction and is a vector quantity. Furthermore, more than one force can act on a single object. We call the individual forces on an object **interaction forces** because they always involve other objects interacting with the object in question. In Fig. 4.2a, for example, the interaction forces are exerted by the people pushing the car. In Fig. 4.2b the interaction forces include the force of air on the plane, the engine force from the hot exhaust gases, and Earth's gravitational force.

We now explore in more detail the relation between force and change in motion. Experiment shows that what matters is the **net force**, meaning the vector sum of all individual interaction forces acting on an object. If the net force on an object is not zero, then the object's motion must be changing—in direction or speed or both (Fig. 4.2a). If the net force on an object is zero—no matter what individual interaction forces contribute to the net force—then the object's motion is unchanging (Fig. 4.2b).

Newton's First Law

The basic idea that force causes change in motion is the essence of **Newton's first law**:

> **Newton's first law of motion:** A body in uniform motion remains in uniform motion, and a body at rest remains at rest, unless acted on by a nonzero net force.

The word "uniform" here is essential; **uniform motion** means unchanging motion—that is, motion in a straight line at constant speed. The phrase "a body at rest" isn't really necessary because rest is just the special case of uniform motion with zero speed, but we include it for consistency with Newton's original statement.

The first law says that uniform motion is a perfectly natural state, requiring no explanation. Again, the word "uniform" is crucial. The first law does *not* say that an object moving in a circle will continue to do so without a nonzero net force; in fact, it says that an object moving in a circle—or in any other curved path—*must* be subject to a nonzero net force because its motion is changing.

GOT IT? 4.1 On a horizontal tabletop is a curved barrier that exerts a force on a ball, guiding its motion in a circular path as shown. After the ball leaves the barrier, which of the dashed paths shown does it follow?

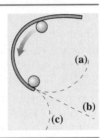

Newton's first law is simplicity itself, but it does seem to violate our Aristotelian preconceptions; after all, your car soon stops when you take your foot off the gas. But because the motion changes, that just means—as the first law says—that there must be a nonzero net force acting. That force is often a "hidden" one, like friction, that isn't as obvious as the push or pull of muscle. Go to an ice show or hockey game, where frictional forces are minimized, and the first law becomes a lot clearer.

Newton's Second Law

Newton's second law quantifies the relation between force and change in motion. Newton reasoned that the product of mass and velocity was the best measure of an object's "quantity of motion." The modern term is **momentum**, and we write

$$\vec{p} = m\vec{v} \quad \text{(momentum)} \tag{4.1}$$

for the momentum of an object with mass m and velocity $\vec{v}$. As the product of a scalar (mass) and a vector (velocity), momentum is itself a vector quantity. Newton's second law relates the rate of change of an object's momentum to the net force acting on that object:

> **Newton's second law of motion:** The rate at which a body's momentum changes is equal to the net force acting on the body:
> $$\vec{F}_{\text{net}} = \frac{d\vec{p}}{dt} \quad \text{(Newton's 2}^{\text{nd}}\text{ law)} \tag{4.2}$$

When a body's mass remains constant, we can use the definition of momentum, $\vec{p} = m\vec{v}$, to write

$$\vec{F}_{\text{net}} = \frac{d\vec{p}}{dt} = \frac{d(m\vec{v})}{dt} = m\frac{d\vec{v}}{dt}$$

But $d\vec{v}/dt$ is the acceleration $\vec{a}$, so

$$\vec{F}_{\text{net}} = m\vec{a} \quad \text{(Newton's 2}^{\text{nd}}\text{ law, constant mass)} \tag{4.3}$$

Although Newton originally wrote his second law in the form 4.2, which remains the most general form, the form 4.3 is more widely recognized.

Newton's second law includes the first law as the special case $\vec{F}_{net} = \vec{0}$. In this case Equation 4.3 gives $\vec{a} = \vec{0}$, so an object's velocity doesn't change.

✓**TIP** Understanding Newton

To apply Newton's law successfully, you have to understand the terms in Equation 4.3. On the left is the net force $\vec{F}_{net}$—the vector sum of all real, physical interaction forces acting on an object. On the right is $m\vec{a}$—not a force but the product of the object's mass and acceleration. The equal sign says that they have the same value, not that they're the same thing. So don't go adding an extra force $m\vec{a}$ when you're applying Newton's second law.

Mass, Inertia, and Force

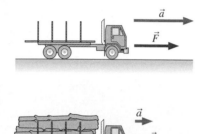

FIGURE 4.3 The loaded truck has greater mass—more inertia—so its acceleration is smaller when the same force is applied.

Because it takes a force to change an object's motion, the first law says that objects naturally resist changes in motion. The term **inertia** describes this resistance, and for that reason the first law is also called the **law of inertia**. Just as we describe a sluggish person as having a lot of inertia, so an object that is hard to start moving—or hard to stop once started—has a lot of inertia. If we solve the second law for the acceleration $\vec{a}$, we find that $\vec{a} = \vec{F}/m$—showing that a given force is *less* effective in changing the motion of a *more massive* object (Fig. 4.3). The mass m that appears in Newton's laws is thus a measure of an object's inertia and determines the object's response to a given force.

By comparing the acceleration of a known and an unknown mass, we can determine the unknown mass. From Newton's second law for a force of magnitude F,

$$F = m_{known}a_{known} \quad \text{and} \quad F = m_{unknown}a_{unknown}$$

where we're interested only in magnitudes so we don't use vectors. Equating these two expressions for the same force, we get

$$\frac{m_{unknown}}{m_{known}} = \frac{a_{known}}{a_{unknown}} \tag{4.4}$$

Equation 4.4 is an operational definition of mass; it shows how, given a known mass and a fixed force, we can determine other masses. This method is used routinely to measure the masses of astronauts in orbiting spacecraft (Fig. 4.4).

The force required to accelerate a 1-kg mass at the rate of 1 m/s^2 is defined to be 1 **newton** (N). Equation 4.3 shows that 1 N is equivalent to 1 kg·m/s^2. Other common force units are the English pound (lb, equal to 4.448 N) and the dyne, a metric unit equal to 10^{-5} N. A 1-N force is rather small; you can readily exert forces measuring hundreds of newtons with your own body.

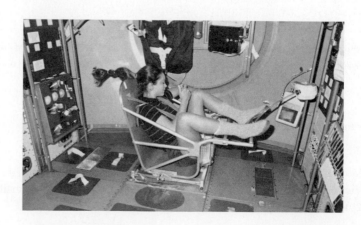

FIGURE 4.4 In a space shuttle, astronaut Tamara Jernigan uses a mass-measuring device based on Newton's second law.

EXAMPLE 4.1 **Force from Newton: A Car Accelerates**

A 1200-kg car accelerates from rest to 20 m/s in 7.8 s, moving in a straight line with constant acceleration. (a) What is the net force acting on the car? (b) If the car then rounds a bend 85 m in radius at a steady 20 m/s, what net force acts on it?

INTERPRET In this problem we're asked to evaluate the net force on a car (a) when it undergoes constant acceleration and (b) when it rounds a turn. In both cases the net force is entirely horizontal, so we need to consider only the horizontal component of Newton's law.

DEVELOP Figure 4.5 shows the horizontal force acting on the car in each case; since this is the net force, it is equal to the car's mass multiplied by its acceleration. We aren't actually given the acceleration in this problem, but for (a) we know the change in speed and the time involved, so we can write $a = \Delta v/\Delta t$. For (b) we're given the speed and the radius of the turn; since the car is in uniform circular motion, Equation 3.16 applies, and we have $a = v^2/r$.

EVALUATE We solve for the unknown acceleration and evaluate the numerical answers for both cases:

(a) $\quad F_{net} = ma = m\dfrac{\Delta v}{\Delta t} = (1200 \text{ kg})\left(\dfrac{20 \text{ m/s}}{7.8 \text{ s}}\right) = 3.1 \text{ kN}$

(b) $\quad F_{net} = ma = m\dfrac{v^2}{r} = (1200 \text{ kg})\dfrac{(20 \text{ m/s})^2}{85 \text{ m}} = 5.6 \text{ kN}.$

ASSESS First, the units worked out; they were actually kg·m/s², but that defines the newton. The answers came out in thousands of N, but we moved the decimal point three places and changed to kN for convenience. And the numbers seem to make sense; we mentioned that 1 newton is a rather small force, so it's not surprising to find forces on cars measured in kilonewtons.

Note that Newton's law doesn't distinguish between forces that change an object's speed—(a) here—and forces that change an object's direction—as in (b). Newton's law relates force, mass, and acceleration in *all* cases. ∎

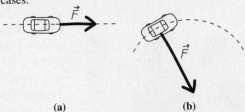

(a)　　　　　(b)

FIGURE 4.5 Our sketch of the net force on the car of Example 4.1.

GOT IT? 4.2 A nonzero net force acts on an object. Does that mean the object necessarily moves in the same direction as the net force?

Inertial Reference Frames

Why don't flight attendants serve beverages when an airplane is accelerating down the runway? For one thing, their beverage cart wouldn't stay put, but would accelerate toward the back of the plane even in the absence of a net force. So is Newton's first law wrong? No, but Newton's laws don't apply in an accelerating airplane. With respect to the ground, in fact, the beverage cart is doing just what Newton says it should: It remains in its original state of motion, while all around it plane and passengers accelerate toward takeoff.

In Section 3.3 we defined a reference frame as a system against which we measure velocities; more generally, a reference frame is the "background" in which we study physical reality. Our airplane example shows that Newton's laws don't work in all reference frames; in particular, they're not valid in accelerating frames. Where they are valid is in reference frames undergoing uniform motion—called **inertial reference frames** because only in these frames does the law of inertia hold. In a noninertial frame like an accelerating airplane, a car rounding a curve, or a whirling merry-go-round, an object at rest doesn't remain at rest, even when no force is acting. A good test for an inertial frame is to check whether Newton's first law is obeyed—that is, whether an object at rest remains at rest, and an object in uniform motion remains in uniform motion, when no force is acting on it.

Strictly speaking, our rotating Earth is not an inertial frame, and therefore Newton's laws aren't exactly valid on Earth. But Earth's rotation has an insignificant effect on most motions of interest, so we can usually treat Earth as an inertial reference frame. An important exception is the motion of oceans and atmosphere; here, scientists must take Earth's rotation into account.

If Earth isn't an inertial frame, what is? That's a surprisingly subtle question, and it pointed Einstein toward his general theory of relativity. The law of inertia is intimately related to questions of space, time, and gravity—questions whose answers lie in Einstein's theory. We'll look briefly at that theory in Chapter 33.

4.3 Forces

The most familiar forces are pushes and pulls you apply with your own body, but passive objects can apply forces, too. A car collides with a parked truck and comes to a stop. Why? Because the truck exerts a force on it. The Moon circles Earth rather than moving in a straight line. Why? Because Earth exerts a gravitational force on it. You sit in a chair and don't fall to the floor. Why not? Because the chair exerts an upward force on you, countering gravity.

When you sit in a chair, the chair compresses and exerts an upward force that balances gravity.

FIGURE 4.6 A compression force.

Some forces, like those you apply with your muscles, can have values that you choose. Other forces take on values determined by the situation. When you sit in the chair shown in Fig. 4.6, the downward force of gravity on you causes the chair to compress slightly. The chair acts like a spring and exerts an upward force. When the chair compresses enough that the upward force is equal in magnitude to the downward force of gravity, there's no net force and you sit without accelerating. The same thing happens with **tension forces** when objects are suspended from ropes or cables—the ropes stretch until the force they exert balances the force of gravity (Fig. 4.7).

Forces like the pull you exert on your rolling luggage, the force of a chair on your body, and the force a baseball exerts on a bat are **contact forces** because the force is exerted through direct contact. Other forces, like gravity and electric and magnetic forces, are **action-at-a-distance forces** because they seemingly act between distant objects, like Earth and the Moon. Actually, the distinction isn't clear-cut; at the microscopic level, contact forces involve action-at-a-distance electric forces between molecules. And the action-at-a-distance concept itself is troubling. How can Earth "reach out" across empty space and pull on the Moon? Later we'll look at an approach to forces that avoids this quandary.

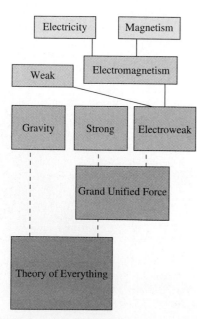

FIGURE 4.7 The climbing rope exerts an upward tension force $\vec{T}$ that balances the force of gravity.

The Fundamental Forces

Gravity, tension forces, compression forces, contact forces, electric forces, friction forces—how many kinds of forces are there? At present, physicists identify three basic forces: the gravitational force, the electroweak force, and the strong force.

Gravity is the weakest of the fundamental forces, but because it acts attractively between all matter, gravity's effect is cumulative. That makes gravity the dominant force in the large-scale universe, determining the structure of planets, stars, galaxies, and the universe itself.

The **electroweak force** subsumes **electromagnetism** and the **weak nuclear force**. Virtually all the nongravitational forces we encounter in everyday life are electromagnetic, including contact forces, friction, tension and compression forces, and the forces that bind atoms into chemical compounds. The weak nuclear force is less obvious, but it's crucial in the Sun's energy production—providing the energy that powers life on Earth.

The **strong force** describes how particles called **quarks** bind together to form protons, neutrons, and a host of less-familiar particles. The force that joins protons and neutrons to make atomic nuclei is a residue of the strong force between their constituent quarks. Although the strong force is not obvious in everyday life, it is ultimately responsible for the structure of matter. If its strength were slightly different, atoms more complex than helium would not be possible, and the universe would be devoid of life!

Unifying the fundamental forces is a major goal of physics. Over the centuries we've come to understand seemingly disparate forces as manifestations of a more fundamental underlying force. Figure 4.8 suggests that the process continues, as physicists attempt first to unify the strong and electroweak forces, and then ultimately to add gravity to give a "Theory of Everything."

FIGURE 4.8 Unification of forces is a major theme in physics.

4.4 The Force of Gravity

Newton's second law shows that mass is a measure of a body's resistance to changes in motion—its inertia. A body's mass is an intrinsic property; it doesn't depend on location. If my mass is 65 kg, it is 65 kg on Earth, in an orbiting spacecraft, or on the Moon. That means no matter where I am, a force of 65 N gives me an acceleration of 1 m/s^2.

We commonly use the term "weight" to mean the same thing as mass. In physics, though, **weight** is the *force* that gravity exerts on a body. Near Earth's surface, a freely falling body accelerates downward at 9.8 m/s^2; we designate this acceleration vector by $\vec{g}$. Newton's second law, $\vec{F} = m\vec{a}$, then says that the force of gravity on a body of mass m is $m\vec{g}$; this force is the body's weight:

$$\vec{w} = m\vec{g} \quad \text{(weight)} \tag{4.5}$$

With my 65-kg mass, my weight near Earth's surface is then $(65 \text{ kg})(9.8 \text{ m/s}^2)$ or 640 N. On the Moon, where the acceleration of gravity is only 1.6 m/s^2, I would weigh only 100 N. And in the remote reaches of intergalactic space, far from any gravitating object, my weight would be essentially zero.

EXAMPLE 4.2   **Mass and Weight: Exploring Mars**

The rovers Spirit and Opportunity that landed on Mars in 2004 each weighed 1.8 kN on Earth. What were their mass and weight on Mars?

INTERPRET Here we're asked about the relation between mass and weight, and the object we're interested in is the Mars rover.

DEVELOP Equation 4.5 describes the relation between mass and weight. Writing this equation in scalar form because we're interested only in magnitudes, we have $w = mg$.

EVALUATE First we want to find mass from weight, so we solve for m using the Earth weight and Earth's gravity:

$$m = \frac{w}{g} = \frac{1.8 \text{ kN}}{9.8 \text{ m/s}^2} = 184 \text{ kg}$$

This mass is the same everywhere, so the weight on Mars is given by $w = mg_{\text{Mars}} = (184 \text{ kg})(3.74 \text{ m/s}^2) = 688 \text{ N}$. Here we found the acceleration of gravity on Mars in Appendix E.

ASSESS Make sense? Sure, Mars's gravitational acceleration is lower than Earth's, and so are the rovers' weights on Mars.

One reason we confuse mass and weight is the common use of the SI unit kilogram to describe "weight." At the doctor's office you may be told that you "weigh" 55 kg. You don't; you have a mass of 55 kg, so your weight is $(55 \text{ kg})(9.8 \text{ m/s}^2)$ or 540 N. The unit of force in the English system is the pound, so giving your weight in pounds is correct.

That we confuse mass and weight at all results from the remarkable fact that the gravitational acceleration of all objects is the same. This makes a body's *weight*, a gravitational property, proportional to its *mass*, a measure of its inertia in terms that have nothing to do with gravity. First inferred by Galileo from his experiments with falling bodies, this relation between gravitation and inertia seemed a coincidence until the early 20th century. Finally Albert Einstein showed how that simple relation reflects the underlying geometry of space and time in a way that intimately links gravitation and acceleration.

Weightlessness

Aren't astronauts "weightless"? Not according to our definition. At typical space-shuttle altitudes, the acceleration of gravity has about 93% of its value at Earth's surface, so the gravitational forces $m\vec{g}$ on the shuttle and its occupants are almost as large as on Earth. But the astronauts *seem* weightless, and indeed they *feel* weightless (Fig. 4.9). What's going on?

Imagine yourself in an elevator whose cable has broken and is dropping freely downward with the gravitational acceleration g. In other words, the elevator and its occupant are in **free fall**, with only the force of gravity acting. If you let go of a book you're holding in the elevator, it too falls freely with acceleration g. But so does everything else around it—and therefore the book

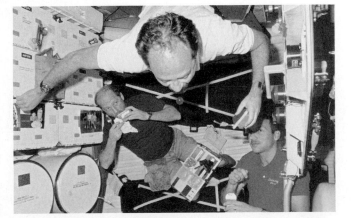

FIGURE 4.9 These astronauts only *seem* weightless.

In a freely falling elevator on Earth, the book and person seem weightless because they fall with the same acceleration as the elevator.

Earth

(a)

Like the elevator in (a), an orbiting spacecraft is falling toward Earth, and because its occupants also fall with the same acceleration, they experience apparent weightlessness.

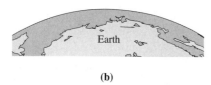

Earth

(b)

FIGURE 4.10 Objects in free fall appear weightless because they all experience the same acceleration.

stays put relative to you (Fig. 4.10a). To you, the book seems "weightless," since it doesn't seem to fall when you let go of it. And you're "weightless" too; if you jump off the elevator's floor, you float to the ceiling rather than falling back down. Of course you, the book, and the elevator are *all* falling, but because all have the same acceleration that isn't obvious to you. The gravitational force is still acting; it's making you fall. So you really do have weight, and your condition is best termed **apparent weightlessness**.

A falling elevator is a dangerous place; your state of apparent weightlessness would end with a deadly smash caused by nongravitational contact forces when you hit the ground. But apparent weightlessness occurs permanently in a state of free fall that doesn't intersect Earth—as in an orbiting spacecraft (Fig. 4.10b). It's not being in outer space that makes astronauts seem weightless; it's that they, like our hapless elevator occupant, are in free fall—moving under the influence of the gravitational force alone. The condition of apparent weightlessness in orbiting spacecraft is sometimes called "microgravity."

GOT IT? 4.3 A popular children's book explains the weightlessness astronauts experience by saying there's no gravity in space. If there were no gravity in space, what would be the motion of a space shuttle, a satellite, or, for that matter, the Moon?

4.5 Using Newton's Second Law

The interesting problems involving Newton's second law are those where more than one force acts on an object. To apply the second law, we then need the net force. For an object of constant mass, the second law relates the net force and the acceleration:

$$\vec{F}_{net} = m\vec{a}$$

Using Newton's second law with multiple forces becomes much easier if we draw a **free-body diagram**, a simple diagram that shows only the object of interest and the forces acting on it.

TACTICS 4.1 **Drawing a Free-Body Diagram**

Drawing a free-body diagram, which shows the forces acting on an object, is the key to solving problems with Newton's laws. To make a free-body diagram:

1 Identify the object of interest and all the forces acting on it.

2 Represent the object as a dot.

3 Draw the vectors for *only* those forces acting *on* the object, with their tails all starting on the dot.

Figure 4.11 shows two examples where we reduce physical scenarios to free-body diagrams. We often add a coordinate system to the free-body diagram so that we can express force vectors in components.

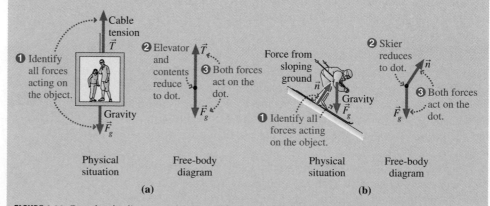

FIGURE 4.11 Free-body diagrams. (a) A one-dimensional situation like those we discuss in this chapter. (b) A two-dimensional situation. We'll deal with such cases in Chapter 5.

Our IDEA strategy applies to Newton's laws as it does to other physics problems. For the second law, we can elaborate on the four IDEA steps:

PROBLEM SOLVING STRATEGY 4.1 **Newton's Second Law**

INTERPRET Interpret the problem to be sure that you know what it's asking and that Newton's second law is the relevant concept. Identify the object of interest and all the individual interaction forces acting on it.

DEVELOP Draw a free-body diagram as described in Tactics 4.1. Develop your solution plan by writing Newton's second law, $\vec{F}_{net} = m\vec{a}$, with $\vec{F}_{net}$ expressed as the sum of the forces you've identified. Then choose a coordinate system so you can express Newton's law in components.

EVALUATE At this point the physics is done, and you're ready to execute your plan by solving Newton's second law and evaluating the numerical answer(s), if called for. Even in the one-dimensional problems of this chapter, remember that Newton's law is a vector equation; that will help you get the signs right. You need to write the components of Newton's law in the coordinate system you chose, and then solve the resulting equation(s) for the quantity(ies) of interest.

ASSESS Assess your solution to see that it makes sense. Are the numbers reasonable? Do the units work out correctly? What happens in special cases—for example, when a mass, a force, an acceleration, or an angle gets very small or very large?

EXAMPLE 4.3 **Newton's Second Law: In the Elevator**

A 740-kg elevator accelerates upward at 1.1 m/s², pulled by a cable of negligible mass. Find the tension force in the cable.

INTERPRET In this problem we're asked to evaluate one of the forces on an object. First we identify the object of interest. Although the problem asks about the cable tension, it's the elevator on which that tension acts, so the elevator is the object of interest. Next, we identify the forces acting on the elevator. There are two: the downward force of gravity $\vec{F}_g$ and the upward cable tension $\vec{T}$.

DEVELOP Figure 4.12a shows the elevator accelerating upward; Fig. 4.12b is a free-body diagram representing the elevator as a dot with the two force vectors acting on it. The applicable equation is Newton's second law, $\vec{F}_{net} = m\vec{a}$, with $\vec{F}_{net}$ given by the sum of the forces we've identified:

$$\vec{F}_{net} = \vec{T} + \vec{F}_g = m\vec{a} \qquad (4.6)$$

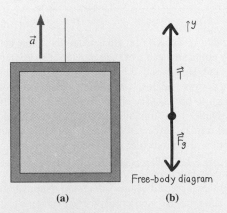

Free-body diagram

(a)　　　　(b)

FIGURE 4.12 The forces on the elevator are the cable tension $\vec{T}$ and gravity $\vec{F}_g$.

✓**TIP** Vectors Tell It All

Don't be tempted to put a minus sign in this equation because one force is downward. You don't have to worry about signs until you write the components of a vector equation in the coordinate system you chose.

Now we need to choose a coordinate system. Here all the forces are vertical, so we choose our y axis pointing upward.

EVALUATE Now we're ready to rewrite Newton's second law—Equation 4.6 in this case—in our coordinate system. Formally, we remove the vector signs and add coordinate subscripts—just y in this case:

$$T_y + F_{gy} = ma_y \qquad (4.7)$$

Still no need to worry about signs. Now, what is T_y? Since the tension points upward and we've chosen that to be the positive direction, the component of tension in the y direction is its magnitude T. What about F_{gy}? Gravity points downward, so this component is negative. Furthermore, we know that the magnitude of the gravitational force is mg. So $F_{gy} = -mg$. Then our Newton's law equation becomes

$$T - mg = ma_y$$

so

$$T = ma_y + mg = m(a_y + g) \qquad (4.8)$$

For the numbers given, this equation yields

$$T = m(a_y + g) = (740 \text{ kg})(1.1 \text{ m/s}^2 + 9.8 \text{ m/s}^2) = 8.1 \text{ kN}$$

ASSESS We can see that this answer makes sense—and learn a lot more about physics—from the algebraic form of the answer in Equation 4.8. Consider some special cases: If the acceleration a_y were zero, then the net force on the elevator would have to be zero. In that case Equation 4.8 gives $T = mg$. Of course: The cable is then supporting the elevator's weight mg but not exerting any additional force to accelerate it.

cont'd.

On the other hand, if the elevator is accelerating upward, then the cable has to provide an extra force in addition to the weight; that's why the tension becomes $ma_y + mg$. Numerically, our answer of 8.1 kN is *greater* than the elevator's weight—and the cable had better be strong enough to handle the extra force.

Finally, if the elevator is accelerating downward, then a_y is negative, and the cable tension is *less* than the weight. In free fall, $a_y = -g$, and the cable tension would be zero.

You might have reasoned out this problem in your head. But we did it very thoroughly because the strategy we followed will let you solve all problems involving Newton's second law, even if they're much more complicated. If you always follow this strategy and don't try to find shortcuts, you'll become confident in using Newton's second law.

■

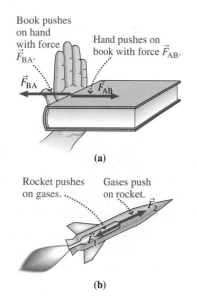

Book pushes on hand with force $\vec{F}_{BA}$.

Hand pushes on book with force $\vec{F}_{AB}$.

$\vec{F}_{BA}$ $\vec{F}_{AB}$

(a)

Rocket pushes on gases.

Gases push on rocket.

$\vec{F}_1$ $\vec{F}_2$

(b)

FIGURE 4.13 Newton's third law says that forces always come in pairs. With objects in contact, both forces act at the contact point. To emphasize that the two forces act on *different* objects, we draw them slightly displaced.

GOT IT? 4.4 For each of the following situations, would the cable tension in Example 4.3 be greater than, less than, or equal to the elevator's weight? (a) elevator starts moving upward, accelerating from rest; (b) elevator decelerates to a stop while moving upward; (c) elevator starts moving downward, accelerating from rest; (d) elevator slows to a stop while moving downward; (e) elevator is moving upward with constant speed.

4.6 Newton's Third Law

Push your book across your desk, and you feel the book push back (Fig. 4.13a). Kick a ball with bare feet, and your toes hurt. Why? You exert a force on the ball, and the ball exerts a force back on you. A rocket engine exerts forces that expel hot gases out of its nozzle—and the hot gases exert a force on the rocket, accelerating it forward (Fig. 4.13b).

Whenever one object exerts a force on a second object, the second object also exerts a force on the first. The two forces are in opposite directions, but they have equal magnitudes. This fact constitutes **Newton's third law** of motion. The familiar expression "for every action there is an equal and opposite reaction" is Newton's 17th-century language. But there's really no distinction between "action" and "reaction"; both are always present. In modern language, the third law states:

> **Newton's third law of motion** If object A exerts a force on object B, then object B exerts an oppositely directed force of equal magnitude on A.

Newton's third law is about forces between objects. It says that such forces always occur in pairs—that it's not possible for object A to exert a force on object B without B ex-

APPLICATION **Hollywood Goes Weightless**

The film *Apollo 13* shows Tom Hanks and his fellow actors floating weightlessly around the cabin of their movie-set spacecraft. What special effects did Hollywood use here? None. The actors' apparent weightlessness was the real thing. But even Hollywood's budget wasn't enough to buy a space-shuttle flight. So the producers rented NASA's weightlessness training aircraft, aptly dubbed the "vomit comet." This airplane executes parabolic trajectories that mimic the free-fall motion of a projectile, so its occupants experience apparent weightlessness. The photo shows not the movie set, but astronaut trainees in the "vomit comet."

Movie critics marveled at how *Apollo 13* "simulated the weightlessness of outer space." Nonsense! The actors were in free fall just like the real astronauts on board the real Apollo 13, and they experienced exactly the same physical phenomenon—apparent weightlessness when moving under the influence of gravity alone.

erting a force back on A. You can see now why we coined the term "interaction forces"—when there's force between two objects, it's always a true *inter*action, with both objects exerting forces and both experiencing forces. We'll use the terms **interaction force pair** and **third-law pair** for the two forces described by Newton's third law.

It's crucial to recognize that the forces of a third-law pair act on *different* objects; the force $\vec{F}_{AB}$ of object A acts on object B, and the force $\vec{F}_{BA}$ of B acts on A. The forces have equal magnitudes and opposite directions, but they don't cancel to give zero net force *because they don't act on the same object*. In Fig. 4.13a, for example, $\vec{F}_{AB}$ is the force the hand exerts on the book. There's no other horizontal force acting on the book, so the net force on the book is nonzero and the book accelerates. Failure to recognize that the two forces of a third-law pair act on different objects leads to a contradiction, embodied in the famous horse-and-cart dilemma illustrated in Fig. 4.14.

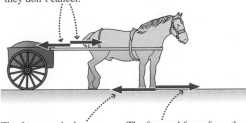

These forces constitute an equal but opposite pair, but they don't act on the same object so they don't cancel.

The force on the horse arises as a reaction to the horse pushing back on the road.

The forward force from the road on the horse is greater than the backward force from the cart so the net force and hence acceleration are forward.

FIGURE 4.14 The horse-and-cart dilemma: The horse pulls on the cart, and the cart pulls back on the horse with a force of equal magnitude. So how can the pair ever get moving? The figure shows the resolution of this dilemma: The *net* force on the horse involves two forces, each from a *different* third-law pair. Their magnitudes aren't equal and the horse experiences a net force in the forward direction.

EXAMPLE 4.4   Newton's Third Law: Pushing Books

On a frictionless horizontal surface, you push with force $\vec{F}$ on a book of mass m_1 that in turn pushes on a book of mass m_2 (Fig. 4.15a). What force does the second book exert on the first?

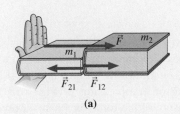

(a)

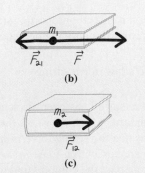

(b)

(c)

FIGURE 4.15 Horizontal forces on the books of Example 4.4. Not shown are the vertical forces of gravity and the normal force from the surface supporting the books.

INTERPRET This problem is about the interaction between two objects, so we identify both books as objects of interest.

DEVELOP In a problem with multiple objects, it's a good idea to draw a separate free-body diagram for each object. We've done that in Fig. 4.15b and c, keeping very light images of the books themselves. Now, we're asked about the force the second book exerts on the first. Newton's third law would give us that force if we knew the force the first book exerts on the second. Since that's the only horizontal force acting on book 2, we could get it from Newton's *second* law if we knew the acceleration of book 2. So here's our plan: (1) Find the acceleration of book 2; (2) use Newton's second law to find the net force on book 2, which in this case is the single force $\vec{F}_{12}$; and (3) apply Newton's third law to get $\vec{F}_{21}$, which is what we're looking for.

EVALUATE (1) The total mass of the two books is $m_1 + m_2$, and the net force applied to the combination is $\vec{F}$. Newton's second law, $\vec{F} = m\vec{a}$, gives

$$\vec{a} = \frac{\vec{F}}{m} = \frac{\vec{F}}{m_1 + m_2}$$

for the acceleration of both books, including book 2. (2) Now that we know book 2's acceleration, we use Newton's second law to find $\vec{F}_{12}$, which we recognize as the net force on book 2:

$$\vec{F}_{12} = m_2\vec{a} = m_2\frac{\vec{F}}{m_1 + m_2} = \frac{m_2}{m_1 + m_2}\vec{F}$$

cont'd.

(3) Finally, the forces the books exert on each other constitute a third-law pair, so we have

$$\vec{F}_{21} = -\vec{F}_{12} = -\frac{m_2}{m_1 + m_2}\vec{F}$$

ASSESS You can see that this result makes sense by considering the first book. It too undergoes acceleration $\vec{a} = \vec{F}/(m_1 + m_2)$, but there are *two* forces acting on it: the applied force $\vec{F}$ and the force $\vec{F}_{21}$ from

the second book. So the net force on the first book is

$$\vec{F} + \vec{F}_{21} = \vec{F} - \frac{m_2}{m_1 + m_2}\vec{F} = \frac{m_1}{m_1 + m_2}\vec{F} = m_1\vec{a}$$

consistent with Newton's second law. Our result shows that Newton's second and third laws are both necessary for a fully consistent description of the motion. ∎

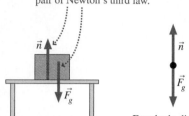

The upward normal force from the table and the downward force due to gravity are not an action/reaction pair of Newton's third law.

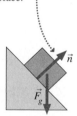

Free-body diagram

(a)

GOT IT? 4.5 The figure shows two blocks with two forces acting on the pair. Is the net force on the larger block (a) greater than 2 N, (b) equal to 2 N, or (c) less than 2 N?

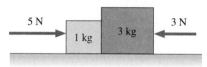

5 N   1 kg   3 kg   3 N

The normal force acts perpendicular to the surface.

The normal force and gravitational force do not balance, so the block slides down the slope.

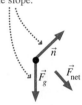

Free-body diagram

(b)

FIGURE 4.16 Normal forces. Also shown in each case is the gravitational force.

A contact force such as the force between the books in Example 4.4 is called a **normal force** (symbol $\vec{n}$) because it acts at right angles to the surfaces in contact. Other examples of normal forces are the upward force that a table or bridge exerts on objects it supports, and the force perpendicular to a sloping surface supporting an object (Fig. 4.16.)

Newton's third law also applies to forces like gravity that don't involve direct contact. Since Earth exerts a downward force on you, the third law says that you exert an equal upward force on Earth (Fig. 4.17). If you're in free fall, then Earth's gravity causes you to accelerate toward Earth. Earth, too, accelerates toward you—but it's so massive that this acceleration is negligible.

Measuring Force

Newton's third law provides a convenient way to measure forces using the tension or compression force in a spring. A spring stretches or compresses in proportion to the force exerted on it. By Newton's third law, the force *on* the spring is equal and opposite to the force the spring exerts on whatever is stretching or compressing it (Fig. 4.18). The spring's stretch or compression thus provides a measure of the force on whatever object is attached to the spring.

In an **ideal spring**, the stretch or compression is directly proportional to the force exerted by the spring. **Hooke's law** expresses this proportionality mathematically:

$$F_{sp} = -kx \qquad \text{(Hooke's law, ideal spring)} \qquad (4.9)$$

Here F_{sp} is the spring force, x is the distance the spring has been stretched or compressed from its normal length, and k is the **spring constant**, which measures the "stiffness" of the spring. Its units are N/m. The minus sign shows that the spring force is *opposite* the distortion of the spring: Stretch it, and the spring responds with a force *opposite* the stretching force; compress it, and the spring pushes back against the compressing force. Real springs obey Hooke's law only up to a point; stretch it too much, and a spring will deform and eventually break.

FIGURE 4.17 Gravitational forces on you and on Earth form a third-law pair. Figure is obviously not to scale!

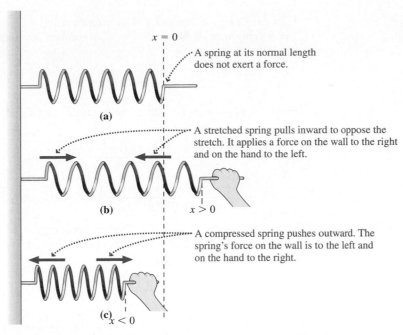

$x = 0$

A spring at its normal length does not exert a force.

(a)

A stretched spring pulls inward to oppose the stretch. It applies a force on the wall to the right and on the hand to the left.

(b) $x > 0$

A compressed spring pushes outward. The spring's force on the wall is to the left and on the hand to the right.

(c) $x < 0$

FIGURE 4.18 A spring responds to stretching or compression with an oppositely directed force.

A **spring scale** is a spring with an indicator and a scale calibrated in force units (Fig. 4.19). Common examples include many bathroom scales, hanging scales in supermarkets, and laboratory spring scales. Even electronic scales are spring scales, with their "springs" materials that produce electrical signals when deformed by an applied force.

When we hang an object to be weighed on a spring scale, the scale spring stretches until the spring force counters the gravitational force on the object. At that point the spring force is equal in magnitude to the weight mg, and thus the spring indicator provides a measure of weight. Given g, this procedure can also provide the object's mass.

Be careful, though: A spring scale provides the true weight only if the scale isn't accelerating; otherwise, the scale reading is only an **apparent weight**. Weigh yourself in an accelerating elevator and you may be horrified or delighted, depending on the direction of the acceleration. Example 4.5 illustrates this point.

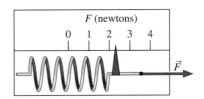

F (newtons)

0 1 2 3 4

$\vec{F}$

FIGURE 4.19 A spring scale.

EXAMPLE 4.5   **True and Apparent Weight: A Helicopter Ride**

A helicopter rises vertically, carrying a load of concrete for the foundation of a ski lift. A 35-kg bag of concrete sits in the helicopter on a spring scale whose spring constant is 3.4 kN/m. By how much does the spring compress (a) when the helicopter is at rest and (b) when it's accelerating upward at 1.9 m/s^2?

INTERPRET This problem is about concrete, a spring scale, and a helicopter. Ultimately, that means it's about mass, force, and acceleration—the content of Newton's laws. We're clearly interested in the spring and the concrete mass resting on it, which share the mo-

tion of the helicopter. We identify two forces acting on the concrete: gravity and the spring force $\vec{F}_{sp}$.

DEVELOP As with any Newton's law problem, we start with a free-body diagram (Fig. 4.20). We then write Newton's second law in its vector form

$$\vec{F}_{net} = \vec{F}_{sp} + \vec{F}_g = m\vec{a}$$

Vectors tell it all; don't worry about signs at this point. Our equation expresses all the physics of the situation, but before we can move on to the solution, we need to choose a coordinate system. Here it's convenient to take the y axis vertically upward.

cont'd.

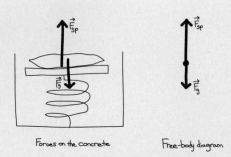

Forces on the concrete   Free-body diagram

FIGURE 4.20 Our drawings for Example 4.5.

$F_{spy} = kx$. Gravity is downward with magnitude mg, so $F_{gy} = -mg$. The y component of Newton's law then becomes $kx - mg = ma_y$, which we solve to get

$$x = \frac{m(a_y + g)}{k}$$

Putting in the numbers (a) with the helicopter at rest $(a_y = 0)$ and (b) with $a_y = 1.9$ m/s² gives

(a) $x = \dfrac{m(a_y + g)}{k} = \dfrac{(35 \text{ kg})(0 + 9.8 \text{ m/s}^2)}{3400 \text{ N/m}} = 10 \text{ cm}$

(b) $x = \dfrac{(35 \text{ kg})(1.9 \text{ m/s}^2 + 9.8 \text{ m/s}^2)}{3400 \text{ N/m}} = 12 \text{ cm}$

EVALUATE The forces are in the vertical direction, so we're concerned with only the y component of Newton's law: $F_{spy} + F_{gy} = ma_y$. The spring force is upward and, from Hooke's law, it has magnitude kx, so

ASSESS Why is the answer to (b) larger? Because, just as with the cable in Example 4.3, the spring needs to provide an additional force to accelerate the concrete upward. ∎

GOT IT? 4.6   (a) Would the answer to (a) in Example 4.5 change if the helicopter were not at rest but moving upward at constant speed? (b) Would the answer to (b) change if the helicopter were moving *downward* but still accelerating *upward*?

Big Picture

The huge idea of this chapter—and of all Newtonian mechanics—is that **force** causes *change* in motion, not motion itself. Uniform motion—straight line, constant speed—needs no cause or explanation. Any deviation, in speed or direction, requires a **net force**. This idea is the essence of Newton's first and second laws. Combined with Newton's third law, these laws provide a consistent description of motion.

Newton's First Law
A body in uniform motion remains in uniform motion, and a body at rest remains at rest, unless acted on by a nonzero net force.

This law is implicit in Newton's second law.

Newton's Second Law
The rate at which a body's momentum changes is equal to the net force acting on the body.

Here **momentum** is the "quantity of motion," the product of mass and velocity.

Newton's Third Law
If object A exerts a force on object B, then object B exerts an oppositely directed force of equal magnitude on A.

Newton's third law says that forces come in pairs.

Solving Problems with Newton's Laws

INTERPRET Interpret the problem to be sure that you know what it's asking and that Newton's second law is the relevant concept. Identify the object of interest and all the individual **interaction forces** acting on it.

DEVELOP Draw a **free-body diagram** as described in Tactics 4.1. Develop your solution plan by writing Newton's second law, $\vec{F}_{net} = m\vec{a}$, with $\vec{F}_{net}$ expressed as the sum of the forces you've identified. Then choose a coordinate system so you can express Newton's law in components.

EVALUATE At this point the physics is done, and you're ready to execute your plan by solving Newton's second law and evaluating the numerical answer(s), if called for. Even in the one-dimensional problems of this chapter, remember that Newton's law is a vector equation; that will help you get the signs right. You need to write the components of Newton's law in the coordinate system you chose, and then solve the resulting equation(s) for the quantity(ies) of interest.

ASSESS Assess your solution to see that it makes sense. Are the numbers reasonable? Do the units work out correctly? What happens in special cases—for example, when a mass, a force, an acceleration, or an angle gets very small or very large?

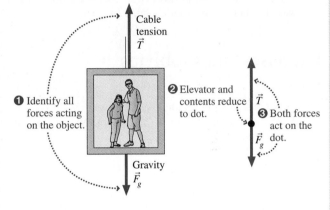

Cable tension $\vec{T}$

❶ Identify all forces acting on the object.

❷ Elevator and contents reduce to dot.

❸ Both forces act on the dot.

Gravity $\vec{F}_g$

Key Concepts and Equations

Mathematically, Newton's second law is $\vec{F}_{net} = d\vec{p}/dt$, where $\vec{p} = m\vec{v}$ is an object's momentum, and $\vec{F}_{net}$ is the sum of all the individual forces acting on the object. When an object has constant mass, the second law takes the familiar form

$$\vec{F}_{net} = m\vec{a} \qquad \text{(Newton's second law)}$$

Newton's second law is a *vector* equation. To use it correctly, you must write the components of the equation in a chosen coordinate system. In one-dimensional problems the result is a single equation.

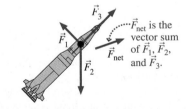

$\vec{F}_{net}$ is the vector sum of $\vec{F}_1$, $\vec{F}_2$, and $\vec{F}_3$.

Cases and Uses

The force of gravity on an object is its **weight**. Since all objects at a given location experience the same gravitational acceleration, weight is proportional to mass:

$$\vec{w} = m\vec{g} \qquad \text{(weight on Earth)}$$

In an accelerated reference frame, an object's **apparent weight** differs from its actual weight; in particular, an object in free fall experiences **apparent weightlessness**.

Springs are convenient force-measuring devices, stretching or compressing in response to the applied force. For an ideal spring, the stretch or compression is directly proportional to the force:

$$F_{sp} = -kx \qquad \text{(Hooke's law)}$$

where k is the **spring constant,** with units of N/m.

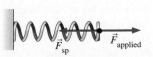

$\vec{F}_{sp}$ $\vec{F}_{applied}$

For Thought and Discussion

1. Distinguish between the Aristotelian and Galilean/Newtonian view of the natural state of motion.
2. A ball bounces off a wall with the same speed it had before it hit the wall. Has its momentum changed? Has a force acted on the ball? Has a force acted on the wall? Relate your answers to Newton's laws of motion.
3. We often use the term "inertia" to describe human sluggishness. How is this usage related to the meaning of "inertia" in physics?
4. My high school physics teacher defined mass as "inverse pushability aroundness." Comment.
5. Does a body necessarily move in the direction of the net force acting on it?
6. A truck crashes into a stalled car. A student trying to explain the physics of this event claims that no forces are involved; the car was just "in the way" so it got hit. Comment.
7. A barefoot astronaut kicks a ball across the recreation area of a space station. Does the ball's apparent weightlessness mean the astronaut's toes don't get hurt? Explain.
8. The surface gravity of Jupiter's moon Io is one-fifth that of Earth. What would happen to your weight and to your mass if you were to travel to Io?
9. In paddling a canoe, you push water backward with your paddle. What force actually propels the canoe forward?
10. Is it possible for a nonzero net force to act on an object without the object's speed changing? Explain.
11. As your plane accelerates down the runway, you take your keys from your pocket and suspend them by a thread. Do they hang vertically? Explain.
12. I tell passengers in my car to buckle their seatbelts, explaining that I believe in the law of inertia. What's that got to do with seatbelts?

Exercises and Problems

Exercises

Section 4.2 Newton's First and Second Laws

13. A subway train has a mass of 1.5×10^6 kg. What force is required to accelerate the train at 2.5 m/s²?
14. A railroad locomotive with a mass of 6.1×10^4 kg can exert a force of 1.2×10^5 N. At what rate can it accelerate (a) by itself and (b) when pulling a 1.4×10^6-kg train?
15. A small plane starts down the runway with acceleration 7.2 m/s². If the force provided by its engine is 1.1×10^4 N, what is the plane's mass?
16. A car leaves the road traveling at 110 km/h and hits a tree, coming to a complete stop in 0.14 s. What average force does a seatbelt exert on a 60-kg passenger during this collision?
17. By how much does the force required to stop a car increase if the initial speed is doubled and the stopping distance remains the same?
18. A 3800-kg jet touches down at 240 km/h on the deck of an aircraft carrier and immediately deploys a parachute to slow itself down. If the plane comes to a stop in 170 m, what is the average force of air on the parachute? Assume the parachute provides essentially all the stopping force.
19. Starting from rest, a 940-kg racing car covers 400 m in 4.95 s. What is the average force acting on the car?
20. In an egg-dropping contest, a student encases an 85-g egg in a styrofoam block. If the force on the egg is not to exceed 1.5 N, and if the block hits the ground at 1.2 m/s, by how much must the styrofoam crush?
21. In a front-end collision, a 1300-kg car with shock-absorbing bumpers can withstand a maximum force of 65 kN before damage occurs. If the maximum speed for a nondamaging collision is 10 km/h, by how much must the bumper be able to move relative to the car?

Section 4.4 The Force of Gravity

22. Show that the units of acceleration can be written as N/kg. Why does it make sense to give g as 9.8 N/kg when talking about mass and weight?
23. My spaceship crashes on one of the Sun's nine planets. Fortunately, the ship's scales are intact and show that my weight is 532 N. If I know my mass is 60 kg, where am I? *Hint:* Consult Appendix E.

24. If I can barely lift a 35-kg concrete block on Earth, how massive a block can I lift on the Moon?
25. A cereal box says "net weight 340 grams." What is the actual weight (a) in SI units and (b) in ounces?
26. A bridge specifies a maximum load of 10 tons. What is the maximum mass, in kilograms, that the bridge can carry?
27. The gravitational acceleration at a typical space-shuttle altitude is about 93% of its surface value. What is the weight of a 68-kg astronaut in a shuttle at this altitude?

Section 4.5 Using Newton's Second Law

28. A 50-kg parachute jumper descends at a steady 40 km/h. What is the force of air on the parachute?
29. A 930-kg motorboat accelerates away from a dock at 2.3 m/s². Its propeller provides a thrust force of 3.9 kN. What drag force is exerted by the water on the boat?
30. An elevator accelerates downward at 2.4 m/s². What force does the floor of the elevator exert on a 52-kg passenger?
31. What is the vertical lifting force on a 747 jetliner when the plane is (a) flying at constant altitude and (b) accelerating upward at 1.1 m/s²? The aircraft's mass is 4.5×10^5 kg.
32. At liftoff, a space shuttle with 2.0×10^6-kg total mass undergoes an upward acceleration of $0.60g$. (a) What is the total thrust force developed by its engines? (b) What force does the seat exert on a 60-kg astronaut during liftoff?
33. You step into an elevator, and it accelerates to a downward speed of 9.2 m/s in 2.1 s. How does your apparent weight during this acceleration time compare with your actual weight?

Section 4.6 Newton's Third Law

34. What upward gravitational force does a 5600-kg elephant exert on Earth?
35. I have a mass of 65 kg. If I jump off a 120-cm-high table, how far toward me does Earth move during the time I fall?
36. What force is necessary to stretch a spring 48 cm, if the spring constant is 270 N/m?
37. A 35-N force is applied to a spring with spring constant $k = 220$ N/m. How much does the spring stretch?
38. A spring with spring constant $k = 340$ N/m is used to weigh a 6.7-kg fish. How far does the spring stretch?

Problems

39. A 1.25-kg object is moving in the x direction at 17.4 m/s. Just 3.41 s later, it is moving at 26.8 m/s at 34.0° to the x axis. What are the magnitude and direction of the force applied during this time?

40. An airplane encounters sudden turbulence, and you feel momentarily lighter. If your apparent weight seems to be about 70% of your normal weight, what are the magnitude and direction of the plane's acceleration?

41. A 74-kg tree surgeon rides a "cherry picker" lift to reach the upper branches of a tree. What force does the bucket of the lift exert on the surgeon when the bucket is (a) at rest; (b) moving upward at a steady 2.4 m/s; (c) moving downward at a steady 2.4 m/s; (d) accelerating upward at 1.7 m/s²; (e) accelerating downward at 1.7 m/s²?

42. A ballet dancer executes a vertical jump during which the floor pushes up on his feet with a force 50% greater than his weight. What is his upward acceleration?

43. An elevator moves upward at 5.2 m/s. What is the minimum stopping time it can have if the passengers are to remain on the floor?

44. A 2.50-kg object is moving along the x axis at 1.60 m/s. As it passes the origin, two forces $\vec{F}_1$ and $\vec{F}_2$ are applied, both in the y direction (plus or minus); $\vec{F}_1 = 15\hat{j}$ N. The forces are applied for 3.00 s, after which the object is at the point $x = 4.80$ m, $y = 10.8$ m. Find $\vec{F}_2$.

45. Blocks of 1.0, 2.0, and 3.0 kg are lined up on a frictionless table, as shown in Fig. 4.21. A rightward-pointing 12-N force is applied to the leftmost block. What force does the middle block exert on the rightmost one?

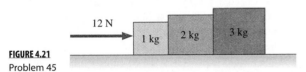

12 N

FIGURE 4.21
Problem 45

46. A child pulls an 11-kg wagon with a horizontal handle whose mass is 1.8 kg, giving the wagon and handle an acceleration of 2.3 m/s². (a) The tension at each end of the handle is different. Why? (b) Find the tension at each end of the handle.

47. A 2200-kg airplane is pulling two gliders, the first of mass 310 kg and the second of mass 260 kg, down the runway with an acceleration of 1.9 m/s² (Fig. 4.22). Neglecting the mass of the two ropes and any frictional forces, determine (a) the horizontal thrust of the plane's propeller; (b) the tension force in the first rope; (c) the tension force in the second rope; and (d) the net force on the first glider.

$\vec{a}$

FIGURE 4.22 Problem 47

48. In a tractor pulling contest, a 2300-kg tractor pulls a 4900-kg sledge with an acceleration of 0.61 m/s². If the tractor exerts a horizontal force of 7700 N on the ground, determine the magnitudes of (a) the force of the tractor on the sledge; (b) the force of the sledge on the tractor; and (c) the frictional force exerted on the sledge by the ground.

49. A biologist is studying the growth of rats in an orbiting space station. To determine a rat's mass, she puts it in a 320-g cage, attaches a spring scale, and pulls so that the scale reads 0.46 N. If the resulting acceleration of the rat and cage is 0.40 m/s², what is the rat's mass?

50. An elastic tow rope has a spring constant of 1300 N/m. It is connected between a truck and a 1900-kg car. As the truck tows the car, the rope stretches 55 cm. Starting from rest, how far do the truck and the car move in 1 min? Assume the car experiences negligible friction.

51. A 2.0-kg mass and a 3.0-kg mass are on a horizontal frictionless surface, connected by a massless spring with spring constant $k = 140$ N/m. A 15-N force is applied to the larger mass, as shown in Fig. 4.23. How much does the spring stretch from its equilibrium length?

2 kg   3 kg $\vec{F}$ 15N

FIGURE 4.23 Problem 51

52. Two large crates, with masses 640 kg and 490 kg, are connected by a stiff, massless spring $(k = 8.1$ kN/m$)$ and propelled along an essentially frictionless, level factory floor by a force applied horizontally to the more massive crate. If the spring compresses 5.1 cm from its equilibrium length, what is the applied force?

53. What downward force is exerted on the air by the blades of a 4300-kg helicopter when it is (a) hovering at constant altitude; (b) dropping at 21 m/s with speed decreasing at 3.2 m/s²; (c) rising at 17 m/s with speed increasing at 3.2 m/s²; (d) rising at a steady 15 m/s; (e) rising at 15 m/s with speed decreasing at 3.2 m/s²?

54. What engine thrust (force) is needed to accelerate a rocket of mass m (a) downward at 1.40g near Earth's surface; (b) upward at 1.40g near Earth's surface; (c) at 1.40g in interstellar space far from any star or planet?

55. An elevator cable can withstand a maximum tension of 19,500 N before breaking. The elevator has a mass of 490 kg and a maximum acceleration of 2.24 m/s². Engineering safety standards require that the cable tension never exceed two-thirds of the breaking tension. How many 65-kg people can the elevator safely accommodate?

56. An F-14 jet fighter has a mass of 1.6×10^4 kg and an engine thrust of 2.7×10^5 N. A 747 jumbo jet has a mass of 3.6×10^5 kg and a total engine thrust of 7.7×10^5 N. Is it possible for either plane to climb vertically with no lift from its wings? If so, what vertical acceleration could it achieve?

57. Two springs have the same unstretched length but different spring constants, k_1 and k_2. (a) If they are connected side by side and stretched a distance x, as shown in Fig. 4.24a, show that the force exerted by the combination is $(k_1 + k_2)x$. (b) If they are connected end to end and the combination is stretched a distance x (Fig. 4.24b), show that they exert a force $k_1 k_2 x/(k_1 + k_2)$.

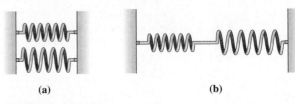

(a)   (b)

FIGURE 4.24 Problem 57

58. Although we usually write Newton's second law for one-dimensional motion in the form $F = ma$, the most basic version of the law is $F = \dfrac{d(mv)}{dt}$. The simpler form holds only when the

mass is constant. Consider an object whose mass may be changing, and show that the rule for the derivative of a product (see Appendix A) can be used to write Newton's law in the form $F = ma + v\dfrac{dm}{dt}$.

59. A railroad car is being pulled beneath a grain elevator that dumps grain at the rate of 450 kg/s. Use the result of Problem 58 to find the force that must be applied to keep the car moving at a constant 2.0 m/s.

60. A block 20% more massive than you hangs from a rope. The other end of the rope goes over a massless, frictionless pulley, and dangles freely. With what acceleration must you climb the rope to keep the block from falling?

61. A crane lifts a 1200-kg bucket of concrete at a construction site, giving it an initial acceleration of 2.6 m/s² upward. Find the tension in the cable supporting the bucket.

62. Your airplane is caught in a brief, violent downdraft. To your amazement, the pretzels on your seatback tray rise vertically, and you estimate their upward acceleration relative to the plane at 2 m/s². What's the downward acceleration of the plane?

63. Airport runways are generally designed with a buffer zone around 300 m long beyond the runway end, to accommodate aircraft that land too fast or too far down the runway and go off the end. Where that's not possible, airports are increasingly installing so-called Engineered Material Arresting Systems (EMAS) to prevent runaway aircraft from entering nearby roads, neighborhoods, or waterways. One such system, at New York's JFK airport, consists of a 132-m-long bed of crushable cement blocks. What average force must this system exert on a 55-Mg jetliner that enters the arrestor bed at 36 m/s if the jet is to stop 120 m into the bed?

64. Two masses are joined together by a massless, inextensible string. A vertical force of 30 N applied to the upper mass gives the system a constant upward acceleration of 3.2 m/s². If the tension in the connecting string is 18 N, what are the two masses?

65. A 65-kg person stands on a scale in a moving elevator while holding a 5-kg mass suspended from a massless spring with spring constant 1.08 kN/m. None of the objects in the elevator is moving relative to the elevator, but the spring is stretched by 5 cm. (a) Is the elevator accelerating, and if so, what are the magnitude and direction of its acceleration? (b) What is the reading of the scale on which the person stands?

66. A block of mass M hangs from a uniform rope of length L and mass m. Find an expression for the tension in the rope as a function of the distance y measured vertically downward from the top of the rope.

67. In Einstein's special theory of relativity, the momentum of a particle is given by $p = mu/\sqrt{1 - u^2/c^2}$, where the mass m and speed of light c are constants and u is the particle's speed. For such a particle moving in one dimension along a straight line, Equation 4.3 no longer applies. How does the general form of Newton's second law, Equation 4.2, relate the net force in the direction of motion to the acceleration, $a = du/dt$?

68. Your science fiction–writing friend is working on a screenplay concerning a civilization that lives on a neutron star (10 km in diameter and extremely massive). Your friend decided the people would have an average mass of 75 kg but does not know how much they should weigh in pounds. An astrophysics book tells you the acceleration due to gravity on a typical neutron star is 5.73×10^{12} m/s². What is the weight of the inhabitants of this strange world?

69. You read about a car accident in the newspaper. A car traveling at 70 km/h collided with a concrete bridge support. The front end of the car was compressed 0.94 m, and the car came to a complete stop within this distance. You wonder, How many "g's" of acceleration did the occupant of the car experience? *Hint:* The "g's" of acceleration are the ratio of the acceleration to 9.8 m/s².

70. You and your lab partner devise a clever experiment to measure the force applied to a ball when it is thrown. You throw a 200-g ball vertically upward. A high-speed video image of the throw indicates your hand pushed on the ball for 0.32 s. The ball rises to a maximum height of 7.23 m above your hand. What force did you apply?

Answers to Chapter Questions

Answer to Chapter Opening Question

The human body exerts a contact force; wind and water are fluids that exert pressure forces, and gravity is an action-at-a-distance force between Earth and the sailboard.

Answers to GOT IT? Questions

4.1 (b).

4.2 No. Look at Fig. 4.5b.

4.3 All would move in straight lines.

4.4 (a) greater; (b) less; (c) less; (d) greater; (e) equal.

4.5 (c) less than 2 N.

4.6 (a) No, the acceleration is still 0; (b) no, the direction of velocity is irrelevant (this situation would occur if the helicopter were moving downward but slowing).

Using Newton's Laws

■ Why doesn't the roller coaster fall off its loop-the-loop track?

Chapter 4 introduced Newton's three laws of motion and used them in one-dimensional applications. Now we apply Newton's laws in two dimensions. This material is at the heart of Newtonian physics, from textbook problems to systems that guide spacecraft to distant planets. The chapter consists largely of examples, to help you learn to apply Newton's laws and also to appreciate their wide range of applicability. We also introduce frictional forces and elaborate on circular motion. As you study the diverse examples, keep in mind that they all follow from the underlying principles embodied in Newton's laws.

5.1 Using Newton's Second Law

Newton's second law, $\vec{F}_{\text{net}} = m\vec{a}$, is the cornerstone of mechanics. We can use it to develop faster skis, engineer skyscrapers, design safer roads, compute a rocket's thrust, and solve myriad other practical problems.

We'll work Example 5.1 in great detail, applying Problem Solving Strategy 4.1. Follow this example closely, and try to understand how our strategy is grounded in Newton's basic statement that the net force on an object determines that object's acceleration.

▶ **To Learn**

By the end of this chapter you should be able to

■ Use Newton's second law to solve problems involving the motion of a single object in two dimensions under the influence of multiple forces (5.1).

■ Solve Newton's law problems involving multiple objects (5.2).

■ Explain that circular motion is just a special case of Newton's second law, and solve circular-motion problems involving multiple forces (5.3).

■ Describe the force of friction, both static and kinetic, and solve Newton's law problems in which one of the forces is friction (5.4).

■ Explain drag forces qualitatively (5.5).

◀ **To Know**

■ You first met Newton's laws in Chapter 4, and you should be familiar with all three of them (4.2).

■ This chapter builds especially on applications of Newton's second law, now generalizing to the case where the forces acting on an object no longer lie along a line (4.5).

EXAMPLE 5.1 **Newton's Law in Two Dimensions: Skiing**

A skier of mass $m = 65$ kg glides down a slope at angle $\theta = 32°$, as shown in Fig. 5.1. Find (a) the skier's acceleration and (b) the force the snow exerts on the skier. The snow is so slippery that you can neglect friction.

$\theta = 32°$

FIGURE 5.1 What's the skier's acceleration?

INTERPRET This problem is about the skier's motion, so we identify the skier as the object of interest. Next, we identify the forces acting on the object. In this case there are just two: the downward force of gravity and the normal force the ground exerts on the skier. As always, the normal force is perpendicular to the surfaces in contact—in this case perpendicular to the slope.

DEVELOP Our strategy for using Newton's second law calls for drawing a free-body diagram that shows only the object and the forces acting on it; that's Fig. 5.2. Determining the relevant equation is straightforward here: It's Newton's second law, $\vec{F}_{net} = m\vec{a}$. We write Newton's law explicitly for the forces we've identified:

$$\vec{F}_{net} = \vec{n} + \vec{F}_g = m\vec{a}$$

To apply Newton's law in two dimensions, we need to choose a coordinate system so that we can write this vector equation in components. Since the coordinate system is just a mathematical construct, you're free to choose any coordinate system you like—but a smart choice can make the problem a lot easier. In this example, the normal force is perpendicular to the slope and the skier's acceleration is along the slope.

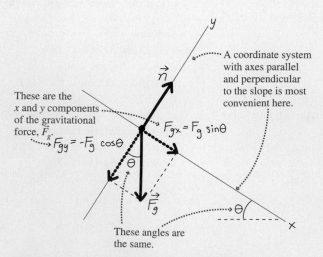

These are the x and y components of the gravitational force, $\vec{F}_g$.

$F_{gy} = -F_g \cos\theta$

$F_{gx} = F_g \sin\theta$

A coordinate system with axes parallel and perpendicular to the slope is most convenient here.

These angles are the same.

FIGURE 5.2 Our free-body diagram for the skier.

That means two of the three vectors in Newton's law will have only a single nonzero component if we choose a coordinate system with axes parallel and perpendicular to the slope. In the standard horizontal/vertical system, gravity would have a single component but we would have to break both the acceleration and the normal force into two components each (see Problem 32). The choice doesn't affect physical reality and you could work the problem either way, but the "tilted" coordinate system makes the math easier. We sketched this coordinate system on the free-body diagram in Fig. 5.2.

EVALUATE The rest is math. First, we write the components of Newton's law in our coordinate system. That means writing a version of the equation for each coordinate direction by removing the arrows indicating vector quantities and adding subscripts for the coordinate directions:

$$x \text{ component: } \quad n_x + F_{gx} = ma_x$$
$$y \text{ component: } \quad n_y + F_{gy} = ma_y$$

Don't worry about signs until the next step, when we actually evaluate the individual terms in these equations. Let's begin with the x equation. With the x axis parallel and the y axis perpendicular to the slope, the normal force has only a y component, so $n_x = 0$. Meanwhile, the acceleration points downslope—that's the positive x direction—so $a_x = a$, the magnitude of the acceleration. Only gravity has two nonzero components and, as Fig. 5.2 shows, trigonometry gives $F_{gx} = F_g \sin\theta$. But F_g, the magnitude of the gravitational force, is just mg, so $F_{gx} = mg \sin\theta$. This component has a positive sign because our x axis slopes downward. Then, with $n_x = 0$, the x equation becomes

$$x \text{ component: } \quad mg \sin\theta = ma$$

On to the y equation. The normal force points in the positive y direction, so $n_y = n$, the magnitude of the normal force. The acceleration has no component perpendicular to the slope, so $a_y = 0$. Figure 5.2 shows that $F_{gy} = -F_g \cos\theta = -mg \cos\theta$, so the y equation is

$$y \text{ component: } \quad n - mg \cos\theta = 0$$

Now we can evaluate to get the answers. The x equation solves directly to give

$$a = g \sin\theta = (9.8 \text{ m/s}^2)(\sin 32°) = 5.2 \text{ m/s}^2$$

which is the acceleration we were asked to find in (a). Next, we solve the y equation to get $n = mg \cos\theta$. Putting in the numbers gives $n = 540$ N. This is the answer to (b), the force the snow exerts on the skier.

ASSESS A look at two special cases shows that these results make sense. First, suppose $\theta = 0°$, so the surface is horizontal. Then the x equation gives $a = 0$, as expected. The y equation gives $n = mg$, showing that a horizontal surface exerts a force that just balances the skier's weight. At the other extreme, consider $\theta = 90°$, so the slope is a vertical cliff. Then the skier falls freely with acceleration g, as expected. In this case $n = 0$ because there's no contact between skier and slope. At intermediate angles, the slope's normal force lessens the effect of gravity, resulting in a lower acceleration. As the x equation shows, that acceleration is independent of the skier's mass—just as in the case of a vertical fall. The force exerted by the snow—here $mg \cos\theta$, or 540 N—is less than the skier's weight mg because the slope has to balance only the perpendicular component of the gravitational force.

If you understand this example, you should be able to apply Newton's second law confidently in other problems involving motion with forces in two dimensions.

Sometimes we're interested in finding the conditions under which an object won't accelerate. Examples are engineering problems, such as ensuring that bridges and buildings don't fall down, and physiology problems involving muscles and bones. Next we give a wilder example.

EXAMPLE 5.2 **Objects at Rest: Taking Bear Precautions**

To protect her 17-kg pack from bears, a camper hangs it from ropes between two trees (Fig. 5.3). What is the tension in each rope?

FIGURE 5.3 Bear precautions.

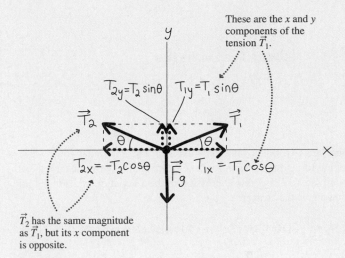

These are the x and y components of the tension $\vec{T}_1$.

$\vec{T}_2$ has the same magnitude as $\vec{T}_1$, but its x component is opposite.

FIGURE 5.4 Our free-body diagram for the pack.

INTERPRET Here the pack is the object of interest. The only forces acting on it are gravity and tension forces in the two halves of the rope. To keep the pack from accelerating, they must sum to zero net force.

DEVELOP Figure 5.4 is our free-body diagram for the pack. The relevant equation is again Newton's second law, $\vec{F}_{net} = m\vec{a}$—this time with $\vec{a} = \vec{0}$. For the three forces acting on the pack, Newton's law is then $\vec{T}_1 + \vec{T}_2 + \vec{F}_g = \vec{0}$. Next, we need a coordinate system. The two rope tensions point in different directions that aren't perpendicular, so it doesn't make sense to align a coordinate axis with either of them. Instead, a horizontal/vertical system is simplest.

EVALUATE First we need to write Newton's law in components. Formally, we have $T_{1x} + T_{2x} + F_{gx} = 0$ and $T_{1y} + T_{2y} + F_{gy} = 0$ for the component equations. Figure 5.4 shows the components of the tension forces, and we see that $F_{gx} = 0$ and $F_{gy} = -F_g = -mg$. So our component equations become

x component: $\quad T_1 \cos\theta - T_2 \cos\theta = 0$

y component: $\quad T_1 \sin\theta + T_2 \sin\theta - mg = 0$

The x equation tells us something that's apparent from the symmetry of the situation: Since the angle θ is the same for both halves of the rope, the magnitudes T_1 and T_2 of the tension forces are the same. Let's just call the magnitude T: $T_1 = T_2 = T$. Then the first two terms in the y equation are equal, and the equation becomes $2T \sin\theta - mg = 0$, which gives

$$T = \frac{mg}{2 \sin\theta} = \frac{(17 \text{ kg})(9.8 \text{ m/s}^2)}{2 \sin 22°} = 220 \text{ N}$$

ASSESS Make sense? Let's look at some special cases. With $\theta = 90°$, the rope hangs vertically, $\sin\theta = 1$, and the tension in each half of the rope is $\frac{1}{2}mg$. Of course: Each piece of the rope supports half the pack's weight. But as θ gets smaller, the ropes become more horizontal and the tension increases. That's because the vertical tension components together still have to support the pack's weight—but now there's a horizontal component as well, increasing the overall tension. Ropes break if the tension becomes too great, and in this example that means the rope's so-called breaking tension must be considerably greater than the pack's weight. If $\theta = 0$, in fact, the tension would become infinite—demonstrating that it's impossible to support a weight with a purely horizontal rope. ∎

EXAMPLE 5.3 **Objects at Rest: Restraining a Ski Racer**

A starting gate acts horizontally to restrain a 60-kg ski racer on a frictionless 30° slope (Fig. 5.5). What horizontal force does the starting gate apply to the skier?

INTERPRET Again, we want the skier to remain unaccelerated. The skier is the object of interest, and we identify three forces acting: gravity, the normal force from the slope, and a horizontal restraining force $\vec{F}_h$ that we're asked to find.

DEVELOP Figure 5.6 is our free-body diagram. The applicable equation is Newton's second law. Again, we want $\vec{a} = \vec{0}$, so with the forces we identified, $\vec{F}_{net} = m\vec{a}$ becomes $\vec{F}_h + \vec{n} + \vec{F}_g = \vec{0}$. Developing our solu-

tion strategy, we choose a coordinate system. With two forces now either horizontal or vertical, a horizontal/vertical system makes the most sense; we've shown this on Fig. 5.6.

EVALUATE As usual, the component equations follow directly from the vector form of Newton's law: $F_{hx} + n_x + F_{gx} = 0$ and $F_{hy} + n_y + F_{gy} = 0$. Figure 5.6 gives the components of the normal force and shows that $F_{hx} = -F_h$, $F_{gy} = -F_g = -mg$, and $F_{gx} = F_{hy} = 0$. Then the component equations become

x: $\quad -F_h + n \sin\theta = 0 \qquad y$: $\quad n \cos\theta - mg = 0$

cont'd.

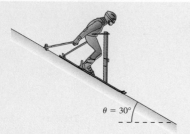

FIGURE 5.5 Restraining a skier.

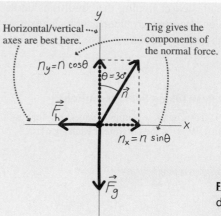

FIGURE 5.6 Our free-body diagram for the restrained skier.

There are two unknowns here—namely, the horizontal force we're looking for and the normal force n. We can solve the y equation to get $n = mg/\cos\theta$. Using this expression in the x equation and solving for F_h then give the answer:

$$F_h = \frac{mg}{\cos\theta}\sin\theta = mg\tan\theta = (60\text{ kg})(9.8\text{ m/s}^2)(\tan 30°) = 340\text{ N}$$

ASSESS Again, let's look at the extreme cases. With $\theta = 0$, we have $F_h = 0$. Of course! It doesn't take any force to restrain a skier on flat ground. But as the slope becomes more vertical, $\tan\theta \to \infty$, and in the vertical limit, it becomes impossible to restrain the skier with a purely horizontal force. ∎

GOT IT? 5.1   A roofer's toolbox rests on an essentially frictionless metal roof with a 45° slope, secured by a horizontal rope as shown. Is the rope tension (a) greater than, (b) less than, or (c) equal to the box's weight?

How does the rope tension compare with the toolbox weight?

45°

5.2 Multiple Objects

In the preceding examples there was a single object of interest. But often we have several objects whose motion is linked. Our Newton's law strategy still applies, with extensions to handle multiple objects.

PROBLEM SOLVING STRATEGY 5.1   **Newton's Second Law and Multiple Objects**

INTERPRET Interpret the problem to be sure that you know what it's asking and that Newton's second law is the relevant concept. Identify the *multiple* objects of interest and all the individual interaction forces acting on *each* object. Finally, identify *connections* between the objects and the resulting *constraints* on their motions.

DEVELOP Draw a *separate* free-body diagram showing all the forces acting on *each* object. Develop your solution plan by writing Newton's law, $\vec{F}_{net} = m\vec{a}$, separately for each object, with $\vec{F}_{net}$ expressed as the sum of the forces acting on that object. Then choose a coordinate system appropriate to each object, so you can express each Newton's law equation in components. The coordinate systems for different objects don't need to have the same orientation.

cont'd.

EVALUATE At this point the physics is done, and you're ready to execute your plan by solving the equations and evaluating the numerical answer(s), if called for. Write the components of Newton's law for each object in the coordinate system you chose for each. You can then solve the resulting equations for the quantity(ies) you're interested in, using the connections you identified to relate the quantities that appear in the equations for the different objects.

ASSESS Assess your solution to see whether it makes sense. Are the numbers reasonable? Do the units work out correctly? What happens in special cases—for example, when a mass, a force, an acceleration, or an angle gets very small or very large?

EXAMPLE 5.4   **Multiple Objects: Rescuing a Climber**

A 70-kg climber finds himself dangling over the edge of an ice cliff, as shown in Fig. 5.7. Fortunately, he's roped to a 940-kg rock located 51 m from the edge of the cliff. Unfortunately, the ice is frictionless, and the climber accelerates downward. What is his acceleration, and how much time does he have before the rock goes over the edge? Neglect the rope's mass.

INTERPRET We need to find the climber's acceleration, and from that we can get the time before the rock goes over the edge. We identify two objects of interest: the climber and the rock, and we note that the rope connects them. There are two forces on the climber: gravity and the upward rope tension. There are three forces on the rock: gravity, the normal force from the surface, and the rightward-pointing rope tension.

DEVELOP Figure 5.8 shows a free-body diagram for each object. Newton's law applies to each, so we write two vector equations:

$$\text{climber:} \quad \vec{T}_c + \vec{F}_{gc} = m_c \vec{a}_c$$
$$\text{rock:} \quad \vec{T}_r + \vec{F}_{gr} + \vec{n} = m_r \vec{a}_r$$

where the subscripts c and r stand for climber and rock, respectively. All forces are either horizontal or vertical, so we can use the same horizontal/vertical coordinate system for both objects, as shown in Fig. 5.8.

EVALUATE Again, the component equations follow directly from the vector forms. There are no horizontal forces on the climber, so only the y equation is significant. We're skilled enough now to skip the intermediate step of writing the components without their actual expressions, and we see from Fig. 5.8a that the y component of Newton's law for the climber becomes $T_c - m_c g = m_c a_c$. For the rock, the only horizontal force is the tension, pointing to the right or positive x direction,

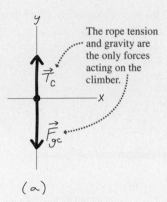

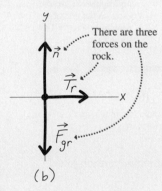

FIGURE 5.8 Our free-body diagrams for (a) the climber and (b) the rock.

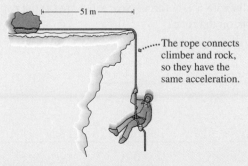

FIGURE 5.7 A climber in trouble.

so the rock's x equation is $T_r = m_r a_r$. Since it's on a horizontal surface, the rock has no vertical acceleration, so its y equation is $n - m_r g = 0$. In writing these equations, we haven't added the subscripts x and y because each vector has only a single nonzero component. Now we need to consider the connection between rock and climber. That's the rope, and its presence means that the magnitude of both accelerations is the same. Calling that magnitude a, we can see from Fig. 5.8 that $a_r = a$ and $a_c = -a$. The value for the rock is positive because $\vec{T}_r$ points to the right, which we defined as the positive x direction; the value for the climber is negative because he's accelerating downward, which we defined as the negative y direction. The rope, furthermore, has negligible

cont'd.

mass, so the tension throughout it must be the same (more on this point just after the example). Therefore the tension forces on rock and climber have equal magnitude T, so $T_c = T_r = T$. Putting this all together gives us three equations:

$$\text{climber, } y: \quad T - m_c g = -m_c a$$
$$\text{rock, } x: \quad T = m_r a$$
$$\text{rock, } y: \quad n - m_r g = 0$$

The rock's x equation gives the tension, which we can substitute into the climber's equation to get $m_r a - m_c g = -m_c a$. Solving for a then gives the answer:

$$a = \frac{m_c g}{m_c + m_r} = \frac{(70 \text{ kg})(9.8 \text{ m/s}^2)}{(70 \text{ kg} + 940 \text{ kg})} = 0.679 \text{ m/s}^2$$

We didn't need the rock's y equation, which just says that the normal force supports the rock's weight.

ASSESS Again, let's look at special cases. Suppose the rock's mass is zero; then our expression gives $a = g$. In this case there's no rope tension and the climber plummets in free fall. Also, acceleration decreases as the rock's mass increases, so with an infinitely massive rock, the climber would dangle without accelerating. You can see physically why our expression for acceleration makes sense. The gravitational force $m_c g$ acting on the climber has to accelerate both rock and climber—whose combined mass is $m_c + m_r$. The result is an acceleration of $m_c g/(m_c + m_r)$.

We're not quite done because we were also asked for the time until the rock goes over the cliff, putting the climber in real trouble. We interpret this as a problem in one-dimensional motion from Chapter 2, and we determine that Equation 2.10, $x = x_0 + v_0 t + \frac{1}{2}at^2$, applies. With $x_0 = 0$ and $v_0 = 0$, we have $x = \frac{1}{2}at^2$. We evaluate by solving for t and using the acceleration we found along with $x = 51$ m for the distance from the rock to the cliff edge:

$$t = \sqrt{\frac{2x}{a}} = \sqrt{\frac{(2)(51 \text{ m})}{0.679 \text{ m/s}^2}} = 12 \text{ s} \qquad ■$$

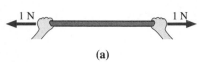

The hand pulls the highlighted section of the rope with a 1-N force to the left.

The net force on the highlighted section is zero, so the rest of the rope must exert a 1-N force to the right.

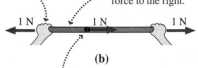

(b)

The dividing point could be anywhere, so there's a 1-N tension force throughout the rope.

FIGURE 5.9 Understanding tension forces.

✓ **TIP** Ropes and Tension Forces

Tension forces can be confusing. In Example 5.4, the rock pulls on one end of the rope and the climber pulls on the other. So why isn't the rope tension the sum of these forces? And why is it important to neglect the rope's mass? The answers lie in the meaning of tension.

Figure 5.9 shows a situation similar to Example 5.4, with two people pulling on opposite ends of a rope with forces of 1 N each. You might think the rope tension is then 2 N, but it's not. To see why, consider the part of the rope that's highlighted in Fig. 5.9b. To the left is the hand pulling leftward with 1 N. The rope isn't accelerating, so there must be a 1-N force pulling to the right on the highlighted piece. The remainder of the rope provides that force. We could have divided the rope anywhere, so we conclude that every part of the rope exerts a 1-N force on the adjacent rope. That 1-N force is what we mean by the rope tension.

As long as the rope isn't accelerating, the net force on it must be zero, so the forces at the two ends have the same magnitude. That conclusion would hold even if the rope were accelerating—provided it had negligible mass. That's often a good approximation in situations involving tension forces. But if a rope, cable, or chain has significant mass and is accelerating, then the tension force differs at the two ends. That difference, according to Newton's second law, is what accelerates the rope.

GOT IT? 5.2 In the figure we've replaced one hand with a hook attaching the rope to a wall. On the right, the hand still pulls with a 1-N force. Now what are (a) the rope tension and (b) the force exerted by the hook on the rope?

5.3 Circular Motion

A car rounds a curve. A satellite circles Earth. A proton whirls around a giant particle accelerator. Since they're not going in straight lines, Newton tells us that a force acts on each (Fig. 5.10). We know from Section 4.4 that the acceleration of an object moving with constant speed v in a circular path of radius r has magnitude v^2/r and points toward the center of the circle. Newton's second law then tells us that the magnitude of the net force on an object of mass m in circular motion is

$$F_{net} = ma = \frac{mv^2}{r} \quad \text{(uniform circular motion)} \quad (5.1)$$

The force is in the same direction as the acceleration—toward the center of the circular path. For that reason it's sometimes called the **centripetal force**, meaning center-seeking (from the Latin *centrum*, "center," and *petere*, "to seek").

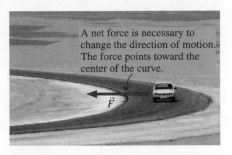

A net force is necessary to change the direction of motion. The force points toward the center of the curve.

$\vec{F}$

FIGURE 5.10 A car rounds a turn on the Trans-Sahara highway.

✓ **TIP** Look for Real Forces

Centripetal force is *not* some new kind of force. It's just the name for *any* forces that keep an object in circular motion—which are always real, physical forces. Common examples of forces involved in circular motion include the gravitational force on a satellite, friction between tires and road, magnetic forces, tension forces, normal forces, and combinations of these and other forces.

Newton's second law describes circular motion exactly as it does any other motion: by relating net force, mass, and acceleration. Therefore we can analyze circular motion with the same strategy we've used in other Newton's law problems.

EXAMPLE 5.5 **Circular Motion: Whirling a Ball on a String**

A ball of mass m whirls around in a horizontal circle at the end of a massless string of length L (Fig. 5.11). The string makes an angle θ with the horizontal. Find the ball's speed and the string tension.

INTERPRET This problem is similar to the Newton's law problems we worked involving force and acceleration. The object of interest is the ball, and only two forces are acting on it: gravity and the string tension.

DEVELOP Figure 5.12 is our free-body diagram showing the two forces we've identified. The relevant equation is Newton's second law, which becomes

$$\vec{T} + \vec{F}_g = m\vec{a}$$

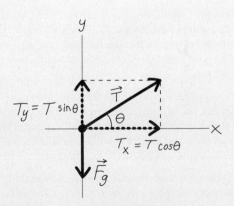

FIGURE 5.12 Our free-body diagram for the whirling ball.

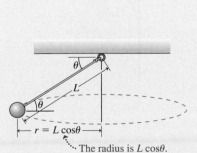

FIGURE 5.11 A ball whirling on a string.

The ball's path is in a horizontal plane, so its acceleration is horizontal. Then two of the three vectors in our problem—$\vec{F}_g$ and $\vec{a}$—are horizontal or vertical, so in developing our strategy, we choose a horizontal/vertical coordinate system.

cont'd.

✓**TIP**   Real Forces Only!

Were you tempted to draw a third force in Fig. 5.12, perhaps pointing outward to balance the other two? *Don't!* Because the ball is accelerating, the net force is nonzero and the individual forces *do not balance*. Or maybe you were tempted to draw an inward-pointing force, mv^2/r. *Don't!* The quantity mv^2/r is *not* another force; it's just the product of mass and acceleration that appears in Newton's law. Students often complicate problems by introducing forces that aren't there. That makes physics seem harder than it is!

EVALUATE We now need the x and y components of Newton's law. Figure 5.12 shows that $F_{gy} = -F_g = -mg$ and gives tension components in terms of trig functions. The acceleration is purely horizontal, so $a_y = 0$, and since the ball is in circular motion, $a_x = v^2/r$. But what's r? Not the string length L, but the radius of the ball's circular path. Figure 5.11 shows that the radius is $L\cos\theta$. With all these expressions, the components of Newton's law become

$$x:\quad T\cos\theta = \frac{mv^2}{L\cos\theta} \qquad y:\quad T\sin\theta - mg = 0$$

We can get the tension directly from the y equation: $T = mg/\sin\theta$. Using this result in the x equation lets us solve for the speed v:

$$v = \sqrt{\frac{TL\cos^2\theta}{m}} = \sqrt{\frac{(mg/\sin\theta)L\cos^2\theta}{m}} = \sqrt{\frac{gL\cos^2\theta}{\sin\theta}}$$

ASSESS In the special case $\theta = 90°$, the string hangs vertically; here $\cos\theta = 0$, so $v = 0$. There's no motion, and the string tension equals the ball's weight. But as the string becomes increasingly horizontal, both speed and tension increase. And, just as in Example 5.2, the tension becomes very great as the string approaches horizontal. Here the string tension has two jobs to do: Its vertical component supports the ball against gravity, while its horizontal component keeps the ball in its circular path. The vertical component is always equal to mg, but as the string approaches horizontal, that becomes an insignificant part of the overall tension—and thus the tension and speed grow very large. ∎

EXAMPLE 5.6   **Circular Motion: Engineering a Road**

Roads designed for high-speed travel have banked curves to give the normal force a component toward the center of the curve. That lets cars turn without relying on friction between tires and road. At what angle should a road with 200-m curvature radius be banked for travel at 90 km/h (25 m/s)?

INTERPRET This is another example involving circular motion and Newton's second law. Although we're asked about the road, a car on the road is the object we're interested in, and we need to design the road so the car can round the curve without needing a frictional force. That means the only forces on the car are gravity and the normal force.

DEVELOP Figure 5.13 shows the physical situation, and Fig. 5.14 is our free-body diagram for the car. Newton's second law is the applicable equation, and here it becomes $\vec{n} + \vec{F}_g = m\vec{a}$. Unlike the skier of Example 5.1, the car isn't accelerating down the slope, so a horizontal/vertical coordinate system makes the most sense.

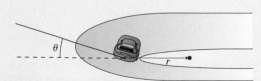

FIGURE 5.13 Car on a banked curve.

EVALUATE First we write Newton's law in components. Gravity has only a vertical component, $F_{gy} = -mg$ in our coordinate system, and Fig. 5.14 shows the two components of the normal force. The acceleration is purely horizontal and points toward the center of the curve; in our coordinate system that's the positive x direction. Since the car is in

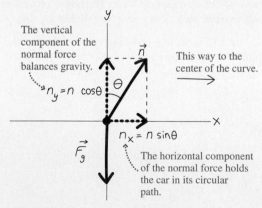

The vertical component of the normal force balances gravity.

$n_y = n\cos\theta$

This way to the center of the curve.

$n_x = n\sin\theta$

The horizontal component of the normal force holds the car in its circular path.

FIGURE 5.14 Our free-body diagram for the car on a banked curve.

cont'd.

circular motion, the magnitude of the acceleration is v^2/r. So the components of Newton's law become

$$x: \quad n \sin\theta = \frac{mv^2}{r} \qquad y: \quad n \cos\theta - mg = 0$$

where the 0 on the right-hand side of the y equation reflects the fact that we don't want the car to accelerate in the vertical direction. Solving the y equation gives $n = mg/\cos\theta$. Then using this result in the x equation gives $mg \sin\theta/\cos\theta = mv^2/r$, or $g \tan\theta = v^2/r$. The mass canceled, which is good news because it means our banked road will work for a vehicle of any mass. Now we can solve for the banking angle:

$$\theta = \tan^{-1}\!\left(\frac{v^2}{gr}\right) = \tan^{-1}\!\left(\frac{(25 \text{ m/s})^2}{(9.8 \text{ m/s}^2)(200 \text{ m})}\right) = 18°$$

ASSESS Make sense? At low speed v or large radius r, the car's motion changes gently and it doesn't take a large force to keep it on its circular path. But as v increases or r decreases, the required force increases and so does the banking angle. That's because the horizontal component of the normal force is what keeps the car in circular motion, and the steeper the angle, the greater that component. ∎

EXAMPLE 5.7 **Circular Motion: Looping the Loop**

The "Great American Revolution" roller coaster at Valencia, California, includes a loop-the-loop section whose radius is 6.3 m at the top (see the chapter opening photo). What is the minimum speed for a roller-coaster car at the top of the loop if it's to stay on the track?

INTERPRET Again, we have circular motion described by Newton's second law. We're asked about the minimum speed for the car to stay on the track. What does it mean to stay on the track? It means there must be a normal force between car and track; otherwise, the two aren't in contact. So we can identify two forces acting on the car: gravity and the normal force from the track.

DEVELOP Figure 5.15 shows the physical situation. The situation is especially simple at the top of the track, where both forces point in the same direction. We show this in our free-body diagram, Fig. 5.16. Since that common direction is downward, it makes sense to choose a coordinate system with the y axis *downward*. The applicable equation is Newton's second law, and with the two forces we've identified, that becomes $\vec{n} + \vec{F}_g = m\vec{a}$.

EVALUATE With both forces in the same direction, we need only the y component of Newton's law. With the downward direction positive, $n_y = n$ and $F_{gy} = mg$. At the top of the loop, the car is in circular motion, so its acceleration is toward the center—downward—and has

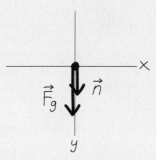

FIGURE 5.16 Our free-body diagram at the top of the loop.

magnitude v^2/r. So $a_y = v^2/r$, and the y component of Newton's law becomes

$$n + mg = \frac{mv^2}{r}$$

Solving for the speed gives $v = \sqrt{(nr/m) + gr}$. Now, the minimum possible speed for contact with the track occurs when n gets arbitrarily small right at the top of the track, so we find this minimum limit by setting $n = 0$. Then the answer is

$$v_{\min} = \sqrt{gr} = \sqrt{(9.8 \text{ m/s}^2)(6.3 \text{ m})} = 7.9 \text{ m/s}$$

ASSESS Do you see what's happening here? With the minimum speed, the normal force vanishes at the top of the loop, and gravity alone provides the force that keeps the object in its circular path. Since the motion is circular, that force must have magnitude mv^2/r. But the force of gravity alone is mg, and $v_{\min} = \sqrt{gr}$ follows directly from equating those two quantities. A car moving any slower than $v_{\min}$ would lose contact with the track and go into the parabolic trajectory of a projectile. For a car moving faster, there would be a nonzero normal force contributing to the downward acceleration at the top of the loop. In the "Great American Revolution," the actual speed at the loop's top is 9.7 m/s to provide a margin of safety. As with many problems involving gravity, the mass cancels. That's a good thing because it means the safe speed doesn't depend on the number or mass of the riders. ∎

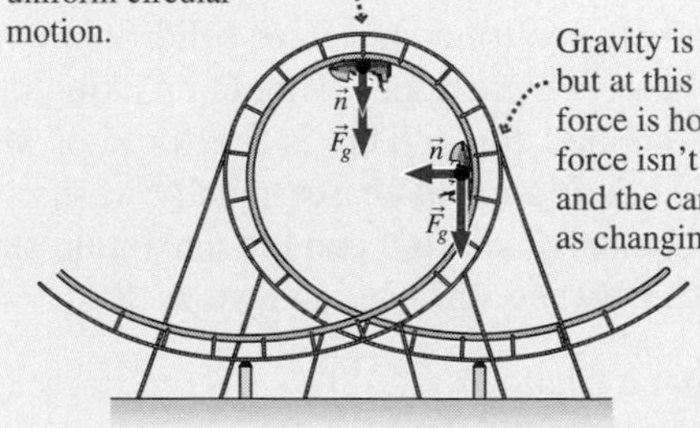

At the top, both forces point downward and the car is momentarily in uniform circular motion.

Gravity is always downward, but at this point the normal force is horizontal. The net force isn't toward the center, and the car is slowing as well as changing direction.

FIGURE 5.15 Forces on the roller-coaster car.

✓**TIP** Force and Motion

We've said this before, but it's worth noting again: Force doesn't cause motion but rather *change* in motion. The direction of an object's motion need not be the direction of the force on the object. That's true in Example 5.7, where the car is moving horizontally at the top of the loop while subject to a downward force. What *is* in the same direction as the force is the *change* in motion, here embodied in the center-directed acceleration of circular motion.

GOT IT? 5.3 Do the riders in the roller coaster of Example 5.7 need seatbelts to keep them from falling out of the car? Explain.

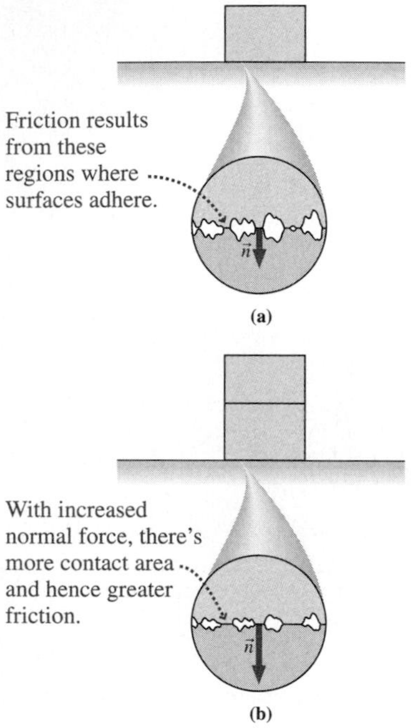

5.4 Friction

Your everyday experience of motion seems inconsistent with Newton's first law. Slide a book across the table, and it stops. Take your foot off the gas, and your car coasts to a stop. But Newton's law is right, so these examples show that some force must be acting. That force is **friction**, a force that opposes the relative motion of two surfaces in contact.

On Earth, we can rarely ignore friction. Some 20% of the gasoline burned in your car is used to overcome friction inside the engine. Friction causes wear and tear on machinery and clothing. But friction is also useful; without it, you couldn't drive or walk.

The Nature of Friction

Friction is ultimately an electrical force between molecules in different surfaces. When two surfaces are in contact, microscopic irregularities adhere, as shown in Fig. 5.17a. At the macroscopic level, the result is a force that opposes any relative movement of the surfaces.

Experiments show that the magnitude of the frictional force depends on the normal force between surfaces in contact. Figure 5.17b shows why this makes sense: As the normal forces push the surfaces together, the actual contact area increases. There's more adherence, and this increases the frictional force.

At the microscopic level, friction is complicated. The simple equations we'll develop here provide approximate descriptions of frictional forces. Friction is important in everyday life, but it's not one of the fundamental physical interactions.

Frictional Forces

Try pushing a heavy trunk across the floor. You start pushing, and at first nothing happens. Push harder; still nothing. Finally, as you push even harder, the trunk starts to slide—and you may notice that once it gets going, you don't have to push quite so hard. Why is that?

With the trunk at rest, microscopic contacts between trunk and floor solidify into relatively strong bonds. As you start pushing, you distort those bonds without breaking them; they respond with a force that opposes your applied force. This is the force of **static friction**, $\vec{f}_s$. As you increase the applied force, static friction increases equally, as shown in Fig. 5.18, and the trunk remains at rest. Experimentally, we find that the maximum static-friction force is proportional to the normal force between surfaces, and we write

$$f_s \leq \mu_s n \quad \text{(static friction)} \tag{5.2}$$

Here the proportionality constant μ_s (lowercase Greek mu, pronounced "myū," with the subscript s for "static") is the **coefficient of static friction**, a quantity that depends on the two surfaces. The $\leq$ sign indicates that the force of static friction ranges from zero up to the maximum value on the right-hand side.

Eventually you push hard enough to break the bonds between trunk and floor, and the trunk begins to move; this is the point in Fig. 5.18 where the force suddenly drops. Now

Friction results from these regions where surfaces adhere.

(a)

With increased normal force, there's more contact area and hence greater friction.

(b)

FIGURE 5.17 Friction originates in the contact between two surfaces.

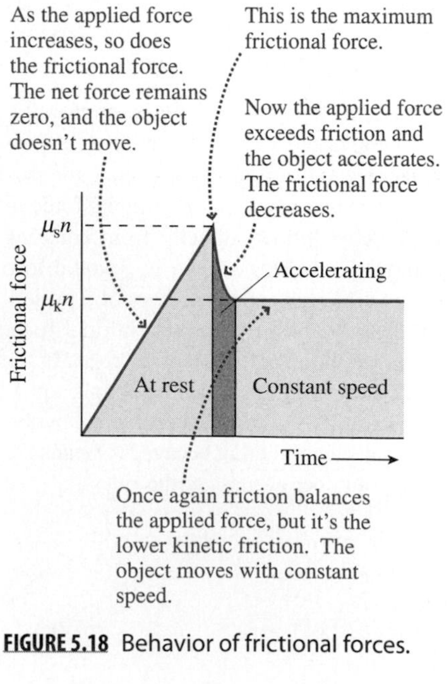

As the applied force increases, so does the frictional force. The net force remains zero, and the object doesn't move.

This is the maximum frictional force.

Now the applied force exceeds friction and the object accelerates. The frictional force decreases.

Accelerating

At rest | Constant speed

Time →

Once again friction balances the applied force, but it's the lower kinetic friction. The object moves with constant speed.

FIGURE 5.18 Behavior of frictional forces.

the microscopic bonds don't have time to strengthen, so the force needed to overcome them isn't so great. This weaker force between surfaces in relative motion is the force of **kinetic friction**, $\vec{f}_k$. Again, it's proportional to the normal force between the surfaces:

$$f_k = \mu_k n \qquad \text{(kinetic friction)} \qquad (5.3)$$

where now the proportionality μ_k is the **coefficient of kinetic friction**. Because kinetic friction is weaker, the coefficient of kinetic friction for a given pair of surfaces is less than the coefficient of static friction. Cross-country skiers exploit that fact by using waxes that provide a high coefficient of static friction for pushing against the snow and for climbing hills, while the lower kinetic friction permits effortless gliding.

Equations 5.2 and 5.3 give only the magnitudes of the frictional forces. The direction of the frictional force is parallel to the two surfaces, in the direction that opposes any applied force (Fig. 5.19a) or the surfaces' relative motion (Fig. 5.19b).

Since they describe proportionality between the magnitudes of two forces, the coefficients of friction are dimensionless. Typical values of μ_k range from less than 0.01 for smooth or lubricated surfaces to about 1.5 for very rough ones. Rubber on dry concrete—vital in driving an automobile—has μ_k about 0.8 and μ_s as high as 1. A waxed ski on dry snow has $\mu_k \approx 0.04$, while the synovial fluid that lubricates your body's joints reduces μ_k to a low 0.003.

If you push a moving object with a force equal to the opposing force of kinetic friction, then the net force is zero and, according to Newton, the object moves at constant speed. Since friction is nearly always present, but not as obvious as the push of a hand or the pull of a rope, you can see why it's so easy to believe that force is needed to make things move—rather than, as Newton recognized, to make them accelerate.

We emphasize that the equations describing friction are empirical expressions that approximate the effects of complicated but more basic interactions at the microscopic level. Our friction equations have neither the precision nor the fundamental character of Newton's laws.

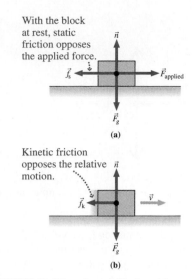

FIGURE 5.19 Direction of frictional forces.

Applications of Friction

Static friction plays a vital role in everyday activities such as walking and driving. As you walk, your foot contacting the ground is momentarily at rest, pushing back against the ground. By Newton's third law, the ground pushes forward, accelerating you forward (Fig. 5.20). Both forces of the third-law pair arise from static friction between foot and ground. On a frictionless surface, walking is impossible.

Similarly, the tires of an accelerating car push back on the road. If they aren't slipping, the bottom of each tire is momentarily at rest (more on this in Chapter 10). Therefore the force is static friction. The third law then requires a frictional force of the road pushing forward on the tires; that's what accelerates the car. Braking is the opposite: The tires push forward, and the road pushes back to decelerate the car (Fig. 5.21). The brakes affect only the wheels; it's friction between tires and road that stops the car. You know this if you've applied your brakes on an icy road!

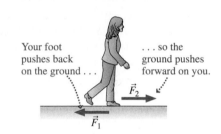

FIGURE 5.20 Walking.

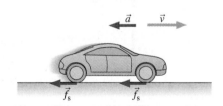

FIGURE 5.21 Friction stops the car.

EXAMPLE 5.8 **Frictional Forces: Stopping a Car**

The kinetic- and static-friction coefficients between a car's tires and a dry road are 0.61 and 0.89, respectively. If the car is traveling at 90 km/h (25 m/s) on a level road, determine the minimum stopping distance and the stopping distance with the wheels fully locked and the car skidding.

INTERPRET Since we're asked about the stopping distance, this is ultimately a question about accelerated motion in one dimension—the subject of Chapter 2. But here friction causes that acceleration, so we have a Newton's law problem. The car is the object of interest, and we identify three forces: gravity, the normal force, and friction.

DEVELOP Figure 5.22 is our free-body diagram. We have a two-part problem here: First, we need to use Newton's second law to find the acceleration, and then we can use Equation 2.11, $v^2 = v_0^2 + 2a\,\Delta x$, to relate distance and acceleration. With the three forces acting on the car, Newton's law becomes $\vec{F}_g + \vec{n} + \vec{f}_f = m\vec{a}$. A horizontal/vertical coordinate system is most appropriate for the components of Newton's law.

EVALUATE The only horizontal force is friction, which points in the $-x$ direction and has magnitude μn, where μ can be either the kinetic- or the static-friction coefficient. The normal force and gravity act in

cont'd.

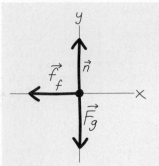

FIGURE 5.22 Our free-body diagram for the braking car.

the vertical direction, so the component equations are

$$x: \quad -\mu n = ma_x \qquad y: \quad -mg + n = 0$$

Solving the y equation for n and substituting in the x equation give the acceleration: $a_x = -\mu g$. We then use this result in Equation 2.11 and solve for the stopping distance Δx. With final speed $v = 0$, this gives

$$\Delta x = \frac{v_0^2}{-2a_x} = \frac{v_0^2}{2\mu g}$$

Using the numbers given, we get $\Delta x = 36$ m for the minimum stopping distance (no skid; static friction) and 52 m for the car skidding with its wheels locked (kinetic friction). The difference could well be enough to prevent an accident.

ASSESS Our result $a_x = -\mu g$ shows that a higher friction coefficient leads to a larger acceleration; this makes sense because friction is what causes the acceleration. What happened to the car's mass? A more massive car requires a larger frictional stopping force for the same acceleration—but friction depends on the normal force, and the latter is greater in proportion to the car's mass. Thus the stopping distance doesn't depend on mass.

This example shows that stopping distance increases as the *square* of the speed. That's one reason high speeds are dangerous: Doubling your speed quadruples your stopping distance! ∎

EXAMPLE 5.9 **Frictional Forces: Steering**

A level road makes a 90° turn with radius 73 m. What is the maximum speed for a car to negotiate this turn when the road is dry $(\mu_s = 0.88)$ and when the road is snow covered $(\mu_s = 0.21)$?

INTERPRET This example is similar to Example 5.8, but now the frictional force acts perpendicular to the car's motion, keeping it in a circular path. Because the car isn't moving in the direction of the force, we're dealing with *static* friction. The car is the object of interest, and again the forces are gravity, the normal force, and friction.

DEVELOP Figure 5.23 is our free-body diagram. Newton's law is the applicable equation, and we're dealing with the acceleration v^2/r that

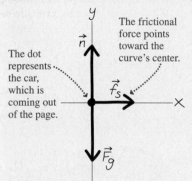

The frictional force points toward the curve's center.

The dot represents the car, which is coming out of the page.

FIGURE 5.23 Our free-body diagram for the cornering car.

occurs in circular motion. With the three forces acting on the car, Newton's law is $\vec{F}_g + \vec{n} + \vec{f}_s = m\vec{a}$. A horizontal/vertical coordinate system is most appropriate, and now it's most convenient to take the x axis in the direction of the acceleration—namely, toward the center of the curve.

EVALUATE Again, the only horizontal force is friction, with magnitude $\mu_s n$. Here it points in the positive x direction, as does the acceleration of magnitude v^2/r. So the x component of Newton's law is $\mu_s n = mv^2/r$. There's no vertical acceleration, so the y component is $-mg + n = 0$. Solving for n and using the result in the x equation give $\mu_s mg = mv^2/r$. Again the mass cancels, and we solve for v to get

$$v = \sqrt{\mu_s g r}$$

Putting in the numbers, we get $v = 25$ m/s (90 km/h) for the dry road and 12 m/s (44 km/h) for the snowy road. Exceed these speeds, and your car inevitably moves in a path with a larger radius—and that means going off the road!

ASSESS Once again, it makes sense that the car's mass doesn't matter. A more massive car needs a larger frictional force, and it gets what it needs because its larger mass results in a larger normal force. The safe speed increases with the curve radius r, and that, too, makes sense: A larger radius means a gentler turn, with less acceleration at a given speed. So less frictional force is needed. ∎

APPLICATION **Antilock Brakes**

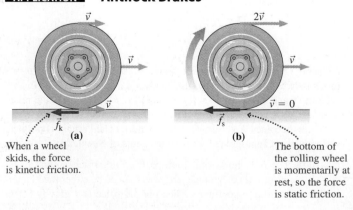

(a) When a wheel skids, the force is kinetic friction.

(b) The bottom of the rolling wheel is momentarily at rest, so the force is static friction.

Today's cars have computer-controlled antilock braking systems (ABS). These systems exploit the fact that static friction is greater than kinetic friction. Slam on the brakes of a non-ABS car and the wheels lock and skid without turning. The force between tires and road is then *kinetic* friction (part a in the figure). But if you pump the brakes to keep the wheels from skidding, then it's the greater force of *static* friction (part b).

ABS improves on this brake-pumping strategy with a computer that independently controls the brakes at each wheel, keeping each just on the verge of slipping. Drivers of ABS cars should slam the brakes hard in an emergency; the ensuing clatter indicates the ABS system is working.

Although ABS can reduce the stopping distance, its real significance is in preventing vehicles from skidding out of control as can happen when you apply the brakes with some wheels on ice and others on pavement. Increasingly, today's cars incorporate their computer-controlled brakes into sophisticated systems that enhance stability during emergency maneuvers.

EXAMPLE 5.10 Friction on a Slope: Avalanche!

A storm dumps new snow on a ski slope. The coefficient of static friction between the new snow and the older layer of snow underneath is 0.46. What is the maximum slope angle to which the new snow can adhere?

INTERPRET The problem asks about an angle, but it's friction that holds the new snow to the old, so this is really a problem about the maximum possible static friction. We aren't given an object, but we can model the new snow as a slab of mass m resting on a slope of unknown angle θ. The forces on the slab are gravity, the normal force, and static friction $\vec{f}_s$.

DEVELOP Figure 5.24 shows the model, and Fig. 5.25 is our free-body diagram. Newton's second law is the applicable equation, here with $\vec{a} = \vec{0}$, giving $\vec{F}_g + \vec{n} + \vec{f}_s = \vec{0}$. We also need the maximum static-friction force, given in Equation 5.2, $f_{s\,max} = \mu_s n$. As in Example 5.1, a tilted coordinate system is simplest and is shown in Fig. 5.25.

EVALUATE With the positive x direction downslope, Fig. 5.25 shows that the x component of gravity is $F_g \sin\theta = mg \sin\theta$, while the frictional force acts upslope ($-x$ direction) and has maximum magnitude $\mu_s n$; therefore, $f_{sx} = -\mu_s n$. So the x component of Newton's law is $mg \sin\theta - \mu_s n = 0$. We can read the y component from Fig. 5.25:

FIGURE 5.24 A layer of snow, modeled as a slab on a sloping surface.

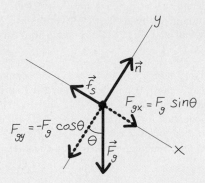

FIGURE 5.25 Our free-body diagram for the snow slab.

$-mg \cos\theta + n = 0$. Solving the y equation gives $n = mg \cos\theta$. Using this result in the x equation then yields $mg \sin\theta - \mu_s mg \cos\theta = 0$. Both m and g cancel, and we have $\sin\theta = \mu_s \cos\theta$ or, since $\tan\theta = \sin\theta/\cos\theta$,

$$\tan\theta = \mu_s$$

For the numbers in this example, we get $\theta = \tan^{-1}\mu_s = \tan^{-1}(0.46) = 25°$.

ASSESS Make sense? Sure: The steeper the slope, the greater the friction needed to keep the snow from sliding. Two effects are at work here: First, as the slope steepens, so does the component of gravity along the slope. Second, as the slope steepens, the normal force gets smaller, and that reduces the frictional force for a given friction coefficient. Note here that the normal force is not simply the weight mg of the snow; again, that's because of the sloping surface.

The real avalanche danger comes at angles slightly smaller than our answer $\tan\theta = \mu_s$, where a thick snowpack can build up. Changes in the snow's composition with temperature may decrease the friction coefficient and unleash an avalanche. ∎

EXAMPLE 5.11 Friction: Dragging a Trunk

You drag a trunk of mass m across a level floor using a massless rope that makes an angle θ with the horizontal (Fig. 5.26). Given a kinetic-friction coefficient μ_k, what rope tension is required to move the trunk at constant speed?

INTERPRET Even though the trunk is moving, it isn't accelerating, so here's another problem involving Newton's law with zero acceleration. The object is the trunk, and now four forces act: gravity, the normal force, friction, and the rope tension.

DEVELOP Figure 5.27 is our free-body diagram showing all four forces acting on the trunk. The relevant equation is Newton's law.

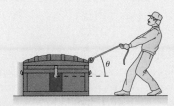

FIGURE 5.26 Dragging a trunk.

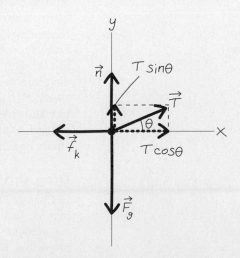

FIGURE 5.27 Our free-body diagram for the trunk.

cont'd.

With no acceleration, it is $\vec{F}_g + \vec{n} + \vec{f}_k + \vec{T} = \vec{0}$, with the magnitude of kinetic friction given by $f_k = \mu_k n$. All vectors except the tension force are horizontal or vertical, so we choose a horizontal/vertical coordinate system.

EVALUATE Looking at Fig. 5.27, we can write the components of Newton's law: $T\cos\theta - \mu_k n = 0$ in the x direction and $T\sin\theta - mg + n = 0$ in the y direction. This time the unknown T appears in both equations. Solving the y equation for n gives $n = mg - T\sin\theta$. Putting this n in the x equation then yields $T\cos\theta - \mu_k(mg - T\sin\theta) = 0$. Factoring terms involving T and solving, we arrive at the answer:

$$T = \frac{\mu_k mg}{\cos\theta + \mu_k \sin\theta}$$

ASSESS Make sense? Without friction, we wouldn't need any force to move the trunk at constant speed, and indeed our expression gives $T = 0$ in this case. If $\theta = 0$, then $\sin\theta = 0$ and we get $T = \mu_k mg$. Of course: In this case the normal force equals the weight, so the frictional force is $\mu_k mg$. Since the frictional force is horizontal and with $\theta = 0$ we're pulling horizontally, this is also the magnitude of the tension force. At intermediate angles, two effects come into play: First, the upward component of tension helps support the trunk's weight, and that means less normal force is needed. With less normal force, there's less friction—making the trunk easier to pull. But as the angle increases, less of the tension is horizontal and that means a larger tension force is needed to overcome friction. In combination, these two effects mean there's an optimum angle at which the rope tension is a minimum. Problem 63 explores this point further. ∎

GOT IT? 5.4 The figure shows a logging vehicle pulling a redwood log. Is the frictional force in this case (a) less than, (b) equal to, or (c) greater than the weight multiplied by the coefficient of friction?

5.5 Drag Forces

Friction is not the only "hidden" force that robs objects of their motion and obscures Newton's first law. Objects moving through fluids like water or air experience **drag forces** that oppose the relative motion of object and fluid. Ultimately, drag forces arise from collisions between fluid molecules and the object. The drag force depends on several factors, including fluid density and the object's cross-sectional area and speed.

Terminal Speed

When an object begins to fall from rest, its speed is low and so is the velocity-dependent drag force. It therefore accelerates downward with nearly the gravitational acceleration g. But as the object gains speed, the drag force increases—until eventually the drag force and gravity have equal magnitudes. At that point the net force on the object is zero, and it falls with constant speed, called its **terminal speed**.

Because the drag force depends on an object's area and the gravitational force depends on its mass, the terminal speed is lower for lighter objects with large areas. A parachute, for example, is designed specifically to have a large surface area that results, typically, in a terminal speed around 5 m/s. A ping-pong ball and a golf ball have about the same size and therefore the same area, but the ping-pong ball's much lower mass leads to a terminal speed of about 10 m/s compared with the golf ball's 50 m/s. For an irregularly shaped object, the drag and thus the terminal speed depend on how large a surface area the object presents to the air. Skydivers exploit this effect to vary their rates of fall prior to their parachutes opening.

Drag and Projectile Motion

In Chapter 3, we consistently neglected air resistance—the drag force of air—in projectile motion. Calculating drag effects on projectiles is not trivial and usually requires computer calculations. The net effect, though, is obvious: Air resistance decreases the range of a projectile. Despite the physicist's need for computer calculations, others—especially athletes—have a feel for drag forces that lets them play their sports by judging correctly the trajectory of a projectile under the influence of drag forces.

Big Picture

The big idea here is the same as in Chapter 4—namely, Newton's laws as a universal description of motion, in which force causes not motion itself but change in motion. Here we focus on Newton's second law, extended to the richer and more complex examples of motion in two dimensions. To use Newton's law, we now sum forces that may point in different directions, but the result is the same: The net force determines an object's acceleration.

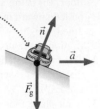

Here's a car on a banked turn. The forces on it don't sum to zero because the car is accelerating toward the center of the turn.

Common forces include gravity, the normal force from surfaces, tension forces, and a force introduced here: friction. Important examples are those where an object is accelerating, including in circular motion, and those where there's no acceleration and therefore the net force is zero.

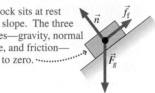

A block sits at rest on a slope. The three forces—gravity, normal force, and friction—sum to zero.

Solving Problems with Newton's Laws

The problem-solving strategy in this chapter is exactly the same as in Chapter 4, except that in two dimensions the choice of coordinate system and the division of forces into components become crucial steps. You usually need both component equations to solve a problem.

A skier on a frictionless slope

Free-body diagram showing the two forces acting

Coordinate system and vector components

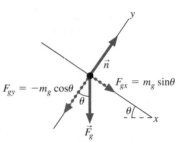

$$\vec{F} = m\vec{a} \rightarrow \vec{n} + \vec{F}_g = m\vec{a} \rightarrow \begin{cases} n_x + F_{gx} = ma_x \\ n_y + F_{gy} = ma_y \end{cases} \rightarrow \begin{cases} mg\sin\theta = ma_x \\ n - mg\cos\theta = 0 \end{cases}$$

Key Concepts and Equations

Newton's second law, $\vec{F}_{net} = m\vec{a}$, is the key equation in this chapter. It's crucial to remember that it's a *vector* equation, representing a pair of scalar equations for its two components in two dimensions.

Cases and Uses

A new force introduced in this chapter is **friction**, which acts between surfaces to oppose their relative motion. The force of friction acts parallel to the surfaces, but its strength depends on the normal force $\vec{n}$ acting perpendicular to them. When surfaces aren't actually in relative motion, the force is **static friction**, whose value ranges from zero to a maximum value $\mu_s n$ as needed to oppose any applied force: $f_s \leq \mu_s n$. Here μ_s is the **coefficient of static friction**, which depends on the nature of the two surfaces. For surfaces in relative motion, the force is **kinetic friction**, given by $f_k = \mu_k n$, where the **coefficient of kinetic friction** is less than the coefficient of static friction.

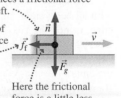

A block moving to the right experiences a frictional force to the left.

The magnitude of the frictional force depends on the normal force: $f_f = \mu n$.

Here the frictional force is a little less than the normal force, so μ is a little less than 1.

For Thought and Discussion

1. Compare the net force on a heavy trunk when it's (a) at rest on the floor; (b) being slid across the floor at constant speed; (c) being pulled upward in an elevator whose cable tension equals the combined weight of the elevator and trunk; and (d) sliding down a frictionless ramp.
2. The force of static friction acts only between surfaces at rest. Yet that force is essential in walking and in accelerating or braking a car. Explain.
3. A jet plane flies at constant speed in a vertical circular loop. At what point in the loop does the seat exert the greatest force on the pilot? The least force?
4. In cross-country skiing, skis should easily glide forward but should remain at rest when the skier pushes back against the snow. What frictional properties should the ski wax have to achieve this goal?
5. Why is it easier for a child to stand nearer the inside of a rotating merry-go-round?
6. Earth's gravity pulls a satellite toward the center of the Earth. So why doesn't the satellite actually fall to Earth?
7. Explain why ABS brakes result in a shorter stopping distance for a car.
8. A fishing line has a breaking strength of 20 lb. Is it possible to break the line while reeling in a 15-lb fish? Explain.
9. Two blocks rest on slopes of unequal angles, connected by a rope passing over a pulley (Fig. 5.28). If the blocks have equal masses, will they remain at rest? Why? Neglect friction.

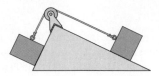

FIGURE 5.28 For Thought and Discussion 9; Exercises 19 and 20

10. Why do airplanes bank when turning?
11. If surfaces are very highly polished, friction actually increases. Why might this be?

Exercises and Problems

Exercises

Section 5.1 Using Newton's Second Law

12. Two forces, both in the x-y plane, act on a 1.5-kg mass that accelerates at 7.3 m/s² in a direction 30° counterclockwise from the x axis. One force has magnitude 6.8 N and points in the $+x$ direction. Find the other force.
13. Two forces act on a 3.1-kg mass that undergoes acceleration $\vec{a} = 0.91\hat{\imath} - 0.27\hat{\jmath}$ m/s². If one of the forces is $\vec{F}_1 = -1.2\hat{\imath} - 2.5\hat{\jmath}$ N, what is the other force?
14. A 3700-kg barge is being pulled along a canal by two mules connected to the barge by horizontal tow ropes. The tension in each rope is 1100 N, and the ropes make 25° angles with the forward direction. What force does the water exert on the barge (a) if it moves with constant velocity and (b) if it accelerates forward at 0.16 m/s²?
15. At what angle should you tilt an air table to simulate motion on the Moon's surface, where $g = 1.6$ m/s²?
16. A skier starts from rest at the top of a 24° slope 1.3 km long. Neglecting friction, how long does it take to reach the bottom?
17. A tow truck is connected to a 1400-kg car by a cable that makes a 25° angle to the horizontal. If the truck accelerates at 0.57 m/s², what is the magnitude of the cable tension? Neglect friction and the mass of the cable.

Section 5.2 Multiple Objects

18. Your 12-kg baby sister is hanging on the bottom of the tablecloth with all her weight. In the middle of the table, 60 cm from each edge, is a 6.8-kg roast turkey. (a) What is the acceleration of the turkey? (b) From the time your sister starts pulling, how long do you have to intervene before the turkey goes over the edge of the table?
19. If the left-hand slope in Fig. 5.28 makes a 60° angle with the horizontal, and the right-hand slope makes a 20° angle, how should the masses compare if the objects are not to slide along the frictionless slopes?
20. Suppose the angles shown in Fig. 5.28 are 60° and 20°. If the left-hand mass is 2.1 kg, what should the right-hand mass be so that (a) it accelerates downslope at 0.64 m/s² and (b) it accelerates upslope at 0.76 m/s²?

21. Two unfortunate climbers, roped together, are sliding freely down an icy mountainside. The upper climber (mass 75 kg) is on a slope at 12° to the horizontal, but the lower climber (mass 63 kg) has gone over the edge to a steeper slope at 38°. (a) Assuming frictionless ice and a massless rope, what is the acceleration of the pair? (b) The upper climber manages to stop the slide with an ice ax. After the climbers have come to a complete stop, what force must the ax exert against the ice?

Section 5.3 Circular Motion

22. Suppose the Moon were held in its orbit not by gravity but by the tension in a massless cable. Estimate the magnitude of the cable tension. *Hint:* See Appendix E for relevant data.
23. Show that the force needed to keep a mass m in a circular path of radius r with period T is $4\pi^2mr/T^2$.
24. A 940-g rock is whirled in a horizontal circle at the end of a 1.3-m-long string. (a) If the breaking strength of the string is 120 N, what is the maximum allowable speed of the rock? (b) At this maximum speed, what angle does the string make with the horizontal?
25. A subway train rounds an unbanked curve at 67 km/h. A passenger hanging onto a strap notices that an adjacent unused strap makes an angle of 15° to the vertical. What is the radius of the turn?
26. A tetherball on a 1.7-m rope is struck so that it goes into circular motion in a horizontal plane, with the rope making a 15° angle to the horizontal. What is the ball's speed?
27. An airplane goes into a turn 3.6 km in radius. If the banking angle required is 28° from the horizontal, what is the plane's speed?

Section 5.4 Friction

28. Movers slide a file cabinet along a floor. The mass of the cabinet is 73 kg, and the coefficient of kinetic friction between cabinet and floor is 0.81. What is the frictional force on the cabinet?
29. A hockey puck is given an initial speed of 14 m/s. If it comes to rest in 56 m, what is the coefficient of kinetic friction?
30. Starting from rest, a skier slides 100 m down a 28° slope. How much longer does the run take if the coefficient of kinetic friction is 0.17 instead of 0?

31. A car moving at 40 km/h negotiates a 130-m-radius banked turn designed for 60 km/h. What coefficient of friction is needed to keep the car on the road?

Problems

32. Repeat Example 5.1, this time using a horizontal/vertical coordinate system.

33. A block is launched up a frictionless ramp that makes an angle of 35° to the horizontal. If the block's initial speed is 2.2 m/s, how far up the ramp does it slide?

34. A 15-kg monkey hangs from the middle of a massless rope, each half of which makes an 8° angle with the horizontal. What is the rope tension? Compare the tension with the monkey's weight.

35. A 10-kg mass is suspended at rest by two strings attached to walls, as shown in Fig. 5.29. Find the tension forces in the two strings.

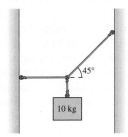

FIGURE 5.29 Problem 35

36. A camper hangs a 26-kg pack between two trees, using two separate pieces of rope of different lengths, as shown in Fig. 5.30. What is the tension in each rope?

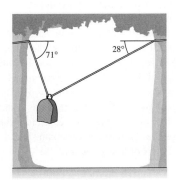

FIGURE 5.30 Problem 36

37. A mass m_1 undergoes circular motion of radius R on a horizontal frictionless table, connected by a massless string through a hole in the table to a second mass m_2 (Fig. 5.31). If m_2 is stationary, find (a) the tension in the string and (b) the period of the circular motion.

FIGURE 5.31 Problem 37

38. Riders on the "Great American Revolution" loop-the-loop roller coaster of Example 5.7 wear seatbelts as the roller coaster negotiates its 6.3-m-radius loop with a speed of 9.7 m/s. At the top of the loop, what are the magnitude and direction of the force exerted on a 60-kg rider (a) by the roller-coaster seat and (b) by the seatbelt? (c) What would happen if the rider unbuckled at this point?

39. A 45-kg skater rounds a 5.0-m-radius turn at 6.3 m/s. (a) What are the horizontal and vertical components of the force the ice exerts on her skate blades? (b) At what angle can she lean without falling over?

40. When a plane turns, it banks as shown in Fig. 5.32 to give the lifting force of the wings a horizontal component that turns the plane. If a plane is flying level at 950 km/h and the banking angle is not to exceed 40°, what is the minimum curvature radius for the turn?

FIGURE 5.32 Problem 40

41. A bucket of water is whirled in a vertical circle of radius 85 cm. What is the minimum speed that will keep the water from falling out?

42. A child sleds down a 12° slope at constant speed. What is the coefficient of friction between slope and sled?

43. The handle of a 22-kg lawnmower makes a 35° angle with the horizontal. If the coefficient of friction between lawnmower and ground is 0.68, what magnitude of force is required to push the mower at constant velocity? Assume the force is applied in the direction of the handle. Compare the force with the mower's weight.

44. Repeat Example 5.4, now assuming that the coefficient of kinetic friction between rock and ice is 0.057.

45. A bat crashes into the vertical front of an accelerating subway train. If the coefficient of friction between bat and train is 0.86, what is the minimum acceleration of the train that will allow the bat to remain in place?

46. The coefficient of static friction between steel train wheels and steel rails is 0.58. The engineer of a train moving at 140 km/h spots a stalled car on the tracks 150 m ahead. If he applies the brakes so that the wheels don't slip, will the train stop in time?

47. A bug crawls outward from the center of a compact disc spinning at 200 revolutions per minute. The coefficient of static friction between the bug's sticky feet and the disc surface is 1.2. How far does the bug get from the center before slipping?

48. A 310-g paperback book rests on a 1.2-kg textbook. A force is applied to the textbook, and the two books accelerate together from rest to 96 cm/s in 0.42 s. The textbook is then brought to a stop in 0.33 s, during which time the paperback slides off. Within what range does the coefficient of static friction between the two books lie?

49. Children sled down a 41-m-long hill inclined at 25°. At the bottom, the slope levels out. If the coefficient of friction is 0.12, how far do the children slide on the level?

50. In a typical front-wheel-drive car, 70% of the car's weight rides on the front wheels. If the coefficient of friction between tires and road is 0.61, what is the maximum acceleration of the car?

51. A police officer investigating an accident estimates from the damage done that a moving car hit a stationary car at 25 km/h. If the moving car left skid marks 47 m long, and if the coefficient of kinetic friction is 0.71, what was the initial speed of the moving car?

52. A slide inclined at 35° takes bathers into a swimming pool. With water sprayed onto the slide to make it essentially frictionless, a bather spends only one-third as much time on the slide as when it is dry. What is the coefficient of friction on the dry slide?

53. You try to move a heavy trunk, pushing down and forward at an angle of 50° below the horizontal. Show that, no matter how hard you try to push, it's impossible to budge the trunk if the coefficient of static friction exceeds 0.84.

54. A block is shoved up a 22° slope with an initial speed of 1.4 m/s. The coefficient of kinetic friction is 0.70. (a) How far up the slope will the block get? (b) Once stopped, will it slide back down?

55. At the end of a factory production line, boxes start from rest and slide down a 30° ramp 5.4 m long. If the slide is to take no more than 3.3 s, what is the maximum allowed frictional coefficient?

56. A space station is in the shape of a hollow ring, 450 m in diameter (Fig. 5.33). At how many revolutions per minute should it rotate in order to simulate Earth's gravity—that is, so that the normal force on an astronaut at the outer edge would be the astronaut's weight on Earth?

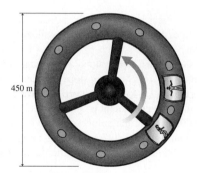

450 m

FIGURE 5.33 Problem 56

57. In a loop-the-loop roller coaster, show that a car moving too slowly would leave the track at an angle ϕ given by $\cos\phi = v^2/rg$, where ϕ is the angle made by a vertical line through the center of the circular track and a line from the center to the point where the car leaves the track.

58. An astronaut is training in an earthbound centrifuge that consists of a small chamber whirled around horizontally at the end of a 5.1-m-long shaft. The astronaut places a notebook on the vertical wall of the chamber and it stays in place. If the coefficient of static friction is 0.62, what is the minimum rate at which the centrifuge must be revolving?

59. You stand on a spring scale at the north pole and again at the equator. (a) Which scale reading will be lower, and why? (b) By what percentage will the lower reading differ from the higher one? *Note:* Assume g has the same value at pole and equator.

60. Driving in thick fog on a horizontal road, a driver spots a tractor-trailer truck jackknifed across the road. To avert a collision, the driver could brake to a stop or swerve in a circular arc, as suggested in Fig. 5.34. Which option offers the greater margin of safety? Assume that the same coefficient of static friction is operative in both cases, and that the car maintains constant speed if it swerves.

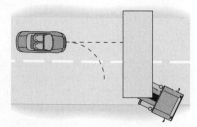

FIGURE 5.34 Problem 60

61. A block is projected up an incline making an angle θ with the horizontal. It returns to its initial position with half its initial speed. Show that the coefficient of kinetic friction is $\mu_k = \frac{3}{5}\tan\theta$.

62. A 2.1-kg mass is connected to a spring with spring constant $k = 150$ N/m and unstretched length 18 cm. The pair are mounted on a frictionless air table, with the free end of the spring attached to a frictionless pivot. The mass is set into circular motion at 1.4 m/s. Find the radius of its path.

63. Take $\mu_k = 0.75$ in Example 5.11, and plot the tension force in units of the trunk's weight, as a function of the rope angle θ (that is, plot T/mg versus θ). Use your plot to determine (a) the minimum tension necessary to move the trunk and (b) the angle at which this minimum tension should be applied.

64. Repeat the preceding problem for an arbitrary value of μ_k, by using calculus to find the minimum force needed to move the trunk with constant speed.

65. Moving through a liquid, an object of mass m experiences a resistive drag force proportional to its velocity, $F_{\text{drag}} = -bv$, where b is a constant. (a) Find an expression for the speed of the object as a function of time, when it starts from rest and falls vertically through the liquid. (b) Show that it reaches a terminal velocity of mg/b.

66. Suppose the object in Problem 65 had an initial velocity in the horizontal direction equal to the terminal speed, $v_{x0} = mg/b$. Show that its horizontal distance is limited by $x_{\max} = mv_{x0}/b$, and find an expression for its trajectory giving y as a function of x.

67. A heavy chain, of length L and uniform mass per unit length m/L, hangs between two supports of the same height. Use the fact that the tension in the chain is along its length (i.e., tangent to the chain) to relate the magnitude of the tension at the lowest point, T_0, to the magnitude of the tension at one of the supports, T_s, given that the angle the tension makes with the horizontal at a support is θ_s. Also, relate the tension at the supports to the weight of the chain, mg. Use similar reasoning to find the slope of the chain at an arbitrary point along its length, dy/dx, where x is horizontal and y is vertical, and differentiate this to eliminate the arc length, $dL = (dx^2 + dy^2)^{1/2}$. The result is a differential equation for the shape of the chain, called a catenary.

68. A florist asks you to make a window display with two hanging pots as shown in Fig. 5.35. The florist is adamant that the strings be as invisible as possible, so you decide to use fishing line but want to use the thinnest line you can. Will fishing line that can withstand 100 N of tension work?

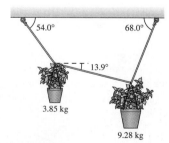

54.0°   68.0°

T 13.9°

3.85 kg

9.28 kg

FIGURE 5.35 Problem 68

69. You're at the state fair. A sideshow barker claims that the star of the show can throw a 7.3-kg Olympic-style hammer "faster than a speeding bullet." You recall that bullets travel at several hundreds of meters per second. The burly hammer thrower whirls the hammer in a circle that you estimate to be 2.4 m in diameter. You guess the chain holding the hammer makes an angle of 10° with the horizontal. When the hammer flies off, is it really moving faster than a bullet?

70. Suppose a construction worker uses a rope and pulley system to hoist construction material to an upper floor, as shown in Fig. 5.36. He knows he should lift the load at a constant speed so as to not break the rope. The rope is rated to withstand 500 N of tension. What mass of materials can he safely hoist?

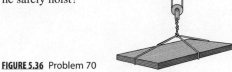

FIGURE 5.36 Problem 70

71. One of the limiting factors in high-performance aircraft is the number of *g*'s to which the pilot can be subjected without blacking out. The new F-22 Raptor fighter can achieve Mach 1.8 (1.8 times the speed of sound, which is about 340 m/s). Suppose a pilot dives in a circle and pulls up. If the pilot cannot exceed 6 *g*'s, or 6 times the normal acceleration of gravity, what is the tightest circle (smallest radius) in which the plane can turn?

Answers to Chapter Questions

Answer to Chapter Opening Question

The roller coaster is "falling," in the sense that it's *accelerating* downward, but that doesn't mean it has to be *moving* downward. The acceleration comes in response to the downward force of gravity and the normal force from the track, both of which are needed to keep the roller coaster moving in its circular path. Quantitatively, the net force is equal to the mass multiplied by the acceleration, so Newton's second law of motion is perfectly satisfied.

Answers to GOT IT? Questions

5.1 (c) Equal—but only because of the 45° slope. At larger angles, the tension would be greater than the weight; at smaller angles, less.

5.2 (a) 1 N; (b) 1 N—the left hand in Fig. 5.9 and the hook in this figure play exactly the same role, balancing the 1-N tension force in the rope.

5.3 No. The riders, too, are in circular motion and therefore experience normal forces from the car as long as the speed exceeds the minimum $v_{min} = \sqrt{gr}$.

5.4 (c) Greater because the chain is pulling downward, making the normal force greater than the log's weight.

6 Work, Energy, and Power

► **To Learn**
By the end of this chapter you should be able to

- Explain the concept of work and evaluate the work done by constant forces (6.1).
- Evaluate the work done by forces that vary with position (6.2).
- Explain the concept of kinetic energy and its relation to work (6.3).
- Describe the relation between energy and power (6.4).

◄ **To Know**

- The idea of work builds on the concept of force. You should have a firm understanding of force from Chapters 4 and 5, and you should be able to distinguish individual forces from the net force on an object (4.1–4.3).
- The relation between work and kinetic energy follows from Newton's second law. Chapters 4 and 5 have given you a solid understanding of this law and its applications (4.2, 4.5, 5.1).

■ Climbing a mountain, these cyclists do work against gravity. Does that work depend on the route chosen?

igure 6.1a shows a skier starting from rest at the top of a uniform slope. What's the skier's speed at the bottom? You can solve this problem by applying Newton's second law to find the skier's constant acceleration and then the speed. But what about the skier in Fig. 6.1b? Here the slope is continuously changing and so is the acceleration. Constant-acceleration equations don't apply, so solving for the details of the skier's motion is difficult.

There are many cases where motion involves changing forces and accelerations. In this chapter, we introduce the important physical concepts of **work** and **energy**. These powerful concepts enable us to "shortcut" the detailed application of Newton's law to analyze these more complex situations. We begin with the concept of work.

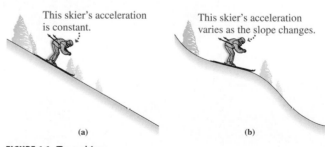

This skier's acceleration is constant.

This skier's acceleration varies as the slope changes.

(a)   (b)

FIGURE 6.1 Two skiers.

6.1 Work

We all have an intuitive sense of the term *work*. Carrying a piece of furniture upstairs in-volves work. The heavier the furniture or the higher the stairs, the greater the work. Push-ing a stalled car involves work. The harder you push or the farther you push, the more work you do. The precise definition of work reflects our intuition:

> For an object moving in one dimension, the work W done on the object by constant applied force $\vec{F}$ is
>
> $$W = F_x \Delta x \tag{6.1}$$
>
> where F_x is the component of the force in the direction of the object's motion and Δx is the object's displacement.

The force $\vec{F}$ need not be the net force. If you're interested, for example, in how much work *you* must do to drag a heavy box across the floor, then $\vec{F}$ is the force *you apply* and W is the work *you do*.

Equation 6.1 shows that the SI unit of work is the newton-meter (N·m). One newton-meter is given the name **joule**, in honor of the 19th-century British physicist and brewer James Joule.

According to Equation 6.1, the person pushing the car in Fig. 6.2a does work equal to the force he applies times the distance the car moves. But the person pulling the suitcase in Fig. 6.2b does work equal to only the horizontal component of the force she applies times the distance the suitcase moves. Furthermore, by our definition, the waiter of Fig. 6.2c does no work on the tray. Why not? Because the force on the tray is vertical while the tray's dis-placement is horizontal; there's no component of force in the direction of the tray's motion.

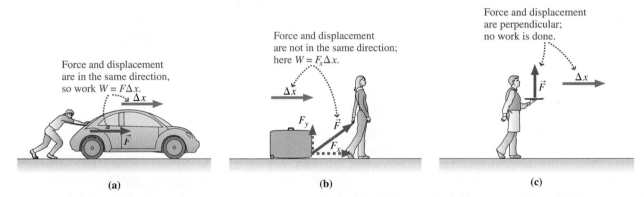

(a)   (b)   (c)

FIGURE 6.2 Work depends on the orientation of force and displacement.

Work can be positive or negative (Fig. 6.3). When a force acts in the same general direction as the motion, it does positive work. A force acting at 90° to the motion does no work. And when a force acts to oppose motion, it does negative work.

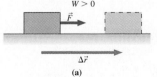
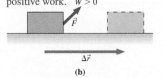
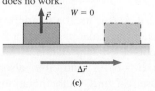
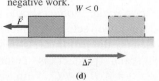
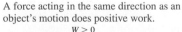

FIGURE 6.3 The sign of the work depends on the relative directions of force and motion. We use $\Delta\vec{r}$ here to indicate that the displacement can be any vector.

EXAMPLE 6.1 **Calculating Work: Pushing a Car**

The person in Fig. 6.2a pushes with a force of 650 N, moving the car a distance of 4.3 m. How much work does he do?

INTERPRET This problem is about work. We identify the car as the object *on* which the work is done and the person as the agent *doing* the work.

DEVELOP Figure 6.2a is our drawing. Equation 6.1, $W = F_x \Delta x$, is the relevant equation, so our plan is to apply that equation. The force is in

the same direction as the displacement, so 650 N is the component we need.

EVALUATE We apply Equation 6.1 to get

$$W = F_x \Delta x = (650 \text{ N})(4.3 \text{ m}) = 2.8 \text{ kJ}$$

ASSESS Make sense? The units work out, with newtons times meters giving joules—here expressed in kilojoules for convenience. ■

EXAMPLE 6.2 **Calculating Work: Pulling a Suitcase**

The airline passenger in Fig. 6.2b exerts a 60-N force on her suitcase, pulling at a 35° angle to the horizontal. How much work does she do in pulling the suitcase 45 m on a level floor?

INTERPRET Again, this example is about work—here done *by* the passenger *on* the suitcase.

DEVELOP Equation 6.1, $W = F_x \Delta x$, applies here, but because the displacement is horizontal while the force isn't, we need to find the horizontal force component. We've redrawn the force vector in Fig. 6.4 to determine F_x.

EVALUATE Applying Equation 6.1 to the x component from Fig. 6.4, we get

$$W = F_x \Delta x = [(60 \text{ N})(\cos 35°)](45 \text{ m}) = 2.2 \text{ kJ}$$

ASSESS The answer of 2.2 kJ is less than the product of 60 N and 45 m, and that makes sense because only the x component of that 60-N force contributes to the work.

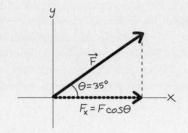

FIGURE 6.4 Our sketch for Example 6.2. ■

Work and the Scalar Product

Work is a *scalar* quantity; it's specified completely by a single number and has no direction. But Fig. 6.2 shows clearly that work involves a relation between two *vectors*: the force $\vec{F}$ and the displacement, which is designated more generally by $\Delta \vec{r}$. If θ is the angle between these two vectors, then the component of the force along the direction of motion is $F \cos \theta$, and the work is

$$W = (F \cos \theta)(\Delta r) = F \Delta r \cos \theta \qquad (6.2)$$

This equation is a generalization of our definition 6.1. If we choose the x axis along $\Delta \vec{r}$, then $\Delta r = \Delta x$ and $F \cos \theta = F_x$, so we recover Equation 6.1.

Equation 6.2 shows that work is the product of the magnitudes of the vectors $\vec{F}$ and $\Delta \vec{r}$ and the cosine of the angle between them. This combination occurs so often that it's given a special name: the **scalar product** of two vectors.

The scalar product of any two vectors $\vec{A}$ and $\vec{B}$ is defined as

$$\vec{A} \cdot \vec{B} = AB \cos \theta \qquad (6.3)$$

where A and B are the magnitudes of the vectors and θ is the angle between them.

The term *scalar product* should remind you that $\vec{A} \cdot \vec{B}$ is itself a *scalar*, even though it's formed from two vectors. A centered dot designates the scalar product; for this reason, it's also called the **dot product**. Figure 6.5 gives a geometric interpretation.

The scalar product is commutative: $\vec{A} \cdot \vec{B} = \vec{B} \cdot \vec{A}$, and it's also distributive: $\vec{A} \cdot (\vec{B} + \vec{C}) = \vec{A} \cdot \vec{B} + \vec{A} \cdot \vec{C}$. With vectors expressed in unit vector notation, Problem 47

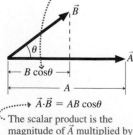

The component of $\vec{B}$ in the direction of $\vec{A}$ is $B \cos\theta$.

$\vec{A} \cdot \vec{B} = AB \cos\theta$

The scalar product is the magnitude of $\vec{A}$ multiplied by the component of $\vec{B}$ in the direction of $\vec{A}$.

FIGURE 6.5 Geometric interpretation of the scalar product.

shows how the distributive law gives a simple form for the scalar product. If $\vec{A} = A_x\hat{i} + A_y\hat{j} + A_z\hat{k}$ and $\vec{B} = B_x\hat{i} + B_y\hat{j} + B_z\hat{k}$, then

$$\vec{A} \cdot \vec{B} = A_xB_x + A_yB_y + A_zB_z \qquad (6.4)$$

Comparing Equation 6.2 with Equation 6.3 shows that the work done by a constant force $\vec{F}$ moving an object through a straight-line displacement $\Delta\vec{r}$ is

$$W = \vec{F} \cdot \Delta\vec{r} \qquad (6.5)$$

As the examples below show, either Equation 6.3 or Equation 6.4 can be used in evaluating the dot product in this expression for work.

EXAMPLE 6.3   **Work and the Scalar Product: A Tugboat**

A tugboat pushes on a cruise ship with a force $\vec{F} = 1.2\hat{i} + 2.3\hat{j}$ MN, moving the ship along a straight path with displacement $\Delta\vec{r} = 380\hat{i} + 460\hat{j}$ m. Find (a) the work done by the tugboat and (b) the angle between the force and the displacement.

INTERPRET Part (a) is about calculating work given force and displacement in unit vector notation. Part (b) is less obvious, but knowing that work involves the angle between force and displacement provides a clue, suggesting that the answer to (a) may lead us to (b).

DEVELOP Figure 6.6 is a sketch of the two vectors that will serve as a check on our final answer. For (a), we want to use Equation 6.5, $W = \vec{F}\cdot\Delta\vec{r}$, with the scalar product given by Equation 6.4. That will give us the work W. We also have the vectors $\vec{F}$ and $\Delta\vec{r}$, so we can find

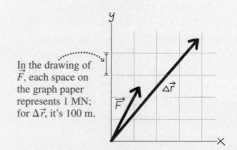

In the drawing of $\vec{F}$, each space on the graph paper represents 1 MN; for $\Delta\vec{r}$, it's 100 m.

FIGURE 6.6 Our sketch of the vectors in Example 6.3.

their magnitudes. That suggests a strategy for (b): Given the work and the vector magnitudes, we can write Equation 6.3 with a single unknown, the angle θ that we're asked to find.

EVALUATE For (a), we use Equations 6.5 and 6.4 to write

$$W = \vec{F} \cdot \Delta\vec{r} = F_x \Delta x + F_y \Delta y$$
$$= (1.2 \text{ MN})(380 \text{ m}) + (2.3 \text{ MN})(460 \text{ m}) = 1510 \text{ MJ}$$

The first equality is from Equation 6.5; the second gives the scalar product in unit vector form from Equation 6.4. Δx and Δy are the components of the displacement $\Delta\vec{r}$. Now that we have the work, we can get the angle. The magnitude of a vector comes from the Pythagorean theorem, as expressed in Equation 3.1. Then we have $F = \sqrt{F_x^2 + F_y^2} = \sqrt{(1.2 \text{ MN})^2 + (2.3 \text{ MN})^2} = 2.59 \text{ MN}$; a similar calculation gives $\Delta r = 597$ m. Now we solve Equation 6.3 for θ:

$$\theta = \cos^{-1}\left(\frac{W}{F\,\Delta r}\right) = \cos^{-1}\left(\frac{1510 \text{ MJ}}{(2.59 \text{ MN})(597 \text{ m})}\right) = 12°$$

ASSESS This small angle is consistent with our sketch in Fig. 6.6. And it makes good physical sense: A tugboat is most efficient when pushing in the direction the ship is supposed to go. Note how the units work out in that last calculation: MJ in the numerator and MN·m in the denominator. But 1 N·m is 1 J, so that's MJ in the denominator, too, giving the dimensionless cosine. ∎

GOT IT? 6.1 Two objects are each displaced the same distance, one by a force F pushing in the direction of motion and the other by a force $2F$ pushing at 45° to the direction of motion. Which force does more work?

6.2 Forces That Vary

Often the force applied to an object varies with position. Important examples include electric and gravitational forces, which vary with the distance between interacting objects. The force of a spring that we encountered in Chapter 4 provides another example; as the spring stretches, the force increases.

Figure 6.7 is a plot of a force F that varies with position x. We want to find the work done as an object moves from x_1 to x_2. We can't simply write $F(x_2 - x_1)$; since the force varies, there's no single value for F. What we can do, though, is divide the region into

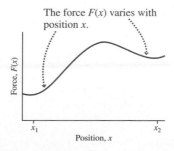

The force $F(x)$ varies with position x.

FIGURE 6.7 A varying force.

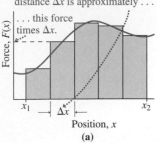

The work done in moving this distance Δx is approximately . . .

. . . this force times Δx.

Force, $F(x)$

x_1 Δx x_2

Position, x

(a)

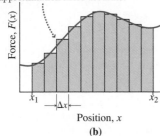

Making the rectangles smaller makes the approximation more accurate.

Force, $F(x)$

x_1 Δx x_2

Position, x

(b)

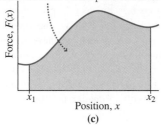

The exact value for the work is the area under the force-versus-position curve.

Force, $F(x)$

x_1 x_2

Position, x

(c)

FIGURE 6.8 Work done by a varying force.

rectangles of width Δx, as shown in Fig. 6.8a. If we make Δx small enough, the force will be nearly constant over the width of each rectangle (Fig. 6.8b). Then the work ΔW done in moving the width Δx of one such rectangle is approximately $F(x)\,\Delta x$, where $F(x)$ is the force at the midpoint x of that rectangle. We write $F(x)$ to show explicitly that the force is a function of position. Note that the quantity $F(x)\,\Delta x$ is the area of the rectangle expressed in the appropriate units (N·m, or, equivalently, J).

Suppose there are N rectangles. Let x_i be the midpoint of the ith rectangle. Then the total work done in moving from x_1 to x_2 is given approximately by the sum of the individual amounts of work ΔW_i associated with each rectangle, or

$$W \simeq \sum_{i=1}^{N} W_i = \sum_{i=1}^{N} F(x_i)\,\Delta x \qquad (6.6)$$

How good is this approximation? That depends on how small we make the rectangles. Suppose we let them get arbitrarily small. Then the number of rectangles must grow arbitrarily large. In the limit of infinitely many infinitesimally small rectangles, the approximation in Equation 6.6 becomes exact (Fig. 6.8c). Then we have

$$W = \lim_{\Delta x \to 0} \sum_i F(x_i)\,\Delta x \qquad (6.7)$$

where the sum is over all the infinitesimal rectangles between x_1 and x_2. The quantity on the right-hand side of Equation 6.7 is the **definite integral** of the function $F(x)$ over the interval from x_1 to x_2. We introduce special symbolism for the limiting process of Equation 6.7:

$$W = \int_{x_1}^{x_2} F(x)\,dx \qquad \left(\begin{array}{l}\text{work done by a varying}\\\text{force in one dimension}\end{array}\right) \qquad (6.8)$$

Equation 6.8 means exactly the same thing as Equation 6.7: It tells us to divide the interval from x_1 to x_2 into many small rectangles of width Δx, to multiply the value of the function $F(x)$ at each rectangle by the width Δx, and to sum those products. As we take arbitrarily many arbitrarily small rectangles, the result of this process gives us the value of the definite integral. You can think of the symbol $\int$ in Equation 6.8 as standing for "sum" and the symbol dx as a limiting case of arbitrarily small Δx. The definite integral has a simple geometric interpretation: It's the area under the curve $F(x)$ between the limits x_1 and x_2 (Fig. 6.8c).

Computers approximate the infinite sum implied in Equation 6.8 using a large number of very small rectangles. But calculus often provides a better way.

TACTICS 6.1   **Integrating**

In your calculus course you've learned, or will soon learn, that integrals and derivatives are inverses. In Section 2.2, you saw that the derivative of x^n is nx^{n-1}; therefore, the integral of x^n is $(x^{n+1})/(n+1)$, as you can verify by differentiating. We determine the value of a definite integral by evaluating this expression at upper and lower limits and subtracting:

$$\int_{x_1}^{x_2} x^n\,dx = \frac{x^{n+1}}{n+1}\bigg|_{x_1}^{x_2} = \frac{x_2^{n+1}}{n+1} - \frac{x_1^{n+1}}{n+1} \qquad (6.9)$$

where the middle term is a shorthand notation for the difference given in the rightmost term. A review of integration and a table of common integrals are given in Appendix A.

Stretching a Spring

A spring provides an important example of a force that varies with position. We've seen that an ideal spring exerts a force proportional to its displacement from equilibrium: $F = -kx$, where k is the spring constant and the minus sign shows that the spring force is opposite the direction of the displacement. It's not just coiled springs that we're interested in here; many physical systems, from molecules to skyscrapers to stars, behave as though they contain springs. The work and energy considerations we develop here apply to those systems as well.

The force exerted *by* a stretched spring is $-kx$, so the force exerted *on* the spring *by* the external stretching force is $+kx$. If we let $x = 0$ be one end of the spring at equilibrium and if we hold the other end fixed and pull the spring until its free end is at a new position x, as shown in Fig. 6.9, then Equation 6.8 shows that the work done on the spring by the external force is

$$W = \int_0^x F(x)\, dx = \int_0^x kx\, dx = \tfrac{1}{2}kx^2 \Big|_0^x = \tfrac{1}{2}kx^2 - \tfrac{1}{2}k(0)^2 = \tfrac{1}{2}kx^2 \qquad (6.10)$$

where we used Equation 6.9 to evaluate the integral. Does this result make sense? The more we stretch the spring, the greater the force we must apply—and that means we must do more work for a given amount of additional stretch. Figure 6.10 shows graphically why the work depends quadratically on the displacement. Although we used the word *stretch* in developing Equation 6.10, the result applies equally to compressing a spring a distance x from equilibrium.

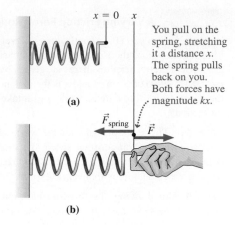

You pull on the spring, stretching it a distance x. The spring pulls back on you. Both forces have magnitude kx.

FIGURE 6.9 Stretching a spring.

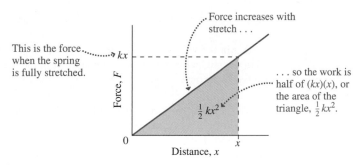

This is the force when the spring is fully stretched.

Force increases with stretch . . .

. . . so the work is half of $(kx)(x)$, or the area of the triangle, $\tfrac{1}{2}kx^2$.

FIGURE 6.10 Work done in stretching a spring.

EXAMPLE 6.4 **The Spring Force: Bungee Jumping**

An elastic cord used in bungee jumping is 20 m long and has spring constant $k = 11$ N/m. At the lowest point in the jump, the cord length has doubled. (a) How much work has been done on the cord? (b) How does the work done in the last meter of stretch compare with the work done in the first meter?

INTERPRET The bungee cord behaves like a spring—as we can tell because we're given its spring constant. So this example is about the work done in stretching a spring, which we just calculated for the general case. We're told the 20-m-long cord doubles in length, so that means it's stretched another 20 m.

DEVELOP A drawing is helpful for understanding (b). Figure 6.11 is the force-distance curve with the first and last meters highlighted.

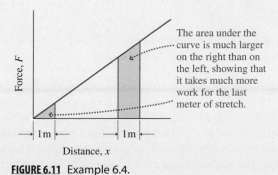

The area under the curve is much larger on the right than on the left, showing that it takes much more work for the last meter of stretch.

FIGURE 6.11 Example 6.4.

Equation 6.10, $W = \tfrac{1}{2}kx^2$, assumes the spring starts out unstretched, so it applies to the entire stretch and to the first meter. For the last meter, we can redo the calculation leading to the final result of Equation 6.10, but this time going from 19 m to 20 m.

EVALUATE For the full stretch, Equation 6.10 gives $W = \tfrac{1}{2}kx^2 = \left(\tfrac{1}{2}\right)(11 \text{ N/m})(20 \text{ m})^2 = 2.2$ kJ. For the first meter, we use the same equation: $W_1 = \tfrac{1}{2}kx^2 = \left(\tfrac{1}{2}\right)(11 \text{ N/m})(1 \text{ m})^2 = 5.5$ J. For the last meter, we start with the middle term in the steps leading to Equation 6.10, using 19 instead of 0 as the lower limit:

$$W_{19\text{-}20} = \tfrac{1}{2}kx^2 \Big|_{19 \text{ m}}^{20 \text{ m}}$$

$$= \left(\tfrac{1}{2}\right)(11 \text{ N/m})(20 \text{ m})^2 - \left(\tfrac{1}{2}\right)(11 \text{ N/m})(19 \text{ m})^2 = 214 \text{ J}$$

We could also get this result by subtracting the work done in the first 19 m from the total work done in the whole 20 m.

ASSESS The work done in the last meter is nearly 40 times that done in the first meter! Figure 6.11 shows why this makes sense: During the last meter, the force is much greater, so much more work is involved. ∎

EXAMPLE 6.5 **A Varying Friction Force: Rough Sliding**

Workers pushing a 180-kg trunk across a level floor encounter a 10-m-long region where the floor becomes increasingly rough. The coefficient of kinetic friction here is given by $\mu_k = \mu_0 + ax^2$, where $\mu_0 = 0.17$, $a = 0.0062$ m^{-2}, and x is the distance from the beginning of the rough region. How much work does it take to push the trunk across the region?

INTERPRET This example asks for the work needed to push the trunk. To move the trunk at constant speed, the workers must apply a force equal in magnitude to the frictional force. That force varies with position, so we're dealing with a varying force.

DEVELOP Our drawing, the force-position curve in Fig. 6.12, emphasizes that we have a varying force. Therefore, we have to integrate

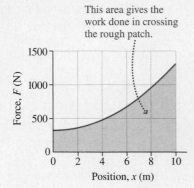

This area gives the work done in crossing the rough patch.

FIGURE 6.12 Force versus position for Example 6.5.

using Equation 6.8, $W = \int_{x_1}^{x_2} F(x)\,dx$. And we need to know the frictional force, which is given by Equation 5.3: $f_k = \mu_k n$. On a level floor, the normal force is equal in magnitude to the weight, mg, so Equation 6.8 becomes $W = \int_{x_1}^{x_2} \mu_k mg\,dx = \int_{x_1}^{x_2} mg(\mu_0 + ax^2)\,dx$.

EVALUATE We evaluate the integral using Equation 6.9. Actually, we have two integrals here: one of dx alone and the other of $x^2\,dx$. According to Equation 6.9, the former gives x and the latter $x^3/3$. So the result is

$$W = \int_{x_1}^{x_2} mg(\mu_0 + ax^2)\,dx = mg\left(\mu_0 x + \tfrac{1}{3}ax^3\right)\Big|_{x_1}^{x_2}$$

$$= mg\left[\left(\mu_0 x_2 + \tfrac{1}{3}ax_2^3\right) - \left(\mu_0 x_1 + \tfrac{1}{3}ax_1^3\right)\right]$$

Putting in the values given for μ_0, a, and m, using $g = 9.8$ m/s^2, and taking $x_1 = 0$ and $x_2 = 10$ m for the endpoints of the rough interval, we get 6.6 kJ for our answer.

ASSESS Is this answer reasonable? Figure 6.12 shows that the maximum force is approximately 1.3 kN. If this force acted over the entire 10-m interval, the work would be about 13 kJ. But it's approximately half that because the coefficient of kinetic friction and therefore the force start out quite low. You can see that the area under the curve in Fig. 6.12 is about half the area of the full rectangle, so our answer of 6.6 kJ makes sense. ∎

✓**TIP** Don't Just Multiply!

When force depends on position, there's no single value for the force, so you can't just multiply force by distance to get work. You need either to integrate, as in Example 6.5, or to use a result that's been derived by integration, as with the equation $W = \tfrac{1}{2}kx^2$ used in Example 6.4.

Force and Work in Two and Three Dimensions

Sometimes a force varies in both magnitude and direction or an object moves on a curved path; either way, the angle between force and motion may vary. Then we have to take the scalar products of the force $\vec{F}$ with small displacements $\Delta\vec{r}$, writing $\Delta W = \vec{F} \cdot \Delta\vec{r}$ for the work involved in one such small displacement. Adding them all gives the total work, which in the limit of very small displacements becomes a **line integral**:

$$W = \int_{\vec{r}_1}^{\vec{r}_2} \vec{F} \cdot d\vec{r} \tag{6.11}$$

where the integral is taken over a specific path between positions $\vec{r}_1$ and $\vec{r}_2$. We won't pursue line integrals further here, but they'll be useful in later chapters.

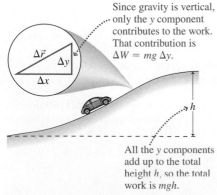

Since gravity is vertical, only the y component contributes to the work. That contribution is $\Delta W = mg\,\Delta y$.

All the y components add up to the total height h, so the total work is mgh.

FIGURE 6.13 A car climbs a hill with varying slope.

Work Done Against Gravity

When an object moves upward or downward on an arbitrary path, the angle between its displacement and the gravitational force varies. But here we don't really need the line integral of Equation 6.11 because we can consider any path as consisting of small horizontal and vertical steps (Fig. 6.13). Only the vertical steps contribute to the work, which then becomes simply $W = mgh$, where h is the total height the object rises—a result that is independent of the particular path taken. (As in our earlier work with gravity, this result holds only near Earth's surface, where we can neglect the variation in gravity with height.)

GOT IT? 6.2 Three forces have magnitudes in newtons that are numerically equal to these quantities: (a) x, (b) x^2, and (c) $\sqrt{x}$, where x is the position in meters. Each force acts on an object as it moves from $x = 0$ to $x = 1$ m. Notice that all three forces have the same values at the two endpoints—namely, 0 N and 1 N. Which force does the most work? Which does the least?

6.3 Kinetic Energy

Closely related to work is **energy**—one of the most important concepts in all of physics. Here we introduce the energy associated with motion, or **kinetic energy**. Our goal is to relate kinetic energy and work. We'll start by evaluating the net work done on an object, and then apply Newton's second law. The net work is done by all the forces acting on an object, so we use the net force in our expression for work. We'll consider one-dimensional motion, with force and displacement along the same line. In that case, Equation 6.8 gives the net work:

$$W_{\text{net}} = \int F_{\text{net}} \, dx$$

But the net force can be written in terms of Newton's second law: $F_{\text{net}} = ma$, or $F_{\text{net}} = m \, dv/dt$, so

$$W_{\text{net}} = \int m \frac{dv}{dt} \, dx$$

The quantities dv, dt, and dx arose as the limits of small numbers Δv, Δt, and Δx. In calculus, you've seen that the limit of a product or quotient is the product or quotient of the individual terms involved. For these reasons, we can rearrange the symbols dv, dt, and dx to rewrite our expression in the form

$$W_{\text{net}} = \int m \, dv \, \frac{dx}{dt}$$

But $dx/dt = v$, so we have

$$W_{\text{net}} = \int mv \, dv$$

The integral here is like the form $\int x \, dx$, which we integrate by raising the exponent and dividing by the new exponent. What about the limits? Suppose our object starts at some speed v_1 and ends at v_2. Then we have

$$W_{\text{net}} = \int_{v_1}^{v_2} mv \, dv = \tfrac{1}{2}mv^2 \Big|_{v_1}^{v_2} = \tfrac{1}{2}mv_2^2 - \tfrac{1}{2}mv_1^2 \qquad (6.12)$$

Equation 6.12 shows that an object has associated with it a quantity $\tfrac{1}{2}mv^2$ that changes when, and only when, net work is done on the object. This quantity plays a vital role in physics and is called the object's **kinetic energy**:

The kinetic energy K of an object of mass m moving at speed v is

$$K = \tfrac{1}{2}mv^2 \qquad (6.13)$$

Like velocity, kinetic energy is a relative term; its value depends on the reference frame in which it's measured. But unlike velocity, kinetic energy is a *scalar*. And since it depends on the *square* of the velocity, kinetic energy is never negative. All moving objects possess kinetic energy.

Equation 6.12 equates the change in an object's kinetic energy with the net work done on the object, a result known as the **work-energy theorem**:

> **Work-energy theorem** The change in an object's kinetic energy is equal to the net work done on the object:
>
> $$\Delta K = W_{net} \qquad (6.14)$$

Equations 6.12 and 6.14 are equivalent statements of the work-energy theorem.

We've seen that work can be positive or negative; so, therefore, can *changes* in kinetic energy. If I stop a moving object, for example, I reduce its kinetic energy from $\frac{1}{2}mv^2$ to zero—a change $\Delta K = -\frac{1}{2}mv^2$. So I do negative work by applying a force directed opposite to the motion. By Newton's third law, the object exerts an equal but oppositely directed force on me, therefore doing positive work $\frac{1}{2}mv^2$ *on me*. So an object of mass m moving at speed v can do work equal to its initial kinetic energy, $\frac{1}{2}mv^2$, if it's brought to rest.

EXAMPLE 6.6 **Work and Kinetic Energy: Passing Zone**

A 1400-kg car enters a passing zone and accelerates from 70 to 95 km/h. (a) How much work is done on the car? (b) If the car then brakes to a stop, how much work is done on it?

INTERPRET Here we're asked about work, but we aren't given any forces as we were in previous examples. Now we know the work-energy theorem, however, relating work and kinetic energy. Kinetic energy depends on the speed, which we are given. So this is a problem involving the work-energy theorem.

DEVELOP The relevant equation is Equation 6.14 or its more explicit form, Equation 6.12. Since we're given speeds, it's easiest to work with Equation 6.12.

EVALUATE For (a), Equation 6.12 gives

$$W_{net} = \tfrac{1}{2}mv_2^2 - \tfrac{1}{2}mv_1^2 = \tfrac{1}{2}m(v_2^2 - v_1^2)$$
$$= \left(\tfrac{1}{2}\right)(1400\ kg)\left[(26.4\ m/s)^2 - (19.4\ m/s)^2\right] = 220\ kJ$$

where we converted the speeds to meters per second before doing the calculation. The work-energy theorem applies equally to the braking car in (b), for which $v_1 = 26.4$ m/s and $v_2 = 0$. Here we have

$$W_{net} = \tfrac{1}{2}m(v_2^2 - v_1^2) = \left(\tfrac{1}{2}\right)(1400\ kg)\left[0^2 - (26.4\ m/s)^2\right] = -490\ kJ$$

ASSESS Make sense? Yes. There's a greater change in speed and thus in kinetic energy in the braking case, so the magnitude of the work involved is greater. But why is the second answer negative? Because stopping the car means applying a force that *opposes* its motion—and that means negative work is done on the car. ∎

GOT IT? 6.3 For each situation, tell whether the net work done on a soccer ball is positive, negative, or zero. Justify your answers using the work-energy theorem. (a) You carry the ball out to the field, walking at constant speed. (b) You kick the stationary ball, starting it flying through the air. (c) The ball rolls along the field, gradually coming to a halt.

Energy Units

Since work is equal to the change in kinetic energy, the units of energy are the same as those of work. In SI, the unit of energy is therefore the joule, equal to 1 newton-meter. In science, engineering, and everyday life, though, you'll encounter other energy units. Scientific units include the **erg**, used in the centimeter-gram-second system of units and equal to 10^{-7} J; the **electron-volt**, used in nuclear, atomic, and molecular physics; and the **calorie**, used in thermodynamics and to describe the energies of chemical reactions. English units include the **foot-pound** and the **British thermal unit** (Btu); the latter is commonly used in engineering of heating and cooling systems. Your electric company charges you for energy use in **kilowatt-hours** (kW·h); we'll see in the next section how this unit relates to the SI joule. Appendix C contains an extensive table of energy units and conversion factors as well as the energy contents of common fuels.

6.4 Power

Climbing a flight of stairs requires the same amount of work no matter how fast you go. But it's harder to *run* up the stairs than to walk. Harder in what sense? In the sense that you do the same work in a shorter time; the *rate* at which you do the work is greater. We define **power** as the rate of doing work:

> If an amount of work ΔW is done in time Δt, then the average power $\overline{P}$ is
>
> $$\overline{P} = \frac{\Delta W}{\Delta t} \quad \text{(average power)} \qquad (6.15)$$

Often the rate of doing work varies with time. Then we define the **instantaneous power** as the average power taken in the limit of an arbitrarily small time interval Δt:

$$P = \lim_{\Delta t \to 0} \frac{\Delta W}{\Delta t} = \frac{dW}{dt} \quad \text{(instantaneous power)} \qquad (6.16)$$

Equations 6.15 and 6.16 both show that the units of power are joules/second. One J/s is given the name **watt** (W) in honor of James Watt, a Scottish engineer and inventor who was instrumental in developing the steam engine as a practical power source. Watt himself defined another unit, the horsepower. One horsepower (hp) is about 746 J/s or 746 W.

EXAMPLE 6.7   **Power: Climbing Mount Washington**

A 55-kg hiker ascends New Hampshire's Mount Washington, making the vertical rise of 1300 m in 2 h. A 1500-kg car drives up the Mount Washington Auto Road, taking half an hour. Neglecting energy lost to friction, what's the average power output for each?

INTERPRET This example is about power, which we identify as the *rate* at which hiker and car expend energy. So we need to know the work done by each and the corresponding time.

DEVELOP Equation 6.15, $\overline{P} = \Delta W/\Delta t$, is relevant, since we want the *average* power. To use this equation we'll need to find the work done in climbing the mountain. As you learned in Section 6.2, work done against gravity is independent of the path taken and is given by mgh, where h is the total height of the climb.

EVALUATE We apply Equation 6.15 in the two cases:

$$\overline{P}_{\text{hiker}} = \frac{\Delta W}{\Delta t} = \frac{(55 \text{ kg})(9.8 \text{ m/s}^2)(1300 \text{ m})}{(2.0 \text{ h})(3600 \text{ s/h})} = 97 \text{ W}$$

$$\overline{P}_{\text{car}} = \frac{\Delta W}{\Delta t} = \frac{(1500 \text{ kg})(9.8 \text{ m/s}^2)(1300 \text{ m})}{(0.50 \text{ h})(3600 \text{ s/h})} = 11 \text{ kW}$$

ASSESS Do these values make sense? A power of 97 W is typical of the sustained long-term output of the human body, as you can confirm by considering a typical daily diet of 2000 "calories" (actually kilocalories; see Exercise 33). The car's output amounts to 14 hp, which you may find low, given that the car's engine is probably rated at several hundred horsepower. But cars are notoriously inefficient machines, with only a small fraction of the rated horsepower available to do useful work. Most of the rest is lost to friction and heating. ∎

When power is constant, so the average and instantaneous power are the same, then Equation 6.15 shows that the amount of work W done in time Δt is

$$W = P \Delta t \qquad (6.17)$$

When the power is not constant, we can consider small amounts of work ΔW, each taken over so small a time interval Δt that the power is nearly constant. Adding all these amounts of work and taking the limit as Δt becomes arbitrarily small, we have

$$W = \lim_{\Delta t \to 0} \sum P \Delta t = \int_{t_1}^{t_2} P \, dt \qquad (6.18)$$

where t_1 and t_2 are the beginning and end of the time interval over which we calculate the power.

EXAMPLE 6.8 **Energy and Power: Yankee Stadium**

Each of the 500 floodlights at Yankee Stadium uses electrical energy at the rate of 1.0 kW. How much does it cost to run these lights during a 4-h night game, if electricity costs 9.5¢/kW·h?

INTERPRET We're given a single floodlight's power consumption and the cost of electricity per kilowatt-hour, a unit of energy. So this problem is about calculating energy given power and time, with a little economics thrown in.

DEVELOP Since the power is constant, we can calculate the energy used over time with Equation 6.17, $W = P \Delta t$.

EVALUATE Solving for work, we find

$$W = P \Delta t = (500 \text{ kW})(4.0 \text{ h}) = 2000 \text{ kW} \cdot \text{h}$$

The cost is then $(2000 \text{ kW·h})(\$0.095/\text{kW·h}) = \$190$.

ASSESS Do we have the right units here? Yes. With power in kilowatts and time in hours, the energy comes out immediately in kilowatt-hours.

APPLICATION **Energy and Society**

Humankind's rate of energy consumption is a matter of concern, especially given our dependence on fossil fuels that pollute the air and whose carbon dioxide emissions threaten global climate change. Just how rapidly are we using energy?

Example 6.7 suggests that the average power output of the human body is approximately 100 W. Before our species harnessed fire and domesticated animals, that was all the power available to each of us. But in today's high-energy societies, we use energy at a much greater rate. For the average citizen of the United States in the early 21st century, for example, the rate of energy consumption is about 11 kW—the equivalent of more than a hundred human bodies. The rate is lower in most other industrialized countries, but it still amounts to many tens of human bodies' worth.

What do we do with all that energy? And where does it come from? The first pie chart shows that most goes for industry and transportation, with lesser amounts used in the residential and commercial sectors. The second chart is a stark reminder that our energy supply is neither diversified nor renewable, with some 86% coming from the fossil fuels coal, oil, and natural gas. That's going to have to change in the coming decades, as a result of both limited fossil-fuel resources and the environmental consequences of fossil-fuel combustion. Much of what you learn in an introductory physics course has direct relevance to the energy challenges we face today.

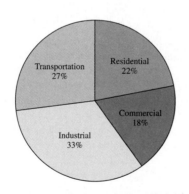

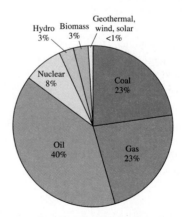

✓ **TIP** Don't Confuse Energy and Power

Is that 11-kW per-capita energy consumption per year, per day, or what? That question reflects a common confusion of energy and power. Power is the *rate* of energy use; it doesn't need any "per time" attached to it. It's just 11 kW, period.

Power and Velocity

We can derive an expression relating power, applied force, and velocity by noting that the work dW done by a force $\vec{F}$ acting on an object that undergoes an infinitesimal displacement $d\vec{r}$ follows from Equation 6.5:

$$dW = \vec{F} \cdot d\vec{r}$$

Dividing both sides by the associated time interval dt gives the power:

$$P = \frac{dW}{dt} = \vec{F} \cdot \frac{d\vec{r}}{dt}$$

But $d\vec{r}/dt$ is the velocity $\vec{v}$, so

$$P = \vec{F} \cdot \vec{v} \qquad\qquad (6.19)$$

EXAMPLE 6.9 **Power and Velocity: Bicycling**

Riding your 14-kg bicycle at a steady 18 km/h (5.0 m/s), you experience a 30-N force from air resistance. If your mass is 68 kg, what power must you supply on level ground and going up a 5° incline?

INTERPRET This example asks about power in two different situations: one with air resistance alone and the other when climbing. We identify the forces involved as air resistance and gravity. You need to exert forces of equal magnitude to overcome them.

DEVELOP Given that we have force and velocity, Equation 6.19, $P = \vec{F} \cdot \vec{v}$, applies. The force you apply to propel the bicycle is in the same direction as its motion, so $\vec{F} \cdot \vec{v}$ in that equation becomes just Fv.

EVALUATE On level ground, we have $P = Fv = (30 \text{ N})(5.0 \text{ m/s}) = 150$ W. Climbing the hill, you have to exert an additional force to overcome the downslope component of gravity, which in Example 5.1 we found to be $mg \sin\theta$. So here we have

$$P = Fv = (F_{air} + mg \sin\theta)v$$
$$= [30 \text{ N} + (82 \text{ kg})(9.8 \text{ m/s}^2)(\sin 5°)](5.0 \text{ m/s}) = 500 \text{ W}$$

where we used your combined mass, body plus bicycle.

ASSESS Both numbers make sense. The values go from a little to a lot more than your body's average power output of around 100 W, and as you've surely experienced, even a modest slope takes much more cycling effort than level ground.

■

GOT IT? 6.4 A newspaper reports that a new power plant will produce "50 megawatts per hour." What's wrong with this statement?

Big Picture

Work and **kinetic energy** are the big ideas here. A force acting on an object does work when the object undergoes a displacement and the force has a component in the direction of that displacement. A force at right angles to the displacement does no work, and a force with a component opposite the displacement does negative work.

Kinetic energy is the energy associated with an object's motion. An object's kinetic energy changes only when net work is done on the object.

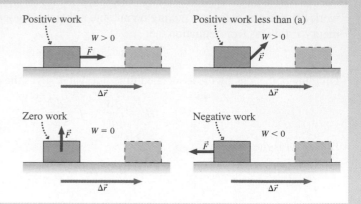

Key Concepts and Equations

Work is the product of force and displacement, but only the component of force in the direction of displacement counts toward the work.

For a constant force and displacement in the x direction,

$$W = F_x \, \Delta x \quad \text{(constant force only)}$$

More generally, for a constant force $\vec{F}$ and arbitrary displacement $\Delta\vec{r}$, the work is

$$W = \vec{F} \cdot \Delta\vec{r} = F \, \Delta r \cos\theta \quad \text{(constant force only)}$$

Here F and Δr are the magnitudes of the force and displacement vectors, and θ is the angle between them. We've written work here using the shorthand notation of the scalar product, defined for any two vectors $\vec{A}$ and $\vec{B}$ as the product of their magnitudes and the cosine of the angle between them:

$$\vec{A} \cdot \vec{B} = AB\cos\theta \quad \text{(scalar product)}$$

When force varies with position, calculating the work involves integrating. In one dimension:

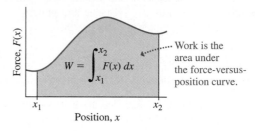

Work is the area under the force-versus-position curve.

$$W = \int_{x_1}^{x_2} F(x)\, dx$$

Most generally, work is the **line integral** of a varying force over an arbitrary path: $W = \displaystyle\int \vec{F} \cdot d\vec{r}$

Kinetic energy is a scalar quantity that depends on an object's mass and speed:

$$K = \tfrac{1}{2} mv^2$$

The **work-energy theorem** states that the change in an object's kinetic energy is equal to the net work done on it:

$$\Delta K = W_{\text{net}} \quad \text{(work-energy theorem)}$$

The unit of energy and work is the joule (J), equal to 1 newton-meter.

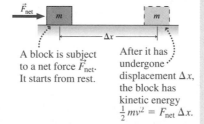

A block is subject to a net force $\vec{F}_{\text{net}}$. It starts from rest. After it has undergone displacement Δx, the block has kinetic energy $\tfrac{1}{2}mv^2 = F_{\text{net}}\,\Delta x$.

Power is the rate at which work is done or energy is used. The unit of power is the **watt** (W), equal to 1 joule/second.

$$P = \frac{dW}{dt} = \vec{F} \cdot \vec{v}$$

Cases and Uses

Common applications of work done against everyday forces are the work mgh needed to raise an object of mass m a distance h against gravity, and the work $\tfrac{1}{2}kx^2$ needed to stretch or compress a spring of spring constant k a distance x from its equilibrium length.

For Thought and Discussion

1. Give two examples of situations in which you might think you are doing work but in which, in the technical sense, you do no work.
2. If the scalar product of two nonzero vectors is zero, what can you conclude about their relative directions?
3. Must you do work to whirl a ball around on the end of a string? Explain.
4. If you pick up a suitcase and put it down, how much total work have *you* done on the suitcase? Does your answer change if you pick up the suitcase and drop it?
5. You want to raise a piano a given height using a ramp. With a fixed, nonzero coefficient of friction, will you have to do more work if the ramp is steeper or more gradual? Explain.
6. Does the gravitational force of the Sun do work on a planet in a circular orbit? On a comet in an elliptical orbit? Explain.
7. Can kinetic energy ever be negative? Explain.
8. A pendulum bob swings back and forth on the end of a string, describing a circular arc. Does the tension force in the string do any work?
9. Does your car's kinetic energy change if you drive at constant speed for 1 hour?
10. A watt-second is a unit of what quantity? Relate it to a more standard SI unit.
11. A truck is moving northward at 55 mi/h. Later, it's moving eastward at the same speed. Has its kinetic energy changed? Has its momentum changed? Has work been done on the truck? Has a force acted on the truck? Explain.

Exercises and Problems

Exercises

Section 6.1 Work

12. How much work do you do as you exert a 75-N force to push a shopping cart through a 12-m-long supermarket aisle?
13. If the coefficient of kinetic friction is 0.21, how much work do you do when you slide a 50-kg box at constant speed across a 4.8-m-wide room?
14. A crane lifts a 650-kg beam vertically upward 23 m and then swings it eastward 18 m. How much work does the crane do? Neglect friction, and assume the beam moves with constant speed.
15. The world's highest waterfall, the Cherun-Meru in Venezuela, has a total drop of 980 m. How much work does gravity do on a cubic meter of water dropping down the Cherun-Meru?
16. A meteorite plunges to Earth, embedding itself 75 cm in the ground. If it does 140 MJ of work in the process, what average force does the meteorite exert on the ground?
17. An elevator of mass m rises a distance h up a vertical shaft with upward acceleration equal to one-tenth g. How much work does the elevator cable do on the elevator?
18. Show that the scalar product obeys the distributive law: $\vec{A} \cdot (\vec{B} + \vec{C}) = \vec{A} \cdot \vec{B} + \vec{A} \cdot \vec{C}$.
19. One vector has magnitude 15 units and another 6.5 units. Find their scalar product if the angle between them is (a) 27° and (b) 78°.
20. Find the work done by a force $\vec{F} = 1.8\hat{i} + 2.2\hat{j}$ N as it acts on an object moving from the origin to the point $\vec{r} = 56\hat{i} + 31\hat{j}$ m.
21. To push a stalled car, you apply a 470-N force at a 17° angle to its direction of motion, doing 860 J of work in the process. How far do you push the car?

Section 6.2 Forces That Vary

22. Find the total work done by the force shown in Fig. 6.14 as the object on which it acts moves (a) from $x = 0$ to $x = 3$ km and (b) from $x = 3$ km to $x = 4$ km.

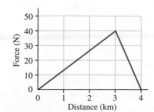

FIGURE 6.14 Exercise 22

23. A spring has spring constant $k = 200$ N/m. How much work does it take to stretch the spring (a) 10 cm from equilibrium and (b) from 10 cm to 20 cm from equilibrium?
24. Uncompressed, the spring for an automobile suspension is 45 cm long. It needs to be fitted into a space 32 cm long. If the spring constant is 3.8 kN/m, how much work does a mechanic have to do to fit the spring?
25. You do 8.5 J of work to stretch a spring with spring constant $k = 190$ N/m, starting with the spring unstretched. How far does the spring stretch?

Section 6.3 Kinetic Energy

26. A 2.4×10^5-kg airplane is cruising at 900 km/h. What is its kinetic energy relative to the ground?
27. A cyclotron accelerates protons from rest to a speed of 2.1×10^7 m/s. How much work does it do on each proton?
28. At what speed must a 950-kg subcompact car be moving to have the same kinetic energy as a 3.2×10^4-kg truck going 20 km/h?
29. A 60-kg skateboarder comes over the top of a hill at 5.0 m/s and reaches 10 m/s at the bottom of the hill. Find the total work done on the skateboarder between the top and bottom of the hill.
30. After a tornado, a 0.50-g drinking straw was found embedded 4.5 cm in a tree. Subsequent measurements showed that the tree would exert a stopping force of 70 N on the straw. What was the straw's speed when it hit the tree?
31. You drop a 150-g baseball from a sixth-story window 16 m above the ground. What are (a) its kinetic energy and (b) its speed when it hits the ground? Neglect air resistance.
32. From what height would you have to drop a car for its impact to be equivalent to a collision at 20 mi/h?

Section 6.4 Power

33. A typical human diet is "2000 calories" per day, where the "calorie" describing food energy is actually 1 kilocalorie. Express 2000 kcal/day in watts.
34. A horse plows a 200-m-long furrow in 5.0 min, exerting a force of 750 N. What is its power output, measured in watts and in horsepower?
35. A typical car battery stores about 1 kW·h of energy. What is its power output if it is drained completely in (a) 1 minute, (b) 1 hour, and (c) 1 day?
36. A sprinter completes a 100-m dash in 10.6 s, doing 22.4 kJ of work. What is her average power output?

37. How much work can a 3.5-hp lawnmower engine do in 1 h?
38. A 75-kg long-jumper takes 3.1 s to reach a prejump speed of 10 m/s. What is his power output?
39. Estimate your power output as you do deep knee bends at the rate of one per second.
40. In midday sunshine, solar energy strikes Earth at the rate of about 1 kW/m². How long would it take a perfectly efficient solar collector of 15-m² area to collect 40 kW·h of energy? *Note:* This is roughly the energy content of a gallon of gasoline.
41. It takes about 20 kJ to melt an ice cube. A typical microwave oven produces 900 W of microwave power. How long will it take to melt the ice cube in this oven?
42. Which consumes more energy, a 1.2-kW hair dryer used for 10 min or a 7-W night-light left on for 24 h?

Problems

43. You slide a box of books at constant speed up a 30° ramp, applying a force of 200 N directed up the slope. The coefficient of sliding friction is 0.18. (a) How much work have you done when the box has risen 1 m vertically? (b) What is the mass of the box?
44. Two people push a stalled car at its front doors, each applying a 280-N force at 25° to the forward direction, as shown in Fig. 6.15. How much work does each person do in pushing the car 5.6 m?

FIGURE 6.15 Problem 44

45. A locomotive does 7.9×10^{11} J of work in pulling a 3.4×10^6-kg train 180 km. What is the average force in the coupling between the locomotive and the rest of the train?
46. A rope pulls a box a horizontal distance of 23 m. If the rope tension is 120 N, and if the rope does 2500 J of work on the box, what angle θ in Fig. 6.16 does it make with the horizontal?

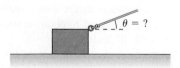

FIGURE 6.16 Problem 46

47. (a) Find the scalar products $\hat{i} \cdot \hat{i}$, $\hat{j} \cdot \hat{j}$, and $\hat{k} \cdot \hat{k}$. (b) Find the scalar products $\hat{i} \cdot \hat{j}$, $\hat{j} \cdot \hat{k}$, and $\hat{k} \cdot \hat{i}$. (c) Use the distributive law to multiply out the scalar product of two arbitrary vectors $\vec{A} = A_x \hat{i} + A_y \hat{j} + A_z \hat{k}$ and $\vec{B} = B_x \hat{i} + B_y \hat{j} + B_z \hat{k}$. Then use the results of (a) and (b) to verify Equation 6.4.
48. (a) Find the scalar product of the vectors $a\hat{i} + b\hat{j}$ and $b\hat{i} - a\hat{j}$. (b) Determine the angle between them. Here a and b are arbitrary constants.
49. Find the angles between all three pairs of the vectors $\vec{A} = 3\hat{i} + 2\hat{j}$, $\vec{B} = -\hat{i} + 6\hat{j}$, and $\vec{C} = 7\hat{i} - 2\hat{j}$.
50. A force $\vec{F} = 67\hat{i} + 23\hat{j} + 55\hat{k}$ N is applied to a body as it moves in a straight line from $\vec{r}_1 = 16\hat{i} + 31\hat{j}$ to $\vec{r}_2 = 21\hat{i} + 10\hat{j} + 14\hat{k}$ m. How much work is done by the force?

51. A force F acts in the x direction, its magnitude given by $F = ax^2$, where x is in meters and a is exactly 5 N/m². (a) Find an exact value for the work done by this force as it acts on a particle moving from $x = 0$ to $x = 6$ m. Now find approximate values for the work by dividing the area under the force curve into rectangles of width (b) $\Delta x = 2$ m, (c) $\Delta x = 1$ m, and (d) $\Delta x = \frac{1}{2}$ m with height equal to the magnitude of the force in the center of the interval. Calculate the percent error in each case.
52. A certain amount of work is required to stretch spring A a certain distance. Twice as much work is required to stretch spring B half that distance. Compare the spring constants of the two springs.
53. A force given by $F = a\sqrt{x}$ acts in the x direction, where $a = 9.5$ N/m$^{1/2}$. Calculate the work done by this force acting on an object as it moves (a) from $x = 0$ to $x = 3$ m; (b) from 3 m to 6 m; and (c) from 6 m to 9 m.
54. The force exerted by a rubber band is given approximately by

$$F = F_0 \left[\frac{L_0 + x}{L_0} - \frac{L_0^2}{(L_0 + x)^2} \right]$$

where L_0 is the unstretched length, x is the stretch, and F_0 is a constant (although F_0 varies with temperature). Find the work needed to stretch the rubber band a distance x.
55. You put your little sister (mass m) on a swing whose chains have length L and pull slowly back until the swing makes an angle ϕ with the vertical. Show that the work you do is $mgL(1 - \cos\phi)$.
56. Two unknown elementary particles pass through a detection chamber. If they have the same kinetic energy and their mass ratio is 4:1, what is the ratio of their speeds?
57. A tractor tows a jumbo jet from its airport gate, doing 8.7 MJ of work. The link from the plane to the tractor makes a 22° angle with the direction of the plane's motion, and the tension in the link is 4.1×10^5 N. How far does the tractor move the plane?
58. A force pointing in the x direction is given by $F = F_0(x/x_0)$, where F_0 and x_0 are constants and x is the position. Find an expression for the work done by this force as it acts on an object moving from $x = 0$ to $x = x_0$.
59. A force pointing in the x direction is given by $F = ax^{3/2}$, where $a = 0.75$ N/m$^{3/2}$. Find the work done by this force as it acts on an object moving from $x = 0$ to $x = 14$ m.
60. Two vectors have equal magnitude, and their scalar product is one-third the square of their magnitude. Find the angle between them.
61. At what rate can a half-horsepower well pump deliver water to a tank 60 m above the water level in the well? Give your answer in kg/s and gal/min.
62. The rate at which the United States imports oil, expressed in terms of the energy content of the imported oil, is about 800 GW. Using the "Energy Content of Fuels" table in Appendix C, convert this figure to gallons per day.
63. By measuring oxygen uptake, sports physiologists have found that the power output of long-distance runners is given approximately by $P = m(bv - c)$, where m and v are the runner's mass and speed, respectively, and b and c are constants given by $b = 4.27$ J/kg·m and $c = 1.83$ W/kg. Determine the work done by a 54-kg runner who runs a 10-km race at a speed of 5.2 m/s.
64. A 1750-kg car delivers energy to its drive wheels at the rate of 35 kW. Neglecting air resistance, what is the greatest speed with which it can climb a 4.5° slope?
65. A 1400-kg car ascends a mountain road at a steady 60 km/h. The force of air resistance on the car is 450 N. If the car's engine supplies energy to the drive wheels at the rate of 38 kW, what is the slope angle of the road?

66. You have to do 2.2 kJ of work to push a 78-kg trunk 3.1 m along a slope inclined upward at 22°, pushing parallel to the slope. What is the coefficient of friction between trunk and slope?

67. (a) How much power is needed to push a 95-kg chest at 0.62 m/s along a horizontal floor where the coefficient of friction is 0.78? (b) How much work is done in pushing the chest 11 m?

68. You mix flour into a thick bread dough, exerting a 45-N force on the stirring spoon. If you move the spoon at 0.29 m/s, (a) what power do you supply? (b) How much work do you do if you stir for 1.0 min?

69. A machine does work at a rate given by $P = ct^2$, where $c = 18$ W/s^2 and t is time. Find the work done by the machine between $t = 10$ s and $t = 20$ s.

70. You're trying to decide whether to buy an energy-efficient, 225-W refrigerator for $1150 or a standard, 425-W model for $850. The standard model will run 20% of the time, but better insulation means the energy-efficient model will run 11% of the time. If electricity costs 9.5¢/kW·h, how long would you have to own the energy-efficient model to make up the difference in cost? Neglect interest you might earn on your money.

71. A machine delivers power at a decreasing rate $P = P_0 t_0^2/(t + t_0)^2$, where P_0 and t_0 are constants. The machine starts at $t = 0$ and runs forever. Show that it nevertheless does only a finite amount of work, equal to $P_0 t_0$.

72. A locomotive accelerates a freight train of total mass M from rest, applying constant power P. Determine the speed and position of the train as functions of time, assuming all the power goes to increasing the train's kinetic energy.

73. A force given by $F = b/\sqrt{x}$ acts in the x direction, where b is a constant with the units N·m$^{1/2}$. Show that even though the force becomes arbitrarily large as x approaches zero, the work done in moving from x_1 to x_2 remains finite even as x_1 approaches zero. Find an expression for that work in the limit $x_1 \to 0$.

74. Figure 6.17 shows the power a baseball bat delivers to the ball as a function of time. Use graphical integration to determine the total work the bat does on the ball.

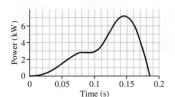

FIGURE 6.17 Problem 74

75. An unusual spring has a force-distance curve described by $F = 100x^2$ for $0 \le x \le 1$ and $F = 100(4x - x^2 - 2)$ for $1 \le x \le 2$, where x is the displacement in meters from the spring's unstretched length and F is in newtons. Find the work done in stretching this spring (a) from $x = 0$ to $x = 1$ m and (b) from $x = 1$ m to $x = 2$ m.

76. You push an object of mass m slowly, partway up a loop-the-loop track of radius R, starting from the bottom, where the normal force to the track is vertically upward, and ending at a point a height $h < R$ above the bottom. The coefficient of friction between the object and the track is a constant μ. Show that the work you do against friction is $\mu mg(2hR - h^2)^{1/2}$. *Hint:* In Problem 55, the work you did against gravity was mgh.

77. A particle moves from the origin to the point $x = 3$ m, $y = 6$ m along the curve $y = ax^2 - bx$, where $a = 2$ m^{-1} and $b = 4$. It is subject to a force $\vec{F} = cxy\hat{\imath} + d\hat{\jmath}$, where $c = 10$ N/m^2 and $d = 15$ N. Calculate the work done by the force.

78. Repeat Problem 77 for the following cases: (a) the particle moves first along the x axis from the origin to the point $(3$ m, $0)$ and then parallel to the y axis until it reaches $(3$ m, 6 m$)$; (b) it moves first along the y axis from the origin to the point $(0, 6$ m$)$ and then parallel to the x axis until it reaches $(3$ m, 6 m$)$.

79. You are investigating the efficiency of airport ground equipment. A tractor that pulls a 747 jet aircraft exerts a force of 4.1×10^5 N at an angle of 22° to the aircraft's direction. The tractor pulls the jet 22.9 m. The tractor should not expend more than 7 MJ of energy if it is to be considered efficient. What do you say in your report?

80. Your friend lifts a 45-kg barbell from the ground to a height of 2.5 m. Your friend claims doing that has "burned" the amount of food calories equivalent to a 43-g (1.55-oz) chocolate bar—namely, 230 food calories (1 food calorie = 4.186 kJ). How many lifts would you need to do to use enough energy to burn off the chocolate bar?

81. Your roommate is studying law and is the plaintiff's attorney in a mock trial. A hospital patient's leg slipped off a stretcher and his heel hit the floor. The defense attorney for the hospital claims the leg, with mass 8 kg, hit the floor with a force equal to the weight of the leg—namely, about 80 N—and the resulting damage was due to a prior injury to the heel. Your roommate's argument will claim the heel hit the floor with a much larger force. Make these assumptions: The leg and heel dropped freely a distance of 0.7 m, and the stopping distance was 2 cm. What data do you give your roommate to convince the jury?

Answers to Chapter Questions

Answer to Chapter Opening Question

No. The work done against gravity in climbing a particular height is independent of the path. A rider on a bicycle with a combined mass of 80 kg does roughly 400 kilojoules or 100 kilocalories of work against gravity regardless of the path up a 500-m mountain. To climb such a mountain in 20 minutes, the rider's power output must exceed 300 watts.

Answers to GOT IT? Questions

6.1 The force $2F$ does $\sqrt{2}$ more work than F does. That's because $2F$'s component along the direction of motion is $2F \cos 45°$, or $2F\sqrt{2}/2 = F\sqrt{2}$.

6.2 (c) $\sqrt{x}$ does the most work. (b) x^2 does the least. You can see this by plotting these two functions from $x = 0$ to $x = 1$ and comparing the areas under each. The case of x is intermediate.

6.3 (a) Kinetic energy doesn't change, so the net work done on the ball is zero. (b) Kinetic energy increases, so the net work is positive. (c) Kinetic energy decreases, so the net work is negative.

6.4 The megawatt is a unit of power; the "per time" is already built in. A correct statement would be that the power plant will produce "energy at the rate of 50 megawatts."

7

Conservation of Energy

■ How many different energy conversions take place as the Yellowstone River plunges over Yellowstone Falls?

The rock climber of Fig. 7.1a does work as she ascends the vertical cliff. So does the mover of Fig. 7.1b, as he pushes a heavy chest across the floor. But there's a difference. If the rock climber lets go, down she goes; the work she put into the climb comes back as the kinetic energy of her fall. If the mover lets go of the chest, though, he and the chest stay right where they are.

This contrast highlights a distinction between two types of forces, called *conservative* and *nonconservative*. From that distinction we'll develop one of the most important principles in all of physics: **conservation of energy**. In this chapter, we consider conservation of **mechanical energy**, which includes kinetic energy, introduced in Chapter 6, and **potential energy**, which we'll introduce here. In later chapters, we'll expand the conservation-of-energy principle to include other forms of energy.

7.1 Conservative and Nonconservative Forces

Both the climber and the mover in Fig. 7.1 are working against external forces—gravity for the climber and friction for the mover. The difference is this: If the climber lets go, the gravitational force "gives back" the work that she did; that work manifests itself as kinetic energy. But the frictional force doesn't "give back" the work that the mover did; that work can't be recovered as kinetic energy.

A **conservative force** is a force like gravity or a spring that "gives back" work done against it. A force like friction is **nonconservative**. A more precise mathematical sense of this distinction comes from considering the work involved in moving an object over a closed path—that is, a path that ends up at the same point where it started. Suppose our rock climber ascends a cliff of height h and then descends to her starting point. How much work has the gravitational force done on her? That force has magnitude mg and points downward. As she climbs up, the force is directed opposite to her motion, so the work done by gravity is $-mgh$. As she descends, the force is in the same direction as her motion, so the work done by gravity is $+mgh$. The total work that gravity does on the climber as she traverses the closed path from the bottom to the top of the cliff and back again is therefore zero.

Now suppose the mover in Fig. 7.1b pushes the chest a distance L across a room, discovers it's the wrong room, and pushes the chest back to the door. Like the climber, the chest describes a closed path. How much work is done by the frictional force acting over this path? That force has magnitude μn, where the normal force in this case of horizontal motion is mg. But the force of kinetic friction *always* acts to oppose the motion, so it *always* does negative work. With frictional force μmg opposing the motion, the work done as the chest crosses the width L of the room is $-\mu mgL$. As the chest comes back, the work is also $-\mu mgL$, so the total work done by the frictional force is $-2\mu mgL$.

The difference between our answers for the work done by the gravitational and frictional forces acting over closed paths provides one precise definition of the distinction between conservative and nonconservative forces:

> When the total work done by a force F acting as an object moves over any closed path is zero, the force is conservative. Mathematically,
>
> $$\oint \vec{F} \cdot d\vec{r} = 0 \qquad \text{(conservative force)} \qquad (7.1)$$

This expression comes from the most general formula for work, Equation 6.11, $W = \int_{\vec{r}_1}^{\vec{r}_2} \vec{F} \cdot d\vec{r}$, which we introduced in Chapter 6. The circle on the integral sign in Equation 7.1 indicates that the integral is to be taken over a *closed* path. A force for which the integral is nonzero, like friction, is nonconservative.

Equation 7.1 suggests a related property of conservative forces. Suppose we move an object along the straight path between points A and B shown in Fig. 7.2, along which a conservative force acts; let the work done by the conservative force be W_{AB}. Since the work done over any closed path is zero, the work W_{BA} done in moving back from B to A must be $-W_{AB}$, whether we return along the straight path or the curved path or any other path. So, going from A to B involves work W_{AB}, regardless of the path taken. In other words:

> The work done by a conservative force in moving between two points is independent of the path taken; mathematically, $\int_A^B \vec{F} \cdot d\vec{r}$ depends only on the endpoints A and B, not on the path between them.

In contrast, the work done by a nonconservative force does depend on the path. On a frictional surface, for example, the least work is done over a straight-line path; any other path involves more work.

Important examples of conservative forces include gravity and the static electric force. The force of an ideal spring—fundamentally an electric force—is also conservative. Nonconservative forces include friction and the electric force in the presence of time-varying magnetic effects, which we'll encounter in Chapter 27.

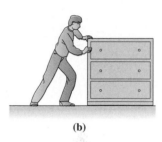

(a)

(b)

FIGURE 7.1 Both the rock climber and the mover do work, but only the climber can recover that work as kinetic energy.

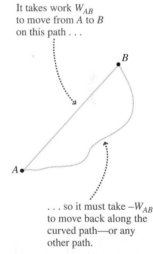

It takes work W_{AB} to move from A to B on this path . . .

B

A

. . . so it must take $-W_{AB}$ to move back along the curved path—or any other path.

FIGURE 7.2 The work done by a conservative force is independent of path.

GOT IT? 7.1   Suppose it takes the same amount of work to push a trunk across a rough floor as it does to lift a weight the same distance straight upward. How do the amounts of work compare if the trunk and weight are moved instead on curved paths between the same starting and ending points?

7.2 Potential Energy

Work done against a conservative force is somehow "stored," in the sense that we can get it back again in the form of kinetic energy. The climber in Fig. 7.1a is acutely aware of that "stored work"; it gives her the potential for a dangerous fall. *Potential* is an appropriate word here: We can consider the "stored work" as **potential energy** U, in the sense that it can be released as kinetic energy.

We define potential energy formally in terms of the work done by a conservative force. Specifically:

> The change ΔU_{AB} in potential energy associated with a conservative force is the negative of the work done by that force as it acts over any path from point A to point B:
>
> $$\Delta U_{AB} = -\int_A^B \vec{F} \cdot d\vec{r} \qquad \text{(potential energy)} \qquad (7.2)$$

Why the minus sign? Because potential energy represents stored work. If a conservative force does *positive* work (as does gravity on a falling object), then potential energy must decrease—and that means ΔU must be *negative*. Conversely, if a conservative force does *negative* work (as does gravity on a weight being lifted), then energy is stored, and ΔU must be *positive*. The minus sign in Equation 7.2 handles both these cases. We'll often drop the subscript AB and write simply ΔU for the potential-energy change. Keeping the subscript is important, though, when we need to be clear about whether we're going from A to B or from B to A.

Changes in potential energy are all that ever matter physically; the actual value of potential energy is meaningless. Often, though, it's convenient to establish a reference point at which the potential energy is defined to be zero. When we say "the potential energy U," we really mean the potential-energy difference ΔU between that reference point and whatever other point we're considering. Our rock climber, for example, might find it convenient to take the zero of potential energy at the base of the cliff. But the choice is purely for convenience; only potential-energy *differences* really matter.

Equation 7.2 is a completely general definition of potential energy, applicable in all circumstances. Often, though, we can consider a path where force and displacement are parallel (or antiparallel). Then Equation 7.2 simplifies to

$$\Delta U = -\int_{x_1}^{x_2} F(x)\, dx \qquad (7.2a)$$

where x_1 and x_2 are the starting and ending points on the x axis, taken to coincide with the path. When the force is constant, this equation simplifies further to

$$\Delta U = -F(x_2 - x_1) \qquad (7.2b)$$

✓**TIP** Understand Your Equations

Equation 7.2b provides a very simple expression for potential-energy changes, but it applies *only* when the force is constant. Equation 7.2b is a special case of Equation 7.2a that follows because a constant force can be taken outside the integral.

Gravitational Potential Energy

We're frequently moving things up and down, causing changes in potential energy. Figure 7.3 shows two possible paths for a book that's lifted from the floor to a shelf of height h. Since the gravitational force is conservative, we can use either path to calculate

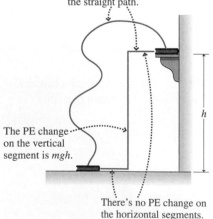

The potential energy (PE) change is the same along either path, but it's calculated more easily for the straight path.

The PE change on the vertical segment is mgh.

There's no PE change on the horizontal segments.

FIGURE 7.3 A good choice of path makes it easier to calculate the potential-energy change.

the potential-energy change. It's easiest to use the path consisting of straight segments. No work or potential-energy change is associated with the horizontal motion, since the gravitational force is perpendicular to the motion. For the vertical lift, the force of gravity is constant and Equation 7.2b gives immediately $\Delta U = mgh$, where the minus sign in Equation 7.2b cancels with the minus sign associated with the *downward* direction of gravity. This result is quite general: When a mass m undergoes a vertical displacement Δy near Earth's surface, its gravitational potential energy changes by

$$\Delta U = mg\,\Delta y \qquad \text{(gravitational potential energy)} \qquad (7.3)$$

The quantity Δy can be positive or negative, depending on whether the object moves up or down; correspondingly, the potential energy can either increase or decrease. We emphasize that Equation 7.3 applies *near Earth's surface*—that is, for distances small compared with Earth's radius. That assumption allows us to treat the gravitational force as constant over the path. We'll explore the more general case in the next chapter.

We've found the *change* in the book's potential energy, but what about the potential energy itself? That depends on where we define the zero of potential energy. If we choose $U = 0$ at the floor, then $U = mgh$ on the shelf. But we could just as well take $U = 0$ at the shelf; then the book's potential energy on the floor would be $-mgh$. Negative potential energies arise frequently, and that's OK because only *differences* in potential energy really matter. Figure 7.4 shows a plot of potential energy versus height with $U = 0$ taken at the floor. The *linear* increase in potential energy with height reflects the *constant* gravitational force.

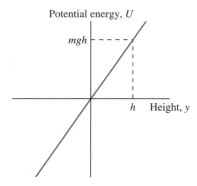

FIGURE 7.4 Gravitational force is constant, so potential energy increases linearly with height.

EXAMPLE 7.1 **Gravitational Potential Energy: Riding the Elevator**

A 55-kg engineer leaves her office on the 33rd floor of a skyscraper and takes an elevator up to the 59th floor. Later she descends to street level. If the engineer takes her office as the zero of potential energy and if the distance from one floor to the next is 3.5 m, what is the engineer's potential energy (a) in her office, (b) on the 59th floor, and (c) at street level?

INTERPRET This is a problem about gravitational potential energy relative to a specified point of zero energy—namely, the engineer's office.

DEVELOP Equation 7.3, $\Delta U = mg\,\Delta y$, gives the change in gravitational energy associated with a change Δy in vertical position. We're given positions in floors, not meters, so we need to convert using the given factor 3.5 m per floor.

EVALUATE (a) In her office, the engineer's potential energy is zero, since she defined it that way. (b) The 59th floor is $59 - 33 = 26$ floors

higher, so the potential energy there is

$$U_{59} = mg\,\Delta y = (55\ \text{kg})(9.8\ \text{m/s}^2)(26\ \text{floors})(3.5\ \text{m/floor})$$
$$= 49\ \text{kJ}.$$

Note that we can write U rather than ΔU because we're calculating the potential-energy change from the place where $U = 0$. (c) The street level is 32 floors *below* the engineer's office, so

$$U_{\text{street}} = mg\,\Delta y = (55\ \text{kg})(9.8\ \text{m/s}^2)(-32\ \text{floors})(3.5\ \text{m/floor})$$
$$= -60\ \text{kJ}.$$

ASSESS Makes sense: When the engineer goes *up*, the potential energy relative to her office is positive; when she goes *down*, it's negative. And the distance down is a bit farther, so the magnitude of the change is greater going down. ∎

APPLICATION **Pumped Storage**

Electricity is a wonderfully versatile form of energy, but it's not easy to store. Large electric power plants are most efficient when operated continuously, yet the demand for power fluctuates. Renewable energy sources like wind and solar vary, not necessarily with demand. Energy storage can help in both cases. Today, the only practical way to store large amounts of excess electrical energy is to convert it to gravitational potential energy. In so-called pumped-storage facilities, surplus electric power pumps water from a lower reservoir to a higher one, thereby increasing the water's gravitational potential energy. When power demand is high, water runs back down, turning the pump motors into generators that produce electricity. The photograph here shows the Northfield Mountain Pumped Storage Project in Massachusetts. You can explore this facility quantitatively in Problem 29.

Elastic Potential Energy

When you stretch or compress a spring or other elastic object, you do work against the spring force, and that work ends up stored as **elastic potential energy**. For an ideal spring, the force is $F = -kx$, where x is the distance the spring is stretched from equilibrium, and the minus sign shows that the force opposes the stretching or compression. Since the force varies with position, we use Equation 7.2a to evaluate the potential energy:

$$\Delta U = -\int_{x_1}^{x_2} F(x)\, dx = -\int_{x_1}^{x_2} (-kx)\, dx = \tfrac{1}{2}kx_2^2 - \tfrac{1}{2}kx_1^2$$

where x_1 and x_2 are the initial and final values of the stretch. If we take $U = 0$ when $x = 0$ —that is, when the spring is neither stretched nor compressed—then we can use this result to write the potential energy at an arbitrary stretch (or compression) x as

$$U = \tfrac{1}{2}kx^2 \qquad \text{(elastic potential energy)} \qquad (7.4)$$

Comparison with Equation 6.10, $W = \tfrac{1}{2}kx^2$, shows that this is equal to the work done in stretching the spring. Of course: That work gets stored as potential energy. Figure 7.5 shows potential energy as a function of the stretch or compression of a spring. The *parabolic* shape of the potential-energy curve reflects the *linear* change of the spring force with stretch or compression.

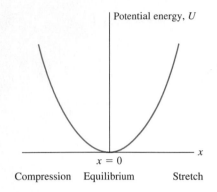

FIGURE 7.5 The potential-energy curve for a spring is a parabola.

EXAMPLE 7.2 **Energy Storage: Springs Versus Gasoline**

A car's suspension consists of springs with an overall effective spring constant of 120 kN/m. How much would you have to compress the springs to store the same amount of energy as in 1 gram of gasoline?

INTERPRET This problem is about the energy stored in a spring, as compared with the chemical energy of gasoline.

DEVELOP Equation 7.4, $U = \tfrac{1}{2}kx^2$, gives a spring's stored energy when it's been compressed a distance x. Here we want that energy to equal the energy in 1 gram of gasoline. We can get that value from the "Energy Content of Fuels" table in Appendix C, which lists 44 MJ/kg for gasoline.

EVALUATE At 44 MJ/kg, the energy in 1 g of gasoline is 44 kJ. Setting this equal to the spring energy $\tfrac{1}{2}kx^2$ and solving for x, we get

$$x = \sqrt{\frac{2U}{k}} = \sqrt{\frac{(2)(44\text{ kJ})}{120\text{ kN/m}}} = 86\text{ cm}$$

ASSESS This answer is absurd. A car's springs couldn't compress anywhere near that far before the underside of the car hit the ground. And 1 g isn't much gasoline. This example shows that springs, though useful energy-storage devices, can't possibly compete with chemical fuels.

∎

EXAMPLE 7.3 **Elastic Potential Energy: A Climbing Rope**

Ropes used in rock climbing are "springy" so that they cushion a fall. A particular rope exerts a force given by $F = -kx + bx^2$, where $k = 223$ N/m, $b = 4.10$ N/m², and x is the stretch. Find the potential energy stored in this rope when it's been stretched 2.62 m, taking $U = 0$ at $x = 0$.

INTERPRET Like Example 7.2, this one is about elastic potential energy. But this one is not so easy because the rope isn't a simple $F = -kx$ spring for which we already have a potential-energy formula.

DEVELOP Because the rope force varies with stretch, we're going to have to integrate. Since force and displacement are in the same direction, we can use Equation 7.2a, $\Delta U = -\int_{x_1}^{x_2} F(x)\, dx$. But that's not so much a formula as a strategy for deriving one.

EVALUATE Applying Equation 7.2 to this particular rope, we have

$$U = -\int_{x_1}^{x_2} F(x)\, dx = -\int_0^x (-kx + bx^2)\, dx = \tfrac{1}{2}kx^2 - \tfrac{1}{3}bx^3 \Big|_0^x$$

$$= \tfrac{1}{2}kx^2 - \tfrac{1}{3}bx^3$$

$$= \left(\tfrac{1}{2}\right)(223\text{ N/m})(2.62\text{ m})^2 - \left(\tfrac{1}{3}\right)(4.1\text{ N/m}^2)(2.62\text{ m})^3$$

$$= 741\text{ J}$$

ASSESS This result is about 3% less than the potential energy $U = \tfrac{1}{2}kx^2$ of an ideal spring with the same spring constant. This shows the effect of the extra term $+bx^2$, whose positive sign reduces the restoring force and thus the work needed to stretch the spring.

∎

GOT IT? 7.2 Gravitational force actually decreases with height, but that decrease is negligible near Earth's surface. To account for the decrease, would the exact value for the potential-energy change associated with a height change h be (a) greater than, (b) less than, or (c) equal to mgh, where g is the gravitational acceleration at Earth's surface?

7.3 Conservation of Mechanical Energy

The work-energy theorem, introduced in Chapter 6, shows that the change ΔK in a body's kinetic energy is equal to the net work done on it:

$$\Delta K = W_{net}$$

Consider separately the work W_c done by conservative forces and the work W_{nc} done by nonconservative forces. Then

$$\Delta K = W_c + W_{nc}$$

We've defined the change in potential energy ΔU as the negative of the work done by conservative forces. So we can write

$$\Delta K = -\Delta U + W_{nc}$$

or

$$\Delta K + \Delta U = W_{nc} \qquad (7.5)$$

We define the sum of the kinetic and potential energy as the **mechanical energy**. Then Equation 7.5 shows that the change in mechanical energy is equal to the work done by non-conservative forces.

In the absence of nonconservative forces, the mechanical energy is unchanged:

$$\Delta K + \Delta U = 0 \qquad (7.6)$$

and, equivalently $\left(\begin{array}{c} \text{conservation of} \\ \text{mechanical energy} \end{array} \right)$

$$K + U = \text{constant} = K_0 + U_0 \qquad (7.7)$$

where K_0 and U_0 are the kinetic and potential energy of a body at some point, and K and U are their values when the body is at any other point. Equations 7.6 and 7.7 express the **law of conservation of mechanical energy**. They show that, in the absence of nonconservative forces, the mechanical energy $K+U$ remains always the same. The kinetic energy K may change, but that change is always compensated by an equal but opposite change in potential energy.

Conservation of mechanical energy is a powerful principle. Throughout physics, from the subatomic realm through practical problems in engineering and on to astrophysics, the principle of energy conservation is widely used in solving problems that would be intractable without it.

PROBLEM SOLVING STRATEGY 7.1   **Conservation of Energy**

Using energy conservation to solve problems is straightforward: Equation 7.7 basically tells it all. Our IDEA problem-solving strategy adapts well to such problems.

INTERPRET First, interpret the problem to be sure that conservation of mechanical energy applies. Are all the forces conservative? If so, mechanical energy is conserved. Next, identify a point at which you know both the kinetic and the potential energy; then you know the total mechanical energy, which is what's conserved. If the problem doesn't do so and it's not implicit in the equations you use, you may need to identify the zero of potential energy—although that's your own arbitrary choice. You also need to identify the quantity the problem is asking for, and the situation in which it has the value you're after. The quantity may be the energy itself or a related quantity like height, speed, or spring compression.

DEVELOP Draw your object first in the situation where you know its energies and then in the situation that contains the unknown. It's helpful to draw simple bar charts suggesting the relative sizes of the potential- and kinetic-energy terms; we'll show you how in several examples. Then you're ready to set up the quantitative statement of mechanical energy conservation, Equation 7.7: $K + U = K_0 + U_0$. Consider which of the four terms you know or can calculate from the given information. You'll probably need secondary equations like the expressions for kinetic energy and for various forms of potential energy. Consider how the quantity you're trying to find is related to an energy.

EVALUATE Write Equation 7.7 for your specific problem, including expressions for kinetic or potential energy that contain the quantity you're after. Solving is then a matter of algebra.

ASSESS As usual, ask whether your answer makes physical sense. Does it have the right units? Are the numbers reasonable? Do the signs make sense? Is your answer consistent with the bar charts in your drawing?

EXAMPLE 7.4 **Energy Conservation: Tranquilizing an Elephant**

A biologist uses a spring-loaded gun to shoot tranquilizer darts into an elephant. The gun's spring has $k = 940$ N/m and is compressed a distance $x_0 = 25$ cm before firing a 38-g dart. Assuming the gun is pointed horizontally, at what speed does the dart leave the gun?

INTERPRET We're dealing with a spring, assumed ideal, so conservation of mechanical energy applies. We identify the initial state—dart at rest, spring fully compressed—as the point where we know both kinetic and potential energy. The state we're then interested in is when the dart just leaves the gun, when its potential energy has been converted to kinetic energy and before gravity has changed its vertical position.

DEVELOP In Fig. 7.6 we've sketched the two states, giving the potential and kinetic energy for each. We've also sketched bar graphs showing the relative sizes of the energies. To use the statement of energy conservation, Equation 7.7, we also need expressions for the kinetic energy $\left(\frac{1}{2}mv^2\right)$ and the spring potential energy $\left(\frac{1}{2}kx^2;$ Equation 7.4). Incidentally, using Equation 7.4 implicitly sets the zero of elastic potential energy when the spring is in its equilibrium position. We might as well set the zero of gravitational energy at the height of the gun, since there's no change in the dart's vertical position between our initial and final states.

EVALUATE We're now ready to write Equation 7.7, $K + U = K_0 + U_0$. We know three of the terms in this equation: The initial kinetic energy K_0 is 0, since the dart is initially at rest. The initial potential energy is that of the compressed spring, $U_0 = \frac{1}{2}kx_0^2$. The final potential energy is $U = 0$ because the spring is now in its equilibrium position and we've taken the gravitational potential energy to be zero. What we don't know is the final kinetic energy, but we do know that it's given by $K = \frac{1}{2}mv^2$. So Equation 7.7 becomes $\frac{1}{2}mv^2 + 0 = 0 + \frac{1}{2}kx^2$, which

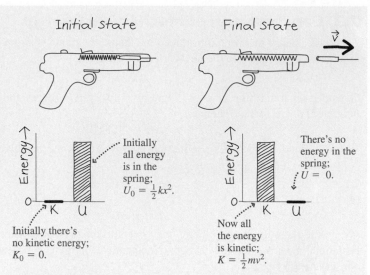

FIGURE 7.6 Our sketches for Example 7.4, showing bar charts for the initial and final states.

solves to give

$$v = \sqrt{\frac{k}{m}}\,x_0 = \left(\sqrt{\frac{940 \text{ N/m}}{0.038 \text{ kg}}}\right)(0.25 \text{ m}) = 39 \text{ m/s}$$

ASSESS Take a look at the answer in algebraic form; it says that a stiffer spring or a greater compression will give a higher dart speed. Increasing the dart mass, on the other hand, will decrease the speed. All this makes good physical sense. And the outcome shows quantitatively what our bar charts suggest—that the dart's energy starts out all potential and ends up all kinetic. ∎

Example 7.4 shows the power of the conservation-of-energy principle. If you had tried to find the answer using Newton's law, you would have been stymied by the fact that the spring force and thus the acceleration of the dart vary continuously. But you don't need to worry about those details; all you want is the final speed, and energy conservation gets you there, shortcutting the detailed application of $\vec{F} = m\vec{a}$.

EXAMPLE 7.5 **Conservation of Energy: A Spring and Gravity**

The spring in Fig. 7.7 has $k = 140$ N/m. A 50-g block is placed against the spring, which is compressed 11 cm. When the block is released, how high up the slope does it rise? Neglect friction.

INTERPRET This example is similar to Example 7.4, but now we have changes in both elastic and gravitational potential energy. Since friction is negligible, we can apply conservation of energy. We identify the initial state as the block at rest against the compressed spring; the final state is the block momentarily at rest at its topmost point on the slope. We might as well take the zero of gravitational potential energy at the bottom.

DEVELOP Figure 7.7 shows the initial and final states, along with bar charts for each. We've drawn separate bars for the spring and gravitational potential energies, U_s and U_g, since both change in different ways between the initial and final states. Now apply Equation 7.7, $K + U = K_0 + U_0$.

EVALUATE In both states the block is at rest, so kinetic energy is zero. In the initial state we know the potential energy U_0: It's the spring energy $\frac{1}{2}kx^2$. We don't know the final-state potential energy, but we do know that it's gravitational energy—and with the zero of potential energy at the bottom, it's $U = mgh$. With $K = K_0 = 0$, $U_0 = \frac{1}{2}kx_0^2$, and

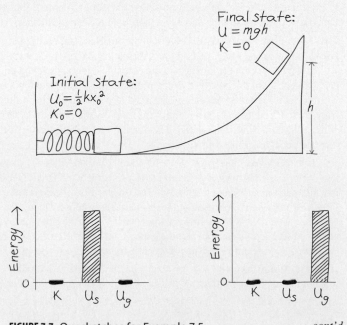

FIGURE 7.7 Our sketches for Example 7.5.

cont'd.

$U = mgh$, Equation 7.7 reads $0 + mgh = 0 + \frac{1}{2}kx^2$. We then solve for the unknown h to get

$$h = \frac{kx^2}{2mg} = \frac{(140 \text{ N/m})(0.11 \text{ m})^2}{(2)(0.050 \text{ kg})(9.8 \text{ m/s}^2)} = 1.7 \text{ m}$$

ASSESS Again, the answer in algebraic form makes sense; the stiffer the spring or the more it's compressed, the higher the block will go. But if the block is more massive or gravity is stronger, then the block won't get as far.

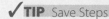 **TIP** Save Steps

You might be tempted to solve first for the block's speed when it leaves the spring and then equate $\frac{1}{2}mv^2$ to mgh to find the height. But conservation of energy shortcuts all the details, getting you right from the initial to the final state. As long as energy is conserved, you don't need to worry about what happens in between.

GOT IT? 7.3 A bowling ball is tied to the end of a long rope and suspended from the ceiling. A student stands at one side of the room and holds the ball to her nose, then releases it from rest. Should she duck as it swings back?

Nonconservative Forces

In these examples we've assumed that energy is strictly conserved. In the everyday world of friction and other nonconservative forces, conservation of energy is sometimes a good approximation and sometimes not. When it's not, we can still apply our strategy, but now Equation 7.5 shows that we need to subtract any energy lost to nonconservative forces.

EXAMPLE 7.6 Nonconservative Forces: A Sliding Block

A block of mass m is launched from a spring of constant k that's initially compressed a distance x_0. After leaving the spring, the block slides on a horizontal surface with frictional coefficient μ. Find an expression for the distance the block slides before coming to rest.

INTERPRET The presence of friction means that mechanical energy isn't conserved. But we can still identify the kinetic and potential energy in the initial state: The kinetic energy is zero and the potential energy is that of the spring. In the final state, there's no mechanical energy at all. The nonconservative frictional force does negative work on the block, reducing its total energy. The block comes to rest when all its initial energy is gone.

DEVELOP Figure 7.8 shows the situation. With $K_0 = 0$, we determine the total initial energy from Equation 7.4, $U_0 = \frac{1}{2}kx_0^2$. The work W_f done by friction follows from Equation 6.1, $W = F_x \Delta x$. Here the frictional force has magnitude $f_f = \mu n = \mu mg$ and so, with its direction opposite the displacement, the frictional work is negative: $W_f = -\mu mg \Delta x$. This is the work W_{nc} in Equation 7.5, $\Delta K + \Delta U = W_{nc}$. The initial and final

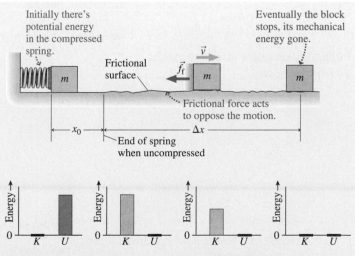

FIGURE 7.8 Intermediate bar charts show gradual loss of mechanical energy.

cont'd.

states here have no kinetic energy, so $\Delta K = 0$. Then the block will have lost all its initial energy when $\Delta U = -U_0$. Therefore Equation 7.5 becomes $-\frac{1}{2}kx_0^2 = -\mu mg\,\Delta x$.

EVALUATE We solve this equation for the unknown distance Δx: $\Delta x = kx_0^2/2\mu mg$. Since we weren't given numbers, there's nothing further to evaluate.

ASSESS Make sense? The stiffer the spring or the more it's compressed, the farther the block goes. The greater the friction or the normal force mg, the sooner the block stops. If $\mu = 0$, mechanical energy is once again conserved; then our result shows that the block would slide forever.

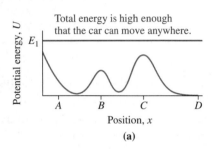

FIGURE 7.9 A roller-coaster track.

7.4 Potential-Energy Curves

Figure 7.9 shows a frictionless roller-coaster track. How fast must a car be coasting at point A if it's to reach point D? Conservation of energy provides the answer. To get to D, the car must clear peak C. Clearing C requires that the car's total energy exceed its potential energy at C; that is, $\frac{1}{2}mv_A^2 + mgh_A > mgh_C$, where we've taken the zero of potential energy at the bottom of the track. Solving for v_A gives $v_A > \sqrt{2g(h_C - h_A)}$. If v_A satisfies this inequality, the car will reach C with some kinetic energy remaining and will coast over the peak.

What if the car is moving a little more slowly at A? Then it will stop at some point to the left of C where its energy is entirely potential—that is, at a height h such that $\frac{1}{2}mv_A^2 + mgh_A = mgh$. Then it will head back, clearing peak B and again coming to a momentary stop, to the left of A, at height h. So it will run back and forth between two **turning points** set by the value of its total energy. Lowering the speed still further would confine the car to the leftmost valley.

Figure 7.9 is a drawing of the actual roller-coaster track. But because gravitational potential energy is directly proportional to height, it's also a plot of potential energy versus position: a **potential-energy curve.** We can study the car's motion by plotting its total energy on the same graph as the potential-energy curve. Since total energy is constant, the total-energy plot is a straight line. Figure 7.10 shows the potential energy and total energy for three values of the total energy. In Fig. 7.10a, the total energy exceeds the potential energy at C; therefore the car will reach C with kinetic energy to spare and will make it all the way to D. In Fig. 7.10b, the total energy is less than the potential energy at C, and the motion is confined between two turning points where the total-energy and potential-energy curves intersect. In Fig. 7.10c the total energy is still lower, and now the car is confined to the leftmost valley.

Even though the car in Figs. 7.10b and c cannot get to D, its total energy still exceeds the potential energy at D. But it's blocked from reaching D by the **potential barrier** of peak C. We say that it's **trapped** in a **potential well** between its turning points.

Potential-energy curves are useful even with nongravitational forces where there's no direct correspondence with hills and valleys. The terminology used here—potential barriers, wells, and trapping—remains appropriate in such cases and indeed is widely used throughout physics.

Figure 7.11 shows the potential energy of a pair of hydrogen atoms as a function of their separation. This energy is associated with attractive and repulsive electrical forces involv-

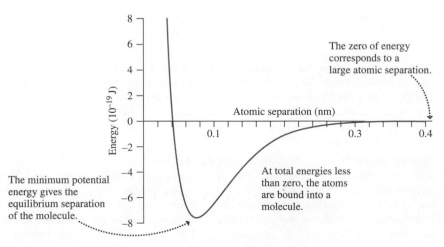

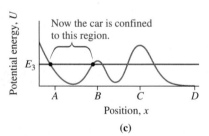

FIGURE 7.10 Potential and total energy for a roller-coaster car.

FIGURE 7.11 Potential-energy curve for two hydrogen atoms.

ing the electrons and the nuclei of the two atoms. The potential-energy curve exhibits a potential well, showing that the atoms can form a **bound system** in which they're unable to separate fully. That bound system is a hydrogen molecule (H_2). The minimum energy, -7.6×10^{-19} J, corresponds to the molecule's equilibrium separation of 0.074 nm. It's convenient to define the zero of potential energy when the atoms are infinitely far apart; Fig. 7.11 then shows that any total energy less than zero results in a bound system. But if the total energy is greater than zero, the atoms are free to move arbitrarily far apart, so they don't form a molecule.

EXAMPLE 7.7 **Molecular Energy: Finding Atomic Separation**

Very near the bottom of the potential well in Fig. 7.11, the potential energy of the two hydrogen atoms is given approximately by $U = U_0 + a(x - x_0)^2$, where $U_0 = -0.760$ aJ, $a = 286$ aJ/nm^2, and $x_0 = 0.0741$ nm is the equilibrium separation. What range of atomic separations is allowed if the total energy is -0.717 aJ?

INTERPRET This problem sounds complicated, with strange units and talk of molecular energies. But it's about just what's shown in Figs. 7.10 and 7.11. Specifically, we're given the total energy and asked to find the turning points—the points where the line representing total energy intersects the potential-energy curve. If the units look strange, remember the SI prefixes (there's a table inside the front cover), which we use to avoid writing large powers of 10. Here 1 aJ = 10^{-18} J and 1 nm = 10^{-9} m.

DEVELOP Figure 7.12 is a plot of the potential-energy curve from the function we've been given. The straight line represents the total energy E. The turning points are the values of atomic separation where the two curves intersect. We could read them off the graph, or we can solve algebraically by setting the total energy equal to the potential energy.

EVALUATE With the potential energy given by $U = U_0 + a(x - x_0)^2$ and the total energy E, we have turning points when $E = U_0 + a(x - x_0)^2$. We could solve directly for x, but then we'd have to use the quadratic formula. Solving for $x - x_0$ is easier:

$$x - x_0 = \pm\sqrt{\frac{E - U_0}{a}} = \pm\sqrt{\frac{-0.717\ \text{aJ} - (-0.760\ \text{aJ})}{286\ \text{aJ/nm}^2}}$$

$$= \pm 0.0123\ \text{nm}$$

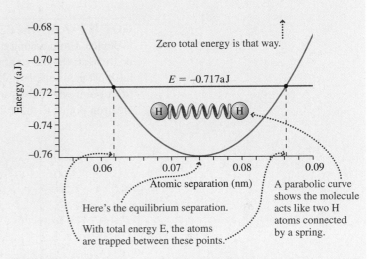

FIGURE 7.12 Analyzing the hydrogen molecule.

Then the turning points are at $x_0 \pm 0.0123$ nm—namely, 0.0864 nm and 0.0618 nm.

ASSESS Make sense? A look at Fig. 7.12 shows that we've correctly located the turning points. The fact that its potential-energy curve is parabolic (like a spring's $U = \frac{1}{2}kx^2$) shows that the molecule can be modeled approximately as two atoms joined by a spring. Chemists frequently use such models and even talk of the "spring constant" of the bond joining atoms into a molecule. ∎

Force and Potential Energy

The roller-coaster track in Fig. 7.9 traces the potential-energy curve for a car on the track. But it also shows the force acting to accelerate the car: where the graph is steep—that is, where the potential energy is changing rapidly—the force is greatest. At the peaks and valleys, the force is zero. So it's the *slope* of the potential-energy curve that tells us about the force (Fig. 7.13).

Just how strong is this force? Consider a small change Δx, so small that the force is essentially constant over this distance. Then we can use Equation 7.2b to write $\Delta U = -F_x \Delta x$, or $F_x = -\Delta U/\Delta x$. In the limit $\Delta x \to 0$, $\Delta U/\Delta x$ becomes the derivative, and we have

$$F_x = -\frac{dU}{dx} \tag{7.8}$$

This equation makes mathematical as well as physical sense. We've already written potential energy as the *integral* of force over distance, so it's no surprise that force is the

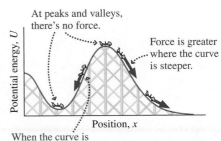

FIGURE 7.13 Force depends on the *slope* of the potential-energy curve.

derivative of potential energy. Equation 7.8 gives the force component in the x direction only. In a three-dimensional situation, we'd have to take derivatives of potential energy with respect to y and z to find the full force vector.

Why the minus sign in Equation 7.8? You can see the answer in the molecular energy curve of Fig. 7.11, where pushing the atoms too close together—moving to the *left* of equilibrium—results in a repulsive force to the *right*, and pulling them apart—moving to the *right*—gives an attractive force to the *left*. You can see the same thing for the roller coaster in Fig. 7.13. In both cases the forces tend to drive the system back toward a minimum-energy state. We'll explore such minimum-energy equilibrium states further in Chapter 12.

GOT IT? 7.4 The figure shows the potential energy for an electron in a microelectronic device. From among the labeled points, find (a) the point where the force on the electron is greatest; (b) the rightmost position possible if the electron has total energy E_1; (c) the leftmost position possible if the electron has total energy E_2 and starts out to the right of D; (d) a point where the force on the electron is zero; and (e) a point where the force on the electron points to the left. In some cases there may be more than one answer.

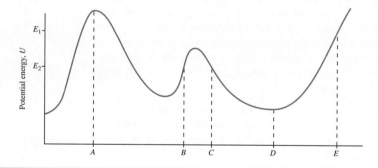

Big Picture

The big idea here is conservation of mechanical energy—the principle that the total energy of a system subject to only conservative forces cannot change. Energy may change from kinetic to potential, and vice versa, but the total remains constant. Applying conservation of energy means understanding the concept of potential energy as stored energy that results when work is done against a conservative force.

A block is against a compressed spring; its energy is all potential.

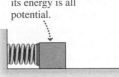

Later, the block is moving. It still has the same energy but now it's kinetic.

A ball is rolling on a slope. At the bottom its energy is all kinetic.

Later, the ball is still moving but more slowly. The sum of its kinetic and potential energy equals its initial energy.

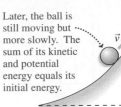

Key Concepts and Equations

The important new concept here is potential energy, defined as the negative of the work done by a conservative force. Only the change ΔU has physical significance. Expressions for potential energy include:

$$\Delta U_{AB} = -\int_A^B \vec{F} \cdot d\vec{r}$$

This one is the most general, but it's mathematically involved. The force can vary over an arbitrary path between points A and B.

$$\Delta U = -\int_{x_1}^{x_2} F(x)\, dx$$

This is a special case, when force and displacement are in a line and force may vary with position.

$$\Delta U = -F(x_2 - x_1)$$

This is the most specialized case, where the force is constant.

Given the concept of potential energy, the principle of conservation of mechanical energy follows from the work-energy theorem of Chapter 6. Here's the mathematical statement of energy conservation:

$$K + U = K_0 + U_0$$

K and U are the kinetic and potential energy at some point where we don't know one of these quantities.

The total energy is conserved, as indicated by the equal sign.

K_0 and U_0 are the kinetic and potential energy at some point where both are known. $K_0 + U_0$ is the *total energy*.

We can describe a wide range of systems—from molecules to roller coasters and on to planets—in terms of **potential-energy curves**. Knowing the total energy then lets us find **turning points** that determine the range of motion available to the system.

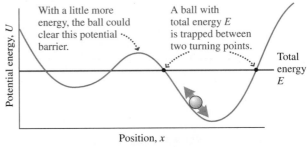

With a little more energy, the ball could clear this potential barrier.

A ball with total energy E is trapped between two turning points.

Total energy E

Position, x

Cases and Uses

Two important cases of potential energy are the elastic potential energy of a spring, $U = \frac{1}{2}kx^2$, and the gravitational potential energy change, $\Delta U = mgh$, associated with lifting an object of mass m through a height h.

The former is limited to ideal springs for which $F = -kx$; the latter to the proximity of Earth's surface, where the variation of gravity with height is negligible.

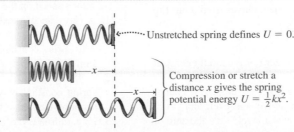

Unstretched spring defines $U = 0$.

Compression or stretch a distance x gives the spring potential energy $U = \frac{1}{2}kx^2$.

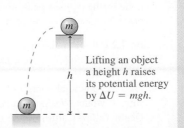

Lifting an object a height h raises its potential energy by $\Delta U = mgh$.

For Thought and Discussion

1. Figure 7.14 shows force vectors at different points in space for two forces. Which is conservative and which nonconservative? Explain.

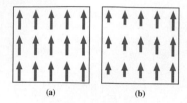

FIGURE 7.14 For Thought and Discussion 1

2. Is the conservation-of-energy principle related to Newton's laws, or is it an entirely separate physical principle? Discuss.
3. Why can't we define a potential energy associated with friction?
4. Can potential energy be negative? Can kinetic energy? Can total mechanical energy? Explain.
5. If the potential energy is zero at a point, must the force also be zero at that point? Give an example.
6. If the force is zero at a point, must the potential energy also be zero at that point? Give an example.
7. If the difference in potential energy between two points is zero, does that necessarily mean that an object moving between those points experiences no force?
8. A tightrope walker follows an essentially horizontal rope between two mountain peaks of equal altitude. A climber descends from one peak and climbs the other. Compare the work done by the gravitational force on the tightrope walker and on the climber.
9. If conservation of energy is a law of nature, why do we have programs—like mileage requirements for cars or insulation standards for buildings—designed to encourage energy conservation?
10. Figure 7.15 shows the potential-energy curve for a particle. At which of the labeled points (a) does the force have the greatest magnitude; (b) does the force point in the negative x direction; (c) is the force zero; (d) are the force and potential energy both zero?

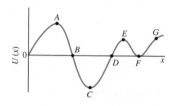

FIGURE 7.15 For Thought and Discussion 10

Exercises and Problems

Exercises

Section 7.1 Conservative and Nonconservative Forces

11. Determine the work done by the frictional force in moving a block of mass m from point 1 to point 2 over the two paths shown in Fig. 7.16. The coefficient of friction has the constant value μ over the surface. *Note:* The diagram lies in a horizontal plane.

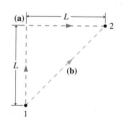

FIGURE 7.16 Exercises 11 and 12

12. Now take Fig. 7.16 to lie in a vertical plane, and find the work done by the gravitational force as an object moves from point 1 to point 2 over each of the paths shown.

Section 7.2 Potential Energy

13. Rework Example 7.1, now taking the zero of potential energy at street level.
14. Find the potential energy of a 70-kg hiker (a) atop New Hampshire's Mount Washington, 1900 m above sea level, and (b) in Death Valley, California, 86 m below sea level. Take the zero of potential energy at sea level.
15. The top of the volcano Haleakala on Maui, Hawaii, is 3050 m above sea level and 18 km inland from the sea. By how much does your gravitational potential energy change as you come down from the mountaintop observatory to swim in the ocean? Assume your mass is 75 kg.
16. You fly from Boston's Logan Airport, at sea level, to Denver, at an altitude of 1.6 km. Taking your mass as 65 kg and the zero of potential energy at Boston, what is your gravitational potential energy (a) at the plane's 11-km cruising altitude and (b) in Denver?
17. A 60-kg hiker ascending 1250-m-high Camel's Hump mountain in Vermont has potential energy -2.4×10^5 J; the zero of potential energy is taken at the mountaintop. What is her altitude?
18. How much energy can be stored in a spring with $k = 320$ N/m if the maximum allowed stretch is 18 cm?
19. How far would you have to stretch a spring of spring constant $k = 1.4$ kN/m for it to store 210 J of energy?

Section 7.3 Conservation of Mechanical Energy

20. A skier starts down a frictionless 32° slope. After a vertical drop of 25 m, the slope temporarily levels out and then drops at 20° an additional 38 m vertically before leveling out again. What is the skier's speed on the two level stretches?
21. A Navy jet of mass 10,000 kg lands on an aircraft carrier and snags a cable to slow it down. The cable is attached to a spring with spring constant 40,000 N/m. If the spring stretches 25 m to stop the plane, what was the landing speed of the plane?
22. A 120-g arrow is shot vertically from a bow whose effective spring constant is 430 N/m. If the bow is drawn 71 cm before shooting the arrow, to what height does the arrow rise?
23. In a railroad yard, a 35,000-kg boxcar moving at 7.5 m/s is brought to a stop by a spring-loaded bumper mounted at the end of the level track. If the spring constant $k = 2.8$ MN/m, how far does it compress in stopping the boxcar?

24. You're designing a toy rocket to be launched by a spring. The launching apparatus has room for a spring that can be compressed 14 cm, and the rocket's mass is 65 g. If the rocket is to reach an altitude of 35 m, what should be the spring constant?

Section 7.4 Potential-Energy Curves

25. A particle slides along the frictionless track shown in Fig. 7.17, starting at rest from point A. Find (a) its speed at B, (b) its speed at C, and (c) the approximate location of its right-hand turning point.

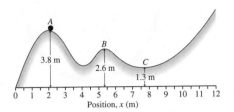

FIGURE 7.17 Exercise 25

26. A particle slides back and forth on a frictionless track whose height as a function of horizontal position x is given by $y = ax^2$, where $a = 0.92 \text{ m}^{-1}$. If the particle's maximum speed is 8.5 m/s, find the turning points of its motion.

27. Figure 7.18 shows the potential-energy curve for a certain particle. Find the force on the particle at each of the labeled curve segments.

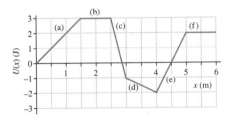

FIGURE 7.18 Exercise 27

28. A particle is trapped in a potential well described by $U(x) = 1.6x^2 - b$, where U is in joules, x is in meters, and $b = 4.0$ J. Find the force on the particle when it's at (a) $x = 2.1$ m, (b) $x = 0$ m, and (c) $x = -1.4$ m.

Problems

29. The reservoir at Northfield Mountain Pumped Storage Project is 270 m above the pump/generators and holds 1.6×10^{10} kg of water (see Application on p. 105). The generators can produce electrical energy at the rate of 1.08 GW. Find (a) the total gravitational energy stored in the reservoir, taking zero potential energy at the generators, and (b) the length of time the station can generate power before the reservoir is drained.

30. The force in Fig. 7.14a is given by $\vec{F} = F_0\hat{\jmath}$, where F_0 is a constant. The force in Fig. 7.14b is given by $\vec{F} = F_0(x/a)\hat{\jmath}$, where the origin is taken at the lower left corner of the box, a is the width of the square box, and the distance x increases horizontally to the right. Determine the work done by $\vec{F}$ on an object moved counterclockwise around each box, starting at the lower left corner.

31. An incline makes an angle θ with the horizontal. Find the gravitational potential energy associated with a mass m located a distance x measured along the incline. Take the zero of potential energy at the bottom of the incline.

32. Show using Equation 7.2b that the potential-energy difference between the ground and a distance h above the ground is mgh regardless of whether you choose the y axis upward or downward.

33. A 1.50-kg brick measures 20.0 cm $\times$ 8.00 cm $\times$ 5.50 cm. Taking the zero of potential energy when the brick lies on its broadest face, what is the potential energy (a) when the brick is standing on end and (b) when it's balanced on its 8-cm edge, with its center directly above that edge? *Note:* You can treat the brick as though all its mass is concentrated at its center.

34. A carbon monoxide molecule can be modeled as a carbon atom and an oxygen atom connected by a spring. If a displacement of the carbon by 1.6×10^{-12} m from its equilibrium position relative to the oxygen increases the molecule's potential energy by 0.015 eV, what is the spring constant?

35. A more accurate expression for the force law of the rope in Example 7.3 is $F = -kx + bx^2 - cx^3$, where k and b have the values given in Example 7.3 and $c = 3.1 \text{ N/m}^3$. Find the energy stored in stretching the rope 2.62 m. By what percentage does your result differ from that of Example 7.3?

36. The force exerted by an unusual spring when it's compressed a distance x from equilibrium is given by $F = -kx - cx^3$, where $k = 220$ N/m and $c = 3.6 \text{ kN/m}^3$. Find the energy stored in this spring when it's been compressed 15 cm.

37. The force on a particle is given by $\vec{F} = A\hat{\imath}/x^2$, where A is a positive constant. (a) Find the potential-energy difference between two points x_1 and x_2, where $x_1 > x_2$. (b) Show that the potential-energy difference remains finite even when $x_1 \to \infty$.

38. A particle moves along the x axis under the influence of a force $F = ax^2 + b$, where a and b are constants. Find its potential energy as a function of position, taking $U = 0$ at $x = 0$.

39. A runaway truck lane heads uphill at 30° to the horizontal. If a 16,000-kg truck goes out of control and enters the lane going 110 km/h, how far along the ramp does it go? Neglect friction.

40. A spring of constant k, compressed a distance x, is used to launch a mass m up a frictionless slope that makes an angle θ with the horizontal. Find an expression for the maximum distance along the slope that the mass moves after leaving the spring.

41. A child is on a swing whose 3.2-m-long chains make a maximum angle of 50° with the vertical. What is the child's maximum speed?

42. With $x - x_0 = h$ and $a = g$, Equation 2.11 gives the speed of an object thrown downward with initial speed v_0 after it has dropped a distance h: $v = \sqrt{v_0^2 + 2gh}$. Use conservation of energy to derive the same result.

43. Two clever kids use a huge spring with $k = 890$ N/m to launch their toboggan at the top of a 9.5-m-high hill (Fig. 7.19). The mass of the kids plus toboggan is 80 kg. If the kids manage to compress the spring 2.6 m, (a) what will be their speed at the bottom of the hill? (b) What fraction of their final kinetic energy was initially stored in the spring? Neglect friction.

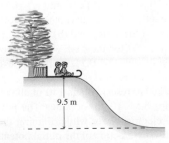

FIGURE 7.19 Problem 43

44. A 200-g block slides back and forth on a frictionless surface between two springs, as shown in Fig. 7.20. The left-hand spring
has $k = 130$ N/m and its maximum compression is 16 cm. The
right-hand spring has $k = 280$ N/m. Find (a) the maximum compression of the right-hand spring and (b) the speed of the block as
it moves between the springs.

FIGURE 7.20 Problem 44

45. A low-damage bumper on a 1400-kg car is mounted on springs
whose effective spring constant is 7.0×10^5 N/m. The springs can
undergo a maximum compression of 16 cm without damage to
the car. What is the maximum speed at which the car can collide
with a stationary object without sustaining damage?

46. A block slides on the frictionless loop-the-loop track shown in
Fig. 7.21. What is the minimum height h at which it can start
from rest and still make it around the loop?

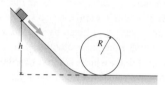

FIGURE 7.21 Problem 46

47. The maximum speed of the pendulum bob in a grandfather clock
is 0.55 m/s. If the pendulum makes a maximum angle of 8.0°
with the vertical, what is the length of the pendulum?

48. A mass m is dropped from a height h above the top of a spring
of constant k that is mounted vertically on the floor. Show
that the maximum compression of the spring is given by
$(mg/k)\left(1 + \sqrt{1 + 2kh/mg}\right)$.

49. A particle with total energy 3.5 J is trapped in a potential well described by $U = 7.0 - 8.0x + 1.7x^2$, where U is in joules and x in meters. Find its turning points.

50. (a) Derive an expression for the potential energy of an object subject to a force $F_x = ax - bx^3$, where $a = 5$ N/m and $b = 2$ N/m³,
taking $U = 0$ at $x = 0$. (b) Graph the potential-energy curve for
$x > 0$, and use it to find the turning points for an object whose
total energy is -1 J.

51. In ionic solids such as NaCl (salt), the potential energy of a pair
of ions takes the form $U = (b/r^n) - (a/r)$, where r is the separation of the ions. For NaCl, a and b have the SI values 4.04×10^{-28}
and 5.52×10^{-98}, respectively, and $n = 8.22$. Find the equilibrium
separation between ions in NaCl.

52. The potential energy of a spring is given by $U = ax^2 - bx + c$,
where $a = 5.20$ N/m, $b = 3.12$ N, and $c = 0.468$ J, and where x is
the *overall* length of the spring (not the stretch). Find (a) the equilibrium length of the spring and (b) the spring constant.

53. Repeat Exercise 20 for the case when the coefficient of kinetic
friction on both slopes is 0.11, while the level stretches remain
frictionless.

54. A pumped-storage reservoir sits 140 m above its generating station and holds 8.5×10^9 kg of water. The power plant generates
330 MW of electric power while draining the reservoir over an
8.0-h period. What fraction of the initial potential energy is lost to
nonconservative forces (i.e., does not emerge as electricity)?

55. A spring of constant $k = 340$ N/m is used to launch a 1.5-kg
block along a horizontal surface whose coefficient of sliding friction is 0.27. If the spring is compressed 18 cm, how far does the
block slide?

56. A bug slides back and forth in a bowl 11 cm deep, starting from
rest at the top, as shown in Fig. 7.22. The bowl is frictionless except for a 1.5-cm-wide sticky patch on its flat bottom, where the
coefficient of friction is 0.61. How many times does the bug cross
the sticky region?

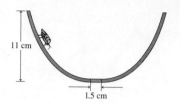

FIGURE 7.22 Problem 56

57. A 190-g block is launched by compressing a spring of constant
$k = 200$ N/m a distance of 15 cm. The spring is mounted horizontally, and the surface directly under it is frictionless. But beyond
the equilibrium position of the spring end, the surface has coefficient of friction $\mu = 0.27$. This frictional surface extends 85 cm,
followed by a frictionless curved rise, as shown in Fig. 7.23. After
launch, where does the block finally come to rest? Measure from
the left end of the frictional zone.

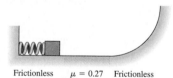

Frictionless $\mu = 0.27$ Frictionless

FIGURE 7.23 Problem 57

58. A block slides down a frictionless incline that terminates in a
ramp pointing up at a 45° angle, as shown in Fig. 7.24. Find an
expression for the horizontal range x shown in the figure as a
function of the heights h_1 and h_2 shown.

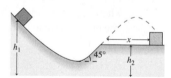

FIGURE 7.24 Problem 58

59. An 840-kg roller-coaster car is launched from a giant spring
of constant $k = 31$ kN/m into a frictionless loop-the-loop track
of radius 6.2 m, as shown in Fig. 7.25. What is the minimum
amount that the spring must be compressed if the car is to stay on
the track?

FIGURE 7.25 Problem 59

60. A particle slides back and forth in a frictionless bowl whose height is given by $h(x) = 0.18x^2$, where x and h are both in meters. If the particle's maximum speed is 47 cm/s, find the x coordinates of its turning points.

61. A child sleds down a frictionless hill whose vertical drop is 7.2 m. At the bottom is a level but rough stretch where the coefficient of kinetic friction is 0.51. How far does she slide across the level stretch?

62. A bug lands on top of the frictionless, spherical head of a bald man. It begins to slide down his head (Fig. 7.26). Show that the bug leaves the head when it has dropped a vertical distance one-third the radius of the head.

FIGURE 7.26 Problem 62

63. A particle of mass m is subject to a force $\vec{F} = (a\sqrt{x})\hat{\imath}$, where a is a constant. The particle is initially at rest at the origin and is given a slight nudge in the positive x direction. Find an expression for the particle's speed as a function of position x.

64. A 17-m-long vine hangs vertically from a tree on one side of a 10-m-wide gorge, as shown in Fig. 7.27. Tarzan runs up, hoping to grab the vine, swing over the gorge, and drop vertically off the vine to land on the other side. At what minimum speed must he be running?

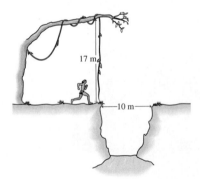

FIGURE 7.27 Problem 64

65. A block of weight 4.5 N is launched up a 30° inclined plane of length 2.0 m by a spring with spring constant 2.0 kN/m and max-

imum compression 0.1 m. The coefficient of kinetic friction is 0.5. (a) Does the block reach the top of the incline? If so, how much kinetic energy does it have there; if not, how close to the top, along the incline, does it get? (b) Answer part (a) for a block of twice the weight.

66. Assume that 30% of the power from the 250-hp engine of a 1500-kg high-performance automobile can be converted to mechanical energy, and that the power delivered is constant, independent of the velocity. How long would it take for this car to accelerate from rest to 60 mi/h on a level road?

67. A uranium nucleus ejects an alpha particle $(m = 6.7 \times 10^{-27}$ kg$)$. After overcoming the nuclear attraction, the particle is at 6.09×10^{-14} m from the center of the nucleus and moving with negligible speed. Your professor theorizes that it is then subject to a repulsive force $F = A/x^2$, where x is the distance from the center of the nucleus and A is a constant. At a great distance from the nucleus $(x \to \infty)$, she measures the particle's speed to be 1.42×10^7 m/s. What is A?

68. Your science fiction writer friend has another harebrained idea. She wants to launch ore mined on the Moon to a processing station in orbit around the Earth. Bins with 1000 kg of ore will be launched by means of a large spring. Her spring cannon will be compressed 15 m. What advice do you give her on the number to use for the spring constant? It takes a speed of 2.4 km/s to escape the Moon's gravity.

69. You have been hired by the college zoology department for the summer. You are excited because you will be working with a world-famous expert on animal behavior. She has assigned you to study several videotapes of African animals leaping into the air from the ground. Your task is to compare the power output of several animals when they jump. You will be able to determine the mass m of each animal from data on the animals collected in the field. By viewing the videotape, you will be able to measure both the vertical distance d over which the animal accelerates when it pushes off the ground and the maximum height h it reaches. Your task is to find an algebraic expression for the power in terms of these other parameters.

70. You and a friend are on vacation in Canada. As you are driving around Vancouver, a child darts out in front of the car; you slam on the brakes and skid to a stop. Being very concerned about possible liability, you return to the scene the next day and collect some data: Your car skidded 20 m before coming to a stop. The posted speed limit is 55 km/h. You call the manufacturer of your tires and learn the coefficient of kinetic friction is 0.5 and the coefficient of static friction is 0.7. Were you speeding?

Answers to Chapter Questions

Answer to Chapter Opening Question
Potential energy turns into kinetic energy, sound, and heat.

Answers to GOT IT? Questions

7.1 On the curved paths, the work is greater for the trunk. The gravitational force is conservative, so the work is independent of path. The frictional force is nonconservative, however.

7.2 (b) The potential-energy change will be slightly less because at greater heights, the gravitational force is lower and so, therefore, is the work done in traversing a given distance.

7.3 No. Mechanical energy is conserved, so if the ball is released from rest, it cannot climb higher than its initial height.

7.4 (a) B; (b) E; (c) C; (d) A or D; (e) B or E.

8 Gravity

■ This TV dish points at a satellite in a fixed position in the sky. How does the satellite manage to stay at that position?

Gravity is the most obvious of nature's fundamental forces. Theories of gravity have brought us new understandings of the nature and evolution of the universe. We've used our knowledge of gravity to explore the solar system and to engineer a host of space-based technologies. In nearly all applications we still use the theory of gravity that Isaac Newton developed in the 1600s. Only in the most extreme astrophysical situations or where—as with Global Positioning System satellites—we need exquisite precision do we use the successor to Newtonian gravitation—namely, Einstein's general theory of relativity.

8.1 Toward a Law of Gravity

Newton's theory of gravity was the culmination of two centuries of scientific revolution that began in 1543 with Polish astronomer Nicolaus Copernicus's radical suggestion that the planets orbit not Earth but the Sun. Fifty years after Copernicus's work was published, the Danish noble Tycho Brahe began a program of accurate planetary observations. After Tycho's death in 1601, his assistant Johannes Kepler worked to make sense of the observations. Success came when Kepler took a radical step: He gave up the long-standing idea that the planets moved in perfect circles. Kepler summarized his new insights in three laws, described in Fig. 8.1. Kepler based his laws solely on observation and gave no theoretical explanation. So Kepler knew *how* the planets moved, but not *why*.

Shortly after Kepler published his first two laws, Galileo trained his first telescopes on the heavens. Among his discoveries were four moons orbiting Jupiter, sunspots that blemished the supposedly perfect sphere of the Sun, and the phases of Venus (Fig. 8.2). His observations called into question the notion that all celestial objects were perfect and also lent credence to the Copernican view of the Sun as the center of planetary motion.

By Newton's time the intellectual climate was ripe for the culmination of the revolution that had begun with Copernicus. Legend has it that Newton was sitting under an apple tree when an apple struck him on the head, causing him to discover gravity. That story is probably a myth, but if it were true the other half would be that Newton was staring at the Moon when the apple struck. Newton's genius was to recognize that *the motion of the apple and the motion of the Moon were the same, that both were "falling" toward Earth under the influence of the same force.* Newton called this force **gravity**, from the Latin *gravitas*, "heaviness." In one of the most sweeping syntheses in human thought, Newton inferred that everything in the universe, on Earth and in the celestial realm, obeys the same physical laws.

8.2 Universal Gravitation

Newton generalized his new understanding of gravity to suggest that any two particles in the universe exert attractive forces on each other, with magnitude given by

$$F = \frac{Gm_1m_2}{r^2} \quad \text{(universal gravitation)} \tag{8.1}$$

Here m_1 and m_2 arc thc particle masses, r the distance between them, and G the **constant of universal gravitation**, whose value—determined after Newton's time—is 6.67×10^{-11} N·m²/kg². The constant G is truly universal; observation and theory suggest that it has the same value throughout the universe.

The force of gravity acts *between* two particles; that is, m_1 exerts an attractive force on m_2, and m_2 exerts an equal but oppositely directed force on m_1. The two forces therefore obey Newton's third law.

Newton's law of universal gravitation applies strictly only to point particles that have no extent. But, as Newton showed using his newly developed calculus, it also holds for spherically symmetric objects of any size if the distance r is measured from their centers. It also applies approximately to arbitrarily shaped objects provided the distance between them is large compared with their sizes. For example, the gravitational force of Earth on the space station is given accurately by Equation 8.1 because (1) Earth is essentially spherical and (2) the station, though irregular in shape, is vastly smaller than its distance from Earth's center.

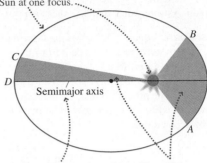

First law: The orbit is elliptical, with the Sun at one focus.

Third law: The square of the orbital period is proportional to the cube of the semimajor axis.

Second law: If the shaded areas are equal, so is the time to go from A to B and from C to D.

FIGURE 8.1 Kepler's laws.

FIGURE 8.2 Phases of Venus. In an Earth-centered system, Venus would always appear the same size because of its constant distance from Earth.

EXAMPLE 8.1 **The Acceleration of Gravity: On Earth and in Space**

Use the law of universal gravitation to find the acceleration of gravity at Earth's surface, at the 380-km altitude of the International Space Station, and on the surface of Mars.

INTERPRET The problem statement tells us this is about universal gravitation, but what's that got to do with the acceleration of gravity? The gravitational force is what causes that acceleration, so we can interpret

this problem as being about the force between Earth (or Mars) and some arbitrary mass.

DEVELOP Since the problem involves universal gravitation, Equation 8.1 applies. But we're asked about acceleration, not force. Newton's second law, $F = ma$, relates the two. So our plan is to use Equation 8.1, $F = Gm_1m_2/r^2$, to find the gravitational force on an arbitrary mass and then use Newton's second law to get the acceleration. There's another bit of planning: We need to find the masses of Earth and Mars and their radii. Astrophysical data like these are given in Appendix E.

EVALUATE Equation 8.1 gives the force a planet of mass M exerts on an arbitrary mass m a distance r from the planet's center: $F = GMm/r^2$. (Here we set m_1 in Equation 8.1 to the large planetary mass M, and m_2 to the smaller mass m.) But Newton's second law says that this force is equal to the product of mass and acceleration for a body in free fall, so we can write $ma = GMm/r^2$. The mass m cancels, and we're left with the acceleration:

$$a = \frac{GM}{r^2} \qquad (8.2)$$

The distance r is measured from the *center* of the object providing the gravitational force, so to find the acceleration at Earth's surface we use R_E, the radius of the Earth, for r. Taking R_E and M_E from Appendix E, we have

$$a = \frac{GM_E}{R_E^2}$$

$$= \frac{(6.67 \times 10^{-11}\,\text{N} \cdot \text{m}^2/\text{kg}^2)(5.97 \times 10^{24}\,\text{kg})}{(6.37 \times 10^6\,\text{m})^2} = 9.81\,\text{m/s}^2$$

This, of course, is the value of g—the acceleration due to gravity at Earth's surface. Note the difference between the two numbers G and g commonly associated with gravity: G is a universal constant, while g describes the gravitational acceleration at a particular place—namely, Earth's surface—and happens to have the value it does because of the size and mass of the Earth.

At the space station's altitude, we have $r = R_E + 380$ km, so

$$a = \frac{GM_E}{r^2}$$

$$= \frac{(6.67 \times 10^{-11}\,\text{N} \cdot \text{m}^2/\text{kg}^2)(5.97 \times 10^{24}\,\text{kg})}{(6.37 \times 10^6\,\text{m} + 380 \times 10^3\,\text{m})^2} = 8.74\,\text{m/s}^2$$

A similar calculation using Appendix E data yields 3.7 m/s² for the acceleration of gravity at the surface of Mars.

ASSESS As we've seen, our result for Earth is just what we expect. The acceleration at the space station is lower but still about 90% of the surface value. This confirms Chapter 4's point that weightlessness doesn't mean the absence of gravity. Rather, as Equation 8.2 shows, an object's gravitational acceleration is independent of its mass—so all objects "fall" together. Finally, our answer for Mars is lower than for Earth, as befits its smaller mass—although not as much lower as mass alone would imply. That's because Mars is also smaller, so r in the denominator of Equation 8.2 is lower. ∎

The variation of gravitational acceleration with distance from Earth's center provided Newton with a clue that the gravitational force should vary as the inverse square of the distance. Newton knew the Moon's orbital period and distance from Earth; from these he could calculate its orbital speed and thus its acceleration v^2/r. Newton found—as you can in Exercise 13—that the Moon's acceleration is about 1/3600 the gravitational acceleration g at Earth's surface. The Moon is about 60 times farther from Earth's center than is Earth's surface; since $60^2 = 3600$, the decrease in gravitational acceleration with distance from Earth's center is consistent with a gravitational force that varies as $1/r^2$.

TACTICS 8.1   Understanding "Inverse Square"

Newton's universal gravitation is the first of several inverse-square force laws you'll encounter, and it's important to understand what this term means. In Equation 8.1 the distance r between the two masses is *squared*, and it occurs in the *denominator*; hence the force depends on the *inverse square* of the distance. Double the distance and the force drops to $1/2^2$, or 1/4 of its original value. Triple the distance and the force drops to $1/3^2$, or 1/9. Although you can always grind through the arithmetic of Equation 8.1, you should use these simple ratio calculations whenever possible. The same considerations apply to gravitational acceleration, since it's proportional to force (Fig. 8.3).

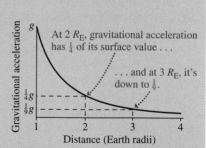

At 2 R_E, gravitational acceleration has ¼ of its surface value . . .

. . . and at 3 R_E, it's down to ⅑.

FIGURE 8.3 Meaning of the inverse-square law.

GOT IT? 8.1   Suppose the distance between two objects is cut in half. Is the gravitational force between them (a) quartered, (b) halved, (c) doubled, or (d) quadrupled?

The Cavendish Experiment: Weighing the Earth

Given the mass and radius of the Earth and the measured value of g, we could use Equation 8.1 to determine the universal constant G. Unfortunately, the only way to determine Earth's mass accurately is to measure its gravitational effect and then use Equation 8.1. But that requires knowing G.

To determine G, we need to measure the gravitational force of a *known* mass. Given the weak gravitational force of normal-size objects, this is a challenging task. It was accomplished in 1798 through an ingenious experiment by the British physicist Henry Cavendish. Cavendish mounted two 5-cm-diameter lead spheres on the ends of a rod suspended from a thin fiber. He then brought two 30-cm lead spheres nearby (Fig. 8.4). Their gravitational attraction caused a slight movement of the small spheres, twisting the fiber. Knowing the properties of the fiber, Cavendish could determine the force. With the known masses and their separation, he then used Equation 8.1 to calculate G. His result determined the mass of the Earth; indeed, his published paper was entitled "On Weighing the Earth."

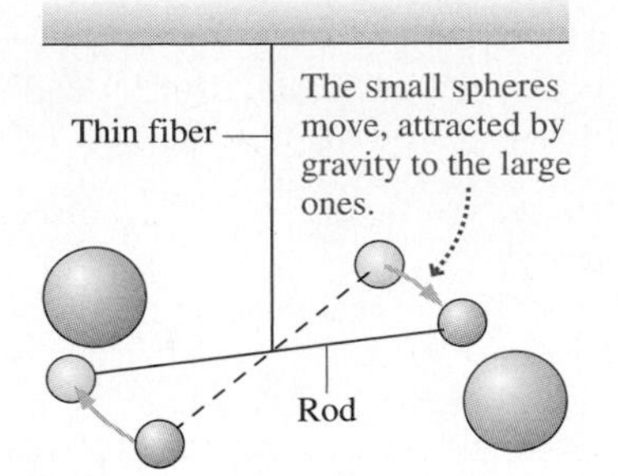

Gravity is the weakest of the fundamental forces, and, as the Cavendish experiment suggests, the gravitational force between everyday objects is negligible. Yet gravity shapes the large-scale structure of matter and indeed the entire universe. Why, if it's so weak? The answer is that gravity, unlike the stronger electric force, is always attractive; there's no such thing as "negative mass." So large concentrations of matter produce substantial gravitational effects. Electric charge, in contrast, can be positive or negative, and electric effects in normal-sized objects tend to cancel out. We'll explore this fundamental distinction further in Chapter 20.

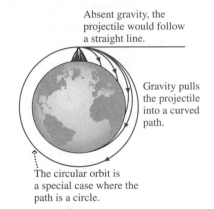

FIGURE 8.4 The Cavendish experiment to determine G.

8.3 Orbital Motion

Orbital motion occurs when gravity is the dominant force acting on a body. It's not just planets and spacecraft that are in orbit. An individual astronaut floating outside the space station is orbiting Earth. The Sun itself orbits the center of the galaxy, taking about 200 million years to complete one revolution. If we neglect air resistance, even a baseball is temporarily in orbit. Here we discuss quantitatively the special case of circular orbits; then we describe qualitatively the general case.

Newton's genius was to recognize that the Moon is held in its circular orbit by the same force that pulls an apple to the ground. From there, it was a short step for Newton to realize that human-made objects could be put into orbit. Nearly 300 years before the first artificial satellites, he imagined a projectile launched horizontally from a high mountain (Fig. 8.5). The projectile moves in a curve, as gravity pulls it from the straight-line path it would follow if no force were acting. As its initial speed is increased, the projectile travels farther before striking Earth. Finally, there comes a speed for which the projectile's path bends in a way that exactly follows Earth's curvature. It's then in **circular orbit**, continuing forever unless a nongravitational force acts.

Why doesn't an orbiting object fall toward Earth? It does! Under the influence of gravity, it gets ever closer to Earth than it would be on a straight-line path. It's behaving exactly as Newton's second law requires of an object under the influence of a force—by accelerating. For a *circular* orbit, that acceleration amounts to a change in the direction, but not the magnitude, of the orbiting object's velocity.

Remember that Newton's laws aren't so much about *motion* as they are about *changes* in motion. To ask why a satellite doesn't fall to Earth is to adopt the archaic Aristotelian view. The Newtonian question is this: Why doesn't the satellite move in a straight line? And the answer is simple: because a force is acting. That force—gravity—is exactly analogous to the tension force that keeps a ball on a string whirling in its circular path.

We can analyze circular orbits quantitatively because we know that a force of magnitude mv^2/r is required to keep an object of mass m and speed v in a circular path of radius r. In the case of an orbit, that force is gravity, so we have

$$\frac{GMm}{r^2} = \frac{mv^2}{r}$$

where m is the mass of the orbiting object and M the mass of the object about which it's orbiting. We assume here that $M \gg m$, so the gravitating object can be considered essentially at rest—a reasonable approximation with Earth satellites or planets orbiting the much more massive Sun. Solving for the orbital speed gives

$$v = \sqrt{\frac{GM}{r}} \quad \text{(speed, circular orbit)} \tag{8.3}$$

FIGURE 8.5 Newton's "thought experiment" showing that projectile and orbital motions are essentially the same.

Often we're interested in the **orbital period**, or the time to complete one orbit. In one period T, the orbiting object moves the orbital circumference $2\pi r$, so its speed is $v = 2\pi r/T$. Squaring Equation 8.3 then gives

$$\left(\frac{2\pi r}{T}\right)^2 = \frac{GM}{r}$$

or

$$T^2 = \frac{4\pi^2 r^3}{GM} \qquad \text{(orbital period, circular orbit)} \qquad (8.4)$$

In deriving Equation 8.4, we've proved Kepler's third law—that the square of the orbital period is proportional to the cube of the semimajor axis—for the special case of a circular orbit whose semimajor axis is its radius.

Note that orbital speed and period are independent of the mass m of the orbiting object. This is another indication that all objects experience the same gravitational acceleration. Astronauts, for example, have the same orbital parameters as the space station. That's why astronauts seem weightless inside the station and why they don't float away if they step outside.

EXAMPLE 8.2 **Orbital Speed and Period: The Space Station**

The International Space Station is in a circular orbit at an altitude of 380 km. What are its orbital speed and period?

INTERPRET This problem involves the speed and period of a circular orbit about Earth.

DEVELOP We can compute the orbit's radius and then use Equation 8.3, $v = \sqrt{GM/r}$, to find the speed and Equation 8.4, $T^2 = 4\pi^2 r^3/GM$, to find the period because the orbit is circular.

EVALUATE As always, the distance is measured from the center of the gravitating body, so r in these equations is Earth's 6.37-Mm radius plus the station's 380-km altitude. So we have

$$v = \sqrt{\frac{GM_E}{r}} = \sqrt{\frac{(6.67 \times 10^{-11}\,\text{N} \cdot \text{m}^2/\text{kg}^2)(5.97 \times 10^{24}\,\text{kg})}{6.37 \times 10^6\,\text{m} + 380 \times 10^3\,\text{m}}} = 7.7\,\text{km/s}$$

or about 17,000 mi/h. We can get the orbital period from the speed and radius, or directly from Equation 8.4, $T = \sqrt{4\pi^2 r^3/GM_E}$. Using the numbers in the calculation for v gives $T = 5.5 \times 10^3$ s, or about 90 min.

ASSESS Make sense? Both answers have the correct units, and 90 min seems reasonable for the period of an orbit at a small fraction of the Moon's distance from Earth. Astronauts who want a circular orbit 380 km up have no choice but this speed and period. In fact, for any "near-Earth" orbit, with altitude much less than Earth's radius, the orbital period is about 90 min. If there were no air resistance and if you could throw a baseball fast enough, it too would go into orbit, skimming Earth's surface with a roughly 90-min period. ∎

Example 8.2 shows that the near-Earth orbital period is about 90 min. The Moon, on the other hand, takes 27 days to complete its nearly circular orbit. So there must be a distance where the orbital period is 24 h—the same as Earth's rotation. A satellite at this distance will remain fixed with respect to Earth's surface provided its orbit is parallel to the equator. TV, weather, and communication satellites are often placed in this **geosynchronous orbit**.

EXAMPLE 8.3 **Geosynchronous Orbit: Finding the Altitude**

What altitude is required for geosynchronous orbit?

INTERPRET Here we're given an orbital period—24 h or 86,400 s—and asked to find the corresponding altitude for a circular orbit.

DEVELOP Equation 8.4, $T^2 = 4\pi^2 r^3/GM$, relates the period T and distance r from Earth's center. Our plan is to solve for r and then subtract Earth's radius to find the altitude (distance from the surface).

EVALUATE Solving for r, we get

$$r = \left(\frac{GM_E T^2}{4\pi^2}\right)^{1/3}$$

$$= \left[\frac{(6.67 \times 10^{-11}\,\text{N} \cdot \text{m}^2/\text{kg}^2)(5.97 \times 10^{24}\,\text{kg})(8.64 \times 10^4\,\text{s})}{4\pi^2}\right]^{1/3}$$

$$= 4.22 \times 10^7\,\text{m}$$

or 42,200 km from Earth's center. Subtracting Earth's radius then gives an altitude of about 36,000 km, or 22,000 miles.

ASSESS Make sense? This is a lot higher than the 90-min low-Earth orbit, but a lot lower than the Moon's 385,000 km distance. Our answer defines one of the most valuable pieces of "real estate" in space—a place where satellites appear suspended over a fixed spot on Earth. The dish antenna in this chapter's opening photo points to such a satellite, positioned 22,000 mi over the equator. A more careful calculation would use Earth's so-called sidereal rotation period, measured with respect to the distant stars rather than the Sun. Because Earth isn't a perfect sphere, geosynchronous satellites drift slightly and therefore fire small rockets every few weeks to stay in position. ∎

Elliptical Orbits

Using his laws of motion and gravity, Newton was able to prove Kepler's assertion that the planets move in elliptical paths with the Sun at one focus. Circular orbits represent the special case where the two foci of the ellipse coincide, so the distance from the gravitating center remains constant. Most planetary orbits are nearly, but not quite, circular; Earth's distance from the Sun, for example, varies by about 3% throughout the year. But the orbits of comets and other smaller bodies are often highly elliptical (Fig. 8.6). Their orbital speeds vary, as they gain speed "falling" toward the Sun, whip quickly around the Sun at the point of closest approach (**perihelion**), and then "climb" ever more slowly to their most distant point (**aphelion**) before returning to the Sun's vicinity.

In Chapter 3, we showed that the trajectory of a projectile is a parabola. But our derivation neglected Earth's curvature and the associated variation in g with altitude. In fact, a projectile is just like any orbiting body. If we neglect air resistance, it too describes an elliptical orbit with Earth's center at one focus. Only for trajectories small compared with Earth's radius are the true elliptical path and the parabola of Chapter 3 essentially indistinguishable (Fig. 8.7).

Are missiles and baseballs really in orbit? Yes. But their orbits happen to intersect the Earth. At that point, nongravitational forces put an end to orbital motion. If Earth suddenly shrank to the size of a grapefruit (but kept the same mass), a baseball would continue happily in orbit, as Fig. 8.7 suggests. Newton's ingenious intuition was correct: Barring air resistance, there's truly no difference between the motion of everyday objects near Earth and the motion of celestial objects.

Open Orbits

With elliptical and circular orbits, the motion repeats indefinitely because the orbit is a closed path. But closed orbits aren't the only possibility. Imagine again Newton's thought experiment—only now fire the projectile faster than necessary for a circular orbit (Fig. 8.8). The projectile rises higher than before, describing an ellipse with its low point at the launch site. Faster, and the ellipse gets more elongated. But with great enough initial speed, the projectile describes a trajectory that takes it ever farther from Earth. We'll see in the next section how energy determines the type of orbit.

GOT IT? 8.2 Suppose the paths in Fig. 8.8 are the paths of four projectiles. Rank each path (circular, elliptical, parabolic, and hyperbolic) according to the initial speed of the corresponding projectile. Assume all are launched from their common point at the top of the figure.

8.4 Gravitational Energy

How much energy does it take to boost a satellite to geosynchronous altitude? Our simple answer mgh won't do here, since g varies substantially over the distance involved. So, as we found in Chapter 7, we have to integrate to determine the potential energy.

Figure 8.9 shows two points at distances r_1 and r_2 from the center of a gravitating mass M, in this case Earth. Equation 7.2 gives the change in potential energy associated with moving a mass m from r_1 to r_2:

$$\Delta U_{12} = -\int_{r_1}^{r_2} \vec{F} \cdot d\vec{r}$$

Here the force points radially inward and has magnitude GMm/r^2, while the path element $d\vec{r}$ points radially outward. Then $\vec{F} \cdot d\vec{r} = -(GMm/r^2)\,dr$, where the minus sign comes

FIGURE 8.6 Orbits of most known comets, like the one shown here, are highly elliptical.

This section approximates a parabola.

Focus is Earth's center.

FIGURE 8.7 Projectile trajectories are actually elliptical.

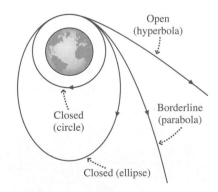

Open (hyperbola)

Closed (circle)

Borderline (parabola)

Closed (ellipse)

FIGURE 8.8 Closed and open orbits.

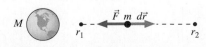

M $\vec{F}$ m $d\vec{r}$
         r_1 r_2

FIGURE 8.9 Finding the potential-energy change requires integration.

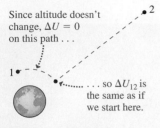

Since altitude doesn't change, $\Delta U = 0$ on this path . . .

. . . so ΔU_{12} is the same as if we start here.

FIGURE 8.10 Gravity is conservative, so we can use any path to evaluate the potential-energy change. Only the radial part of the path contributes to ΔU.

from the factor cos 180° in the dot product of oppositely directed vectors. Then the potential energy difference is

$$\Delta U_{12} = \int_{r_1}^{r_2} \frac{GMm}{r^2}\, dr = GMm \int_{r_1}^{r_2} r^{-2}\, dr = GMm \frac{r^{-1}}{-1}\Big|_{r_1}^{r_2} = GMm\left(\frac{1}{r_1} - \frac{1}{r_2}\right) \quad (8.5)$$

Does this make sense? Yes: For $r_1 < r_2$, ΔU_{12} is positive, showing that potential energy increases with height—consistent with our simpler result $\Delta U = mgh$ near Earth's surface. Although we derived Equation 8.5 for two points on a radial line, Fig. 8.10 shows that it holds for any two points at distances r_1 and r_2 from the gravitating center.

EXAMPLE 8.4 **Gravitational Potential Energy: Steps to the Moon**

Materials to construct an 11,000-kg lunar observatory are boosted from Earth to geosynchronous orbit. There they are assembled and then launched to the Moon, 385,000 km from Earth. Compare the work that must be done against Earth's gravity on the two legs of the trip.

INTERPRET This problem asks about work done against gravity, a conservative force.

DEVELOP As we saw in Chapter 7, the work done against a conservative force is equal to the change in potential energy; here that change is given by Equation 8.5. For the first leg, we have $r_1 = R_E$ and then, from Example 8.3, $r_2 = 42,200$ km.

EVALUATE Since the quantity GM_Em that appears in Equation 8.5 will be used in both steps, we calculate it first: $GM_Em = 4.38 \times 10^{18}$ N·m². Then for the first step we have

$$W = \Delta U_{12} = GM_Em\left(\frac{1}{r_1} - \frac{1}{r_2}\right)$$

$$= (4.38 \times 10^{18}\ \text{N}\cdot\text{m}^2)\left(\frac{1}{6.37 \times 10^6\ \text{m}} - \frac{1}{4.22 \times 10^7\ \text{m}}\right)$$

$$= 5.8 \times 10^{11}\ \text{J}$$

From geosynchronous orbit to the Moon a similar calculation gives

$$W = (4.38 \times 10^{18}\ \text{N}\cdot\text{m}^2)\left(\frac{1}{4.22 \times 10^7\ \text{m}} - \frac{1}{3.85 \times 10^8\ \text{m}}\right)$$

$$= 9.2 \times 10^{10}\ \text{J}$$

ASSESS Make sense? Even though the second leg is much longer, the rapid drop-off in the gravitational force means that less work is required than for the shorter boost to geosynchronous altitude. Our calculations here include only the work done against Earth's gravity; additional energy would be required to attain a circular geosynchronous orbit. On the other hand, the Moon's gravitational attraction would lower the required energy somewhat. ∎

The Zero of Potential Energy

Equation 8.5 has an interesting feature: The potential-energy difference remains finite even when the points are infinitely far apart, as you can see by setting either r_1 or r_2 to infinity. Although the gravitational force always acts, it weakens so rapidly that its effect is finite over even infinite distances. This property makes it convenient to set the zero of potential energy at infinity. Setting $r_1 = \infty$ and dropping the subscript on r_2, we then have an expression for the potential energy at an arbitrary distance r from a gravitating center:

$$U(r) = -\frac{GMm}{r} \qquad \text{(gravitational potential energy)} \qquad (8.6)$$

The potential energy is negative because we chose $U = 0$ at $r = \infty$. Any other point is closer to the gravitating center and therefore has lower potential energy.

Knowing the gravitational potential energy allows us to apply the powerful conservation-of-energy principle. Figure 8.11 shows the potential-energy curve given by Equation 8.6. Superposing three values of total energy E shows that orbits with $E < 0$ have a turning point where they intersect the potential-energy curve, and are therefore closed. Orbits with $E > 0$, in contrast, are open because they never intersect the curve and therefore extend to infinity.

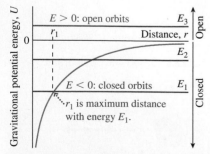

$E > 0$: open orbits $\quad E_3$

r_1 Distance, r

E_2

$E < 0$: closed orbits $\quad E_1$

r_1 is maximum distance with energy E_1.

Gravitational potential energy, U

Open

Closed

FIGURE 8.11 A gravitational potential-energy curve.

EXAMPLE 8.5   Conservation of Energy: Blast Off!

A rocket is launched vertically upward at 3.1 km/s. How high does it go?

INTERPRET This sounds like a problem from Chapter 2, but here gravity changes, so the acceleration isn't constant. The conservation-of-energy principle lets us cut through those details, so we can apply the methods of Chapter 7. "How high does it go?" in the problem statement means we're dealing with the initial launch state and a final state where the rocket is momentarily at rest at the top of its trajectory.

DEVELOP Equation 7.7 states the conservation of energy: $K + U = K_0 + U_0$. Here we're given speed v at the bottom, so $K_0 = \frac{1}{2}mv^2$. We're going to be using Equation 8.6, $U(r) = -GMm/r$, for potential energy, and that's already established the zero of potential energy at infinity. So U_0 isn't zero but is given by Equation 8.6 with r equal to Earth's radius. Finally, at the top, $K = 0$ and U is also given by Equation 8.6, but now we don't know r. Our plan is to solve for that r and from it get the rocket's altitude. Figure 8.12 shows "before" and "after" diagrams with bar graphs, like those we introduced in Chapter 7.

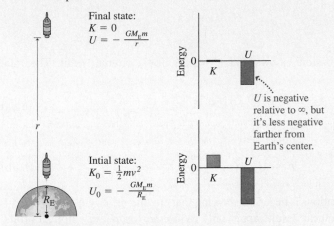

Final state:
$K = 0$
$U = -\dfrac{GM_{\mathrm{E}}m}{r}$

U is negative relative to ∞, but it's less negative farther from Earth's center.

Intial state:
$K_0 = \frac{1}{2}mv^2$
$U_0 = -\dfrac{GM_{\mathrm{E}}m}{R_{\mathrm{E}}}$

FIGURE 8.12 Diagrams for Example 8.5.

EVALUATE With our values of kinetic and potential energy, $K + U = K_0 + U_0$ becomes

$$-\frac{GM_{\mathrm{E}}m}{r} = \tfrac{1}{2}mv_0^2 - \frac{GM_{\mathrm{E}}m}{R_{\mathrm{E}}}$$

where m is the rocket's mass, r is the distance from Earth's center at the peak, and Earth's radius R_{E} is the distance at launch. Solving for r gives

$$r = \left(\frac{1}{R_{\mathrm{E}}} - \frac{v_0^2}{2GM_{\mathrm{E}}} \right)^{-1}$$

$$= \left(\frac{1}{6.37 \times 10^6\ \mathrm{m}} - \frac{(3100\ \mathrm{m/s})^2}{2(6.67 \times 10^{-11}\ \mathrm{N \cdot m^2/kg^2})(5.97 \times 10^{24}\ \mathrm{kg})} \right)^{-1}$$

$$= 6.90\ \mathrm{Mm}$$

Again, this is the distance from Earth's center; subtracting Earth's radius then gives a peak altitude of 530 km.

ASSESS Make sense? Yes. Our answer of 530 km is significantly greater than the 490 km you'd get assuming a potential-energy change of $\Delta U = mgh$. That's because the decreasing gravitational force lets the rocket go higher before all its kinetic energy becomes potential.

✓**TIP** All Conservation-of-Energy Problems Are the Same

This problem is essentially the same as throwing a ball straight up and solving for its maximum height using $U = mgh$ for the potential energy. The only difference is the more complicated potential-energy function $U = -GMm/r$, used here because the variation in gravity is significant over the rocket's trajectory. Recognize what's common to all similar problems, and you'll begin to see how physics is a powerful description based on a few simple principles.

Escape Speed

What goes up comes down, right? Not always! Figure 8.11 shows that an object with total energy zero or greater can escape infinitely far from a gravitating body, never to return. Consider an object of mass m at the surface of a gravitating body of mass M and radius r. It has gravitational potential energy given by Equation 8.6, $U = -GMm/r$. Toss it upward with speed v, and it's also got kinetic energy $\frac{1}{2}mv^2$. Its total energy will be zero if

$$0 = K + U = \tfrac{1}{2}mv^2 - \frac{GMm}{r}$$

The speed v here that makes the total energy zero is called the **escape speed** because an object with this speed or greater has enough energy to escape forever from the gravitating body. Solving for v in the preceding equation gives the escape speed:

$$v_{\mathrm{esc}} = \sqrt{\frac{2GM}{r}} \qquad \text{(escape speed)} \tag{8.7}$$

At Earth's surface, $v_{\mathrm{esc}} = 11.2$ km/s. Earth-orbiting spacecraft have lower speeds. Moon-bound astronauts go at just under v_{esc}, so if anything goes wrong (as with Apollo 13), they can return to Earth. Planetary spacecraft have speeds greater than v_{esc}. The Pioneer and Voyager missions to the outer planets gained enough additional energy in their encounters with Jupiter that they now have escape speed relative to the Sun and will coast indefinitely through interstellar space.

Energy in Circular Orbits

In the special case of a circular orbit, kinetic and potential energies are related in a simple way. In Section 8.3, we found that the speed in a circular orbit is given by

$$v^2 = \frac{GM}{r}$$

where r is the distance from a gravitating center of mass M. So the kinetic energy of the orbiting object is

$$K = \tfrac{1}{2}mv^2 = \frac{GMm}{2r}$$

while the potential energy is given by Equation 8.6:

$$U = -\frac{GMm}{r}$$

Comparing these two expressions shows that $U = -2K$ for a circular orbit. The total energy is therefore

$$E = U + K = -2K + K = -K \tag{8.8a}$$

or, equivalently,

$$E = \tfrac{1}{2}U = -\frac{GMm}{2r} \tag{8.8b}$$

The total energy in these equations is negative, showing that circular orbits are—obviously— bound orbits. We stress that these results apply only to *circular* orbits; in elliptical orbits, there's a continuous interchange between kinetic and potential energy as the orbiting object moves relative to the gravitating center.

Equation 8.8a shows that *higher* kinetic energy corresponds to *lower* total energy. This surprising result occurs because *higher* orbital speed corresponds to a *lower* orbit, with lower potential energy. The implications for space flight are counterintuitive. To get into a faster orbit, a spacecraft must *lose* energy—as if a car, to speed up, had to apply its brakes. Astronauts attempting orbital rendezvous actually fire braking rockets in order to catch up with their target (Fig. 8.13).

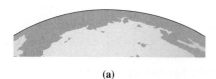

(a)

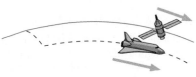

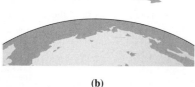

(b)

FIGURE 8.13 To catch the satellite, the shuttle needs to lose energy. It does so by turning to fire its engine opposite its direction of motion. It drops lower, turns again and fires its engine to achieve a circular orbit, now faster and lower than before.

GOT IT? 8.3 Spacecraft A and B are in circular orbits about Earth, with B at a higher altitude. Which of the following statements are true? (a) B has greater total energy; (b) B is moving faster; (c) B takes longer to complete its orbit; (d) B has greater potential energy; (e) a larger proportion of B's total energy is potential energy.

8.5 The Gravitational Field

Our description of gravity so far suggests that a massive body like Earth somehow "reaches out" across empty space to pull on objects like falling apples, satellites, or the Moon. This view—called **action-at-a-distance**—has bothered both physicists and philosophers for centuries. How can the Moon, for example, "know" about the presence of the distant Earth?

An alternative view holds that Earth creates a **gravitational field** and that objects respond to the field in their immediate vicinity. The field is described by vectors that give the force per unit mass that would arise at each point if a mass were placed there. Near Earth's surface, for instance, the gravitational field vectors point vertically downward and have magnitude 9.8 N/kg. We can express this field vectorially by writing

$$\vec{g} = -g\hat{j} \qquad \text{(gravitational field near Earth's surface)} \qquad (8.9)$$

More generally, the field points toward a spherical gravitating center, and its strength decreases inversely with the square of the distance:

$$\vec{g} = -\frac{GM}{r^2}\hat{r} \qquad \text{(gravitational field of a spherical mass } M) \qquad (8.10)$$

where $\hat{r}$ is a unit vector that points radially outward. Figure 8.14 shows pictorial representations of Equations 8.9 and 8.10. You can show that the units of gravitational field (N/kg) are equivalent to those of acceleration (m/s^2), so the field is really just a vectorial representation of g, the local acceleration of gravity.

What do we gain by this field description? As long as we deal with situations where nothing changes, the action-at-a-distance and field descriptions are equivalent. But what if, for example, Earth suddenly gains mass. How does the Moon know to adjust its orbit? Under the field view, its orbit doesn't change immediately; instead, it takes a small but nonzero time for the information about the more massive Earth to propagate out to the Moon. The Moon always responds to the gravitational field *in its immediate vicinity*, and it takes a short time for the field itself to change. That description is consistent with Einstein's notion that instantaneous transmission of information is impossible; the action-at-a-distance view is not.

More generally, the field view provides a powerful way of describing interactions in physics. We'll see fields again when we study electricity and magnetism, and you'll find that fields aren't just mathematical or philosophical conveniences but are every bit as real as matter itself.

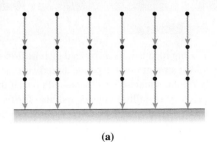

(a)

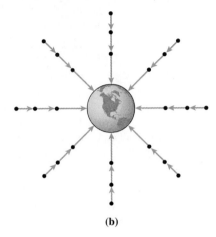

(b)

FIGURE 8.14 Gravitational field vectors at points (a) near Earth's surface and (b) on a larger scale.

APPLICATION   **Tides**

If the gravitational field were uniform, all parts of a freely falling object would experience exactly the same acceleration. But gravity always varies over large enough scales. The result is a force—arising not from gravity itself but from *changes* in gravity with position—that tends to stretch or compress an object. Ocean tides result from this **tidal force**, as the nonuniform gravitational forces of the Sun and Moon stretch the oceans and create bulges that move across Earth as the planet rotates. The figure shows that the greatest force is on the ocean nearest the Moon, causing one tidal bulge. The solid Earth experiences an intermediate force, pulling it away from the ocean on the far side. The water that's "left behind" forms a second bulge opposite the Moon. The bulges shown are highly exaggerated. Furthermore, shoreline effects and the differing relative positions of the Moon and

Sun complicate this simple picture that suggests two equal high tides and two equal low tides a day. Tidal forces also cause internal heating of satellites like Jupiter's moon Io and contribute to the formation of planetary rings.

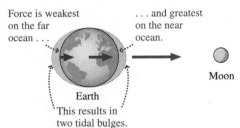

Force is weakest on the far ocean . . .   . . . and greatest on the near ocean.

Moon

Earth

This results in two tidal bulges.

Big Picture

The big idea here is **universal gravitation**—an attractive force that acts between all matter with a strength that depends directly on the product of two interacting masses and inversely on the square of the distance between them. Gravitation is responsible for the familiar behavior of falling objects and also for the orbits of planets and satellites. Depending on energy, orbits may be closed—ellipses or circles—or open.

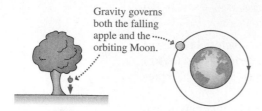

Gravity governs both the falling apple and the orbiting Moon.

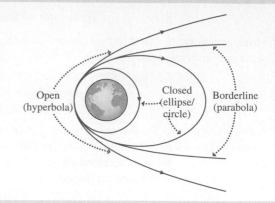

Open (hyperbola)   Closed (ellipse/circle)   Borderline (parabola)

Key Concepts and Equations

Mathematically, Newton's law of universal gravitation describes the attractive force F between two masses m_1 and m_2 located a distance r apart:

$$F = \frac{Gm_1 m_2}{r^2} \quad \text{(universal gravitation)}$$

This equation applies to point masses of negligible size and to spherically symmetric masses of any size. It's an excellent approximation for any objects whose size is much smaller than their separation. In all cases, r is measured from the centers of the gravitating objects.

Because the strength of gravity varies with distance, potential-energy changes over large distances aren't just a product of force and distance. Integration shows that the potential energy change ΔU involved in moving a mass m originally a distance r_1 from the center of a mass M to a distance r_2 is

$$\Delta U = GMm\left(\frac{1}{r_1} - \frac{1}{r_2}\right) \quad \text{(change in potential energy)}$$

With gravity, it's convenient to choose the zero of potential energy at infinity; then

$$U = -\frac{GMm}{r} \quad \text{(potential energy, } U = 0 \text{ at infinity)}$$

for the potential energy of a mass m located a distance r from the center of a mass M.

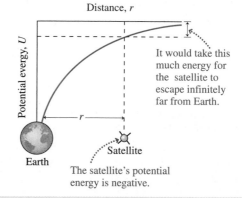

Distance, r

Potential energy, U

It would take this much energy for the satellite to escape infinitely far from Earth.

Earth   Satellite

The satellite's potential energy is negative.

Cases and Uses

A total energy—kinetic plus potential—of zero marks the dividing line between closed and open orbits. An object located a distance r from a gravitating mass M must have at least the **escape speed** to achieve an open orbit and escape M's vicinity forever:

$$v_{\text{esc}} = \sqrt{\frac{2GM}{r}}$$

The **gravitational field** concept provides a way to describe gravity that avoids the troublesome action-at-a-distance. A gravitating mass creates a field in the space around it, and a second mass responds to the field in its immediate vicinity.

Gravitational field

Force arises from field at Moon's location.

Circular orbits are readily analyzed using Newton's laws and concepts from circular motion. A circular orbit of radius r about a mass M has a period given by

$$T^2 = \frac{4\pi^2 r^3}{GM}$$

Its kinetic and potential energies are related by $U = -2K$. Total energy is negative, as appropriate for a closed orbit, and the object actually moves faster the lower its total energy.

A special orbit is the **geosynchronous orbit**, parallel to Earth's equator at an altitude of about 36,000 km. Here the orbital period is 24 h, so a satellite in geosynchronous orbit appears from Earth's surface to be fixed in the sky. TV, communications, and weather satellites use geosynchronous orbits.

For Thought and Discussion

1. Why is Newton's assertion that planetary orbits are elliptical more satisfying than Kepler's?
2. What do Newton's apple and the Moon have in common?
3. Explain the difference between G and g.
4. When you stand on Earth, the distance between you and Earth is zero. So why isn't the gravitational force infinite?
5. The force of gravity on an object is proportional to the object's mass, yet all objects fall with the same gravitational acceleration. Why?
6. A friend who knows nothing about physics asks what keeps an orbiting space station from falling to Earth. Give an answer that will satisfy your friend.
7. Could you put a satellite in an orbit that keeps it stationary over the south pole? Explain.
8. Why are satellites generally launched eastward from low latitudes? *Hint:* Think about Earth's rotation.
9. Given the mass of the Earth, the distance to the Moon, the period of the Moon's orbit, and the value of G, could you calculate the Moon's mass? How, or why not?
10. How should a satellite be launched so that its orbit takes it over every point on the (rotating) Earth?
11. Does the gravitational force of the Sun ever do work on a planet in a circular orbit? Explain.

Exercises and Problems

Exercises

Section 8.2 Universal Gravitation

12. Space explorers land on a planet that has the same mass as Earth, but they find they weigh twice as much as they would on Earth. What is the radius of the planet?
13. Use data for the Moon's orbit from Appendix E to compute the Moon's acceleration in its circular orbit, and verify that the result is consistent with Newton's law of universal gravitation.
14. To what fraction of its current radius would Earth have to be shrunk (with no change in mass) for the gravitational acceleration at its surface to triple?
15. Calculate the gravitational acceleration at the surface of (a) Mercury and (b) Saturn's moon Titan.
16. Two identical lead spheres are 14 cm apart and attract each other with a force of 0.25 μN. What is their mass?
17. What is the approximate value of the gravitational force between a 67-kg astronaut and a 73,000-kg space shuttle when they're 84 m apart?
18. A sensitive gravimeter is carried to the top of Chicago's Sears Tower, where its reading for the acceleration of gravity is 0.00136 m/s² lower than at street level. Find the height of the building.

Section 8.3 Orbital Motion

19. At what altitude will a satellite complete a circular orbit of the Earth in 2.0 h?
20. Find the speed of a satellite in geosynchronous orbit.
21. Mars's orbit has a diameter 1.52 times that of Earth's orbit. How long does it take Mars to orbit the Sun?
22. Calculate the orbital period for Jupiter's moon Io, which orbits 4.22×10^5 km from the center of the 1.9×10^{27}-kg planet.
23. An astronaut hits a golf ball horizontally from the top of a lunar mountain so fast that it goes into circular orbit. What is its orbital period?
24. Find the period of a spacecraft in a circular orbit 150 km above Mars's surface.

Section 8.4 Gravitational Energy

25. Earth's distance from the Sun varies from 1.47×10^{11} m at perihelion to 1.52×10^{11} m at aphelion because its orbit is not quite circular. Find the change in potential energy as Earth goes from perihelion to aphelion.
26. How much energy does it take to launch a 230-kg instrument package on a vertical trajectory that peaks at an altitude of 1800 km?
27. A rocket is launched vertically upward from Earth's surface at a speed of 5.1 km/s. What is its maximum altitude?

28. What vertical launch speed is necessary to get a rocket to an altitude of 1100 km?
29. Find the energy necessary to put 1 kg, initially at rest on Earth's surface, into geosynchronous orbit.
30. What is the total mechanical energy associated with Earth's orbital motion?
31. The escape speed from a planet of mass 2.9×10^{24} kg is 7.1 km/s. What is the planet's radius?
32. Determine the escape speed from (a) Jupiter's moon Callisto, with mass 1.07×10^{23} kg and radius 2.40 Mm, and (b) a neutron star, with the Sun's mass crammed into a sphere with radius 6.0 km.
33. To what radius would Earth have to be shrunk, with no loss of mass, for escape speed at its surface to be 30 km/s?

Problems

34. The gravitational acceleration at the surface of a planet is 22.5 m/s². Find the acceleration at a height above the surface equal to half the planet's radius.
35. Astronauts put their spaceship into orbit about a planet. They find that the acceleration of gravity at their orbital altitude is half that at the planet's surface. How far above the planet's surface are they orbiting? Answer in terms of the planet's radius.
36. If you're standing on the ground 15 m directly below the center of a spherical water tank containing 4×10^6 kg of water, by what fraction is your weight reduced due to the gravitational attraction of the water?
37. Given the orbital radius of 384,400 km and period of 27.3 days, calculate the acceleration of the Moon in its circular orbit, and compare it with the acceleration of gravity at Earth's surface. Show that the acceleration of the Moon is lower by the ratio of the square of Earth's radius to the square of the Moon's orbital radius, thus confirming the inverse-square law for the gravitational force.
38. During the Apollo Moon landings, one astronaut remained with the command module in lunar orbit, about 130 km above the surface. For half of each orbit, this astronaut was completely cut off from the rest of humanity, as the spacecraft rounded the far side of the Moon. How long did this period last?
39. A white dwarf is a collapsed star with roughly the mass of the Sun compressed into the size of the Earth. What would be (a) the orbital speed and (b) the orbital period for a spaceship in orbit just above the surface of a white dwarf?
40. Given that our Sun orbits the galaxy with a period of 200 My at a distance of 2.6×10^{20} m from the galactic center, estimate the mass of the galaxy. Assume (incorrectly) that the galaxy is essentially spherical and that most of its mass lies interior to the Sun's orbit. To how many Sun-mass stars is your estimate equivalent?

41. Satellites A and B are in circular orbits, with A twice as far from Earth's center as B. How do their orbital periods compare?

42. The asteroid Pasachoff orbits the Sun with a period of 1417 days. What is the semimajor axis of its orbit? Determine from Kepler's third law, using Earth's orbital radius and period, respectively, as your units of distance and time.

43. One proposal for dealing with radioactive waste is to shoot it into the Sun. Suppose a waste canister were simply dropped, starting from rest in the vicinity of Earth's orbit. At what speed would it hit the Sun?

44. At its perihelion in February 1986, Comet Halley was 8.79×10^7 km from the Sun and was moving at 54.6 km/s. What was its speed when it crossed Neptune's orbit in 2006?

45. Neglecting air resistance, to what height would you have to fire a rocket for the constant-acceleration equations of Chapter 2 to give a height that is in error by 1%? Would those methods over- or underestimate the height?

46. Show that an object released from rest very far from Earth $(r \gg R_\text{E})$ reaches Earth's surface at essentially escape speed.

47. By what factor must the speed of an object in circular orbit be increased to reach escape speed from its orbital altitude?

48. Astronomers discover a new comet. As it crosses Earth's orbit, the comet is moving at 45 km/s. Is the comet in an open or closed orbit?

49. Two meteoroids are 250,000 km from Earth and moving at 2.1 km/s. One is headed straight for Earth, while the other is on a path that will come within 8500 km of Earth's center (Fig. 8.15). (a) What is the speed of the first meteoroid when it strikes Earth? (b) What is the speed of the second meteoroid at its closest approach to Earth? (c) Will the second meteoroid ever return to Earth's vicinity?

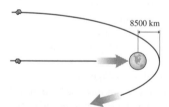

8500 km

FIGURE 8.15 Problem 49

50. Neglecting Earth's rotation, show that the energy needed to launch a satellite of mass m into circular orbit at altitude h is

$$\left(\frac{GM_\text{E}m}{R_\text{E}}\right)\left(\frac{R_\text{E} + 2h}{2(R_\text{E} + h)}\right)$$

51. A projectile is launched vertically upward from a planet of mass M and radius R; its initial speed is twice the escape speed. Derive an expression for its speed as a function of the distance r from the center of the planet.

52. A spacecraft is in a circular Earth orbit at an altitude of 5500 km. By how much will its altitude decrease if it moves to a new circular orbit where (a) its orbital speed is 10% higher or (b) its orbital period is 10% shorter?

53. Two meteoroids are 160,000 km from Earth's center and heading straight toward Earth. One is moving at 10 km/s, the other at 20 km/s. At what speeds will they strike Earth?

54. Two rockets are launched from Earth's surface, one at 12 km/s and the other at 18 km/s. How fast is each moving when it crosses the Moon's orbit?

55. A satellite is in an elliptical orbit at altitudes ranging from 230 to 890 km. At the high point it's moving at 7.23 km/s. How fast is it moving at the low point?

56. A missile's trajectory takes it to a maximum altitude of 1200 km. If its launch speed is 6.1 km/s, how fast is it moving at the peak of its trajectory?

57. A 720-kg spacecraft has total energy -5.3×10^{11} J and is in circular orbit about the Sun. Find (a) its orbital radius, (b) its kinetic energy, and (c) its speed.

58. Mercury's orbital speed varies from 38.8 km/s at aphelion to 59.0 km/s at perihelion. If the planet is 6.99×10^{10} m from the Sun's center at aphelion, how far is it at perihelion?

59. Show that the form $\Delta U = mg\,\Delta r$ follows from Equation 8.5 when $r_1 \simeq r_2$. *Hint:* Write $r_2 = r_1 + \Delta r$, and apply the binomial approximation (see Appendix A).

60. Two satellites are in geosynchronous orbit but in diametrically opposite positions (Fig. 8.16). Into how much lower a circular orbit should one spacecraft descend if it is to catch up with the other after 10 complete orbits? Neglect rocket firing times and time spent moving between the two circular orbits.

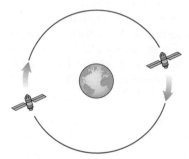

FIGURE 8.16 Problem 60

61. We derived Equation 8.4 on the assumption that the massive gravitating center remains fixed. Now consider the case of two objects of equal mass M orbiting each other as shown in Fig. 8.17. Show that the orbital period is given by

$$T^2 = \frac{16\pi^2 r^3}{GM}$$

where r is the orbital radius (half the distance between the objects).

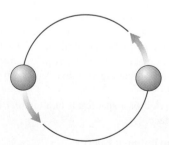

FIGURE 8.17 Problem 61

62. Tidal effects in the Earth-Moon system are causing the Moon's orbital period to increase at a current rate of about 35 ms per century. Assuming the Moon's orbit around the Earth is circular, to what rate of change in the Earth-Moon distance does this correspond? *Hint:* Differentiate Kepler's third law, Equation 8.4, and consult Appendix E.

63. Tide-producing forces are proportional to the change in the force of gravity with distance. By differentiating Equation 8.1, estimate the ratio of the tide-producing forces on the Earth due to the Sun and the Moon. Compare your answer with the ratio of the gravitational force that the Sun and the Moon exert on the Earth. Use data from Appendix E.

64. For a circular orbit around a massive gravitating body, the speed depends on the radius according to Equation 8.3; for elliptical orbits, the speed varies according to the equation $v^2 = 2GM[(1/r) - (1/2a)]$, where r is the distance from the massive body and a is the semimajor axis of the ellipse (i.e., half the sum of the closest and farthest distances). A satellite can be transferred from one circular orbit (at radius r_1) to a higher orbit (at radius r_2) by boosting the circular speed v_1 at r_1 to the appropriate speed for an elliptical orbit whose distance varies between r_1 and r_2, and then boosting the speed in the elliptical orbit at r_2 to the circular speed v_2. This is called a Hohmann transfer. (a) How much energy is required for the first boost in such a transfer to take a 250-kg satellite from a circular orbit at a 400-km altitude to the altitude of a geosynchronous orbit? (b) How much energy is required for the second boost? (c) Check that the sum of your answers to parts (a) and (b) equals the difference in the total energies in the two circular orbits.

65. In the 2004 Summer Olympic Games, the men's high-jump medalists cleared a height of at least 2.34 m. There are numerous small asteroids (radii of just a few kilometers; assume them to be spherical) orbiting the Sun between Jupiter and Mars. Could one of these athletes actually jump *off* an asteroid, never to return? What size is the largest asteroid from which an Olympian could jump off? Based on your calculations, a new event may be initiated for the 2040 Summer Games to be hosted by the Mars Colony. You may make these assumptions: The high jumper can jump at the same initial velocity on the asteroid as on Earth, and the asteroid has the same average density as Earth.

66. Astronaut Alan Shepard took a golf ball to the Moon. He hit what is thought to be the longest golf drive in history. (Recall that the range of a projectile is inversely proportional to the acceleration due to gravity, so the range on the Moon is about six times longer than on Earth.) What speed would an astronaut need to give a lunar golf ball in order to put the ball in orbit around the Moon? Do you think it could be done?

67. When people on a cruise ship cross Earth's equator for the first time, they are awarded a certificate to commemorate the experience. Imagine that you are the tour director for a trip to the Moon. Transplanetary Tours promises tour participants a certificate to commemorate their passage from the stronger influence of Earth's gravitational pull to the stronger pull of the Moon at the point where the two forces on your spaceship are equal. Where on the trip should you award the certificate? (You may express your answer as the distance from either the Earth or the Moon.)

68. You are reading a newspaper article about asteroids. Astronomers studying the impact of asteroids with Earth used 11 km/s for an impact speed. An assumption they made is that the asteroid came from a very great distance from Earth. Why did they use this speed?

Answers to Chapter Questions

Answer to Chapter Opening Question

The satellite orbits Earth in 24 hours, so from Earth's surface it appears at a fixed position in the sky.

Answers to GOT IT? Questions

8.1 (d) Quadrupled. If the original distance were r, the original force would be proportional to $1/r^2$. At half that distance, the force is proportional to $1/(r/2)^2 = 4/r^2$.

8.2 Hyperbolic > parabolic > elliptical > circular.

8.3 (a), (c), and (d). Since B has higher total energy, it must have lower kinetic energy and is therefore moving slower. B is farther from the gravitating body, so its potential energy is higher—still negative, but less so than A's. For circular orbits, the ratio of potential energy to total energy is always the same—namely, $U = 2E$.

Systems of Particles

■ As the skier flies through the air, most parts of his body follow complex trajectories. But one special point follows a parabola. What's that point, and why is it special?

So far we've treated objects as point particles, ignoring the fact that most objects are composed of smaller parts. Here we deal explicitly with systems of many particles. These include **rigid bodies**—objects such as baseballs, cars, and planets whose constituent particles are stuck together in fixed orientations—as well as systems like human bodies, exploding fireworks, or flowing rivers, whose parts move relative to one another. In subsequent chapters we'll look at specific instances of many-particle systems, including the rotational motion of rigid bodies (Chapter 10) and the behavior of fluids (Chapter 15).

9.1 Center of Mass

The motion of the skier in the photo above is complex, with each part of his body moving on a different path. But the superimposed curve shows one point following the parabola we expect of a projectile (Section 3.5). This point is the **center of mass**, an average position of all the mass making up the skier. Since the net force on the skier as a whole is gravity, the photo suggests that the center of mass obeys Newton's second law, $\vec{F}_{net} = M\vec{a}_{cm}$, where M is the skier's total mass and $\vec{a}_{cm}$ the acceleration of the center of mass. (We'll use the subscript cm for quantities associated with the center of mass.) To find the center of mass, we therefore need to locate a point whose acceleration obeys $\vec{F}_{net} = M\vec{a}_{cm}$, with $\vec{F}_{net}$ the net force on the entire system.

Consider a system of many particles. To find the center of mass, we want an equation that looks like Newton's second law but involves the total mass of the system and the net force on the entire system. If we apply Newton's second law to the *i*th particle in the system, we have

$$\vec{F}_i = m_i \vec{a}_i = m_i \frac{d^2 \vec{r}_i}{dt^2} = \frac{d^2 m_i \vec{r}_i}{dt^2}$$

where $\vec{F}_i$ is the net force on the particle, m_i is its mass, and we've written the acceleration $\vec{a}_i$ as the second derivative of the position $\vec{r}_i$. The total force on the system is the sum of the forces acting on all N particles. We write this sum compactly using the summation symbol Σ:

$$\vec{F}_{\text{total}} = \sum_{i=1}^{N} \vec{F}_i = \sum_{i=1}^{N} \frac{d^2 m_i \vec{r}_i}{dt^2}$$

where the sum runs over all particles composing the system, from $i = 1$ to N. But the sum of derivatives is the derivative of the sum, so

$$\vec{F}_{\text{total}} = \frac{d^2 \left(\sum m_i \vec{r}_i \right)}{dt^2}$$

We can now put this equation in the form of Newton's second law. Multiplying and dividing the right-hand side by the total mass $M = \sum m_i$, and distributing this constant M through the differentiation, we have

$$\vec{F}_{\text{total}} = M \frac{d^2}{dt^2} \left(\frac{\sum m_i \vec{r}_i}{M} \right) \tag{9.1}$$

Equation 9.1 has a form like Newton's law applied to the total mass if we define

$$\vec{r}_{\text{cm}} = \frac{\sum m_i \vec{r}_i}{M} \quad \text{(center of mass)} \tag{9.2}$$

Then the derivative in Equation 9.1 becomes $d^2 \vec{r}_{\text{cm}}/dt^2$, which we recognize as the center-of-mass acceleration, $\vec{a}_{\text{cm}}$. So now Equation 9.1 reads $\vec{F}_{\text{total}} = M \vec{a}_{\text{cm}}$. This is almost Newton's law—but not quite, because the force here is the sum of all the forces acting on all the particles of the system, and we want just the net **external force**—the net force applied from *outside* the system. We can write the force $\vec{F}_{\text{total}}$ as

$$\vec{F}_{\text{total}} = \sum \vec{F}_{\text{ext}} + \sum \vec{F}_{\text{int}}$$

where $\sum \vec{F}_{\text{ext}}$ is the sum of all the external forces and $\sum \vec{F}_{\text{int}}$ the sum of the internal forces. According to Newton's third law, each of the internal forces has an equal but oppositely directed force that itself acts on a particle of the system and is therefore included in the sum $\sum \vec{F}_{\text{int}}$. (Each external force is also part of a third-law pair, but forces paired with the external forces act *outside* the system and therefore aren't included in the sum.) Added vectorially, the internal forces therefore cancel in pairs, so $\sum \vec{F}_{\text{int}} = \vec{0}$, and the force $\vec{F}_{\text{total}}$ in Equation 9.1 is just the net *external* force applied to the system. So the point $\vec{r}_{\text{cm}}$ defined in Equation 9.2 does obey Newton's law, written in the form

$$\vec{F}_{\text{net ext}} = M \vec{a}_{\text{cm}} = M \frac{d^2 \vec{r}_{\text{cm}}}{dt^2} \tag{9.3}$$

where $\vec{F}_{\text{net ext}}$ is the net external force applied to the system and M is the total mass.

We've defined the center of mass $\vec{r}_{\text{cm}}$ so we can apply Newton's second law to the entire system rather than to each individual particle. As far as its overall motion is concerned, a complex system acts as though all its mass were concentrated at the center of mass.

Finding the Center of Mass

Equation 9.2 shows that the center-of-mass position is an average of the positions of the individual particles, weighted by their masses. For a one-dimensional system, Equation 9.2 becomes a single equation: $x_{cm} = \sum m_i x_i/M$; in two and three dimensions, there are similar equations for the center-of-mass coordinates y_{cm} and z_{cm}. Finding the center of mass (CM) is a matter of establishing a coordinate system and then using the components of Equation 9.2.

EXAMPLE 9.1 The CM in One Dimension: Weightlifting

Find the center of mass of a barbell consisting of 50-kg and 80-kg weights at the opposite ends of a 1.5-m-long bar of negligible mass.

INTERPRET This is a problem about center of mass. We identify the system as consisting of two "particles"—namely, the two weights.

DEVELOP Figure 9.1 shows the barbell. Here, with just two particles, we have a one-dimensional situation and Equation 9.2, $\vec{r}_{cm} = \sum m_i \vec{r}_i/M$, becomes $x_{cm} = (m_1 x_1 + m_2 x_2)/(m_1 + m_2)$. Before we can apply this equation, however, we need to set up a coordinate system. Choosing the origin at one of the constituent particles makes $x = 0$ for that particle and thus eliminates one term in the equation. Of course, any coordinate system will do—but a smart choice makes the math easier. So let's take $x = 0$ at the 50-kg mass. Our plan is then to find the center-of-mass coordinate x_{cm} using our one-dimensional version of Equation 9.2.

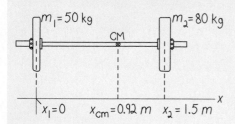

FIGURE 9.1 Our sketch of the barbell.

EVALUATE With $x = 0$ at the left end of the barbell, the coordinate of the 80-kg mass is $x_2 = 1.5$ m. So our equation becomes

$$x_{cm} = \frac{m_1 x_1 + m_2 x_2}{m_1 + m_2} = \frac{m_2 x_2}{m_1 + m_2} = \frac{(80 \text{ kg})(1.5 \text{ m})}{(50 \text{ kg} + 80 \text{ kg})} = 0.92 \text{ m}$$

where the equation simplified because of our choice $x_1 = 0$.

ASSESS As Fig. 9.1 shows, this result makes sense: The center of mass is closer to the heavier weight. If the weights had been equal, the center of mass would have been right in the middle.

✔ **TIP** Choosing the Origin

Choosing the origin at one of the masses here conveniently makes one of the terms in the sum $\sum m_i x_i$ zero. But, as always, the choice of origin is purely for convenience and doesn't influence the actual physical location of the center of mass. Exercise 14 demonstrates this point, repeating Example 9.1 with a different origin.

EXAMPLE 9.2 The CM in Two Dimensions: A Space Station

Figure 9.2 shows a space station consisting of three modules arranged in an equilateral triangle, connected by struts of length L and of negligible mass. Two modules have mass m, the other $2m$. Find the center of mass.

INTERPRET We're after the center of mass of the system consisting of the three modules.

DEVELOP Figure 9.2 is our drawing. We'll use Equation 9.2, $\vec{r}_{cm} = \sum m_i \vec{r}_i/M$, to find the center-of-mass coordinates x_{cm} and y_{cm}. A sensible coordinate system has the origin at the module with mass $2m$ and the y axis downward, as shown in Fig. 9.2.

EVALUATE Labeling the modules from left to right, we see that $x_1 = -L\sin 30° = -\frac{1}{2}L$, $y_1 = L\cos 30° = L\sqrt{3}/2$; $x_2 = y_2 = 0$; and $x_3 = -x_1 = \frac{1}{2}L$, $y_3 = y_1 = L\sqrt{3}/2$. Writing explicitly the x and y components of Equation 9.2 for this case gives

$$x_{cm} = \frac{mx_1 + mx_3}{4m} = \frac{m(x_1 - x_1)}{4m} = 0$$

$$y_{cm} = \frac{my_1 + my_3}{4m} = \frac{2my_1}{4m} = \frac{1}{2}y_1 = \frac{\sqrt{3}}{4}L \approx 0.43L$$

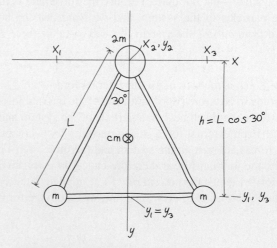

FIGURE 9.2 Our sketch of the space station.

cont'd.

Although there are three "particles" here, our choice of coordinate system left only two nonzero terms in the numerator, both associated with the same mass m. The more massive module is still in the problem, though; its mass $2m$ contributes to make the total mass M in the denominator equal to $4m$.

ASSESS That $x_{cm} = 0$ is apparent from symmetry (more on this in the following Tip). How about the result for y_{cm}? We have $2m$ at the top of the triangle, and $m + m = 2m$ at the bottom—so shouldn't the center of mass lie midway up the triangle? It does! Expressing the center of mass in terms of the triangle side L obscures this fact. The triangle's height is $h = L \cos 30° = L\sqrt{3}/2$, and our answer for y_{cm} is indeed half this value. We marked the CM on Fig. 9.2.

✔ **TIP** Exploit Symmetries

It's no accident that x_{cm} here lies on the vertical line that bisects the triangle; after all, the triangle is symmetric about that line, so its mass is distributed evenly on either side. Exploit symmetry whenever you can; that can save you a lot of computation throughout physics!

■

Continuous Distributions of Matter

We've expressed the center of mass as a sum over individual particles. Ultimately, matter *is* composed of individual particles. But it's often a convenient approximation to consider that matter is continuously distributed; we don't want to deal with 10^{23} or so atoms to find the center of mass of a book or other macroscopic object. We can consider continuous matter to be composed of individual pieces of mass Δm_i, with position vectors $\vec{r}_i$; we call these pieces **mass elements** (Fig. 9.3). The center of mass of the entire chunk is then given by Equation 9.2: $\vec{r}_{cm} = \left(\sum \Delta m_i \vec{r}_i \right)/M$, where $M = \sum \Delta m_i$ is the total mass. In the limit as the mass elements become arbitrarily small, this expression becomes an integral:

$$\vec{r}_{cm} = \lim_{\Delta m_i \to 0} \frac{\sum \Delta m_i \vec{r}_i}{M} = \frac{\int \vec{r}\, dm}{M} \quad \left(\begin{array}{c} \text{center of mass,} \\ \text{continuous matter} \end{array} \right) \quad (9.4)$$

where the integration is over the entire volume of the object. Like the sum in Equation 9.2, the integral of the vector $\vec{r}$ here stands for three separate integrals for the three components of the center-of-mass position.

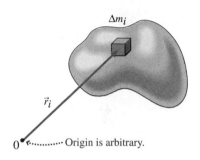

FIGURE 9.3 A chunk of continuous matter, showing one mass element Δm_i and its position vector $\vec{r}_i$.

EXAMPLE 9.3 **The CM of Continuous Matter: An Aircraft Wing**

A supersonic aircraft wing is an isosceles triangle of length L, width w, and negligible thickness. It has mass M, distributed uniformly over the wing. Where's its center of mass?

INTERPRET Here the matter is distributed continuously, so we need to integrate to find the center of mass. We identify an axis of symmetry through the wing, which we designate the x axis. By symmetry, the center of mass lies along this x axis, so $y_{cm} = 0$ and we'll need to calculate only x_{cm}.

DEVELOP Figure 9.4 shows the wing. Equation 9.4 applies, and we need only the x component because the y component is evident from symmetry. The x component of Equation 9.4 is $x_{cm} = \left(\int x\, dm \right)/M$. Developing a plan for dealing with an integral like this requires some thought; we'll first do the work and then summarize the general steps involved.

Our goal is to find an appropriate mass element dm in terms of the infinitesimal coordinate interval dx. As shown in Fig. 9.4, here it's easiest to use a vertical strip of width dx. Each such strip has a different height h, depending on its position x. If we choose a coordinate system with origin at the wing apex, then, as you can see from the figure, the height grows linearly from 0 at $x = 0$ to w at $x = L$. So $h = (w/L)x$. Now the strip is infinitesimally narrow, so the sloping edges don't

matter and its area is that of a very thin rectangle—namely, $h\, dx = (w/L)x\, dx$. The strip's mass dm is then the same fraction of the total wing mass M as its area is of the total wing area $\frac{1}{2}wL$; that is,

$$\frac{dm}{M} = \frac{(w/L)x\, dx}{\frac{1}{2}wL} = \frac{2x\, dx}{L^2}$$

so $dm = 2Mx\, dx/L^2$.

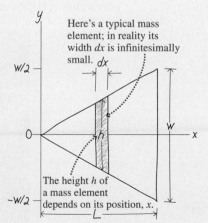

Here's a typical mass element; in reality its width dx is infinitesimally small.

The height h of a mass element depends on its position, x.

FIGURE 9.4 Our sketch of the supersonic aircraft wing.

cont'd.

In the integral we weight each mass element dm by its distance x from the origin, and then sum—that is, integrate—over all mass elements. So, from Equation 9.4, we have

$$x_{cm} = \frac{1}{M} \int x \, dm = \frac{1}{M} \int_0^L x \left(\frac{2Mx}{L^2} dx \right) = \frac{2}{L^2} \int_0^L x^2 \, dx$$

As always, constants come outside the integral because they're shared by all terms in the infinite sum that the integral represents. We set the limits to 0 and L to cover all the mass elements in the wing. Now we're finally ready to find x_{cm}.

EVALUATE The hard part of our plan is done. All that's left is to evaluate the integral:

$$x_{cm} = \frac{2}{L^2} \int_0^L x^2 \, dx = \frac{2}{L^2} \left. \frac{x^3}{3} \right|_0^L = \frac{2L^3}{3L^2} = \frac{2}{3} L$$

ASSESS Make sense? Yes. Our answer puts the center of mass toward the back of the wing where, because of its increasing width, most of the mass lies. And in a complicated calculation like this one, it's reassuring to see that the answer is a quantity with the units of length. ∎

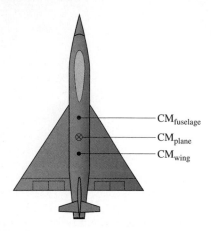

FIGURE 9.5 The center of mass of the airplane is found by treating the wing and fuselage as point particles located at their respective centers of mass.

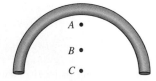

FIGURE 9.6 Got it? The center of mass lies outside the semicircular wire, but which point is it?

TACTICS 9.1 **Setting Up an Integral**

An integral like $\int x \, dm$ can be confusing because you see both x and dm after the integral sign and they don't seem related. But they are, and here's how to proceed:

1 Find a suitable shape for your mass elements, preferably one that exploits any symmetry in the situation. One dimension of the elements should involve an infinitesimal interval in one of the coordinates x, y, or z. In Example 9.3, the mass elements were strips, symmetric about the wing's centerline and with width dx.

2 Find an expression for the infinitesimal area of your mass elements (in a one-dimensional problem it would be the length; in a three-dimensional problem, the volume). In Example 9.3, the infinitesimal area of each mass element was the strip height h multiplied by the width dx.

3 Form ratios that relate the infinitesimal coordinate interval to the physical quantity in the integral—which in Example 9.3 is the mass element dm. Here we formed the ratio of the area of a mass element to the total area, and equated that to the ratio of dm to the total mass M.

4 Solve your ratio statement for the infinitesimal quantity, in this case dm, that appears in your integral. Then you're ready to evaluate the integral.

Sometimes you'll be given a density—mass per volume, per area, or per length—and then in place of steps 3 and 4 you find dm by multiplying the density by the infinitesimal volume, area, or length you identified in step 2.

Although we described this procedure in the context of Example 9.3, it also applies to other integrals you'll encounter in different areas of physics.

With still more complex objects, it's convenient to find the centers of mass of sub-parts and then treat those parts as point particles to find the center of mass of the entire object; see Fig. 9.5.

The center of mass need not lie within an object, as Fig. 9.6 makes clear. High jumpers exploit this fact as they straddle the bar with their arms and legs dangling on either side (Fig. 9.7). Although the jumper's entire body may clear the bar, his center of mass doesn't need to!

GOT IT? 9.1 A thick wire is bent into a semicircle, as shown in Fig. 9.6. Which of the points shown is the center of mass?

FIGURE 9.7 A high jumper clears the bar, but his center of mass doesn't!

Motion of the Center of Mass

We defined the center of mass so its motion would obey Newton's law $\vec{F}_{\text{net ext}} = M\vec{a}_{\text{cm}}$, with $\vec{F}_{\text{net ext}}$ the net external force on the system and M the total mass. When gravity is the only external force, the center of mass follows the trajectory of a point particle. But if the net external force is zero, then the center-of-mass acceleration $\vec{a}_{\text{cm}}$ is also zero, and the center of mass moves with constant velocity. In the special case of a system at rest, the center of mass remains at rest despite any motions of its internal parts.

EXAMPLE 9.4 **CM Motion: Circus Train**

Jumbo, a 4.8-t elephant, is standing near one end of a 15-t railcar, which is at rest, all by itself, on a frictionless horizontal track. (Here t is the abbreviation for tonne, or metric ton, equal to 1000 kg.) Jumbo walks 19 m toward the other end of the car. How far does the car move?

INTERPRET We're asked about the car's motion, but we can interpret this problem as being fundamentally about the center of mass. We identify the relevant system as comprising Jumbo and the car. Because there's no net external force acting on the system, its center of mass can't move.

DEVELOP Figure 9.8a shows the initial situation. The symmetric car has its CM at its center (here we care only about the x component). Let's take a coordinate system with $x = 0$ at this point—that is, at the *initial* location of the car's center. After the car moves, its center will be somewhere else! Equation 9.2 applies—here in the simpler

one-dimensional, two-object form we used in Example 9.1: $x_{\text{cm}} = (m_J x_J + m_c x_c)/M$, where we use the subscripts J and c for Jumbo and the car, respectively, and where M is the total mass. We have a before/after situation in which the CM position can't change, so our plan is to write two versions of this expression, before and after Jumbo's walk. We'll then set them equal to state mathematically that the CM itself doesn't move.

We chose the coordinates so that $x_{\text{ci}} = 0$, where i designates the initial state, so our initial expression is $x_{\text{cm}} = m_J x_{Ji}/M$. After Jumbo's walk, our final expression is $x_{\text{cm}} = (m_J x_{Jf} + m_c x_{cf})/M$, where f stands for final. We don't know either coordinate here, but we do know that Jumbo walks 19 m *with respect to the car*. The elephant's final position x_{Jf} is therefore 19 m to the right of the initial x_{Ji}, adjusted by the car's displacement. Therefore Jumbo ends up at $x_{Jf} = x_{Ji} + 19\text{ m} + x_{cf}$. You might think we need a minus sign here because the car must move to the left. That's true, but the sign of x_{cf} will take care of that. Trust algebra! So our final expression is

$$x_{\text{cm}} = \frac{m_J x_{Jf} + m_c x_{cf}}{M} = \frac{m_J(x_{Ji} + 19\text{ m} + x_{cf}) + m_c x_{cf}}{M}.$$

EVALUATE Finally, we execute our plan, equating the two expressions for the unchanging position of the center of mass. The total mass M cancels, and we're left with $m_J x_{Ji} = m_J(x_{Ji} + 19\text{ m} + x_{cf}) + m_c x_{cf}$. We aren't given x_{Ji}, but the term $m_J x_{Ji}$ is on both sides of this equation, so it cancels, leaving $0 = m_J(19\text{ m} + x_{cf}) + m_c x_{cf}$. We solve for the unknown x_{cf} to get

$$x_{cf} = -\frac{(19\text{ m})m_J}{(m_J + m_c)} = -\frac{(19\text{ m})(4.8\text{ t})}{(15\text{ t} + 4.8\text{ t})} = -4.6\text{ m}$$

The minus sign here indicates a displacement to the left, as we anticipated (Fig. 9.8b). Because the masses appear only in ratios, we didn't need to convert to kilograms.

ASSESS The car's 4.6-m displacement is quite a bit less than Jumbo's (which is 19 m − 4.6 m, or 14.4 m relative to the ground). That makes sense because Jumbo is considerably less massive than the car. ∎

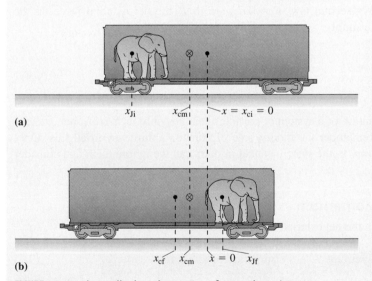

(a)

(b)

FIGURE 9.8 Jumbo walks, but the center of mass doesn't move.

9.2 Momentum

In Chapter 4 we defined the linear momentum $\vec{p}$ of a particle as $\vec{p} = m\vec{v}$, and we first wrote Newton's law in the form $\vec{F} = d\vec{p}/dt$. We suggested that this general form would play an important role in many-particle systems. We're now ready to explore that role.

The momentum of a system of particles is the vector sum of the individual momenta: $\vec{P} = \sum \vec{p}_i = \sum m_i \vec{v}_i$, where m_i and $\vec{v}_i$ are the masses and velocities of the individual particles. But we really don't want to keep track of all the particles in the system. Is there a simpler way to express the total momentum? There is, and it comes from writing the individual velocities as time derivatives of position: $\vec{v} = d\vec{r}/dt$. Then

$$\vec{P} = \sum m_i \frac{d\vec{r}_i}{dt} = \frac{d}{dt} \sum m_i \vec{r}_i$$

where the last step follows because the individual particle masses are constant and because the sum of derivatives is the derivative of the sum. In Section 9.1, we defined the center-of-mass position $\vec{r}_{cm}$ as $\sum m_i \vec{r}_i / M$, where M is the total mass. So the total momentum becomes

$$\vec{P} = \frac{d}{dt} M \vec{r}_{cm}$$

or, assuming the system mass M remains constant,

$$\vec{P} = M \frac{d\vec{r}_{cm}}{dt} = M\vec{v}_{cm} \tag{9.5}$$

where $\vec{v}_{cm} = d\vec{r}_{cm}/dt$ is the center-of-mass velocity. So the momentum of a system is given by an expression similar to that of a single particle; it's the product of the system mass and the system velocity—that is, the velocity of the center of mass. If this seems so obvious as not to need deriving, watch out! We'll see soon that the same is *not* true for the system's total energy.

If we differentiate Equation 9.5 with respect to time, we have

$$\frac{d\vec{P}}{dt} = M \frac{d\vec{v}_{cm}}{dt} = M\vec{a}_{cm}$$

where $\vec{a}_{cm}$ is the center-of-mass acceleration. But we defined the center of mass so that its motion obeyed Newton's second law, $\vec{F} = M\vec{a}_{cm}$, with $\vec{F}$ the net external force on the system. So we can write simply

$$\vec{F}_{net\ ext} = \frac{d\vec{P}}{dt} \tag{9.6}$$

showing that the momentum of a system of particles changes only if there's a net external force on the system. Remember the hidden role of Newton's third law in all this: Only because the forces *internal* to the system cancel in pairs can we ignore them and consider just the external force.

Conservation of Momentum

In the special case when the net external force is zero, Equation 9.6 gives $d\vec{P}/dt = \vec{0}$, so

$$\vec{P} = \text{constant} \qquad \text{(conservation of linear momentum)} \tag{9.7}$$

Equation 9.7 describes the **conservation of linear momentum**, one of the most fundamental laws of physics:

Conservation of linear momentum: When the net external force on a system is zero, the total momentum $\vec{P}$ of the system—the vector sum of the individual momenta $m\vec{v}$ of its constituent particles—remains constant.

Momentum conservation holds no matter how many particles are involved and no matter how they're moving. It applies to systems ranging in size from atomic nuclei to pool balls, from colliding cars to galaxies. Although we derived Equation 9.7 from Newton's laws, momentum conservation is even more basic than those laws, since it applies to subatomic and nuclear systems where the laws and even the language of Newtonian physics are hopelessly inadequate. The following examples show the wide range and power of the momentum conservation principle.

GOT IT! 9.2 A 500-g fireworks rocket is moving with velocity $\vec{v} = 60\hat{j}$ m/s at the instant it explodes. If you were to add the momentum vectors of all its fragments just after the explosion, what would be the result?

EXAMPLE 9.5 **Conservation of Momentum: Kayaking**

Jess (mass 53 kg) and Nick (mass 72 kg) sit in a 26-kg kayak at rest on frictionless water. Jess tosses Nick a 17-kg pack, giving it a horizontal speed of 3.1 m/s relative to the water. What's the kayak's speed while the pack is in the air and after Nick catches it?

INTERPRET The water is frictionless, so there's no net external force acting on the system, which we identify as comprising Jess, Nick, the kayak, and the pack. Since there's no net force, the concept behind this problem is that the system's momentum is conserved. Everything is initially at rest, so the momentum is zero.

DEVELOP Figure 9.9 shows the situation before and during the back-pack transfer. The relevant equation is Equation 9.7, $\vec{P} = $ constant, with the constant momentum in this case being zero. Our plan is to

Initially all momenta are zero . . .

$\vec{P}_p$

$\vec{P}_k + \vec{P}_J + \vec{P}_N$ . . . while the pack is in the air, momenta still sum to zero.

FIGURE 9.9 Our sketch for Example 9.5.

write an expression for the total momentum while the pack is in the air, and then use it to find the kayak's speed. Since the kayakers stay put in the kayak, they share the kayak's velocity v_k, and the combined momentum of Jess, Nick, and the kayak is $(m_J + m_N + m_k)\vec{v}_k$. The pack, meanwhile, has momentum $m_p \vec{v}_p$. We're interested only in what's happening in the horizontal direction, so we can treat this as a one-dimensional problem and drop the vector signs.

EVALUATE We sum to find the total momentum and set it equal to the initial value, zero, because momentum is conserved: $m_p v_p + (m_J + m_N + m_k)v_k = 0$. Solving for the unknown kayak velocity while the pack is in the air, we get

$$v_k = -\frac{m_p v_p}{m_J + m_N + m_k} = \frac{(17\ \text{kg})(3.1\ \text{m/s})}{53\ \text{kg} + 72\ \text{kg} + 26\ \text{kg}} = -0.35\ \text{m/s}$$

The minus sign shows that the kayak moves opposite to the pack. There's no need to calculate the speed after Nick catches the pack; kayak, passengers, and pack all have the same velocity then. For the total momentum to remain zero, that velocity must be zero.

ASSESS Make sense? The kayak and passengers together are much more massive than the pack, so their speed is lower. The momentum approach cuts through a lot of details, like the force and acceleration of the pack—but if you want to think more deeply, consider that Newton's third law requires the force on the pack to be of the same magnitude as the force the pack exerts on Jess and the rest of the system. That means less acceleration and therefore lower speed for the kayak and passengers. Momentum conservation and the third law are intimately related! ∎

EXAMPLE 9.6 **Conservation of Momentum: Radioactive Decay**

A lithium-5 nucleus (^{5}Li) is moving at 1.6 Mm/s when it decays into a proton (^{1}H, or p) and an alpha particle (^{4}He, or α). [The superscripts are the total numbers of nucleons and give the approximate masses in unified atomic mass units (u)]. The alpha particle is detected moving at 1.4 Mm/s, at 33° to the original velocity of the ^{5}Li nucleus. What are the magnitude and direction of the proton's velocity?

INTERPRET Although the physical situation here is entirely different from the preceding example, we can interpret this one, too, as being about momentum conservation. But there are two differences: First, in this case the total momentum is not zero, and, second, this situation involves two dimensions. The fundamental principle is the same, however: In the absence of external forces, a system's total momentum is unchanged before and after some event. Whether that event is the tossing of a pack or the decay of a nucleus makes no difference.

DEVELOP Figure 9.10 shows what we know: the velocities for the Li and He nuclei. You can probably guess that the proton must emerge with a downward momentum component, but we'll let the math confirm that. We determine that Equation 9.7, $\vec{P} = $ constant, applies, with the constant equal to the ^{5}Li momentum. After the decay, we have two momenta to account for, so Equation 9.7 becomes

$$m_{Li}\vec{v}_{Li} = m_p \vec{v}_p + m_\alpha \vec{v}_\alpha$$

FIGURE 9.10 Our sketch for Example 9.6: what we're given.

cont'd.

Let's choose the x axis along the direction of $\vec{v}_{Li}$. Then the two components of the momentum conservation equation become

x component: $\quad m_{Li}v_{Li} = m_pv_{px} + m_\alpha v_{\alpha x}$

y component: $\quad 0 = m_pv_{py} + m_\alpha v_{\alpha y}$

Our plan is to solve these equations for the unknowns v_{px} and v_{py}. From these we can get the magnitude and direction of the proton's velocity.

EVALUATE From Fig. 9.10 it's evident that $v_{\alpha x} = v_\alpha \cos\phi$ and $v_{\alpha y} = v_\alpha \sin\phi$. So we can solve our two equations to get

$$v_{px} = \frac{m_{Li}v_{Li} - m_\alpha v_{\alpha x}}{m_p} = \frac{m_{Li}v_{Li} - m_\alpha v_\alpha \cos\phi}{m_p}$$

$$= \frac{(5.0\,\text{u})(1.6\,\text{Mm/s}) - (4.0\,\text{u})(1.4\,\text{Mm/s})(\cos 33°)}{1.0\,\text{u}}$$

$$= 3.30\,\text{Mm/s}$$

$$v_{py} = -\frac{m_\alpha v_{\alpha y}}{m_p} = -\frac{m_\alpha v_\alpha \sin\phi}{m_p}$$

$$= \frac{(4.0\,\text{u})(1.4\,\text{Mm/s})(\sin 33°)}{1.0\,\text{u}} = -3.05\,\text{Mm/s}$$

Thus the proton's speed $v_p = \sqrt{v_{px}^2 + v_{py}^2} = 4.5\,\text{Mm/s}$, and its direction is $\theta = \tan^{-1}(v_{py}/v_{px}) = -43°$. Note that here, as in Example 9.4, the masses appear only in ratios so we don't need to change units.

ASSESS Make sense? That negative θ tells us the proton's velocity is downward, as we anticipated. Figure 9.11 makes our result clear. Here we multiplied the velocities by the masses to get momentum vectors. The two momenta after the decay event have equal but opposite vertical components, reflecting that the total momentum of the system never had a vertical component. And the two horizontal components sum to give the initial momentum of the lithium nucleus. Momentum is indeed conserved.

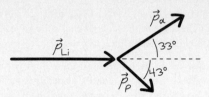

FIGURE 9.11 Our momentum diagram for Example 9.6.

Defining the System: Internal and External Forces

Whether a force is internal or external to a system depends on how we choose to define the system. Sometimes we're interested in the force on a particular object, so it's convenient to define that object as our system. If an external force is acting, then Newton's law says that the system's momentum will change at a rate equal to the net external force. Example 9.7 makes this point.

EXAMPLE 9.7 **Changing Momentum: Fighting a Fire**

A firefighter directs a stream of water against the window of a burning building, hoping to break the window so water can get to the fire. The hose delivers water at the rate of 45 kg/s, and the water hits the window moving horizontally at 32 m/s. After hitting the window, the water drops vertically. What horizontal force does the water exert on the window?

INTERPRET We're asked about the window, but we're told a lot more about the water. The water stops at the window, so clearly the window exerts a force on the water—and by Newton's third law, that force is equal in magnitude to the force we're after—namely, the force of the water on the window. So we identify the water as our system.

DEVELOP Newton's law in the form $\vec{F} = d\vec{P}/dt$ applies to the water. So our plan is to find the rate at which the water's momentum changes. By Newton's second law, that's equal to the window's force on the water, and by Newton's third law, that's equal to the water's force on the window.

EVALUATE The water strikes the window at 32 m/s, so each kilogram of water loses 32 kg·m/s of momentum. Water strikes the window at the rate of 45 kg/s, so the rate at which it loses momentum to the window is

$$\frac{dP}{dt} = (45\,\text{kg/s})(32\,\text{m/s}) = 1400\,\text{kg}\cdot\text{m/s}^2$$

By Newton's second law, that's equal to the force on the water, and by the third law, that in turn is equal in magnitude to the force on the window. So the window experiences a 1400-N force from the water. Since the window is rigidly attached to the building and the Earth, it doesn't experience significant acceleration—until it breaks and the glass fragments accelerate violently.

ASSESS 1400 N is about twice the weight of a typical person, and a fire hose produces quite a blast of water, so this number seems reasonable. Check the units, too: 1 kg·m/s² is equal to 1 N, so our answer does have the units of force.

GOT IT? 9.3 Two skaters toss a basketball back and forth on frictionless ice. Which of the following does not change: (a) the momentum of an individual skater; (b) the momentum of the basketball; (c) the momentum of the system consisting of one skater and the basketball; (d) the momentum of the system consisting of both skaters and the basketball?

9.3 Kinetic Energy of a System

We've seen how the momentum of a many-particle system is determined entirely by the motion of its center of mass; the detailed behavior of the individual particles doesn't matter. For example, a firecracker sliding on ice has the same total momentum before and after it explodes.

The same, however, is *not* true of a system's kinetic energy. Energetically, that firecracker is very different after it explodes; internal potential energy has become kinetic energy of the fragments. Nevertheless, the center-of-mass concept remains useful in categorizing the kinetic energy associated with a system of particles.

The total kinetic energy of a system is the sum of the kinetic energies of the constituent particles: $K = \sum \frac{1}{2}m_i v_i^2$. But the velocity $\vec{v}_i$ of a particle can be written as the vector sum of the center-of-mass velocity $\vec{v}_{cm}$ and a velocity $\vec{v}_{i\,rel}$ of that particle relative to the center of mass: $\vec{v}_i = \vec{v}_{cm} + \vec{v}_{i\,rel}$. Then the total kinetic energy of the system is

$$K = \sum \tfrac{1}{2}m_i(\vec{v}_{cm} + \vec{v}_{i\,rel}) \cdot (\vec{v}_{cm} + \vec{v}_{i\,rel}) = \sum \tfrac{1}{2}m_i v_{cm}^2 + \sum m_i \vec{v}_{cm} \cdot \vec{v}_{i\,rel} + \sum \tfrac{1}{2}m_i v_{i\,rel}^2 \quad (9.8)$$

Let's examine the three sums making up the total kinetic energy. Since the center-of-mass speed v_{cm} is common to all particles, it can be factored out of the first sum, so $\sum \frac{1}{2}m_i v_{cm}^2 = \frac{1}{2}v_{cm}^2 \sum m_i = \frac{1}{2}M v_{cm}^2$, where M is the total mass. This is the kinetic energy of a particle with mass M moving at speed v_{cm}, so we call it K_{cm}, the **kinetic energy of the center of mass**.

APPLICATION   **Rockets**

Rockets provide our means of escaping Earth's gravity. And in the vacuum of space, where there's nothing for a wheel or propeller to push against, rockets are our most practical means of propulsion.

A rocket engine ejects hot gases through its exhaust nozzle. If no external forces act, the total momentum of the (rocket + fuel) system stays constant—and as a result momentum carried by the exhaust results in equal but opposite momentum gained by the rocket. So momentum conservation explains why the rocket moves opposite its exhaust.

The rate of momentum change is the force on the rocket; engineers call this the *thrust*. As in the fire hose of Example 9.7, thrust is the product of the exhaust rate (dM/dt, in kg/s) and the exhaust speed v_{ex}: $F = v_{ex}\,dM/dt$. Because a rocket has to carry—and accelerate—the mass that it's going to exhaust, the most efficient rockets provide a given thrust with the lowest fuel mass. That calls for the highest possible exhaust velocity.

What actually propels the rocket? The figure shows that it's ultimately the hot gases inside the rocket engine pushing on the *front* of the engine chamber. It's not that the rocket "pushes against" anything outside itself; all the pushing is done inside the rocket engine, accelerating the rocket forward. That's why rockets work just fine in the vacuum of space.

High-pressure gas pushes equally in opposite directions, and this "rocket" goes nowhere.

(a)

Open an exhaust port . . .

Exhaust port

(b)

. . . and there's now an unbalanced force on the front of the rocket.

The center-of-mass velocity can also be factored out of the second term in Equation 9.8, giving $\sum m_i \vec{v}_{cm} \cdot \vec{v}_{i\,rel} = \vec{v}_{cm} \cdot \sum m_i \vec{v}_{i\,rel}$. Because the $\vec{v}_{i\,rel}$'s are the particle velocities relative to the center of mass, the sum here is the total momentum relative to the center of mass. But that's zero, so the entire second term in Equation 9.8 is zero.

The third term in Equation 9.8, $\sum \frac{1}{2} m_i v_{i\,rel}^2$, is the sum of the individual kinetic energies measured in a frame of reference moving with the center of mass. We call this term K_{int}, the **internal kinetic energy**.

With the middle term gone, Equation 9.8 shows that the kinetic energy of a system breaks into two terms:

$$K = K_{cm} + K_{int} \qquad \text{(kinetic energy of a system)} \qquad (9.9)$$

The first term, the kinetic energy of the center of mass, depends only on the center-of-mass motion. In our firecracker example, K_{cm} doesn't change when the firecracker explodes. The second term, the internal kinetic energy, depends only on the motions of the individual particles relative to the center of mass. Explosion of the firecracker dramatically increases the internal kinetic energy.

9.4 Collisions

A **collision** is a brief, intense interaction between objects. Examples abound: automobile collisions; collisions among balls on a pool table; the collision of a tennis ball and racket, baseball and bat, or football and foot; an asteroid colliding with a planet; and the collisions among high-energy elementary particles that probe the fundamental structure of matter. Less obvious are the stately collisions among galaxies that last a few hundred million years, the interaction of a spacecraft with a planet as the craft gains energy for a voyage to the outer solar system, and the repulsive interaction of two protons that approach and reverse direction without ever touching. But all these are collisions because they meet two criteria.

First, they're brief—meaning they last for what is a short time in the overall context of the colliding objects' motions. On the pool table, the collision time is short compared with the time it takes a ball to roll across the table. An automobile collision lasts a fraction of a second. A baseball spends far more time coming from the pitcher or heading to left field than it does interacting with the bat. And even a hundred million years is short compared with the lifetime of a galaxy.

Second, collisions are intense. That means the forces among the interacting objects are far larger than any external forces that may be acting on the system. The effect of external forces is therefore negligible over the brief collision time.

Momentum in Collisions

With external forces negligible, the total momentum of the colliding objects is essentially unchanged during the collision. Even though the forces of the collision are large, they're *internal* to the system comprising the colliding objects, so they can't alter the total momentum. We'll therefore assume that momentum is conserved during a collision—and that means we can relate before- and after-collision states without considering the often-complicated details of the collision and the forces involved.

However, the force of a collision alters the motions of the individual colliding particles. How much depends not only on the magnitude of the force but also on how long it's applied. If $\overline{F}$ is the average force acting on one particle during a collision that lasts for time Δt, then Newton's second law reads $\overline{F} = \Delta \vec{p} / \Delta t$ or

$$\Delta \vec{p} = \overline{F} \, \Delta t \qquad (9.10a)$$

The product of average force and time that appears on the right-hand side of this equation is the **impulse** $\vec{J}$ associated with the collision. Its units are newton-seconds (N·s). A given impulse produces the same momentum change regardless of whether it involves a larger force exerted over a shorter time or a smaller force exerted over a longer time.

The force involved in a collision is usually not constant but can fluctuate wildly throughout the collision. In that case we find the impulse by integrating the force over time: $\vec{J} = \int_{\substack{\text{collision} \\ \text{time}}} \vec{F}(t)\, dt$, so the change in momentum becomes

$$\Delta \vec{p} = \vec{J} = \int \vec{F}(t)\, dt \qquad \text{(impulse)} \qquad (9.10\text{b})$$

Figure 9.12 shows the relation between the actual force $\vec{F}(t)$ and its average $\overline{\vec{F}}$.

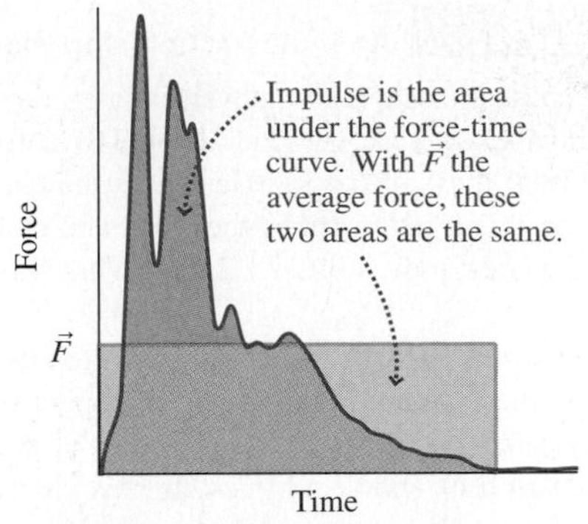

FIGURE 9.12 The force measured during an automobile crash test is a complicated function of time. Average force is the value that gives the same impulse—that is, the same area under the force-time curve.

Energy in Collisions

Kinetic energy may or may not be conserved in a collision. If it is, then the collision is **elastic**; if not, the collision is **inelastic**. The fundamental criterion for an elastic collision is that the forces between the interacting objects be conservative; then kinetic energy is stored briefly as potential energy, and released again by the time the collision is over. Interactions at the atomic and nuclear scales are often truly elastic. In the macroscopic realm, nonconservative forces produce heat or permanently deform the colliding objects, either way robbing the colliding system of kinetic energy. But even many macroscopic collisions are close enough to elastic that we can neglect energy loss during the collision (Fig. 9.13).

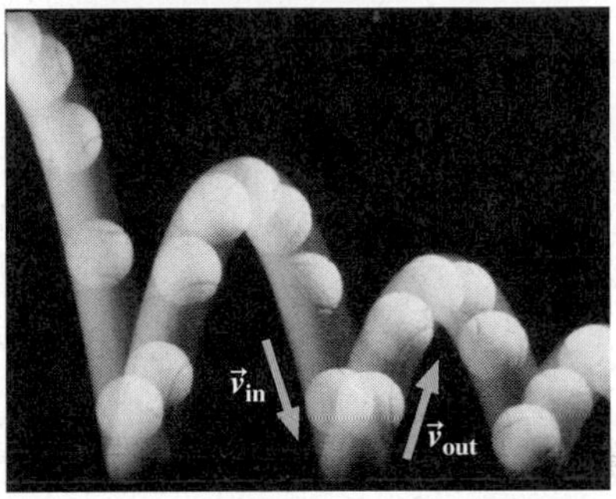

FIGURE 9.13 The collisions of a rubber ball with the floor are nearly elastic, with incoming and outgoing velocities nearly equal.

GOT IT? 9.4 Which of the following qualifies as a collision? Of the collisions, which are nearly elastic and which inelastic? (a) a basketball rebounds off the backboard; (b) two magnets approach, their north poles facing; they repel and reverse direction without touching; (c) a basketball flies through the air on a parabolic trajectory; (d) a truck strikes a parked car and the two slide off together, crumpled metal hopelessly intertwined; (e) a snowball splats against a tree, leaving a lump of snow adhering to the bark

---

9.5 Totally Inelastic Collisions

In a **totally inelastic collision**, the colliding objects stick together to form a single object. Even then, kinetic energy is usually not all lost. But a totally inelastic collision entails the maximum energy loss consistent with momentum conservation. The motion after a totally inelastic collision is determined entirely by momentum conservation, and that makes totally inelastic collisions easy to analyze.

Consider two masses m_1 and m_2 with initial velocities $\vec{v}_1$ and $\vec{v}_2$ that undergo a totally inelastic collision. After colliding, they stick together to form a single object of mass $m_1 + m_2$ and final velocity $\vec{v}_f$. Conservation of momentum states that the initial and final momenta of this system must be the same:

$$m_1 \vec{v}_1 + m_2 \vec{v}_2 = (m_1 + m_2)\vec{v}_f \quad \text{(totally inelastic collision)} \qquad (9.11)$$

Given four of the five quantities m_1, $\vec{v}_1$, m_2, $\vec{v}_2$, and $\vec{v}_f$, we can solve for the fifth.

EXAMPLE 9.8 An Inelastic Collision: Hockey

The hockey captain, a physics major, decides to measure the speed of the hockey puck. He loads a small Styrofoam chest with sand, giving it a total mass of 6.4 kg. He places the chest at rest on frictionless ice. The 160-g puck strikes the chest and embeds itself in the Styrofoam. The chest moves off at 1.2 m/s. What was the puck's speed?

INTERPRET This is an example of a totally inelastic collision. We identify the system as consisting of the puck and the chest. Initially, all the system's momentum is in the puck; after the collision, it's in the combination of chest and puck. In this case of a single nonzero velocity before collision and a single velocity after, momentum conservation requires that both motions be in the same direction. Therefore we have a one-dimensional problem.

DEVELOP Figure 9.14 is a sketch of the situation before and after the collision. With a totally inelastic collision, Equation 9.11—the statement of momentum conservation—tells it all. In our one-dimensional situation, this equation becomes $m_p v_p = (m_p + m_c)v_c$, where the subscripts p and c stand for puck and chest, respectively.

EVALUATE Here we want the initial puck velocity, so we solve for v_p:

$$v_p = \frac{(m_p + m_c)v_c}{m_p} = \frac{(0.16 \text{ kg} + 6.4 \text{ kg})(1.2 \text{ m/s})}{0.16 \text{ kg}} = 49 \text{ m/s}$$

ASSESS Make sense? Yes: The puck's mass is small, so it needs a much higher speed to carry the same momentum as the much more massive chest. Variations on the hockey captain's technique are often used to determine speeds that would be difficult to measure directly.

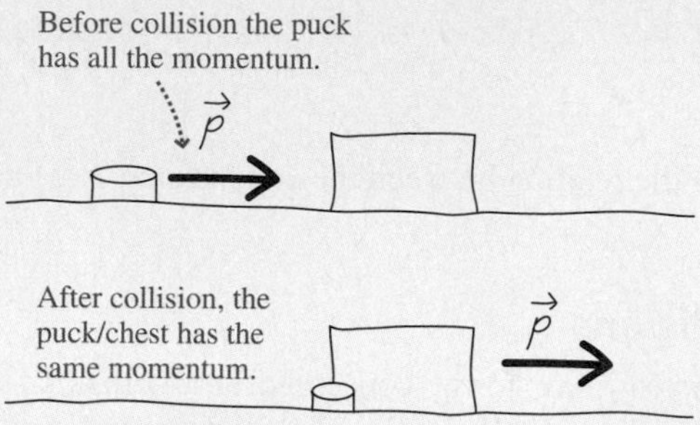

Before collision the puck has all the momentum.

After collision, the puck/chest has the same momentum.

FIGURE 9.14 Our sketch for Example 9.8. ▪

EXAMPLE 9.9 Conservation of Momentum: Fusion

In a fusion reaction, two deuterium nuclei (^{2}H) join to form helium (^{4}He). Initially, one of the deuterium nuclei is moving at 3.5 Mm/s, the second at 1.8 Mm/s at a 64° angle to the velocity of the first. Find the speed and direction of the helium nucleus.

INTERPRET Although the context is very different, this is another example of a totally inelastic collision. But here both objects are initially moving, and in different directions, so we have a two-dimensional situation. We identify the system as consisting of initially the two deuterium nuclei and finally the single helium nucleus. We're asked for the final velocity of the helium, expressed as magnitude (speed) and direction.

DEVELOP Figure 9.15 shows the situation. Momentum is conserved, so Equation 9.11 applies; solving that equation for $\vec{v}_f$ gives $\vec{v}_f = (m_1\vec{v}_1 + m_2\vec{v}_2)/(m_1 + m_2)$. In two dimensions, this represents two equations for the two components of $\vec{v}_f$. We need a coordinate system, and Fig. 9.15 shows our choice, with the x axis along the motion of the first deuterium nucleus. We need the components of the initial velocities in order to apply our equation for $\vec{v}_f$.

EVALUATE With $\vec{v}_1$ in the x direction, we have $v_{1x} = 3.5$ Mm/s and $v_{1y} = 0$. Figure 9.15 shows that $v_{2x} = (1.8 \text{ Mm/s})(\cos 64°) = 0.789$ Mm/s and $v_{2y} = (1.8 \text{ Mm/s})(\sin 64°) = 1.62$ Mm/s. So the components of our equation become

$$v_{fx} = \frac{m_1 v_{1x} + m_2 v_{2x}}{m_1 + m_2}$$

$$= \frac{(2 \text{ u})(3.5 \text{ Mm/s}) + (2 \text{ u})(0.789 \text{ Mm/s})}{2 \text{ u} + 2 \text{ u}} = 2.14 \text{ Mm/s}$$

$$v_{fy} = \frac{m_1 v_{1y} + m_2 v_{2y}}{m_1 + m_2}$$

$$= \frac{0 + (2 \text{ u})(1.62 \text{ Mm/s})}{2 \text{ u} + 2 \text{ u}} = 0.809 \text{ Mm/s}$$

As in Example 9.6, the superscripts are the nuclear masses in u, and because the mass units cancel, there's no need to convert to kilograms.

From these velocity components we can get the speed and direction: $v_f = \sqrt{v_{fx}^2 + v_{fy}^2} = 2.3$ Mm/s and $\theta = \tan^{-1}(v_{fy}/v_{fx}) = 21°$. We show this final velocity on the diagram in Fig. 9.15.

ASSESS In this example the two incident particles have the same masses, so their velocities are proportional to their momenta. Figure 9.15 shows that the total initial momentum is largely horizontal, with a smaller vertical component, so the 21° angle of the final velocity makes sense. The magnitude of $\vec{v}_f$ also makes sense: Now the total momentum is contained in a single, more massive particle, so we expect a final speed comparable to the initial speeds. ▪

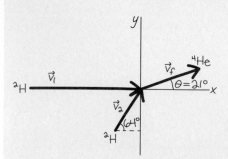

FIGURE 9.15 Our sketch of the velocity vectors for Example 9.9.

EXAMPLE 9.10 **The Ballistic Pendulum: Conserving Momentum, then Energy**

The ballistic pendulum measures the speeds of fast-moving objects like bullets. It consists of a wooden block of mass M suspended from vertical strings (Fig. 9.16). A bullet of mass m strikes and embeds itself in the block, and the block swings upward through a vertical distance h. Find an expression for the bullet's speed.

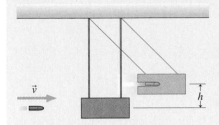

FIGURE 9.16 A ballistic pendulum (Example 9.10).

INTERPRET Interpreting this example is a bit more involved. We actually have two separate events: the bullet striking the block and the subsequent rise of the block. We can interpret the first event as a one-dimensional totally inelastic collision, as in Example 9.8. Momentum is conserved during this event but, because the collision is inelastic, energy is not. Then the block rises, and now a net external force—resulting from string tension and gravity—acts to change the momentum. But gravity is conservative, and the string tension does no work, so here mechanical energy is conserved.

DEVELOP Figure 9.16 is our drawing. Our plan is to separate the two parts of the problem and then to combine the results to get our final answer. First is the inelastic collision; here momentum is conserved, so

Equation 9.11 applies. In one dimension, that reads $mv = (m + M)V$, where v is the initial bullet speed and V is the speed of the block with embedded bullet just after the collision. Solving gives $V = mv/(m + M)$. Now the block swings upward. Momentum isn't conserved, but mechanical energy is. Setting the zero of potential energy in the block's initial position, we have $U_0 = 0$ and—using the situation just after the collision as the initial state—$K_0 = \frac{1}{2}(m + M)V^2$. At the peak of its swing the block is momentarily at rest, so $K = 0$. But it's risen a height h, so its potential energy is $U = (m + M)gh$. Conservation of mechanical energy reads $K_0 + U_0 = K + U$—in this case, $\frac{1}{2}(m + M)V^2 = (m + M)gh$.

EVALUATE Now we've got two equations describing the two parts of the problem. Using our expression for V from momentum conservation in the energy conservation equation, we get

$$\frac{1}{2}\left(\frac{mv}{m + M}\right)^2 = gh$$

Solving for the bullet speed v then gives our answer:

$$v = \left(\frac{m + M}{m}\right)\sqrt{2gh}$$

ASSESS Make sense? Yes: The smaller the bullet mass m, the higher velocity it must have to carry a given momentum; that's reflected by the factor m alone in the denominator. The higher the rise h, obviously, the greater the bullet speed. But the speed scales not as h itself but as $\sqrt{h}$. That's because kinetic energy—which turned into potential energy of the rise—depends on velocity *squared*. ∎

9.6 Elastic Collisions

We've seen that momentum is essentially conserved in any collision. In an elastic collision, kinetic energy is conserved as well. In the most general case of a two-body collision, we consider two objects of masses m_1 and m_2, moving initially with velocities $\vec{v}_{1i}$ and $\vec{v}_{2i}$, respectively. Their final velocities after collision are $\vec{v}_{1f}$ and $\vec{v}_{2f}$. Then the conservation statements for momentum and kinetic energy become

$$m_1\vec{v}_{1i} + m_2\vec{v}_{2i} = m_1\vec{v}_{1f} + m_2\vec{v}_{2f} \tag{9.12}$$

and

$$\frac{1}{2}m_1v_{1i}^2 + \frac{1}{2}m_2v_{2i}^2 = \frac{1}{2}m_1v_{1f}^2 + \frac{1}{2}m_2v_{2f}^2 \tag{9.13}$$

Given initial velocities, we'd like to predict the outcome of a collision. In the totally inelastic two-dimensional collision, we had enough information to solve the system. Here, in the two-dimensional elastic case, we have the two components of the momentum conservation equation 9.12 and the single scalar equation for energy conservation 9.13. But we have four unknowns—the magnitudes and directions of both final velocities. With three equations and four unknowns, we don't have enough information to solve the general two-dimensional elastic collision. Later in this section we'll see how other information can help solve such problems. First, though, we look at the special case of a one-dimensional elastic collision.

Elastic Collisions in One Dimension

When two objects collide head-on, the internal forces act along the same line as the incident motion, and the objects' subsequent motion must therefore be along that same line (Fig. 9.17a). Although such one-dimensional collisions are a special case, they do occur and they provide much insight into the more general case.

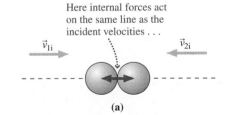

Here internal forces act on the same line as the incident velocities . . .

(a)

. . . but here they don't, so the motion involves two dimensions.

(b)

FIGURE 9.17 Only a head-on collision is one-dimensional.

In the one-dimensional case, the momentum conservation equation 9.12 has only one nontrivial component:

$$m_1 v_{1i} + m_2 v_{2i} = m_1 v_{1f} + m_2 v_{2f} \tag{9.12a}$$

where the v's stand for velocity components, rather than magnitudes, and can therefore be positive or negative. If we collect together the terms in Equations 9.12a and 9.13 that are associated with each mass, we have

$$m_1(v_{1i} - v_{1f}) = m_2(v_{2f} - v_{2i}) \tag{9.12b}$$

and

$$m_1(v_{1i}^2 - v_{1f}^2) = m_2(v_{2f}^2 - v_{2i}^2) \tag{9.13a}$$

But $a^2 - b^2 = (a+b)(a-b)$, so Equation 9.13a can be written

$$m_1(v_{1i} - v_{1f})(v_{1i} + v_{1f}) = m_2(v_{2f} - v_{2i})(v_{2f} + v_{2i}) \tag{9.13b}$$

Dividing the left and right sides of Equation 9.13b by the corresponding sides of Equation 9.12b then gives

$$v_{1i} + v_{1f} = v_{2f} + v_{2i}$$

Rearranging shows that

$$v_{1i} - v_{2i} = v_{2f} - v_{1f} \tag{9.14}$$

What does this equation tell us? Both sides describe the relative velocity between the two particles; the equation therefore shows that the relative speed remains unchanged after the collision, although the direction reverses. If the two objects are approaching at a relative speed of 5 m/s, then after collision they'll separate at 5 m/s.

Continuing our search for the final velocities, we solve Equation 9.14 for v_{2f}:

$$v_{2f} = v_{1i} - v_{2i} + v_{1f}$$

and use this result in Equation 9.12a:

$$m_1 v_{1i} + m_2 v_{2i} = m_1 v_{1f} + m_2(v_{1i} - v_{2i} + v_{1f})$$

Solving for v_{1f} then gives

$$v_{1f} = \frac{m_1 - m_2}{m_1 + m_2} v_{1i} + \frac{2m_2}{m_1 + m_2} v_{2i} \tag{9.15a}$$

Problem 63 asks you to show similarly that

$$v_{2f} = \frac{2m_1}{m_1 + m_2} v_{1i} + \frac{m_2 - m_1}{m_1 + m_2} v_{2i} \tag{9.15b}$$

Equations 9.15 are our desired result, expressing the final velocities in terms of the initial velocities alone.

To see that these results make sense, we suppose that $v_{2i} = 0$. (This is really not a special case, since we can always work in a frame of reference in which m_2 is initially at rest.) We then consider the three special cases of one-dimensional elastic collisions illustrated in Fig. 9.18.

Case 1: $m_1 \ll m_2$ (Fig. 9.18a) For this case, picture a ping-pong ball colliding with a bowling ball, or any object colliding elastically with a perfectly rigid surface. If we set $v_{2i} = 0$ in Equations 9.15, and drop m_1 as being negligible compared with m_2, Equations 9.15 become simply

$$v_{1f} = -v_{1i}$$

and

$$v_{2f} = 0$$

That is, the lighter object rebounds with no change in speed, while the heavier object remains at rest. Does this make sense in light of the conservation laws that Equations 9.15

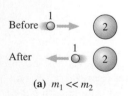

(a) $m_1 \ll m_2$

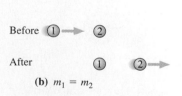

(b) $m_1 = m_2$

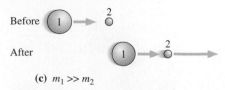

(c) $m_1 \gg m_2$

FIGURE 9.18 Special cases of elastic collisions in one dimension.

are supposed to reflect? Clearly energy is conserved: The kinetic energy of m_2 remains zero and the kinetic energy $\frac{1}{2}m_1v_1^2$ is unchanged. But what about momentum? The momentum of the lighter object has changed, from m_1v_{1i} to $-m_1v_{1i}$. But momentum *is* conserved; the momentum given up by the lighter object is absorbed by the heavier object. In the limit of an arbitrarily large m_2, the heavier object can absorb huge amounts of momentum mv without acquiring significant speed. If we "back off" from the extreme case that m_1 can be neglected altogether compared with m_2, we would find that a lighter object striking a heavier one rebounds with reduced speed and that the heavier object begins moving slowly in the opposite direction.

Case 2: $m_1 = m_2$ (Fig. 9.18b) Again with $v_{2i} = 0$, Equations 9.15 now give

$$v_{1f} = 0$$

and

$$v_{2f} = v_{1i}$$

So the first object stops abruptly, transferring all its energy and momentum to the second. For purposes of energy transfer, two equal-mass particles are perfectly "matched." We'll encounter analogous instances of energy transfer "matching" when we discuss wave motion and again in connection with electric circuits.

Case 3: $m_1 \gg m_2$ (Fig. 9.18c) Now Equations 9.15 give

$$v_{1f} = v_{1i}$$

and

$$v_{2f} = 2v_{1i}$$

where we've neglected m_2 compared with m_1. So here the more massive object barrels right on with no change in motion, while the lighter one heads off with twice the speed of the massive one. This result is entirely consistent with our earlier claim that the relative speed remains unchanged in a one-dimensional elastic collision. How are momentum and energy conserved in this case? In the extreme limit where we neglect the mass m_2, its energy and momentum are negligible. Essentially all the energy and momentum remain with the more massive object, and both these quantities are essentially unchanged in the collision. In the less extreme case where an object of finite mass strikes a less massive object initially at rest, both objects move off in the initial direction of the incident object, with the lighter one moving faster.

EXAMPLE 9.11 **Elastic Collisions: Nuclear Engineering**

Nuclear power reactors include a substance called a *moderator*, whose job is to slow the neutrons liberated in nuclear fission, making them more likely to induce additional fission and thus sustain a nuclear chain reaction. A reactor design developed in Canada uses so-called *heavy water* as its moderator. In heavy water, ordinary hydrogen atoms are replaced by deuterium, the rare form of hydrogen whose nucleus consists of a proton and a neutron. The mass of this so-called deuteron is thus about 2 u, compared with a neutron's 1 u. Find the fraction of a neutron's kinetic energy that's transferred to an initially stationary deuteron in a head-on elastic collision.

INTERPRET This example is about a head-on collision, so we're dealing with a one-dimensional situation. The system of interest consists of the neutron and the deuteron. We're not told much else except the masses of the two particles. That should be enough, though, because we're not asked for the final velocities but rather for a ratio of related quantities—namely, kinetic energies.

DEVELOP Since we have a one-dimensional elastic collision, Equations 9.15 apply. We're asked for the fraction of the neutron's kinetic energy that gets transferred to the deuteron, so we need to express the deuteron's final velocity in terms of the neutron's initial velocity. If we take the neutron to be particle 1, then we want Equation 9.15b. With the deuteron initially at rest, $v_{2i} = 0$ and the equation becomes $v_{2f} = 2m_1v_{1i}/(m_1 + m_2)$. Our plan is to use this equation to determine the kinetic-energy ratio.

EVALUATE The kinetic energies of the two particles are given by $K_1 = \frac{1}{2}m_1v_1^2$ and $K_2 = \frac{1}{2}m_2v_2^2$. Using our equation for v_{2f} gives

$$K_2 = \frac{1}{2}m_2\left(\frac{2m_1v_1}{m_1 + m_2}\right)^2 = \frac{2m_2m_1^2v_1^2}{(m_1 + m_2)^2}$$

We want to compare this with K_1:

$$\frac{K_2}{K_1} = K_2\left(\frac{1}{K_1}\right) = \left(\frac{2m_2m_1^2v_1^2}{(m_1 + m_2)^2}\right)\left(\frac{1}{\frac{1}{2}m_1v_1^2}\right) = \frac{4m_1m_2}{(m_1 + m_2)^2} \quad (9.16)$$

cont'd.

In this case $m_1 = 1$ u and $m_2 = 2$ u, so we have $K_2/K_1 = 8/9 \approx 0.89$. Thus 89% of the incident energy is transferred in a single collision, leaving the neutron with 11% of its initial energy.

ASSESS Let's take a look at Equation 9.16 in the context of our three special cases. We numbered this equation because it's a general result for the fractional energy transfer in any one-dimensional elastic collision. In case 1, $m_1 \ll m_2$, so we neglect m_1 compared with m_2 in the denominator; then our energy ratio is approximately $4m_1/m_2$. This becomes zero in the extreme limit where m_1's mass is negligible— consistent with our case 1 where the massive object didn't move at all. In case 2, $m_1 = m_2$, and Equation 9.16 becomes $4m^2/(2m)^2 = 1$, where

m is the mass of both objects. That too agrees with our earlier analysis: The incident object stops and transfers all its energy to the struck object. Finally, in case 3, $m_1 \gg m_2$, so we neglect m_2 in the denominator. Now the energy ratio becomes $4m_2/m_1$. As in case 1, this approaches zero as the mass ratio gets extremely large. So the maximum energy transfer occurs with two equal masses, and tails off toward zero if the mass ratio becomes extreme in either direction.

For the particles in this example, the mass ratio 1:2 is close enough to equality that the energy transfer is nearly 90% efficient. Problem 73 explores further this energy transfer. ∎

GOT IT? 9.5 One ball is at rest on a level floor. A second ball collides elastically with the first, and the two move off separately but in the same direction. What can you conclude about the masses of the two balls?

Elastic Collisions in Two Dimensions

Analyzing an elastic collision in two dimensions requires the full vector statement of momentum conservation (Equation 9.12), along with the statement of energy conservation (Equation 9.13). But these equations alone don't provide enough information to solve a problem. In a collision between reasonably simple macroscopic objects, that information may be provided by the so-called **impact parameter**, a measure of how much the collision differs from being head-on (Fig. 9.19). More typically—especially with atomic and nuclear interactions—the necessary information must be supplied by measurements done after the collision. Knowing the direction of motion of one particle after collision, for example, provides enough information to analyze a collision if the masses and initial velocities are also known.

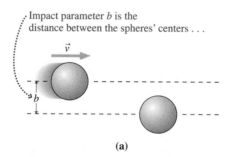

Impact parameter b is the distance between the spheres' centers . . .

. . . b determines where they hit and thus the direction of the collision forces.

$\vec{F}$

$\vec{v}$

b

b

$\vec{F}$

FIGURE 9.19 The impact parameter b determines the directions of the collision forces.

(a)   (b)

EXAMPLE 9.12   **A Two-Dimensional Elastic Collision: Croquet**

A croquet ball strikes a stationary ball of equal mass. The collision is elastic, and the incident ball goes off at 30° to its original direction. In what direction does the other ball move?

INTERPRET We've got an elastic collision, so both momentum and kinetic energy are conserved. The system consists of the two croquet balls. We aren't given a lot of information, but since we're asked only for a direction, the magnitudes of the velocities won't matter. Thus we've got what we need to know about the initial velocities, and we've got one other piece of information, so we have enough to solve the problem.

DEVELOP Figure 9.20 shows the situation, in which we're after the unknown angle θ. Since the collision is elastic, Equations 9.12 (momentum conservation) and 9.13 (energy conservation) both apply. The masses are equal, so they cancel from both equations. With $v_{2i} = 0$, we then have $\vec{v}_{1i} = \vec{v}_{1f} + \vec{v}_{2f}$ for momentum conservation and $v_{1i}^2 = v_{1f}^2 + v_{2f}^2$ for energy conservation. The rest will be algebra.

EVALUATE Solving for one unknown in terms of another is going to get messy here, with some velocities squared and some not. Here's a more

cont'd.

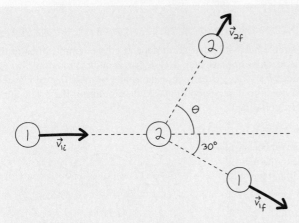

FIGURE 9.20 Our sketch of the collision between croquet balls of equal mass (Example 9.12).

clever approach: Rather than write the momentum equation in two components, let's take the dot product of each side with itself. That will bring in velocity-squared terms, letting us combine the momentum and energy equations. And the dot product includes an angle—which is what we're asked to find.

The dot product is distributive and commutative, so here's what we get when we dot the momentum equation with itself:

$$\vec{v}_{1i} \cdot \vec{v}_{1i} = (\vec{v}_{1f} + \vec{v}_{2f}) \cdot (\vec{v}_{1f} + \vec{v}_{2f})$$
$$= \vec{v}_{1f} \cdot \vec{v}_{1f} + \vec{v}_{2f} \cdot \vec{v}_{2f} + 2\vec{v}_{1f} \cdot \vec{v}_{2f}$$

Recall that the dot product of two vectors is the product of their magnitudes with the cosine of the angle between them: $\vec{A} \cdot \vec{B} = AB\cos\theta$. Since the angle between a vector and itself is zero, the dot product of a vector with itself is the square of its magnitude: $\vec{A} \cdot \vec{A} = A^2\cos(0) = A^2$. So our equation becomes

$$v_{1i}^2 = v_{1f}^2 + v_{2f}^2 + 2v_{1f}v_{2f}\cos(\theta + 30°)$$

where the argument of the cosine follows because, as Fig. 9.20 shows, the angle between $\vec{v}_{1f}$ and $\vec{v}_{2f}$ is $\theta + 30°$. We now subtract the energy equation from this new equation to get

$$2v_{1f}v_{2f}\cos(\theta + 30°) = 0$$

But neither of the final speeds is zero, so this equation requires that $\cos(\theta + 30°) = 0$. Thus $\theta + 30° = 90°$, and our answer follows: $\theta = 60°$.

ASSESS This result seems reasonable, although we don't have a lot to go on because we haven't calculated the final speeds. But it's intriguing that the two balls go off at right angles to each other. Is this a coincidence? No: It happens in any two-dimensional elastic collision between objects of equal mass, when one is initially at rest. ∎

The Center-of-Mass Frame

Two-dimensional collisions take a particularly simple form in a frame of reference moving with the center of mass of the colliding particles, since the total momentum in such a frame must be zero. That remains true after a collision, which involves only *internal* forces that don't affect the the center of mass. Therefore, both the initial and final momenta form pairs of oppositely directed vectors of equal magnitude, as shown in Fig. 9.21. In an elastic collision, energy conservation requires further that the incident and final momenta have the same values, so a single number—the angle θ in Fig 9.21—completely describes the collision.

It's often easier to analyze a collision by transforming to the center-of-mass frame, doing the analysis, and then transforming the resulting momentum and velocity vectors back to the original or "lab" frame. High-energy physicists routinely make such transformations as they seek to understand the fundamental forces between elementary particles. Those forces are described most simply in the center-of-mass frame of colliding particles, but in some experiments—those where lighter particles slam into massive nuclei or stationary targets—the physicists and their particle accelerators are not in the center-of-mass frame.

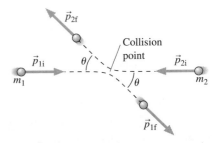

FIGURE 9.21 An elastic collision viewed in the center-of-mass frame, showing that the initial and final momentum vectors form pairs with equal magnitudes and opposite directions.

Big Picture

The big idea of this chapter is that systems consisting of many particles exhibit simple behaviors that don't depend on the complexities of their internal structure or motions. That, in turn, allows us to understand those internal details. In particular, a system responds to external forces as though it were a point particle located at the **center of mass**. If the net external force on a system is zero, then the center of mass does not accelerate and the system's total momentum is conserved. Conservation of momentum holds to a very good approximation during the brief, intense encounters called **collisions**, allowing us to relate particles' motions before and after colliding.

Newton's second and third laws are behind these big ideas. The third law, in particular, says that forces *internal* to a system cancel in pairs, and therefore they don't contribute to the net force on the system. That's what allows us to describe a system's overall motion without having to worry about what's going on internally.

Newton's 3rd law ⟹ Internal forces cancel in pairs ⟹ The center of mass satisfies Newton's 2nd law

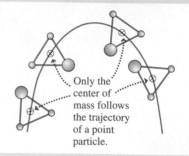

Only the center of mass follows the trajectory of a point particle.

Key Concepts and Equations

The **center of mass** position $\vec{r}_{cm}$ is a weighted average of the positions of a system's constituent particles:

$$\vec{r}_{cm} = \frac{\sum m_i \vec{r}_i}{M} \quad \text{or, with continuous matter,} \quad \vec{r}_{cm} = \frac{\int \vec{r}\, dm}{M}$$

Here M is the system's total mass and the sum or integral is taken over the entire system. The center of mass obeys Newton's second law:

$$\vec{F}_{net\,ext} = M\vec{a}_{cm} = M\frac{d\vec{P}}{dt}$$

where $\vec{F}_{net\,ext}$ is the net external force on the system, $\vec{a}_{cm}$ the acceleration of the center of mass, and $\vec{P}$ the system's total momentum.

A **collision** is a brief, intense interaction between particles involving large internal forces. External forces have little effect during a collision, so to a good approximation the total momentum of the interacting particles is conserved.

In a **totally inelastic collision**, the colliding objects stick together to form a composite; in that case momentum conservation entirely determines the outcome:

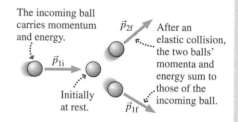

$$m_1\vec{v}_1 + m_2\vec{v}_2 = (m_1 + m_2)\vec{v}_f \quad \text{(conservation of momentum, totally inelastic collision)}$$

An **elastic collision** conserves kinetic energy as well as momentum, and the colliding particles separate after the collision:

$$m_1\vec{v}_{1i} + m_2\vec{v}_{2i} = m_1\vec{v}_{1f} + m_2\vec{v}_{2f} \quad \text{(conservation of momentum, elastic collision)}$$

$$\tfrac{1}{2}m_1 v_{1i}^2 + \tfrac{1}{2}m_2 v_{2i}^2 = \tfrac{1}{2}m_1 v_{1f}^2 + \tfrac{1}{2}m_2 v_{2f}^2 \quad \text{(conservation of energy, elastic collision)}$$

In the special case of a one-dimensional elastic collision, knowledge of the mass and initial velocities is sufficient to determine the outcome. To analyze elastic collisions in two dimensions requires an additional piece of information, such as the impact parameter or the direction of one of the particles after the collision.

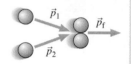

The incoming ball carries momentum and energy.

After an elastic collision, the two balls' momenta and energy sum to those of the incoming ball.

Initially at rest.

Cases and Uses

One-dimensional collisions with one object initially at rest provide insights into the nature of collisions. There are three cases, depending on the relative masses:

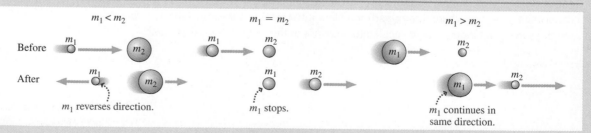

$m_1 < m_2$ — m_1 reverses direction.

$m_1 = m_2$ — m_1 stops.

$m_1 > m_2$ — m_1 continues in same direction.

Rockets provide an important technological application of the principle that a system's momentum is conserved in the absence of external forces. A rocket exhausts matter out the back at high velocity; momentum conservation then requires that the rocket gain momentum in the forward direction. Rocket propulsion requires no interaction with any external material, which is why rockets work in space.

For Thought and Discussion

1. Roughly where is your center of mass when you're standing?
2. Explain why a high jumper's center of mass need not clear the bar.
3. The center of mass of a solid sphere is clearly at its center. If the sphere is cut in half and the two halves are stacked as in Fig. 9.22, is the center of mass at the point where they touch? If not, roughly where is it? Explain.

FIGURE 9.22 For Thought and Discussion 3

4. The momentum of a system of pool balls is the same before and after they are hit by the cue ball. Is it still the same after one of the balls strikes the edge of the table? Explain.

5. An hourglass is inverted and placed on a scale. Compare the scale readings (a) before sand begins to hit the bottom; (b) while sand is hitting the bottom; and (c) when all the sand is on the bottom.
6. Why are cars designed so that their front ends crush during an accident?
7. Give three everyday examples of inelastic collisions.
8. Is it possible to have an inelastic collision in which *all* the kinetic energy of the colliding objects is lost? If not, why not? If so, give an example.
9. If you want to stop the neutrons in a reactor, why not use massive nuclei like lead?
10. A pitched baseball moves no faster than the pitcher's hand. But a batted ball can move much faster than the bat. Why the difference?
11. Two identical satellites are going in opposite directions in the same circular orbit when they collide head-on. Describe their subsequent motion if the collision is (a) elastic or (b) inelastic.

Exercises and Problems

Exercises

Section 9.1 Center of Mass

12. A 28-kg child sits at one end of a 3.5-m-long seesaw. Where should her 65-kg father sit so the center of mass will be at the center of the seesaw?
13. Two particles of equal mass m are at the vertices of the base of an equilateral triangle. The center of mass of the triangle is midway between the base and the third vertex. What is the mass at the third vertex?
14. Rework Example 9.1 with the origin at the center of the barbell, showing that the physical location of the center of mass is independent of the choice of coordinate system.
15. Three equal masses lie at the corners of an equilateral triangle of side L. Where is the center of mass?
16. How far from the center of Earth is the center of mass of the Earth-Moon system? *Hint:* Consult Appendix E.
17. Where is the center of mass of a canoe if a 76-kg paddler is 0.85 m from the stern and the 52-kg passenger is 3.9 m from the stern? Neglect the mass of the canoe itself.

Section 9.2 Momentum

18. A popcorn kernel in a hot pan bursts into two pieces, with masses of 91 mg and 64 mg. The more massive piece moves horizontally at 47 cm/s. Describe the motion of the second piece.
19. A 60-kg ice skater, at rest on frictionless ice, tosses a 12-kg snowball with velocity given by $\vec{v} = 3.0\hat{\imath} + 4.0\hat{\jmath}$ m/s, where the x and y axes are both in the horizontal plane. Find the subsequent velocity of the skater.
20. A plutonium-239 nucleus at rest decays into a uranium-235 nucleus by emitting an alpha particle (^{4}He) with kinetic energy of 5.15 MeV. What is the speed of the uranium nucleus?
21. A runaway toboggan of mass 8.6 kg is moving horizontally at 23 km/h. As it passes under a tree, 15 kg of snow drop onto it. What is its subsequent speed?
22. A 42-kg child stands at rest on ice skates. She catches a 1.1-kg ball moving horizontally at 9.5 m/s. What is her speed after she's caught the ball?

Section 9.3 Kinetic Energy of a System

23. A 150-g trick baseball is thrown at 60 km/h. It explodes in flight into two pieces, with a 38-g piece continuing straight ahead at 85 km/h. How much energy do the pieces gain in the explosion?
24. A 1200-kg car moving at 88 km/h collides with a 7600-kg truck moving in the same direction at 65 km/h. The two stick together, continuing in their original direction at 68 km/h. Determine the center-of-mass and internal energies of the (car + truck) system before and after the collision.
25. An object with kinetic energy K_0 explodes into two pieces, each of which moves with twice the speed of the original object. Compare the internal and center-of-mass energies after the explosion.

Section 9.5 Totally Inelastic Collisions

26. In a railroad switchyard, a 56-ton freight car is sent at 7.0 mi/h toward a 31-ton car that is moving in the same direction at 2.6 mi/h. (a) What is the speed of the pair after they couple together? (b) What fraction of the initial kinetic energy was lost in the collision?
27. In a totally inelastic collision between two equal masses, one of which is initially at rest, show that half the initial kinetic energy is lost.
28. A neutron (mass 1 u) strikes a deuteron (mass 2 u), and the two combine to form a tritium nucleus. If the neutron's initial velocity was $28\hat{\imath} + 17\hat{\jmath}$ Mm/s and if the tritium nucleus leaves the reaction with velocity $12\hat{\imath} + 20\hat{\jmath}$ Mm/s, what was the velocity of the deuteron?
29. Two identical trucks have mass 5500 kg when empty. One truck carries a 9500-kg load and is moving at 65 km/h. It collides inelastically with the second truck, which is initially at rest, and the pair moves off at 40 km/h. What is the load of the second truck?

Section 9.6 Elastic Collisions

30. An alpha particle (^{4}He) strikes a stationary gold nucleus (^{197}Au) head-on. What fraction of the alpha particle's kinetic energy is transferred to the gold nucleus? Assume the collision is totally elastic.
31. While playing ball in the street, a child accidentally tosses a ball at 18 m/s toward the front of a car moving toward him at 14 m/s. What is the speed of the ball after it rebounds elastically from the car?

32. A block of mass m undergoes a one-dimensional elastic collision with a block of mass M initially at rest. If both blocks have the same speed after the collision, how are their masses related?

33. A proton moving at 6.9 Mm/s collides elastically and head-on with a second proton moving in the opposite direction at 11 Mm/s. Find their velocities after the collision.

34. A head-on, elastic collision between two particles with equal initial speed v leaves the more massive particle (mass m_1) at rest. Find (a) the ratio of the particle masses and (b) the final speed of the less massive particle.

Problems

35. Find the center of mass of a pentagon with five equal sides a but with one triangle missing, as shown in Fig. 9.23. *Hint:* See Example 9.3, and treat the pentagon as a group of triangles.

FIGURE 9.23 Problem 35

36. Consider a system of three equal-mass particles moving in a plane; their positions are given by $a_i\hat{i} + b_i\hat{j}$, where a_i and b_i are functions of time with the units of position. Particle 1 has $a_1 = 3t^2 + 5$ and $b_1 = 0$; particle 2 has $a_2 = 7t + 2$ and $b_2 = 2$; and particle 3 has $a_3 = 3t$ and $b_3 = 2t + 6$. Find the position, velocity, and acceleration of the center of mass as functions of time.

37. You are with 19 other people on a boat at rest in frictionless water. Your total mass is 1500 kg, and the mass of the boat itself is 12,000 kg. The entire party walks the 6.5-m distance from bow to stern. How far does the boat move?

38. A hemispherical bowl is at rest on a frictionless kitchen counter. A mouse drops onto the rim of the bowl from a cabinet directly overhead. The mouse climbs down the inside of the bowl to eat the crumbs at the bottom. If the bowl moves along the counter a distance equal to one-tenth of its diameter, how does the mouse's mass compare with the bowl's mass?

39. Find the center of mass of the uniform, solid cone of height h, base radius R, and constant density ρ shown in Fig. 9.24. *Hint:* Integrate over disk-shaped mass elements of thickness dy, as shown in the figure.

FIGURE 9.24 Problem 39

40. A firecracker, initially at rest, explodes into two fragments. The first, of mass 14 g, moves in the positive x direction at 48 m/s. The second moves at 32 m/s. Find its mass and the direction of its motion.

41. An 11,000-kg freight car rests against a spring bumper at the end of a railroad track. The spring has constant $k = 3.2 \times 10^5$ N/m. The car is hit by a second car of 9400-kg mass moving at 8.5 m/s, and the two cars couple together. (a) What is the maximum compression of the spring? (b) What is the speed of the two cars together when they rebound from the spring?

42. On an icy road, a 1200-kg car moving at 50 km/h strikes a 4400-kg truck moving in the same direction at 35 km/h. The combination is soon hit from behind by a 1500-kg car speeding at 65 km/h. If all three vehicles stick together, what is the speed of the wreckage?

43. A car of mass M is initially at rest on a frictionless surface. A jet of water carrying mass at the rate dm/dt and moving horizontally at speed v_0 strikes the rear window of the car, which makes a 45° angle with the horizontal; the water bounces off at the same relative speed with which it hit the window, as shown in Fig. 9.25. (a) Find an expression for the initial acceleration of the car. (b) What is the maximum speed reached by the car?

FIGURE 9.25 Problem 43

44. A 950-kg compact car is moving with velocity $\vec{v}_1 = 32\hat{i} + 17\hat{j}$ m/s. It skids on a frictionless icy patch and collides with a 450-kg hay wagon moving with velocity $\vec{v}_2 = 12\hat{i} + 14\hat{j}$ m/s. If the two stay together, what is their velocity?

45. A ^{238}U nucleus is moving in the x direction at 5.0×10^5 m/s when it decays into an alpha particle (^{4}He) and a ^{234}Th nucleus. If the alpha particle moves off at 22° above the x axis with a speed of 1.4×10^7 m/s, what is the recoil velocity of the thorium nucleus?

46. A cylindrical concrete silo is 4.0 m in diameter and 30 m high. It consists of a 6,000-kg concrete base and 38,000-kg cylindrical concrete walls. Locate the center of mass of the silo (a) when it's empty and (b) when it's two-thirds full of silage whose density is 800 kg/m^3. Neglect the thickness of the silo walls and base.

47. A 42-g firecracker is at rest at the origin when it explodes into three pieces. The first, with mass 12 g, moves along the x axis at 35 m/s. The second, with mass 21 g, moves along the y axis at 29 m/s. Find the velocity of the third piece.

48. A 60-kg astronaut floating in space simultaneously tosses away a 14-kg oxygen tank and a 5.8-kg camera. The tank moves in the x direction at 1.6 m/s, and the astronaut recoils at 0.85 m/s in a direction 200° counterclockwise from the x axis. Find the velocity of the camera.

49. Assume the pieces in Exercise 25 have equal masses, and find the angles their velocities make with the direction of the initial object's motion.

50. A 55-kg sprinter is standing at the left end of a 240-kg cart moving to the left at 7.6 m/s. She runs to the right end and continues horizontally off the cart. What should be her speed relative to the cart in order to leave the cart with no horizontal velocity component relative to the ground?

51. Cookie sheets move along a conveyor belt toward an oven, with mounds of unbaked dough dropping vertically onto the sheets at the rate of one 12-g mound every 2 seconds. What average force must the conveyor belt exert on a cookie sheet to keep it moving at a constant 50 cm/s?

52. While standing on frictionless ice, you (mass 65.0 kg) toss a 4.50-kg rock with initial speed 12.0 m/s. If the rock is 15.2 m from you when it lands, (a) at what angle did you toss it? (b) How fast are you moving?

53. A drunk driver in a 1600-kg car plows into a 1300-kg parked car with its brake set. Police measurements show that the two cars skid together a distance of 25 m before stopping. If the effective coefficient of friction is 0.77, how fast was the drunk going just before the collision?

54. A fireworks rocket is launched vertically upward at 40 m/s. At the peak of its trajectory, it explodes into two equal-mass fragments. One reaches the ground 2.87 s after the explosion. When does the second reach the ground?

55. Explosive bolts separate a 950-kg communications satellite from its 640-kg booster rocket. If the impulse of the explosion is 350 N·s, at what relative speed do the satellite and booster separate?

56. (a) Estimate the impulse imparted by the force shown in Fig. 9.26. (b) What is the average impulsive force?

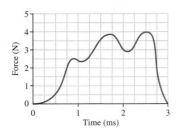

FIGURE 9.26 Problem 56

57. A 1200-kg Toyota and a 2200-kg Buick collide at right angles in an intersection. They lock together and skid 22 m; the coefficient of friction is 0.91. Show that at least one car must have exceeded the 25 km/h speed limit in effect at the intersection.

58. A 400-mg popcorn kernel is skittering across a nonstick frying pan at 8.2 cm/s when it pops and breaks into two equal-mass pieces. If one piece ends up at rest, how much energy was released in the popping?

59. Two identical objects with the same initial speed collide and stick together. If the composite object moves with half the initial speed of either object, what was the angle between the initial velocities?

60. A proton (mass 1 u) moving at 6.90 Mm/s collides elastically and head-on with a second particle moving in the opposite direction at 2.80 Mm/s. After the collision, the proton is moving opposite to its initial direction at 8.62 Mm/s. Find the mass and final velocity of the second particle.

61. Two objects, one initially at rest, undergo a one-dimensional elastic collision. If half the kinetic energy of the initially moving object is transferred to the other object, what is the ratio of their masses?

62. Blocks B and C have masses $2m$ and m, respectively, and are at rest on a frictionless surface. Block A, also of mass m, is heading at speed v toward block B as shown in Fig. 9.27. Determine the final velocity of each block after all subsequent collisions are over. Assume all collisions are elastic.

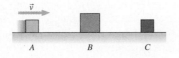

FIGURE 9.27 Problem 62

63. Derive Equation 9.15b.

64. An object collides elastically with an equal-mass object initially at rest. If the collision is not head-on, show that the final velocity vectors are perpendicular.

65. A proton (mass 1 u) collides elastically with a stationary deuteron (mass 2 u). If the proton is deflected through 37°, what fraction of its kinetic energy does it transfer to the deuteron?

66. Two identical billiard balls are initially at rest when they are struck symmetrically by a third identical ball moving with velocity $\vec{v}_0 = v_0\hat{i}$ (Fig. 9.28). Find the velocities of all three balls after they undergo an elastic collision.

FIGURE 9.28 Problem 66

67. A 32-u oxygen molecule (O_2) moving in the $+x$ direction at 580 m/s collides with an oxygen atom (mass 16 u) moving at 870 m/s at 27° to the x axis. The particles stick together to form an ozone molecule. Find the velocity of the ozone.

68. A 114-g Frisbee is lodged on a tree branch 7.65 m above the ground. To free it, you lob a 240-g wad of mud vertically upward. The mud leaves your hand at a point 1.23 m above the ground, moving at 17.7 m/s. It sticks to the Frisbee. Find (a) the maximum height reached by the Frisbee-mud combination and (b) the speed with which the combination hits the ground.

69. You set a small ball of mass m atop a large ball of mass $M \gg m$ and drop the pair from a height h. Assuming the balls are perfectly elastic, show that the smaller ball rebounds to a height $9h$.

70. A car moving at speed v undergoes a one-dimensional collision with an identical car initially at rest. The collision is neither elastic nor fully inelastic; $\frac{5}{18}$ of the initial kinetic energy is lost. Find the velocities of the two cars after the collision.

71. A 200-g block is released from rest at a height of 25 cm on a frictionless 30° incline. It slides down the incline and then along a frictionless surface until it collides elastically with an 800-g block at rest 1.4 m from the bottom of the incline (Fig. 9.29). How much later do the two blocks collide again?

FIGURE 9.29 Problem 71

72. A 14-kg projectile is launched at 380 m/s at a 55° angle to the horizontal. At the peak of its trajectory it collides with a second projectile moving horizontally, in the opposite direction, at 140 m/s. The two stick together and land 9.6 km horizontally downrange from the first projectile's launch point. Find the mass of the second projectile.

73. A block of mass m_1 undergoes a one-dimensional elastic collision with an initially stationary block of mass m_2. Find an expression for the fraction of the initial kinetic energy transferred to the second block, and plot your result for mass ratios m_1/m_2 from 0 to 20. Show also that, for a given mass ratio, the energy transfer is the same no matter which mass is initially at rest.

74. In Figure 9.6, if the radius of the uniform semicircular wire is R, how far above the center of the semicircle is its center of mass?

75. Where is the center of mass of a slice of pizza? Assume the slice is a uniform circular sector with radius R and central angle θ. *Hint:* Use symmetry to help locate the CM and integrate over area using Equation 9.4.

76. In a ballistic pendulum demonstration gone bad, a 0.52-g air gun pellet, fired horizontally with kinetic energy 3.25 J, passes straight through the 400-g Styrofoam pendulum block. If the pendulum swung to a maximum height of 0.50 mm, how much kinetic energy did the pellet have after it emerged from the Styrofoam?

77. The leading theory for the origin of the Moon suggests that a large asteroid the size of Mars struck the Earth. Some of the asteroid material and some of the terrestrial material combined to form the Moon. Assume the combined mass today of the Earth and Moon equals the combined mass of the primitive Earth and the asteroid. Also assume that the asteroid struck the Earth at about 10 km/s (see Problem 68 in Chapter 8). Assume the collision was inelastic. How much did the Earth's orbital speed change as a result of the collision?

78. An astronaut has become detached from the safety line connecting her to the space shuttle. She is carrying a tool kit with a mass of 10 kg. She is not moving with respect to the shuttle. In order to get back to the shuttle, and safety, she throws the tool kit away from her with a velocity of 8 m/s. The 80-kg astronaut is 200 m away from the shuttle when she throws the tool kit, and has 4 min of air remaining in her air tank. Does she make it back to the shuttle in time?

79. You have been called upon to investigate an accident at the corner of University Avenue and Harvard Street. A small compact car of mass 900 kg traveling east on University collided with a 1500-kg-mass maroon pickup truck traveling north on Harvard Street. As a result of the collision, the two vehicles stuck together, brakes locked, and slid in a direction 24° east of north. Skid marks indicate the two vehicles slid 16.33 m before coming to rest. (For the purposes of this problem, ignore the fact that there are buildings on all corners of the street: Assume the intersection is infinitely large! You may also assume there was no skidding prior to the impact.) The coefficient of kinetic friction between the tires and the pavement is found to be 0.8. Your job is to determine whether anyone was speeding. The speed limit on University Avenue is 30 mi/h and the speed limit on Harvard Street is 25 mi/h. And, since your best friend was driving the car, your result is important.

80. Astronomers can detect extrasolar planets by observing the slight wobble of stars due to the gravitational forces between the stars and the planets. The Sun wobbles about the center of mass of the solar system. Find the position of the center of mass by including the Sun, Jupiter, and Saturn, which make up most of the solar system's mass.

81. A friend suggests that everyone in the world simultaneously jumping off a high platform could cause an earthquake. "Yeah, right," you say. Assume all 6.5 billion inhabitants (average mass 70 kg each) jump together off a 4-m-high platform. How much would the Earth "recoil" or move? What energy would be imparted to the Earth by the impact? A 7.0-magnitude earthquake releases about 2×10^{15} J. Is your friend's claim reasonable? (One person pointed out the jumpers would need to jump with their knees locked and the major result would be a very loud cry of pain upon impact.)

Answers to Chapter Questions

Answer to Chapter Opening Question

The skier's center of mass follows the simple path of a projectile because, as Newton's laws show, the skier's mass acts like it's all concentrated at this point.

Answers to GOT IT? Questions

9.1 The CM is the uppermost point A. You can see this by imagining horizontal strips through the loop; the higher the strip the more mass is included, so the CM must lie nearer the top of the loop. The bottommost point would be the CM for a complete circle.

9.2 Momentum is conserved, so the momentum both before and after the explosion is the same: $\vec{P} = m\vec{v} = (0.50 \text{ kg})(60\hat{j} \text{ m/s}) = 30\hat{j}$ kg·m/s.

9.3 Only (d). The individual skaters experience external forces from the ball, as does the ball from the skaters. A system consisting of the ball and one skater experiences external forces from the other skater. Only the system of all three has no net external force and therefore has conserved momentum.

9.4 All but (c) are collisions; (a) and (b) are nearly elastic; (d) and (e) are inelastic.

9.5 The ball initially at rest is less massive; otherwise, the incident ball would have reversed direction (or stopped if the masses were equal).

10 Rotational Motion

■ For a blade of fixed mass, how should you engineer the blades of a wind turbine so it's easiest for the wind to get the turbine rotating?

▶ **To Learn**

By the end of this chapter you should be able to

■ Describe the rotational motion of rigid bodies by analogy with one-dimensional linear motion, using the concepts of angular velocity and acceleration, rotational inertia, torque, and the rotational analog of Newton's law (10.1–10.3).

■ Calculate the rotational inertias of objects with sufficient symmetry, by summing or integrating as appropriate (10.3).

■ Solve problems that combine linear and rotational motion (10.3).

■ Calculate rotational kinetic energy, and explain its relation to torque and work (10.4).

■ Describe rolling motion (10.5).

◀ **To Know**

■ The description of rotational motion here is directly analogous to Chapter 2's material on one-dimensional linear motion (2.1–2.4).

■ This chapter's rotational analog of Newton's law builds on the one-dimensional applications of Newton's second law in Chapter 4 (4.2, 4.5).

■ You should be comfortable with the concepts of work and kinetic energy introduced in Chapter 6 (6.1, 6.3).

You're sitting on a rotating planet. The wheels of your car rotate. Your favorite movie comes from a rotating DVD. A circular saw rotates to rip its way through a board. A dancer pirouettes, and a satellite spins about its axis. Even molecules rotate. Rotational motion is commonplace throughout the physical universe.

In principle, we could treat rotational motion by analyzing the motion of each particle in a rotating object. But that would be a hopeless task for all but the simplest objects. Instead, we'll describe rotational motion in analogy with linear motion as governed by Newton's laws.

This chapter parallels our study of one-dimensional motion in Chapters 2 and 4. In the next chapter we introduce a full vector description to treat multidimensional rotational motion.

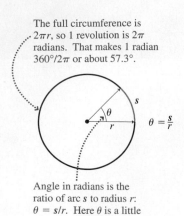

The full circumference is $2\pi r$, so 1 revolution is 2π radians. That makes 1 radian $360°/2\pi$ or about $57.3°$.

$$\theta = \frac{s}{r}$$

Angle in radians is the ratio of arc s to radius r: $\theta = s/r$. Here θ is a little less than 1 radian.

FIGURE 10.1 Radian measure of angles.

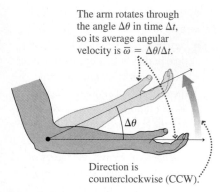

The arm rotates through the angle $\Delta\theta$ in time Δt, so its average angular velocity is $\overline{\omega} = \Delta\theta/\Delta t$.

Direction is counterclockwise (CCW).

FIGURE 10.2 Average angular velocity.

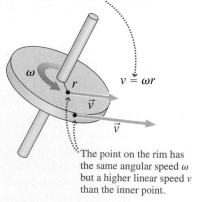

Linear speed is proportional to distance from the rotation axis.

$$v = \omega r$$

The point on the rim has the same angular speed ω but a higher linear speed v than the inner point.

FIGURE 10.3 Linear and rotational speeds.

10.1 Angular Velocity and Acceleration

You slip a DVD into a player, and it starts spinning. You could describe its motion by giving the speed and direction of each point on the disc. But it's much easier just to say that the disc is rotating at 800 revolutions per minute (rpm). As long as the disc is a **rigid body**—one whose parts remain in fixed positions relative to one another—then that single statement suffices to describe the motion of the entire disc.

Angular Velocity

The rate at which a body rotates is its **angular velocity**— the rate at which the angular position of any point on the body changes. With our 800-rpm DVD, the unit of angle was one full revolution (360°, or 2π radians), and the unit of time was the minute. But we could equally well express angular speed in revolutions per second (rev/s), degrees per second (°/s), or radians per second (rad/s or simply s^{-1} since radians are dimensionless). Because of the mathematically simple nature of radian measure, we often use radians in calculations involving rotational motion (Fig. 10.1).

We use the Greek symbol ω (omega) for angular velocity and define **average angular velocity** $\overline{\omega}$ as

$$\overline{\omega} = \frac{\Delta\theta}{\Delta t} \quad \text{(average angular velocity)} \tag{10.1}$$

where $\Delta\theta$ is the **angular displacement**—that is, the change in angular position—occurring in the time Δt (Fig. 10.2). When angular velocity is changing, we define **instantaneous angular velocity** as the limit over arbitrarily short time intervals:

$$\omega = \lim_{\Delta t \to 0} \frac{\Delta\theta}{\Delta t} = \frac{d\theta}{dt} \quad \text{(instantaneous angular velocity)} \tag{10.2}$$

These definitions are analogous to those of average and instantaneous linear velocity introduced in Chapter 2. Just as we use the term *speed* for the magnitude of velocity, so we define **angular speed** as the magnitude of the angular velocity.

Velocity is a vector quantity, with magnitude and direction. Is angular velocity also a vector? Yes, but we'll wait until the next chapter for the full vector description of rotational motion. In this chapter, it's sufficient to know whether an object's rotation is clockwise (CW) or counterclockwise (CCW) about a fixed axis—as suggested by the curved arrow in Fig. 10.2. This restriction to a fixed axis is analogous to Chapter 2's restriction to one-dimensional motion.

Angular and Linear Speed

Individual points on a rotating object undergo circular motion. Each point has an instantaneous linear velocity $\vec{v}$ whose magnitude is the linear speed v. We now relate this linear speed v to the angular speed ω. The definition of angular measure in radians (Fig. 10.1) is $\theta = s/r$. Differentiating this expression with respect to time, we have

$$\frac{d\theta}{dt} = \frac{1}{r}\frac{ds}{dt}$$

because the radius r is constant. But $d\theta/dt$ is the angular velocity, and ds/dt is the linear speed v, so $\omega = v/r$, or

$$v = \omega r \tag{10.3}$$

Thus the linear speed of any point on a rotating object is proportional both to the angular speed of the object and to the distance from that point to the axis of rotation (Fig. 10.3).

✓**TIP** Radian Measure

Equation 10.3 was derived using the definition of angle *in radians* and therefore holds for only angular speed measured in radians per unit time. If you're given other angular measures—degrees or revolutions, for example—you should convert to radians before using Equation 10.3.

EXAMPLE 10.1 **Angular Speed: A Wind Turbine**

A wind turbine's blades are 28 m long and rotate at 21 rpm. Find the angular speed of the blades in radians per second, and determine the linear speed at the tip of a blade.

INTEPRET This problem is about converting between two units of angular speed, revolutions per minute and radians per second, as well as finding linear speed given angular speed and radius.

DEVELOP We'll first convert the units to radians per second and then calculate the linear speed using Equation 10.3, $v = \omega r$.

EVALUATE One revolution is 2π rad, and 1 min is 60 s, so we have

$$\omega = 21 \text{ rpm} = \frac{(21 \text{ rev/min})(2\pi \text{ rad/rev})}{60 \text{ s/min}} = 2.2 \text{ rad/s}$$

The speed at the tip of a 28-m-long blade then follows from Equation 10.3: $v = \omega r = (2.2 \text{ rad/s})(28 \text{ m}) = 62 \text{ m/s}$.

ASSESS With ω in radians per second, multiplying by length in meters gives correct velocity units of meters per second because radians are dimensionless. ∎

Angular Acceleration

If the angular velocity of a rotating object changes, then the object undergoes **angular acceleration** α, defined analogously to linear acceleration:

$$\alpha = \lim_{\Delta t \to 0} \frac{\Delta \omega}{\Delta t} = \frac{d\omega}{dt} \quad (\text{angular acceleration}) \qquad (10.4)$$

Taking the limit gives the instantaneous angular acceleration; if we don't take the limit, then we have an average over the time interval Δt. The SI units of angular acceleration are rad/s², although we sometimes use other units such as rpm/s or rev/s².

Angular acceleration has the same direction as angular velocity—CW or CCW—if the angular speed is increasing, and the opposite direction if it's decreasing. These situations are analogous to a car that's speeding up (acceleration and velocity in the same direction) or braking (acceleration opposite velocity).

When a rotating object undergoes angular acceleration, points on the object speed up or slow down. Therefore they have a **tangential acceleration** dv/dt directed parallel or antiparallel to their linear velocity (Fig. 10.4). We introduced this idea of tangential acceleration back in Chapter 3; here we can recast it in terms of the angular acceleration:

$$a_t = \frac{dv}{dt} = r\frac{d\omega}{dt} = r\alpha \quad (\text{tangential acceleration}) \qquad (10.5)$$

Whether or not there's angular acceleration, points on a rotating object also have **radial acceleration** because they're in circular motion. Radial acceleration is given, as usual, by $a_r = v^2/r$; using $v = \omega r$ from Equation 10.3, we can recast this equation in angular terms as $a_r = \omega^2 r$.

Because angular velocity and acceleration are defined analogously to linear velocity and acceleration, all the relations among linear position, velocity, and acceleration automatically apply among angular position, angular velocity, and angular acceleration. If angular acceleration is constant, then all our constant-acceleration formulas of Chapter 2 apply when we make the substitutions θ for x, ω for v, and α for a. Table 10.1 summarizes this direct analogy between linear and rotational quantities. With Table 10.1, problems involving rotational motion are analogous to the one-dimensional linear problems you solved in Chapter 2.

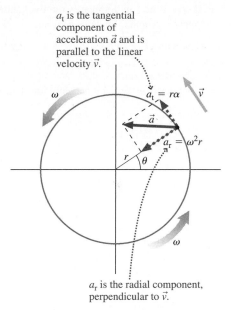

a_t is the tangential component of acceleration $\vec{a}$ and is parallel to the linear velocity $\vec{v}$.

a_r is the radial component, perpendicular to $\vec{v}$.

FIGURE 10.4 Radial and tangential acceleration.

TABLE 10.1 Angular and Linear Position, Velocity, and Acceleration

Linear Quantity	Angular Quantity
Position x	Angular position θ
Velocity $v = \dfrac{dx}{dt}$	Angular velocity $\omega = \dfrac{d\theta}{dt}$
Acceleration $a = \dfrac{dv}{dt} = \dfrac{d^2x}{dt^2}$	Angular acceleration $\alpha = \dfrac{d\omega}{dt} = \dfrac{d^2\theta}{dt^2}$

Equations for Constant Linear Acceleration		Equations for Constant Angular Acceleration	
$\bar{v} = \frac{1}{2}(v_0 + v)$	(2.8)	$\bar{\omega} = \frac{1}{2}(\omega_0 + \omega)$	(10.6)
$v = v_0 + at$	(2.7)	$\omega = \omega_0 + \alpha t$	(10.7)
$x = x_0 + v_0 t + \frac{1}{2}at^2$	(2.10)	$\theta = \theta_0 + \omega_0 t + \frac{1}{2}\alpha t^2$	(10.8)
$v^2 = v_0^2 + 2a(x - x_0)$	(2.11)	$\omega^2 = \omega_0^2 + 2\alpha(\theta - \theta_0)$	(10.9)

EXAMPLE 10.2 **Linear Analogies: Spin-down**

When the wind dies, the turbine of Example 10.1 spins down with constant angular acceleration of magnitude 0.12 rad/s². How many revolutions does the turbine make before coming to a stop?

INTERPRET The key to problems involving rotational motion is to identify the analogous situation for linear motion. This problem is analogous to asking how far a braking car travels before coming to a stop. We identify the number of rotations—the angular displacement—as the analog of the car's linear displacement. The given angular acceleration is analogous to the car's braking acceleration. The initial angular speed (2.2 rad/s, from Example 10.1) is analogous to the car's initial speed. And in both cases the final state we're interested in has zero speed—whether linear or angular.

DEVELOP Our plan is to develop the analogy further so we can find the angular displacement. The easiest way to solve the linear problem would be to use Equation 2.11, $v^2 = v_0^2 + 2a(x - x_0)$, with $v = 0$, v_0 the initial velocity, a the car's acceleration, and $\Delta x = x - x_0$ the distance

we're solving for. In Table 10.1, Equation 10.9 is the analogous equation for rotational motion: $\omega^2 = \omega_0^2 + 2\alpha \, \Delta\theta$, where we've written $\theta - \theta_0 = \Delta\theta$ for the rotational displacement during the spin-down.

EVALUATE We solve for $\Delta\theta$:

$$\Delta\theta = \frac{\omega^2 - \omega_0^2}{2\alpha} = \frac{0 - (2.2 \text{ rad/s})^2}{(2)(-0.12 \text{ rad/s}^2)} = 20 \text{ rad} = 3.2 \text{ revolutions}$$

where the last conversion follows because 1 revolution is 2π radians.

ASSESS The turbine blades are turning rather slowly—less than 1 revolution every second—so it's not surprising that a small angular acceleration can bring them to a halt in a short angular "distance." Note, too, how the units work out. Also, by taking ω as positive, we needed to treat α as negative because the angular acceleration is opposite the angular velocity when the rotation rate is slowing—just as the braking car's linear acceleration is opposite its velocity. ∎

10.2 Torque

Newton's second law, $\vec{F} = m\vec{a}$, proved very powerful in our study of motion. Ultimately Newton's law governs all motion, but its application to every particle in a rotating object would be terribly cumbersome. Can we instead formulate an analogous law that deals with rotational quantities?

To develop such a law, we need rotational analogs of force, mass, and acceleration. Angular acceleration α is the analog of linear acceleration; in the next two sections we develop analogs for force and mass.

How can a small child balance her father on a seesaw? By sitting far from the seesaw's rotation axis; that way, her smaller weight at a greater distance from the pivot is as effective as her father's greater weight closer to the pivot. In general, the effectiveness of a force in bringing about changes in rotational motion—a quantity called **torque**—depends not only on the magnitude of the force but also on how far from the rotation axis it's applied (Fig. 10.5). The effectiveness of the force also depends on the *direction* in which it's applied, as Fig. 10.6 suggests. Based on these considerations, we define torque as the product of the distance r from the rotation axis and the component of force perpendicular to that axis. Torque is given the symbol τ (Greek tau, pronounced to rhyme with "how"). Then we can write

$$\tau = rF \sin\theta \qquad (10.10)$$

The same force is applied at different points on the wrench.

Closest to O, τ is smallest.

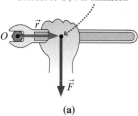

(a)

Farther away, τ becomes larger.

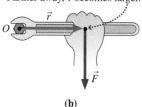

(b)

Farthest away, τ becomes greatest.

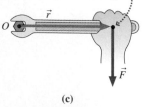

(c)

FIGURE 10.5 Torque increases with the distance r from the rotation axis O to the point where force is applied.

The same force is applied at different angles.

Torque is greatest when $\vec{F}$ is perpendicular to $\vec{r}$.

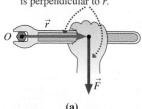

(a)

Torque decreases when $\vec{F}$ is no longer perpendicular to $\vec{r}$.

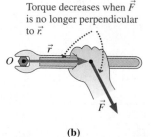

(b)

Torque is zero when $\vec{F}$ is parallel to $\vec{r}$.

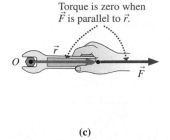

(c)

FIGURE 10.6 Torque is greatest with $\vec{F}$ and $\vec{r}$ at right angles, and diminishes to zero as they become colinear.

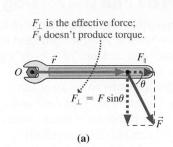

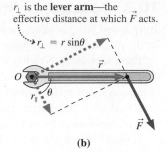

FIGURE 10.7 Two ways of thinking about torque. (a) $\tau = rF_\perp$; (b) $\tau = r_\perp F$. Both give $\tau = rF\sin\theta$.

where θ is the angle between the force vector and the vector $\vec{r}$ from the rotation axis to the force application point. Figure 10.7 shows two interpretations of Equation 10.10. Figure 10.7b also defines the so-called **lever arm**.

Torque, which you can think of as a "twisting force," plays the role of force in the rotational analog of Newton's second law. Equation 10.10 shows that torque is measured in newton-meters. Although this is the same unit as energy, torque is a different physical quantity, so we reserve the term *joule* ($=1$ N·m) for energy.

Does torque have direction? Yes, and we'll extend our notion of torque to provide a vector description in the next chapter. For now we'll specify the direction as either clockwise or counterclockwise.

EXAMPLE 10.3 **Torque: Changing a Tire**

You're tightening your car's wheel nuts after changing a flat tire. The instructions specify a tightening torque of 95 N·m so the nuts won't come loose. If your 45-cm-long wrench makes a 67° angle with the horizontal, with what force must you pull horizontally to produce the required torque?

INTERPRET We need to find the force required to produce a specific torque, given the distance from the rotation axis and the angle the force makes with the wrench.

DEVELOP Figure 10.8 is our drawing, and we'll calculate the torque using Equation 10.10, $\tau = rF\sin\theta$. With the force applied horizontally, comparison of Figs. 10.7a and 10.8 shows that the angle θ in Equation 10.10 is $180° - 67° = 113°$.

EVALUATE We solve Equation 10.10 for the force F:

$$F = \frac{\tau}{r\sin\theta} = \frac{95 \text{ N} \cdot \text{m}}{(0.45 \text{ m})(\sin 113°)} = 230 \text{ N}$$

ASSESS Is a 230-N force reasonable? Yes: It's roughly the same size as the force needed to lift a 23-kg ($\sim$50-lb) suitcase. Tightening torques,

as in this example, are often specified for nuts and bolts in critical applications. Mechanics use specially designed "torque wrenches" that provide a direct indication of the applied torque.

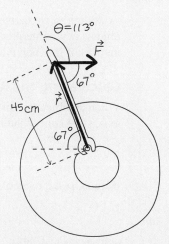

FIGURE 10.8 Our sketch of the wrench and wheel nut.

✓**TIP** Specify the Axis

Torque depends on where the force is applied *relative to some rotation axis*. The same physical force results in different torques about different axes. Be sure the rotation axis is specified before you make a calculation involving torque.

GOT IT? 10.1 The forces in Figs. 10.5 and 10.6 all have the same magnitude. (a) Which of Figs. 10.5a, 10.5b, and 10.6b has the greatest torque? (b) Which of these has the least torque?

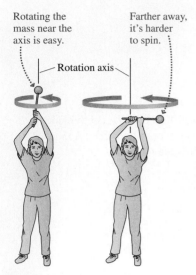

Rotating the mass near the axis is easy.

Farther away, it's harder to spin.

Rotation axis

FIGURE 10.9 It's easier to set an object rotating if the mass is concentrated near the axis.

10.3 Rotational Inertia and the Analog of Newton's Law

Torque and angular acceleration are the rotational analogs of force and linear acceleration. To develop a rotational analog of Newton's law, we still need the rotational analog of mass.

The mass m in Newton's law is a measure of a body's inertia—of its resistance to changes in motion. So we want a quantity that describes resistance to changes in rotational motion. Figure 10.9 shows that it's easier to set an object rotating when its mass is concentrated near the rotation axis. Therefore our rotational analog of inertia must depend not only on mass itself but also on the distribution of mass relative to the rotation axis.

Suppose the object in Fig. 10.9 consists of an essentially massless rod of length R with a ball of mass m on the end. We allow the object to rotate about an axis through the free end of the rod and apply a force $\vec{F}$ to the ball, always at right angles to the rod (Fig. 10.10). The ball undergoes a tangential acceleration given by Newton's law: $F = ma_t$. (There's also a tension force in the rod, but because it acts along the rod, it doesn't contribute to the torque or angular acceleration.) We can use Equation 10.5 to express the tangential acceleration in terms of the angular acceleration α and the distance R from the rotation axis: $F = ma_t = m\alpha R$. We can also express the force F in terms of its associated torque. Since the force is perpendicular to the rod, Equation 10.10 gives $\tau = RF$. Using our expression for F, we have

$$\tau = (mR^2)\alpha$$

Here we have Newton's law, $F = ma$, written in terms of rotational quantities. The torque—analogous to force—is the product of the angular acceleration and the quantity mR^2, which must therefore be the rotational analog of mass. We call this quantity the **rotational inertia** or **moment of inertia** and give it the symbol I. Rotational inertia is measured in kg·m² and accounts for both an object's mass and the distribution of that mass. Like torque, the value of the rotational inertia depends on the location of the rotation axis. Given the rotational inertia I, our rotational analog of Newton's law becomes

$$\tau = I\alpha \quad \text{(Rotational analog of Newton's 2}^{\text{nd}}\text{ law)} \quad (10.11)$$

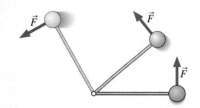

FIGURE 10.10 A force applied perpendicular to the rod results in angular acceleration.

Although we derived Equation 10.11 for a single, localized mass, it applies to extended objects if we interpret τ as the net torque on the object and I as the sum of the rotational inertias of the individual mass elements making up the object.

Calculating the Rotational Inertia

When an object consists of a number of discrete mass points, its rotational inertia about an axis is the sum of the rotational inertias of the individual mass points:

$$I = \sum m_i r_i^2 \quad \text{(rotational inertia)} \quad (10.12)$$

Here m_i is the mass of the ith mass point, and r_i is its distance from the rotation axis.

EXAMPLE 10.4 **Rotational Inertia: A Sum**

A dumbbell-shaped object consists of two equal masses $m = 0.64$ kg on the ends of a massless rod of length $L = 85$ cm. Calculate its rotational inertia about an axis one-fourth of the way from one end of the rod and perpendicular to it.

INTERPRET Here we have two discrete masses, so this problem is asking us to calculate the rotational inertia by summing over the individual masses.

DEVELOP Figure 10.11 is our sketch. We'll use Equation 10.12, $I = \sum m_i r_i^2$, to sum the two individual rotational inertias.

EVALUATE
$$I = \sum m_i r_i^2 = m\left(\tfrac{1}{4}L\right)^2 + m\left(\tfrac{3}{4}L\right)^2 = \tfrac{5}{8}mL^2$$
$$= \tfrac{5}{8}(0.64 \text{ kg})(0.85 \text{ m})^2 = 0.29 \text{ kg} \cdot \text{m}^2$$

ASSESS Make sense? Even though there are two masses, our answer is less than the rotational inertia mL^2 of a single mass rotated about a rod of length L. That's because distance from the rotation axis is *squared*, so it contributes more in determining rotational inertia than does mass.

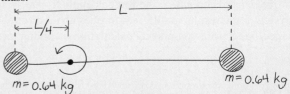

$m = 0.64$ kg $m = 0.64$ kg

FIGURE 10.11 Our sketch for Example 10.4, showing rotation about an axis perpendicular to the page.

GOT IT? 10.2 Would the rotational inertia of the two-mass dumbbell in Example 10.4 (a) increase, (b) decrease, or (c) stay the same if the rotation axis were at the center of the rod? If it were at one end?

With continuous distributions of matter, we consider a large number of very small mass elements dm throughout the object, and sum the individual rotational inertias $r^2\, dm$ over the entire object (Fig. 10.12). In the limit of an arbitrarily large number of infinitesimally small mass elements, that sum becomes an integral:

$$I = \int r^2\, dm \quad \left(\begin{array}{c}\text{rotational inertia,}\\ \text{continuous matter}\end{array}\right) \qquad (10.13)$$

where the limits of integration cover the entire object.

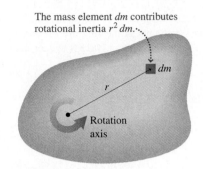

The mass element dm contributes rotational inertia $r^2\, dm$.

FIGURE 10.12 Rotational inertia can be found by integrating the rotational inertias $r^2\, dm$ of the mass elements making up an object.

EXAMPLE 10.5 Rotational Inertia by Integration: A Rod

Find the rotational inertia of a uniform, narrow rod of mass M and length L about an axis through its center and perpendicular to the rod.

INTERPRET The rod is a continuous distribution of matter, so calculating the rotational inertia is going to involve integration. We identify the rotation axis as being in the center of the rod.

DEVELOP Figure 10.13 shows the rod and rotation axis; we added a coordinate system with x axis along the rod and the origin at the rotation axis. With a continuous distribution, Equation 10.13, $I = \int r^2\, dm$, applies. To develop a solution plan, we need to set up the integral in Equation 10.13. That equation may seem confusing because the integral contains both the geometric variable r and the mass element dm. How are they related? At this point you might want to review Tactics 9.1 (page 138); we'll follow its steps here. (1) We're first supposed to find a suitable mass element; here, with a one-dimensional rod, that can be a short section of the rod. We marked a typical mass element in Fig. 10.13. (2) This step is straightforward in this one-dimensional case; the length of the mass element is dx, signifying an infinitesimally short piece of the rod. (3) Now we form ratios to relate dx and the

mass element dm. The total mass of the rod is M, and its total length is L. With the mass distributed uniformly, that means dx is the same fraction of L that dm is of M, or $dx/L = dm/M$. (4) We solve for the mass element: $dm = (M/L)\, dx$.

We're almost done. But the integral in Equation 10.13 contains r, and we've related dm and dx. No problem: On the one-dimensional rod, distances from the rotation axis are just the coordinates x. So r becomes x in our integral, and we have

$$I = \int r^2\, dm = \int_{-L/2}^{L/2} x^2 \frac{M}{L}\, dx$$

We chose the limits to include the entire rod; with the origin at the center, it runs from $-L/2$ to $L/2$.

EVALUATE The constants M and L come outside the integral, so we have

$$I = \int_{-L/2}^{L/2} x^2 \frac{M}{L}\, dx = \frac{M}{L} \int_{-L/2}^{L/2} x^2\, dx = \frac{M}{L} \frac{x^3}{3}\Bigg|_{-L/2}^{L/2} = \tfrac{1}{12} ML^2 \quad (10.14)$$

ASSESS Make sense? In Example 10.4 we found $I = \tfrac{5}{8} mL^2$ for a rod with two masses m on the ends. If you thought about GOT IT? 10.2, you probably realized that the rotational inertia would be $\tfrac{1}{2} mL^2$ for rotation about the rod's center. The total mass for that one was $M = 2m$, so in terms of total mass the rotational inertia about the center would be $I = \tfrac{1}{4} ML^2$—a lot larger than what we've found for the continuous rod. That's because much of the continuous rod's mass is close to the rotation axis, so it contributes less to the rotational inertia. ∎

The mass element has mass dm and length dx.

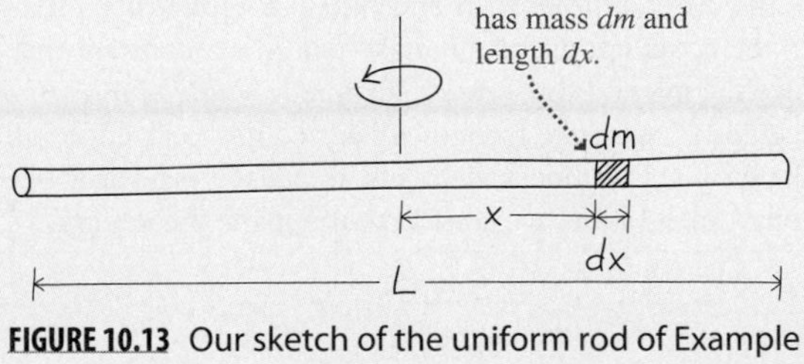

FIGURE 10.13 Our sketch of the uniform rod of Example 10.5.

EXAMPLE 10.6 **Rotational Inertia by Integration: A Ring**

Find the rotational inertia of a thin ring of radius R and mass M about the ring's axis.

INTERPRET This example is similar to Example 10.5, but the geometry has changed from a rod to a ring.

DEVELOP Figure 10.14 shows the ring with a mass element dm. All the mass elements in the ring are the same distance R from the rotation axis, so r in Equation 10.13 is the constant R, and the equation becomes

$$I = \int R^2 \, dm = R^2 \int dm$$

where the integration is over the ring.

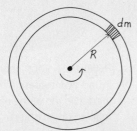

FIGURE 10.14 Our sketch of a thin ring, showing one mass element dm.

EVALUATE Because the sum of the mass elements over the ring is the total mass M, we find

$$I = MR^2 \quad \text{(thin ring)} \qquad (10.15)$$

ASSESS The rotational inertia of the ring is the same as if all the mass were concentrated in one place a distance R from the rotation axis; the angular distribution of the mass about the axis doesn't matter. Notice, too, that it doesn't matter whether the ring is narrow like a loop of wire or long like a section of hollow pipe, as long as it's thin enough that all of it is essentially equidistant from the rotation axis (Fig 10.15).

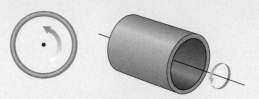

FIGURE 10.15 The rotational inertia is MR^2 for any thin ring, whether it's narrow like a wire loop or long like a pipe.

EXAMPLE 10.7 **Rotational Inertia by Integration: A Disk**

A disk of radius R and mass M has uniform density. Find the rotational inertia of the disk about an axis through its center and perpendicular to the disk.

INTERPRET Again we need to find the rotational inertia for a piece of continuous matter, this time a disk.

DEVELOP Because the disk is continuous, we need to integrate using Equation 10.13, $I = \int r^2 \, dm$. We'll condense the strategy we applied in Example 10.5. The result of Example 10.6 suggests dividing the disk into rings, as shown in Fig. 10.16. Equation 10.15, with $M \to dm$, shows that a ring of radius r and mass dm contributes $r^2 \, dm$ to the rotational inertia of the disk. Then the total inertia will be $I = \int_0^R r^2 \, dm$, where we chose the limits to pick up contributions from all the mass elements on the disk. Again we need to relate r and dm. Think of

"unwinding" the ring shown in Fig. 10.16; it becomes essentially a rectangle whose area dA is its circumference multiplied by its width: $dA = 2\pi r \, dr$. Next, we form ratios. The ring area dA is to the total disk area πR^2 as the ring mass dm is to the total mass M: $2\pi r \, dr / \pi R^2 = dm/M$. Solving for dm gives $dm = (2Mr/R^2) \, dr$.

EVALUATE We now evaluate the integral:

$$I = \int_0^R r^2 \, dm = \int_0^R r^2 \left(\frac{2Mr}{R^2} \right) dr$$

$$= \frac{2M}{R^2} \int_0^R r^3 \, dr = \frac{2M}{R^2} \left. \frac{r^4}{4} \right|_0^R = \tfrac{1}{2} MR^2 \quad \text{(disk)} \qquad (10.16)$$

ASSESS Again, this result makes sense. In the disk, some of the mass is closer to the rotation axis, so the rotational inertia should be less than the value MR^2 for the ring.

✓**TIP** Constants and Variables

Note the different roles of R and r here. R represents a fixed quantity—the actual radius of the disk—and it's a constant that can go outside the integral. In contrast, r is the *variable of integration*, and it changes as we range from the disk's center to its edge, adding up all the infinitesimal mass elements. Because r is a variable over the region of integration, we can't take it outside the integral.

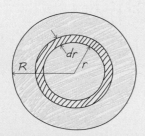

FIGURE 10.16 A disk may be divided into ring-shaped mass elements of mass dm, radius r, and width dr.

TABLE 10.2 Rotational Inertias

Thin rod about center
$I = \frac{1}{12}ML^2$

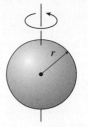

Thin ring or hollow cylinder about its axis
$I = MR^2$

Solid sphere about diameter
$I = \frac{2}{5}MR^2$

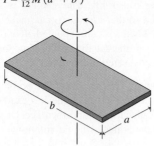

Flat plate about perpendicular axis
$I = \frac{1}{12}M(a^2 + b^2)$

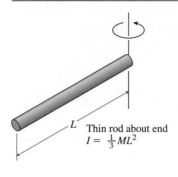

Thin rod about end
$I = \frac{1}{3}ML^2$

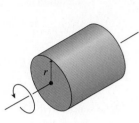

Disk or solid cylinder about its axis
$I = \frac{1}{2}MR^2$

Hollow spherical shell about diameter
$I = \frac{2}{3}MR^2$

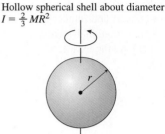

Flat plate about central axis
$I = \frac{1}{12}Ma^2$

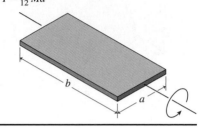

The rotational inertias of other shapes about various axes are found by integration as in these examples. Table 10.2 lists results for some common shapes. Note that more than one rotational inertia is listed for some shapes, since the rotational inertia depends on the rotation axis.

If we know the rotational inertia I_{cm} about an axis through the center of mass of a body, a useful relation called the **parallel-axis theorem** allows us to calculate the rotational inertia I through any parallel axis. The parallel-axis theorem states that

$$I = I_{cm} + Md^2 \tag{10.17}$$

where d is the distance from the center-of-mass axis to the parallel axis and M is the total mass of the object. Figure 10.17 shows the meaning of the parallel-axis theorem.

GOT IT? 10.3 Explain why the rotational inertia of the solid sphere in Table 10.2 is less than that of the spherical shell with the same radius and the same mass.

Rotational Dynamics

Knowing a body's rotational inertia, we can use the rotational analog of Newton's second law (Equation 10.11) to determine its behavior, just as we used Newton's law itself to analyze linear motion. Like the force in Newton's law, the torque in Equation 10.11 is the *net* external torque—the sum of all external torques acting on the body.

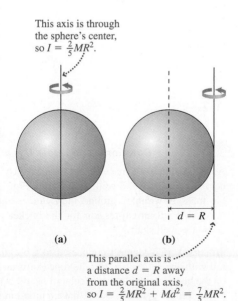

This axis is through the sphere's center, so $I = \frac{2}{5}MR^2$.

(a)   (b)

$d = R$

This parallel axis is a distance $d = R$ away from the original axis, so $I = \frac{2}{5}MR^2 + Md^2 = \frac{7}{5}MR^2$.

FIGURE 10.17 Meaning of the parallel-axis theorem.

EXAMPLE 10.8 **Rotational Dynamics: De-Spinning a Satellite**

A cylindrical satellite is 1.4 m in diameter, with its 940-kg mass distributed uniformly. The satellite is spinning at 10 rpm but must be stopped so that astronauts can make repairs. Two small gas jets, each with 20-N thrust, are mounted on opposite sides of the satellite and fire tangent to the satellite's rim. How long must the jets be fired in order to stop the satellite's rotation?

INTERPRET This is ultimately a problem about angular acceleration, but we're given the forces the jets exert. So it becomes a problem about calculating torque and then acceleration—that is, a problem in rotational dynamics using the rotational analog of Newton's law.

DEVELOP Figure 10.18 shows the situation. We're asked about the time, which we can get from the angular acceleration and initial angular speed. We can find the acceleration using the rotational analog of Newton's law, Equation 10.11, if we know both torque and rotational inertia. So here's our plan: (1) Find the satellite's rotational inertia from Table 10.2, treating it as a solid cylinder. (2) Find the torque due

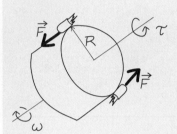

FIGURE 10.18 Torque from the jets stops the satellite's rotation.

to the jets using Equation 10.10, $\tau = rF\sin\theta$. (3) Use the rotational analog of Newton's law—Equation 10.11, $\tau = I\alpha$—to find the angular acceleration. (4) Use the change in angular speed to get the time.

EVALUATE Following our plan, (1) the rotational inertia from Table 10.2 is $I = \frac{1}{2}MR^2$. (2) With the jets tangent to the satellite, $\sin\theta$ in Equation 10.10 is 1, so each jet contributes a torque of magnitude RF, where R is the satellite radius and F the jet thrust force. With two jets, the net torque then has magnitude $\tau = 2RF$. (3) Equation 10.11 gives $\alpha = \tau/I = (2RF)/(\frac{1}{2}MR^2) = 4F/MR$. (4) We want this torque to drop the angular speed from $\omega_0 = 10$ rpm to zero, so the magnitude of the speed change is

$$\Delta\omega = 10\,\text{rev/min} = (10\,\text{rev/min})(2\pi\,\text{rad/rev})/(60\,\text{s/min}) = 1.05\,\text{rad/s}.$$

Since angular acceleration is $\alpha = \Delta\omega/\Delta t$, our final answer is

$$\Delta t = \frac{\Delta\omega}{\alpha} = \frac{MR\,\Delta\omega}{4F}$$

$$= \frac{(940\,\text{kg})(0.70\,\text{m})(1.05\,\text{rad/s})}{(4)(20\,\text{N})} = 8.6\,\text{s}$$

ASSESS Make sense? Yes: The thrust F appears in the denominator, showing that a larger force and hence torque will bring the satellite more rapidly to a halt. Larger M and R contribute to a larger rotational inertia, thus lengthening the stopping time—although a larger R also means a larger torque, an effect that reduces the R dependence from the R^2 that appears in the expression for rotational inertia. ∎

A single problem can involve both rotational and linear motion with more than one object. The strategy for dealing with such problems is similar to the multiple-object strategy we developed in Chapter 5, where we identified the objects whose motions we were interested in, drew a free-body diagram for each, and then applied Newton's law separately to each object. We used the physical connections among the objects to relate quantities appearing in the separate Newton's law equations. Here we do the same thing, except that when an object is rotating, we use Equation 10.11, the rotational analog of Newton's law. Often the physical connection will entail relations between the force on an object in linear motion and the torque on a rotating object, as well as between the objects' linear and rotational accelerations.

EXAMPLE 10.9 **Rotational and Linear Dynamics: Into the Well**

A solid cylinder of mass M and radius R is mounted on a frictionless horizontal axle over a well, as shown in Fig. 10.19. A rope of negligible mass is wrapped around the cylinder and supports a bucket of mass m. Find an expression for the bucket's acceleration as it falls down the well shaft.

INTERPRET If it weren't connected to the cylinder, the bucket would fall with acceleration g. But the rope exerts an upward tension force $\vec{T}$ on the bucket, reducing its acceleration and at the same time exerting a torque on the cylinder. So we have a problem involving both rotational and linear dynamics. We identify the bucket and the cylinder as the objects of interest; the bucket is in linear motion while the cylinder rotates. The connection between them is the rope.

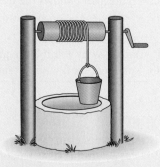

FIGURE 10.19 Example 10.9.

cont'd.

DEVELOP Figure 10.20 shows the free-body diagrams for the two objects; note that both involve the rope tension, $\vec{T}$. We chose the downward direction as positive in the bucket diagram and the clockwise direction as positive in the cylinder diagram. Now we're ready to write Newton's second law and its analog—Equation 10.11, $\tau = I\alpha$—for the two objects. Our plan is to formulate both equations and solve using the connection between them—physically the rope and mathematically the magnitude of the rope tension. We have to express the torque on the cylinder in terms of the tension force, using Equation 10.10, $\tau = rF \sin\theta$. We also need to relate the cylinder's angular acceleration to the bucket's linear acceleration, using Equation 10.5, $a_t = r\alpha$.

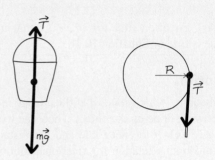

FIGURE 10.20 Our free-body diagrams for the bucket and cylinder.

EVALUATE With the downward direction positive, Newton's second law for the bucket reads $F_{net} = mg - T = ma$. For the cylinder we have the rotational analog of Newton's second law: $\tau = I\alpha$. But here the torque is due to the rope tension, which exerts a force T at right angles to a line from the rotation axis and so produces torque RT. Then the Newton's law analog becomes $RT = I\alpha$. As the rope unwinds, the tangential acceleration of the cylinder's edge must be equal to the bucket's linear acceleration; thus, using Equation 10.5, we have $\alpha = a/R$, and the cylinder equation becomes $RT = Ia/R$ or $T = Ia/R^2$. But the cylinder's rotational inertia, from Table 10.2, is $I = \frac{1}{2}MR^2$, so $T = \frac{1}{2}Ma$. Using this result in the bucket equation gives $ma = mg - T = mg - \frac{1}{2}Ma$; solving for a, we then have

$$a = \frac{mg}{m + \frac{1}{2}M}$$

ASSESS Make sense? If $M = 0$, there would be no rotational inertia and we would have $a = g$. Of course: With no torque needed to accelerate the cylinder, there would be no rope tension and the bucket would fall freely. But as the cylinder's mass M increases, the bucket's deceleration drops as greater torque and thus rope tension are needed to give the cylinder its rotational acceleration. You may be surprised to see that the cylinder radius doesn't appear in our answer. That, too, makes sense: The rotational inertia scales as R^2, but both the torque and the tangential acceleration scale with R. Since the cylinder's tangential acceleration is the same as the bucket's acceleration, the increases in torque and tangential acceleration cancel the effect of a greater rotational inertia. ∎

GOT IT? 10.4 The figure shows two identical masses m connected by a string that passes over a frictionless pulley whose mass M is *not* negligible. One mass rests on a frictionless table; the other hangs vertically, as shown. Is the magnitude of the tension force in the vertical section of the string (a) greater than, (b) equal to, or (c) less than that in the horizontal section? Explain.

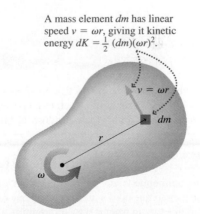

10.4 Rotational Energy

A rotating object clearly has kinetic energy. We define an object's **rotational kinetic energy** as the sum of the kinetic energies of all its mass elements, taken with respect to the rotation axis. Figure 10.21 shows that an individual mass element dm a distance r from the rotation axis has kinetic energy given by $dK = \frac{1}{2}(dm)(v^2) = \frac{1}{2}(dm)(\omega r)^2$. The rotational kinetic energy is given by summing—that is, integrating—over the entire object:

$$K_{rot} = \int dK = \int \frac{1}{2}(dm)(\omega r)^2 = \frac{1}{2}\omega^2 \int r^2 \, dm$$

where we've taken ω^2 outside the integral because it's the same for every mass element in the rigid, rotating object. The remaining integral is just the rotational inertia I, so we have

$$K_{rot} = \frac{1}{2}I\omega^2 \quad \text{(rotational kinetic energy)} \qquad (10.18)$$

This formula should come as no surprise: Since I and ω are the rotational analogs of mass and speed, Equation 10.18 is the rotational equivalent of $K = \frac{1}{2}mv^2$.

A mass element dm has linear speed $v = \omega r$, giving it kinetic energy $dK = \frac{1}{2}(dm)(\omega r)^2$.

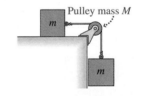

FIGURE 10.21 Kinetic energy of a mass element.

EXAMPLE 10.10 **Rotational Energy: Flywheel Storage**

A flywheel has a 135-kg solid cylindrical rotor with radius 30 cm and spins at 31,000 rpm. How much energy does it store?

INTERPRET We're being asked about kinetic energy stored in a rotating cylinder.

DEVELOP Equation 10.18, $K_{rot} = \frac{1}{2}I\omega^2$, gives the rotational energy. To use it, we need the rotational inertia from Table 10.2, and we need to convert the rotation rate in revolutions per minute to angular speed ω in radians per second.

EVALUATE Table 10.2 gives the rotational inertia, $I = \frac{1}{2}MR^2 = (\frac{1}{2})(135\,kg)(0.30\,m)^2 = 6.1\,kg\cdot m^2$, and 31,000 rpm is equivalent to $(31,000\,rev/min)(2\pi\,rad/rev)/(60\,s/min) = 3246\,rad/s$. Then Equation 10.18 gives

$$K_{rot} = \frac{1}{2}I\omega^2 = \left(\frac{1}{2}\right)(6.1\,kg\cdot m^2)(3246\,rad/s)^2 = 32\,MJ$$

ASSESS That 32 MJ is roughly the energy contained in a liter of gasoline. The advantages of the flywheel over a fuel or a chemical battery are more concentrated energy storage and greater efficiency at getting energy into and out of storage; see Application below. ∎

✓**TIP** When to Use Radians

We derived Equation 10.18, $K = \frac{1}{2}I\omega^2$, using Equation 10.3, $v = \omega r$. Since that equation works only with radian measure, the same is true of Equation 10.18.

Energy and Work in Rotational Motion

In Section 6.3 we proved the work-energy theorem, which states that the change in an object's linear kinetic energy is equal to the net work done on the object. There the work was the product (or the integral, for a changing force) of the net force and the distance the object moves. Not surprisingly, there's an analogous relation for rotational motion: The change in an object's rotational kinetic energy is equal to the net work done on the object. Now the work is, analogously, the product (or the integral) of the torque and the angular displacement:

$$W = \int_{\theta_i}^{\theta_f} \tau\, d\theta = \Delta K_{rot} = \frac{1}{2}I\omega_f^2 - \frac{1}{2}I\omega_i^2 \qquad \left(\begin{array}{c}\text{work-energy theorem,} \\ \text{rotational motion}\end{array}\right) \qquad (10.19)$$

Here the subscripts refer to the initial and final states.

Flywheel Energy Storage

Flywheels provide an attractive alternative to batteries in applications requiring short bursts of power. Examples include acceleration and hill climbing in hybrid vehicles, industrial lifting equipment and amuse-

ment park rides, power management on the electric grid, and uninterruptible power supplies. Flywheel-based hybrid vehicles would achieve high efficiency by storing mechanical energy in the flywheel during braking, rather than dissipating it as heat in conventional brakes or even storing it in a chemical battery as in today's hybrids.

Equation 10.18 shows that the stored energy can be substantial, provided the flywheel has significant rotational inertia and angular speed—the latter being especially important because the energy scales as the *square* of the angular speed. Modern flywheels can supply tens of kilowatts of power for as long as a minute; unlike batteries, their output isn't reduced in cold weather. They achieve rotation rates of 30,000 rpm and more using advanced carbon composite materials that can withstand the forces needed to maintain the radial acceleration of magnitude $\omega^2 r$. Advanced flywheels spin in vacuum, using magnetic bearings to minimize friction. Some even use superconducting materials, which eliminate electrical losses that we'll examine in Chapter 26. The photo shows a high-speed flywheel used in a prototype hybrid bus operating in Austin, Texas. The flywheel helps the bus achieve 30% fuel savings.

EXAMPLE 10.11 **Work and Rotational Energy: Balancing a Tire**

An automobile wheel with tire has rotational inertia 2.7 kg·m². What constant torque does a tire-balancing machine need to apply in order to spin this tire up from rest to 700 rpm in 25 revolutions?

INTERPRET The wheel's rotational kinetic energy changes as it spins up, so the machine must be doing work by applying a torque. Therefore the concept behind this problem is the work-energy theorem for rotational motion. *cont'd.*

DEVELOP The work-energy theorem of Equation 10.19 relates the work to the change in rotational kinetic energy: $W = \int_{\theta_i}^{\theta_f} \tau\, d\theta = \Delta K_{rot} = \frac{1}{2}I\omega_f^2 - \frac{1}{2}I\omega_i^2$. We're given the initial and final angular velocities, although we have to convert them to radians per second. With constant torque, the integral in Equation 10.19 becomes the product $\tau\,\Delta\theta$, so we can solve for the torque.

EVALUATE The initial angular speed ω_i is zero, and the final speed $\omega_f = (700\ \text{rev/min})(2\pi\ \text{rad/rev})/(60\ \text{s/min}) = 73.3\ \text{rad/s}$. The angular

displacement $\Delta\theta$ is $(25\ \text{rev})(2\pi\ \text{rad/rev}) = 157\ \text{rad}$. Then Equation 10.19 becomes $W = \tau\,\Delta\theta = \frac{1}{2}I\omega_f^2$, which gives

$$\tau = \frac{\frac{1}{2}I\omega_f^2}{\Delta\theta} = \frac{\left(\frac{1}{2}\right)(2.7\ \text{kg}\cdot\text{m}^2)(73.3\ \text{rad/s})^2}{157\ \text{rad}} = 46\ \text{N}\cdot\text{m}$$

ASSESS If this torque results from a force applied at the rim of a typical 40-cm-radius tire, then the magnitude of the force would be just under 100 N, about the weight of a 10-kg mass and thus a reasonable value. ∎

10.5 Rolling Motion

A rolling object exhibits both rotational motion and translational motion—the motion of the whole object from place to place. How much kinetic energy is associated with each of these motions?

In Section 9.3, we found that the kinetic energy of a composite object comprises two terms: the kinetic energy of the center of mass and the internal kinetic energy relative to the center of mass: $K = K_{cm} + K_{internal}$. A wheel of mass M moving with speed v has center-of-mass kinetic energy $K_{cm} = \frac{1}{2}Mv^2$. In the center-of-mass frame, the wheel is simply rotating with angular speed ω about the center of mass, so its internal kinetic energy is $K_{internal} = \frac{1}{2}I_{cm}\omega^2$, where the rotational inertia is taken about the center of mass. We now sum K_{cm} and $K_{internal}$ to get the total kinetic energy:

$$K_{total} = \tfrac{1}{2}Mv^2 + \tfrac{1}{2}I_{cm}\omega^2 \tag{10.20}$$

When a wheel is *rolling*—moving without slipping against the ground—its translational speed v and angular speed ω about its center of mass are related. Imagine a wheel that rolls half a revolution and therefore moves horizontally half its circumference (Fig. 10.22). Then the wheel's angular speed is the angular displacement $\Delta\theta$, here half a revolution, or π radians, divided by the time Δt: $\omega = \pi/\Delta t$. Its translational speed is the actual distance the wheel travels divided by the same time interval. But we've just argued that the wheel travels half a circumference, or πR, where R is its radius. So its translational speed is $v = \pi R/\Delta t$. Comparing our expressions for v and ω, we see that

$$v = \omega R \quad \text{(rolling motion)} \tag{10.21}$$

Equation 10.21 looks deceptively like Equation 10.3. But it says more. In Equation 10.3, $v = \omega r$, v is the linear speed of a point a distance r from the center of a rotating object. In Equation 10.21, v is the translational speed of the whole object and R is its radius. The two equations look similar because, as our argument leading to Equation 10.21 shows, an object that rolls without slipping moves with respect to the ground at the same rate that a point on its rim moves in the center-of-mass frame.

Our description of rolling motion leads to a point you may at first find absurd: In a rolling wheel, the point in contact with the ground is, instantaneously, at rest! Figure 10.23 shows how this surprising situation comes about.

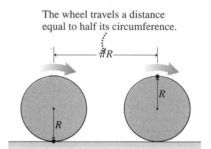

The wheel travels a distance equal to half its circumference.

FIGURE 10.22 A rolling wheel turns through half a revolution.

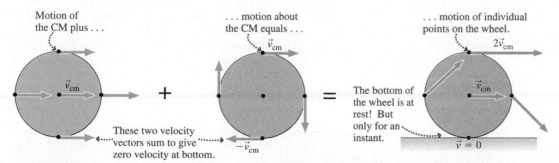

Motion of the CM plus . . .

. . . motion about the CM equals . . .

. . . motion of individual points on the wheel.

These two velocity vectors sum to give zero velocity at bottom.

The bottom of the wheel is at rest! But only for an instant.

FIGURE 10.23 Motion of a rolling wheel, decomposed into translation of the entire wheel plus rotation about the center of mass.

Why would an object roll without slipping? The answer is friction. On an icy slope, a wheel just slides down without rolling. Normally, though, the force of static friction keeps it from sliding. Instead, it rolls (Fig. 10.24). Because the contact point is at rest, the frictional force does no work and therefore mechanical energy is conserved. This lets us use the conservation-of-energy principle to analyze rolling objects.

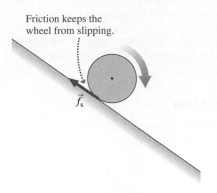

FIGURE 10.24 A wheel rolls down a sloping surface.

EXAMPLE 10.12   **Energy Conservation: Rolling Downhill**

A solid ball of mass M and radius R starts from rest and rolls down a hill. Its center of mass drops a total distance h. Find the ball's speed at the bottom of the hill.

INTERPRET This is similar to conservation-of-energy problems from Chapter 7, but now we identify two types of kinetic energy: translational and rotational. The ball starts on the slope with some gravitational potential energy, which ends up as kinetic energy at the bottom. The frictional force that keeps the ball from slipping does no work, so we can apply conservation of energy.

DEVELOP Figure 10.25 shows the situation, including bar graphs showing the distribution of energy in the ball's initial and final states. We've determined that conservation of energy holds, so

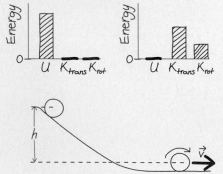

FIGURE 10.25 How fast is the ball moving at the bottom of the hill?

$K_0 + U_0 = K + U$. Here $K_0 = 0$ and, if we take the zero of potential energy at the bottom, then $U_0 = Mgh$ and $U = 0$. Finally, K consists of both translational and rotational kinetic energy as expressed in Equation 10.20, $K_{total} = \frac{1}{2}Mv^2 + \frac{1}{2}I\omega^2$. Our plan is to use this expression in the conservation-of-energy statement and solve for v. It looks like there's an extra variable, ω, that we don't know. But the ball isn't slipping, so Equation 10.21 holds and gives $\omega = v/R$. Then conservation of energy becomes

$$Mgh = \tfrac{1}{2}Mv^2 + \tfrac{1}{2}I\omega^2 = \tfrac{1}{2}Mv^2 + \tfrac{1}{2}\left(\tfrac{2}{5}MR^2\right)\left(\frac{v}{R}\right)^2 = \tfrac{7}{10}Mv^2$$

where we found the rotational inertia of a solid sphere, $\frac{2}{5}MR^2$, from Table 10.2.

EVALUATE Solving for v gives our answer:

$$v = \sqrt{\frac{10}{7}gh}$$

ASSESS This result is less than the speed $v = \sqrt{2gh}$ for an object that slides down a frictionless incline. Make sense? Yes: Some of the energy the rolling object gains goes into rotation, leaving less for translational motion. As often happens with gravitational problems, mass doesn't matter. Neither does radius: That factor $\frac{7}{10}$ results from the distribution of mass that gives the sphere its particular rotational inertia and would be the same for all spheres regardless of radius or mass. ∎

GOT IT? 10.5 A solid ball and a hollow ball roll without slipping down a ramp. Which reaches the bottom first? Explain.

Big Picture

The big idea of this chapter is rotational motion, quantified as the rate of change of angular position of any point on a rotating object. All the quantities used to describe linear motion have analogs in rotational motion. The analogs of force, mass, and acceleration are, respectively, torque, rotational inertia, and angular acceleration—and together they obey the rotational analog of Newton's second law.

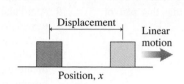

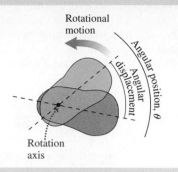

Key Concepts and Equations

The defining relations for rotational quantities are analogous to those for linear quantities, as is the statement of Newton's second law for rotational motion. Key concepts include angular velocity and acceleration, torque, and rotational inertia.

Angular velocity, ω

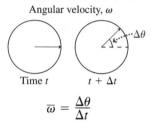

$$\overline{\omega} = \frac{\Delta\theta}{\Delta t}$$

Torque, τ

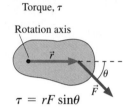

$$\tau = rF\sin\theta$$

Rotational inertia, I

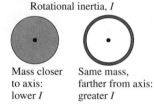

Mass closer to axis: lower I Same mass, farther from axis: greater I

$$I = \sum m_i r_i^2 \longrightarrow \int r^2\,dm$$

Discrete masses   Continuous matter

This table summarizes the analogies between linear and rotational quantities, along with quantitative relations that link rotational and linear quantities. Many of these relations require that angles be measured in radians, and most require explicit specification of a rotation axis.

Linear Quantity or Equation	Angular Quantity or Equation	Relation Between Linear and Angular Quantities
Position x	Angular position θ	
Speed $v = dx/dt$	Angular speed $\omega = d\theta/dt$	$v = \omega r$
Acceleration a	Angular acceleration α	$a_t = \alpha r$
Mass m	Rotational inertia I	$I = \int r^2\,dm$
Force F	Torque τ	$\tau = rF\sin\theta$
Kinetic energy $K_{trans} = \frac{1}{2}mv^2$	Kinetic energy $K_{rot} = \frac{1}{2}I\omega^2$	
Newton's second law (constant mass or rotational inertia):		
$F = ma$	$\tau = I\alpha$	

Cases and Uses

Constant angular acceleration: When angular acceleration is constant, equations analogous to those of Chapter 2 apply.

Equations for Constant Linear Acceleration		Equations for Constant Angular Acceleration	
$\overline{v} = \frac{1}{2}(v_0 + v)$	(2.8)	$\overline{\omega} = \frac{1}{2}(\omega_0 + \omega)$	(10.6)
$v = v_0 + at$	(2.7)	$\omega = \omega_0 + \alpha t$	(10.7)
$x = x_0 + v_0 t + \frac{1}{2}at^2$	(2.10)	$\theta = \theta_0 + \omega_0 t + \frac{1}{2}\alpha t^2$	(10.8)
$v^2 = v_0^2 + 2a(x - x_0)$	(2.11)	$\omega^2 = \omega_0^2 + 2\alpha(\theta - \theta_0)$	(10.9)

Rolling motion: When an object of radius R rolls without slipping, the point in contact with the ground is instantaneously at rest. In this case the object's translational and rotational speeds are related by $v = \omega R$. The object's kinetic energy is shared among translational kinetic energy $\frac{1}{2}mv^2$ and rotational kinetic energy $\frac{1}{2}I\omega^2$, with the division between these forms dependent on the rotational inertia.

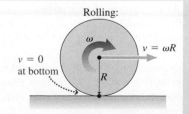

Rolling:

For Thought and Discussion

1. Do all points on a rigid, rotating object have the same angular velocity? Linear speed? Radial acceleration?
2. Part of a train wheel extends below the point of contact with the rails, as shown in Fig. 10.26. It is often said that this part of the train moves backward. Explain.

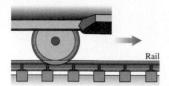

FIGURE 10.26 For Thought and Discussion 2

3. A point on the rim of a rotating wheel has nonzero acceleration, since it's moving in a circular path. Does it necessarily follow that the wheel is undergoing angular acceleration?
4. Why doesn't it make sense to talk about a body's rotational inertia unless a rotation axis is specified?
5. Two different forces act on an object, but the net force is zero. Must the net torque be zero? If so, why? If not, give a counter-example.
6. Is it possible to apply a counterclockwise torque to an object that's rotating clockwise? If not, why not? If so, how will the object's motion change?
7. A solid sphere and a hollow sphere, of the same mass and radius, are rolling along level ground. If they have the same total kinetic energy, which is moving faster?

8. A solid cylinder and a hollow cylinder, of the same mass and radius, are rolling along level ground at the same speed. Which has more kinetic energy?
9. A circular saw takes a long time to stop rotating after the power is turned off. Without the saw blade mounted, the motor stops much more quickly when turned off. Why?
10. A solid sphere and a solid cube have the same mass, and the side of the cube is equal to the diameter of the sphere. Which has greater rotational inertia about an axis through the center of mass? The rotational axis for the cube is perpendicular to two of the cube faces.
11. The lower leg of a horse contains essentially no muscle. How does this help the horse to run fast? Explain in terms of rotational inertia.
12. You wish to store energy in a rotating flywheel. Given a fixed amount of a material, what shape should you make the flywheel so it will store the most energy at a given angular speed?
13. A ball starts from rest and rolls without slipping down a slope, and then starts up a *frictionless* slope (Fig. 10.27). Compare its maximum height on the frictionless slope with its starting height on the first slope.

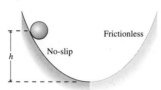

FIGURE 10.27 For Thought and Discussion 13, Problem 64

Exercises and Problems

Exercises

Section 10.1 Angular Velocity and Acceleration

14. Determine the angular speed, in rad/s, of (a) Earth about its axis; (b) the minute hand of a clock; (c) the hour hand of a clock; and (d) an egg beater turning at 300 rpm.
15. What is the linear speed of a point (a) on Earth's equator and (b) at your latitude?
16. Express each of the following in radians per second: (a) 720 rpm; (b) 50°/h; (c) 1000 rev/s; (d) 1 rev/year (which is the angular speed of Earth in its orbit).
17. A 25-cm-diameter circular saw blade spins at 3500 rpm. How fast would you have to push a straight hand saw to have the teeth move through the wood at the same rate as the circular saw teeth?
18. The rotation rate of a compact disc varies from about 200 rpm to 500 rpm. If the disc plays for 74 min, what is its average angular acceleration in (a) rpm/s and (b) rad/s^2?
19. During startup of a power plant, a turbine accelerates from rest at 0.52 rad/s^2. (a) How long does it take to reach its 3600-rpm operating speed? (b) How many revolutions does it make during this time?
20. A merry-go-round starts from rest and accelerates with angular acceleration 0.010 rad/s^2 for 14 s. (a) How many revolutions does it make during this time? (b) What is its average angular speed during the spin-up time?

Section 10.2 Torque

21. A frictional force of 320 N acts on the rim of a 1.0-m-diameter wheel to oppose its rotational motion. What is the torque about the wheel's central axis?
22. A torque of 110 N·m is required to start a revolving door rotating. If a child can push with a maximum force of 90 N, how far from the door's rotation axis must she apply this force?
23. A car tune-up manual calls for tightening the spark plugs to a torque of 35.0 N·m. To achieve this torque, with what force must you pull on the end of a 24.0-cm-long wrench if you pull (a) at right angles to the wrench shaft and (b) at 110° to the wrench shaft?
24. A 55-g mouse runs out to the end of the 17-cm-long minute hand of a grandfather clock when the clock reads 10 minutes past the hour. What torque does the mouse's weight exert about the rotation axis of the clock hand?
25. You have your bicycle upside-down for repairs. The front wheel is free to rotate and is perfectly balanced except for the 25-g valve stem. If the valve stem is 32 cm from the rotation axis and is located 24° below the horizontal, what is the resulting torque about the wheel's axis?

Section 10.3 Rotational Inertia and the Analog of Newton's Law

26. Four equal masses m are located at the corners of a square of side L, connected by essentially massless rods. Find the rotational inertia of this system about an axis (a) that coincides with one side and (b) that bisects two opposite sides.

27. The shaft connecting a turbine and an electric generator in a power plant is a solid cylinder of mass 6.8×10^3 kg and diameter 85 cm. Find its rotational inertia.

28. The chamber of a rock-tumbling machine is a hollow cylinder with mass 65 g and radius 7.1 cm. The chamber is closed off by end caps in the form of uniform circular disks, each of mass 22 g. (a) What is the rotational inertia of the chamber about its central axis? (b) What torque is necessary to give the chamber an angular acceleration of 3.4 rad/s²?

29. The full diameter of a wheel is 92 cm, and its rotational inertia is 7.8 kg·m². (a) What is the minimum mass it could have? (b) How could it have more mass?

30. Three 1.2-kg masses are located at the vertices of an equilateral triangle 88 cm on a side, connected by rods of negligible mass. Find the rotational inertia of this object (a) about an axis through the center of the triangle and perpendicular to its plane and (b) about an axis that passes through one vertex and the midpoint of the opposite side.

31. (a) Estimate the rotational inertia of Earth, assuming it to be a uniform solid sphere. (b) What torque would have to be applied to Earth to cause the length of the day to change by 1 second every century?

32. A neutron star is an extremely dense, rapidly spinning object that results from the collapse of a star at the end of its life. A neutron star of 1.8 times the Sun's mass has an approximately uniform density of 1×10^{18} kg/m³. (a) What is its rotational inertia? (b) The neutron star's spin rate slowly decreases as a result of torque associated with magnetic forces. If the spin-down rate is 5×10^{-5} rad/s², what is the magnetic torque?

33. A 108-g Frisbee is 24 cm in diameter and has about half its mass spread uniformly in a disk and the other half concentrated in the rim. With a quarter-turn flick of the wrist, a student sets the Frisbee rotating at 550 rpm. (a) What is the rotational inertia of the Frisbee? (b) What is the magnitude of the torque, assumed constant, that the student applies?

34. At the MIT Magnet Laboratory, energy is stored in huge solid flywheels of mass 7.7×10^4 kg and radius 2.4 m. The flywheels ride on shafts 41 cm in diameter. If a frictional force of 34 kN acts tangentially on the shaft, how long will it take the flywheel to coast to a stop from its normal rotation rate of 360 rpm?

Section 10.4 Rotational Energy

35. A 25-cm-diameter circular saw blade has a mass of 0.85 kg, distributed uniformly as in a disk. (a) What is its rotational kinetic energy at 3500 rpm? (b) What average power must be applied to bring the blade from rest to 3500 rpm in 3.2 s?

36. Humanity uses energy at the rate of about 10^{13} W. If we found a way to extract this energy from Earth's rotation, how long would it take before the length of the day increased by 1 minute?

37. A 150-g baseball is pitched at 33 m/s, and it's spinning at 42 rad/s. What fraction of its kinetic energy is rotational? Treat the baseball as a uniform solid sphere of radius 3.7 cm.

38. Estimate Earth's rotational energy, approximating the planet as a solid sphere of uniform density.

39. (a) Find the energy stored in the flywheel of Exercise 34 when it's rotating at 360 rpm. (b) The wheel is attached to an electric generator and the rotation rate drops from 360 rpm to 300 rpm in 3.0 s. What is the average power output?

Section 10.5 Rolling Motion

40. A solid 2.4-kg sphere is rolling at 5.0 m/s. Find (a) its translational kinetic energy and (b) its rotational kinetic energy.

41. What fraction of a solid disk's kinetic energy is rotational if it's rolling without slipping?

42. A rolling ball has total kinetic energy of 100 J, 40 J of which is rotational energy. Is the ball solid or hollow?

Problems

43. A wheel turns through 2.0 revolutions while being accelerated from rest at 18 rpm/s. (a) What is the final angular speed? (b) How long does it take to turn the 2.0 revolutions?

44. You switch a food blender from its high to its low setting; the blade speed drops from 3600 rpm to 1800 rpm in 1.4 s. How many revolutions does it make during this time?

45. A compact disc (CD) player varies the rotation rate of the disc in order to keep the part of the disc from which information is being read moving at a constant linear speed of 1.30 m/s. Compare the rotation rates of a 12.0-cm-diameter CD when information is being read (a) from its outer edge and (b) from a point 3.75 cm from the center. Give your answers in rad/s and rpm.

46. You rev your car's engine and watch the tachometer climb steadily from 1200 rpm to 5500 rpm in 2.7 s. (a) What is the angular acceleration of the engine? (b) What is the tangential acceleration of a point on the edge of the engine's 3.5-cm-diameter crankshaft? (c) How many revolutions does the engine make during this time?

47. A piece of machinery is spinning at 680 rpm. When a brake is applied, its rotation rate drops to 440 rpm while it turns through 180 revolutions. What is the magnitude of the angular deceleration?

48. A circular saw blade completes 1200 revolutions in 40 s while coasting to a stop after being turned off. Assuming constant deceleration, what are (a) the angular deceleration and (b) the initial angular speed?

49. A pulley 12 cm in diameter is free to rotate about a horizontal axle. A 220-g mass and a 470-g mass are tied to either end of a massless string, and the string is hung over the pulley. If the string does not slip, what torque must be applied to keep the pulley from rotating?

50. A square frame is made from four thin rods, each of length L and mass m. Calculate its rotational inertia about the three axes shown in Fig. 10.28.

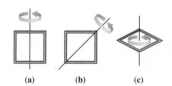

FIGURE 10.28 Problem 50

51. Use integration to show that the rotational inertia of a thick ring of mass M and inner and outer radii R_1 and R_2 is given by $\frac{1}{2}M(R_1^2 + R_2^2)$. *Hint:* See Example 10.7.

52. A uniform rectangular flat plate has mass M and dimensions a by b. Use the parallel-axis theorem in conjunction with Table 10.2 to show that its rotational inertia about the side of length b is $\frac{1}{3}Ma^2$.

53. Each propeller on a King Air twin-engine airplane consists of three blades, each of mass 10 kg and length 125 cm. The blades may be treated approximately as uniform, thin rods. (a) What is the rotational inertia of the propeller? (b) If the propeller is driven by an engine that develops a torque of 2700 N·m, how long will it take to change the propeller's angular speed from 1400 rpm to 1900 rpm?

54. Verify by direct integration the formula given in Table 10.2 for the rotational inertia of a flat plate about a central axis. *Hint:* Divide the plate into strips parallel to the axis.

55. A space station is constructed in the shape of a wheel 22 m in diameter, with essentially all of its 5.0×10^5-kg mass at the rim. Once the station is completed, it is set rotating at a rate that requires an object at the rim to have radial acceleration g, thereby simulating Earth's surface gravity. This is accomplished using two small rockets, each with 100-N thrust, that are mounted on the rim of the station. (a) How long will it take to reach the desired spin rate? (b) How many revolutions will the station make in this time?

56. A motor is connected to a solid cylindrical drum with a diameter of 1.2 m and a mass of 51 kg. A massless rope is attached to the drum and tied at the other end to a 38-kg weight, so the rope will wind onto the drum as it turns. What torque must the motor apply if the weight is to be lifted with an acceleration of 1.1 m/s²?

57. A 2.4-kg block rests on a 30° slope and is attached by a string of negligible mass to a solid drum of mass 0.85 kg and radius 5.0 cm, as shown in Fig. 10.29. When released, the block accelerates down the slope at 1.6 m/s². What is the coefficient of friction between block and slope?

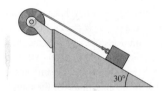

FIGURE 10.29 Problem 57

58. You've got your bicycle upside-down for repairs, with its 66-cm-diameter wheel spinning freely at 230 rpm. The mass of the wheel is 1.9 kg and is concentrated mostly at the rim. You hold a wrench against the tire for 3.1 s, with a normal force of 2.7 N. If the coefficient of friction between the wrench and the tire is 0.46, what is the final angular speed of the wheel?

59. A potter's wheel is a stone disk 90 cm in diameter with a mass of 120 kg. If the potter's foot pushes at the outer edge of the initially stationary wheel with a 75-N force for one-eighth of a revolution, what will be the final speed? *Hint:* Use the work-energy theorem.

60. A ship's anchor weighs 5000 N. Its cable passes over a roller of negligible mass and is wound around a hollow cylindrical drum of mass 380 kg and radius 1.1 m. The drum is mounted on a frictionless axle. The anchor is released and drops 16 m to the water. Use energy considerations to determine the drum's rotation rate when the anchor hits the water. Neglect the mass of the cable.

61. A hollow basketball rolls down a 30° incline. If it starts from rest, what is its speed after it's gone 8.4 m along the incline?

62. A hollow ball is rolling along a horizontal surface at 3.7 m/s when it encounters an upward incline. If it rolls without slipping up the incline, what maximum height will it reach?

63. The rotational kinetic energy of a rolling automobile wheel is 40% of its translational kinetic energy. The wheel is then redesigned to have 10% lower rotational inertia and 20% less mass, while keeping its radius the same. By what percentage does its total kinetic energy at a given speed decrease?

64. A solid ball of mass M and radius R starts at rest at height h above the bottom of the path shown in Fig. 10.27. It rolls without slipping down the left side of the path. The right side of the path, starting at the bottom, is frictionless. To what height does the ball rise on the right?

65. A disk of radius R has an initial mass M. Then a hole of radius $\frac{1}{4}R$ is drilled, with its edge at the disk center (Fig. 10.30). Find the new rotational inertia about the central axis. *Hint:* Find the rotational inertia of the missing piece, and subtract it from that of the whole disk. You'll need to determine what fraction the missing mass is of the total M and use the parallel-axis theorem.

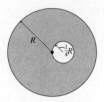

FIGURE 10.30 Problems 65 and 70

66. A 50-kg mass is tied to a massless rope wrapped around a solid cylindrical drum. The drum is mounted on a frictionless horizontal axle. When the mass is released, it falls with acceleration $a = 3.7$ m/s². Find (a) the tension in the rope and (b) the mass of the drum.

67. A 320-kg motorcycle includes two wheels, each of which is 52 cm in diameter and has rotational inertia 2.1 kg·m². The cycle and its 75-kg rider are coasting at 85 km/h on a flat road when they encounter a hill. If the cycle rolls up the hill with no applied power and no significant internal friction, what vertical height will it reach?

68. A solid marble starts from rest and rolls without slipping on the loop-the-loop track shown in Fig. 10.31. Find the minimum starting height of the marble from which it will remain on the track through the loop. Assume the marble radius is small compared with R.

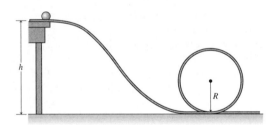

FIGURE 10.31 Problem 68

69. A disk of radius R and thickness w has a mass density that increases from the center outward, given by $\rho = \rho_0 r/R$, where r is the distance from the axis of the disk. (a) Calculate the total mass M of the disk. (b) Calculate the rotational inertia about the disk axis in terms of M and R. Compare with the results for a solid disk of uniform density and for a ring.

70. The wheel shown in Fig. 10.30 is rotating freely about a frictionless horizontal axle. Since the wheel is unbalanced, its angular speed varies as it rotates. If the maximum angular speed is ω_{max}, find an expression for the minimum speed. *Hint:* How does the potential energy change as the wheel rotates?

71. Consider the rotational inertia of a thin, flat object about an axis perpendicular to the plane of the object. Show that this is equal to the sum of the rotational inertias about two perpendicular axes in the plane of the object, passing through the given axis. This is called the perpendicular-axis theorem.

72. Use the perpendicular-axis theorem in Problem 71 to (a) verify the entry in Table 10.2 for a flat rectangular plate about a perpendicular axis, and (b) find the rotational inertia of a uniform thin disk of radius R and mass M about an axis along a diameter.

73. Calculate the rotational inertia of a uniform right circular cone of mass M, height h, and base-radius R, about its axis.

74. Show that the rotational inertia of a uniform solid spheroid, about its axis of revolution, is $\frac{2}{5}MR^2$, where M is its mass and R is the semi-axis perpendicular to the axis of revolution. Why does this result look the same for both a prolate or oblate spheroid and a sphere?

75. Your team is helping to design an orbiting space station. The station will be in the shape of a wheel 100 m in diameter with essentially all of its 6×10^5-kg mass contained in the rim portion of the wheel (the mass of the spokes is negligible). Once the station is completed, it will rotate at a rate such that a person working in the wheel will experience a simulated gravitational force equal to that of Earth; that is, the normal force of the floor on an occupant would be equal to mg ($g = 9.8$ m/s^2). When the station is finished, it will be necessary to set it rotating at the required rotation rate to produce this "artificial gravity." This is accomplished by firing four small thruster rockets attached symmetrically to the outer wall of the wheel. The rockets will fire for 5 hours to bring the station "up to speed." Your task is to determine the thrust, or force, each rocket must have to produce the required rotation rate in this amount of time. You may assume the station starts rotating from rest.

76. You are designing a machine to be used in your company's manufacturing process. The machine spins a wheel at 7.48 Hz and must accelerate the wheel to 9.39 rev/s in 140 s. Your assistant needs to know the necessary acceleration so he can order the proper motor. What do you tell him?

77. The local historical society has asked for your assistance in writing the interpretive material for a display featuring an old steam locomotive. You have information on the torque in a flywheel but need to know the force applied to the flywheel by means of an attached horizontal rod. The rod joins the wheel with a flexible connection 95 cm from the wheel's axis. The maximum torque the rod produces on the flywheel is 10.1 kN. What force is applied via the rod?

78. Hybrid cars are becoming popular due to increasing gasoline costs. You are considering buying one but are skeptical of the manufacturer's claims. One model uses a ring-shaped flywheel to store energy. Supposedly it can store 10 MJ of energy and supply 25 kW of power for about 400 seconds. You find these data on the flywheel: mass of 48 kg, rotation speed of 30,000 rpm, and diameter of 41.1 cm. Are the specs correct?

Answers to Chapter Questions

Answer to Chapter Opening Question

The blade mass should be concentrated toward the rotation axis, thus lowering the turbine's rotational inertia—the rotational analog of mass.

Answers to GOT IT? Questions

10.1 (a) 10.5b; (b) 10.5a.

10.2 (b) Rotational inertia would decrease with the axis at the center $(mL^2/2)$ and (a) increase (mL^2) with the axis at one end.

10.3 The mass of the shell is farther from the rotation axis.

10.4 (a). There must be a net torque acting to increase the pulley's clockwise angular velocity. The difference in the two tension forces provides that torque.

10.5 The solid ball reaches the bottom first. With more of its mass closer to the center, it has lower rotational inertia and thus more of its energy ends up as translational kinetic energy. So it rolls faster.

11 Rotational Vectors and Angular Momentum

■ Earth isn't quite round. How does this affect its rotation axis, and what's this got to do with ice ages? (The deviation from roundness is exaggerated.)

Summer, fall, winter, spring: the cycle of the seasons is ultimately determined by the vector direction of Earth's angular velocity. The changing angular velocity of protons in living tissue produces MRI images that give physicians a noninvasive look inside the human body. Rising and rotating, moist, heated air forms itself into the ominous funnel of a tornado. You ride your bicycle, the rotating wheels helping stabilize what seems a precarious balance. These examples all involve rotational motion in which not only the magnitude but also the *direction* matters. They're best understood in terms of the rotational analog of Newton's law, which we introduce here in full vector form involving a rotational analog of momentum. The transition from Chapter 10 to Chapter 11 is analogous to the leap from Chapter 2's one-dimensional description of motion to the full vector description in Chapter 3. Here, as there, we'll find a new richness of phenomena involving motion.

11.1 Angular Velocity and Acceleration Vectors

So far we've ascribed direction to rotational motion using the terms "clockwise" and "counterclockwise." But that's not enough: To describe rotational motion fully we need to specify the direction of the rotation axis. We therefore define **angular velocity** $\vec{\omega}$ as a vector whose magnitude is the angular speed ω and whose direction is parallel to the rotation axis. There's an ambiguity in this definition, since there are two possible directions parallel to the axis. We resolve the ambiguity with the **right-hand rule**: If you curl the fingers of your right hand to follow the rotation, then your right thumb points in the direction of the angular velocity (Fig. 11.1). This refinement means that $\vec{\omega}$ not only gives the angular speed and the direction of the rotation axis but also distinguishes what we would have described previously as clockwise or counterclockwise rotation.

By analogy with the linear acceleration vector, we define angular acceleration as the rate of change of the angular velocity vector:

$$\vec{\alpha} = \lim_{\Delta t \to 0} \frac{\Delta \vec{\omega}}{\Delta t} = \frac{d\vec{\omega}}{dt} \qquad \text{(angular acceleration vector)} \qquad (11.1)$$

where, as with Equation 10.2, we get the average angular acceleration if we don't take the limit.

Equation 11.1 says that angular acceleration points in the direction of the *change* in angular velocity. If that change is only in magnitude, then $\vec{\omega}$ simply grows or shrinks, and $\vec{\alpha}$ is parallel or antiparallel to the rotation axis (Fig. 11.2a, b). But a change in *direction* is also a change in angular velocity. When the angular velocity $\vec{\omega}$ changes only in direction, then the angular acceleration vector is perpendicular to $\vec{\omega}$ (Fig. 11.2c). More generally, both the magnitude and direction of $\vec{\omega}$ may change; then $\vec{\alpha}$ is neither parallel nor perpendicular to $\vec{\omega}$. These cases are exactly analogous to the situations we treated in Chapter 3, where acceleration parallel to velocity changes only the speed, while acceleration perpendicular to velocity changes only the direction of motion.

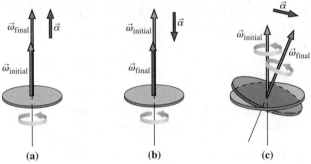

FIGURE 11.2 Angular acceleration can (a) increase or (b) decrease the magnitude of the angular velocity, or (c) change its direction.

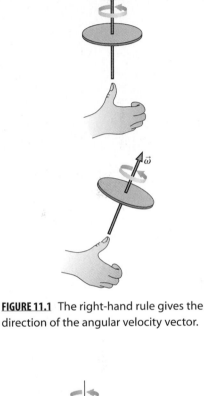

FIGURE 11.1 The right-hand rule gives the direction of the angular velocity vector.

11.2 Torque and the Vector Cross Product

Figure 11.3 shows a wheel, initially stationary, with a force applied at its rim. The torque associated with this force sets the wheel rotating in the direction shown; applying the right-hand rule, we see that angular velocity vector $\vec{\omega}$ points upward. Since the angular speed is increasing, the angular acceleration $\vec{\alpha}$ also points upward. So that our rotational analog of Newton's law—angular acceleration proportional to torque—will hold for directions as well as magnitudes, we'd like the torque to have an upward direction, too.

We already know the magnitude of the torque: From Equation 10.10, it's $\tau = rF \sin\theta$, where θ is the angle between the vectors $\vec{r}$ and $\vec{F}$ in Fig. 11.3. We define the direction of the torque as being perpendicular to both $\vec{r}$ and $\vec{F}$, as given by the right-hand rule shown in Fig. 11.4. You can verify that this rule gives an upward direction for the torque in Fig. 11.3.

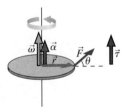

FIGURE 11.3 The torque vector is perpendicular to $\vec{r}$ and $\vec{F}$, and in the same direction as the angular acceleration. Here $\vec{F}$ lies in the plane of the disk.

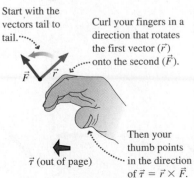

Start with the vectors tail to tail.

Curl your fingers in a direction that rotates the first vector ($\vec{r}$) onto the second ($\vec{F}$).

$\vec{\tau}$ (out of page)

Then your thumb points in the direction of $\vec{\tau} = \vec{r} \times \vec{F}$.

FIGURE 11.4 The right-hand rule for the direction of torque.

The Cross Product

The magnitude of the torque, $\tau = rF\sin\theta$, is determined by the magnitudes of the vectors $\vec{r}$ and $\vec{F}$ and the angle between them; the direction of the torque is determined by the vectors $\vec{r}$ and $\vec{F}$ through the right-hand rule. This operation—forming from two vectors $\vec{A}$ and $\vec{B}$ a third vector $\vec{C}$ of magnitude $C = AB\sin\theta$ and direction given by the right-hand rule—occurs frequently in physics and is called the **cross product**:

> The cross product $\vec{C}$ of two vectors $\vec{A}$ and $\vec{B}$ is written
>
> $$\vec{C} = \vec{A} \times \vec{B}$$
>
> and is a vector with magnitude $AB\sin\theta$, where θ is the angle between $\vec{A}$ and $\vec{B}$, and where the direction of $\vec{C}$ is given by the right-hand rule of Fig. 11.4.

Torque is an instance of the cross product, and we can write the torque vector simply as

$$\vec{\tau} = \vec{r} \times \vec{F} \qquad \text{(torque vector)} \qquad (11.2)$$

Both direction and magnitude are described succinctly in this equation.

TACTICS 11.1 Multiplying Vectors

The cross product $\vec{A} \times \vec{B}$ is the second way of multiplying vectors that you've encountered. The first was the scalar product $\vec{A} \cdot \vec{B} = AB\cos\theta$ introduced in Chapter 6 and also called the dot product. Both depend on the product of the vector magnitudes and on the angle between them. But where the dot product depends on the *cosine* of the angle and is therefore maximum when the two vectors are parallel, the cross product depends on the *sine* and is therefore maximum for perpendicular vectors. There's another crucial distinction between dot product and cross product: the dot product is a *scalar*—a single number, with no direction—while the cross product is a *vector*. That's why $AB\cos\theta$ completely specifies the dot product, but $AB\sin\theta$ gives only the magnitude of the cross product; it's also necessary to specify the direction via the right-hand rule.

The cross product obeys the usual distributive rule: $\vec{A} \times (\vec{B} + \vec{C}) = \vec{A} \times \vec{B} + \vec{A} \times \vec{C}$, but it's *not* commutative; in fact, as you can see by rotating $\vec{F}$ onto $\vec{r}$ instead of $\vec{r}$ onto $\vec{F}$ in Fig. 11.4, $\vec{B} \times \vec{A} = -\vec{A} \times \vec{B}$.

GOT IT? 11.1 The figure shows four pairs of force and radius vectors and eight torque vectors. Which numbered torque vector goes with each pair of force-radius vectors? Consider only direction, not magnitude.

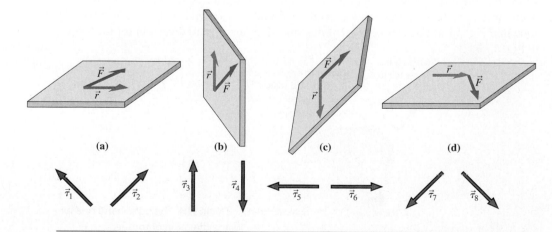

11.3 Angular Momentum

We first used Newton's law in the form $\vec{F} = m\vec{a}$, but later found the form $\vec{F} = d\vec{p}/dt$ especially powerful. The same is true in rotational motion: To explore fully some surprising aspects of rotational dynamics, we need to define angular momentum and develop a relation between its rate of change and the applied torque. Once we've done that, we'll be able to answer questions like why a gyroscope doesn't fall over and how spinning protons lead to MRI images of your body's innards.

Like other rotational quantities, angular momentum is specified with respect to a given point or axis. We begin with the **angular momentum** $\vec{L}$ of a single particle:

> If a particle with linear momentum $\vec{p}$ is at position $\vec{r}$ with respect to some point, then its angular momentum $\vec{L}$ about that point is defined as
>
> $$\vec{L} = \vec{r} \times \vec{p} \qquad \text{(angular momentum)} \qquad (11.3)$$

EXAMPLE 11.1 **Calculating Angular Momentum: A Single Particle**

A particle of mass m moves counterclockwise at speed v around a circle of radius r in the x-y plane. Find its angular momentum about the center of the circle, and express the answer in terms of its angular velocity.

INTERPRET We're given the motion of a particle—namely, uniform motion in a circle—and asked to find the corresponding angular momentum and its relation to angular velocity.

DEVELOP Figure 11.5 is our sketch, showing the particle in its circular path. We added an xyz coordinate system with the circular path in the x-y plane. Equation 11.3, $\vec{L} = \vec{r} \times \vec{p}$, gives the angular momentum in terms of the vector $\vec{r}$ and the linear momentum $\vec{p}$. We know that linear momentum is the product $m\vec{v}$, so we have everything we need to apply Equation 11.3. We'll then express our result in terms of angular velocity using $v = \omega r$.

EVALUATE Figure 11.5 shows that the linear momentum $m\vec{v}$ is perpendicular to $\vec{r}$, so $\sin\theta = 1$ in the cross product, and the magnitude of the angular momentum becomes $L = mvr$. Applying the right-hand rule shows that $\vec{L}$ points in the z direction, so we can write $\vec{L} = mvr\hat{k}$. But $v = \omega r$, and the right-hand rule shows that $\vec{\omega}$, too, points in the z direction. So we can write

$$\vec{L} = mvr\hat{k} = mr^2\omega\hat{k} = mr^2\vec{\omega}$$

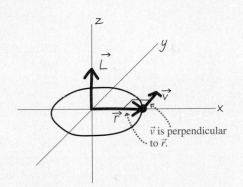

FIGURE 11.5 Finding the angular momentum $\vec{L}$ of a particle moving in a circle.

ASSESS Make sense? The faster the particle is going, the more linear momentum it has. But angular momentum depends on linear momentum and distance from the rotation axis, so at a given angular speed, the angular momentum scales as the *square* of the radius. ∎

Angular momentum is the rotational analog of linear momentum $\vec{p} = m\vec{v}$. Since rotational inertia I is the analog of mass m, and angular velocity $\vec{\omega}$ is the analog of linear velocity $\vec{v}$, you might expect that we could write

$$\vec{L} = I\vec{\omega} \qquad (11.4)$$

The rotational inertia of a single particle is mr^2, so you can see that the result of Example 11.1 can indeed be written $\vec{L} = I\vec{\omega}$. Equation 11.4 also holds for symmetric objects like a wheel or sphere rotating about a fixed axis. But in more complicated cases, Equation 11.4 may not hold; surprisingly, $\vec{L}$ and $\vec{\omega}$ can even have different directions. We'll leave such cases for more advanced courses.

Torque and Angular Momentum

We're now ready to develop the full vector analog of Newton's law in the form $\vec{F} = d\vec{P}/dt$. Recall that $\vec{F}$ here is the *net* external force on a system, and $\vec{P}$ is the system's momentum—the vector sum of the momenta of its constituent particles. Can we write, by analogy, $\vec{\tau} = d\vec{L}/dt$? To see that we can, we write the angular momentum of a system as the sum of the angular momenta of its constituent particles:

$$\vec{L} = \sum \vec{L}_i = \sum (\vec{r}_i \times \vec{p}_i)$$

where the subscript i refers to the ith particle. Differentiating gives

$$\frac{d\vec{L}}{dt} = \sum \left(\vec{r}_i \times \frac{d\vec{p}_i}{dt} + \frac{d\vec{r}_i}{dt} \times \vec{p}_i \right)$$

where we've applied the product rule for differentiation, being careful to preserve the order of the cross product since it's not commutative. But $d\vec{r}_i/dt$ is the velocity of the ith particle, so the second term in the sum is the cross product of velocity $\vec{v}$ and momentum $\vec{p} = m\vec{v}$. Since these two vectors are parallel, their cross product is zero, and we're left with only the first term in the sum:

$$\frac{d\vec{L}}{dt} = \sum \left(\vec{r}_i \times \frac{d\vec{p}_i}{dt} \right) = \sum (\vec{r}_i \times \vec{F}_i)$$

where we've used Newton's law to write $d\vec{p}_i/dt = \vec{F}_i$. But $\vec{r}_i \times \vec{F}_i$ is the torque $\vec{\tau}_i$ on the ith particle, so

$$\frac{d\vec{L}}{dt} = \sum \vec{\tau}_i$$

The sum here includes both external and internal torques—the latter due to interactions among the particles of the system. Newton's third law assures us that internal *forces* cancel in pairs, but what about *torques*? They'll cancel, too, provided the internal forces act along lines joining pairs of particles. This condition is stronger than Newton's third law alone, and it usually but not always holds. When it does, the sum of torques reduces to the net *external* torque, and we have

$$\frac{d\vec{L}}{dt} = \vec{\tau} \qquad \left(\begin{array}{l} \text{rotational analog,} \\ \text{Newton's 2}^{\text{nd}} \text{ law} \end{array} \right) \qquad (11.5)$$

where $\vec{\tau}$ is the net external torque. Thus our analogy between linear and rotational motion holds for momentum as well as for the other quantities we've discussed.

11.4 Conservation of Angular Momentum

When there's no external torque on a system, Equation 11.5 tells us that the angular momentum is constant. This statement—that the angular momentum of an isolated system cannot change—is of fundamental importance in physics, and applies to systems ranging from subatomic particles to galaxies. Because a composite system can change its configuration—and hence its rotational inertia I—conservation of angular momentum requires that angular speed increase if I decreases, and vice versa. The classic example is a figure skater who starts spinning relatively slowly with arms and leg extended and then pulls in her limbs to spin rapidly (Fig. 11.6). A more dramatic example is the collapse of a star at the end of its lifetime, explored in the next example.

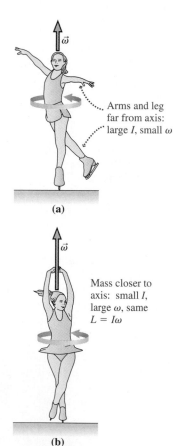

Arms and leg far from axis: large I, small ω

(a)

Mass closer to axis: small I, large ω, same $L = I\omega$

(b)

FIGURE 11.6 As the skater's rotational inertia decreases, her angular speed increases to conserve angular momentum.

EXAMPLE 11.2 **Conservation of Angular Momentum: Pulsars**

A star rotates once every 45 days. At the end of its life, it undergoes a supernova explosion, hurling much of its mass into the interstellar medium. But the inner core of the star, whose radius is initially 20 Mm, collapses into a neutron star only 6 km in radius. As it rotates, the neutron star emits regular pulses of radio waves, making it a *pulsar*. Calculate the rotation rate, which is the same as the pulse rate that radio astronomers detect. Consider the core to be a uniform sphere, and assume that no external torques act during the collapse.

INTERPRET Here we're given the radius and rotation rate of the stellar core before collapse and asked for the rotation rate afterward. That kind of "before and after" question often calls for the application of a conservation law. In this case there's no external torque, so it's angular momentum that's conserved.

DEVELOP The magnitude of the angular momentum is $I\omega$, so our plan is to write this expression before and after collapse, and then equate the two to find the new rotation rate: $I_1\omega_1 = I_2\omega_2$. We need to use Table 10.2's expression for the rotational inertia of a solid sphere: $I = \frac{2}{5}MR^2$.

EVALUATE Given I, our statement of angular momentum conservation becomes $\frac{2}{5}MR_1^2\omega_1 = \frac{2}{5}MR_2^2\omega_2$, or

$$\omega_2 = \omega_1\left(\frac{R_1}{R_2}\right)^2 = \left(\frac{1\ \text{rev}}{45\ \text{day}}\right)\left(\frac{2\times 10^7\ \text{m}}{6\times 10^3\ \text{m}}\right) = 2.5\times 10^5\ \text{rev/day}$$

ASSESS Our answer is huge, about 3 revolutions per second. But that makes sense. This neutron star is a fantastic thing—an object with more mass than the entire Sun, crammed into a diameter of about 8 miles. It's because of that dramatic reduction in radius—and thus in rotational inertia—that the pulsar's rotation rate is so high. Note that in a case like this, where ω appears on both sides of the equation, it isn't necessary to convert to radian measure. ∎

EXAMPLE 11.3 **Conserving Angular Momentum: On the Playground**

A merry-go-round of radius $R = 1.3$ m has rotational inertia $I = 240$ kg·m² and is rotating freely at $\omega_1 = 11$ rpm. A boy of mass $m_b = 28$ kg runs straight toward the center of the merry-go-round at $v_b = 2.5$ m/s, and leaps on. At the same time a girl of mass $m_g = 32$ kg, running tangentially at speed $v_g = 3.7$ m/s in the same direction as the merry-go-round's tangential velocity, also leaps on. Find the new angular speed ω_2 once both children are seated on the merry-go-round's rim.

INTERPRET This case is more complicated and requires some thought to identify the system. We're interested in what happens after the boy and girl leap onto the merry-go-round, so we identify the system as consisting of both children and the merry-go-round. Then there's no *external* torque on the system, so angular momentum is conserved.

DEVELOP Figure 11.7 shows the situation. We determined that angular momentum is conserved, so our plan is to write expressions for the angular momentum before and after the children board the merry-go-round, and then equate them and solve for the final angular speed. We're given the merry-go-round's initial angular speed ω_1 and its rotational inertia I, so we know its initial angular momentum: $I\omega_1$. For the children, we know only the linear velocities, so we have to use Equation 11.3, the fundamental definition of angular momentum: $\vec{L} = \vec{r} \times \vec{p}$. The boy is running straight toward the rotation axis, so for him the momentum $\vec{p}_b$ and radius vector $\vec{r}_b$ in Equation 11.3 are parallel. Since the cross product of parallel vectors is zero, he carries no angular momentum. So although we're given his speed, we don't really need it. For the girl, running tangentially to the rim of the merry-go-round, the vectors in Equation 11.3 are at right angles at the moment she reaches the rim. Then $\sin\theta = 1$ in the cross product and so, with $p_g = m_g v_g$ and $r_g = R$, she carries angular momentum $m_g v_g R$.

After the children are aboard, they and the merry-go-round have the same angular velocity ω_2. The merry-go-round's angular momentum is then $I\omega_2$. We can treat both children as particles in

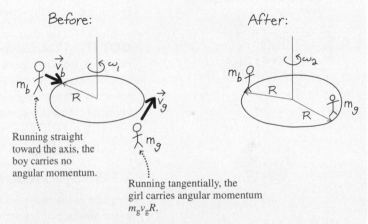

FIGURE 11.7 Our diagrams for Example 11.3.

circular motion of radius R. Then the result of Example 11.1 shows that their angular momenta are $m_b R^2 \omega_2$ and $m_g R^2 \omega_2$.

EVALUATE With expressions for the angular momenta before and after the children board the merry-go-round, we're ready to write the mathematical statement of angular momentum conservation:

$$I\omega_1 + m_g v_g R = I\omega_2 + m_b R^2\omega_2 + m_g R^2\omega_2$$

Solving for the unknown ω_2 gives

$$\omega_2 = \frac{I\omega_1 + m_g v_g R}{I + (m_b + m_g)R^2}$$

cont'd.

We have to be a little careful with units here. The initial angular momentum of the merry-go-round is $I\omega_1 = (240 \text{ kg·m}^2)(11 \text{ rpm})$ $= 2640 \text{ kg·m}^2\text{·rpm}$. But the girl's initial angular momentum is in SI units; it works out to $m_g v_g R = 154 \text{ kg·m}^2\text{·s}^{-1}$ or $154 \text{ kg·m}^2\text{·rad/s}$. Let's stay with rpm for a change; then we can convert the girl's value with the factor $(1 \text{ rev}/2\pi \text{ rad})(60 \text{ s/min})$; the result is $m_g v_g R = 1470 \text{ kg·m}^2\text{·rpm}$. These two angular momenta are in the numerator of our expression for ω_2, so we have

$$\omega_2 = \frac{(2640 \text{ kg} \cdot \text{m}^2 \cdot \text{rpm}) + (1470 \text{ kg} \cdot \text{m}^2 \cdot \text{rpm})}{(240 \text{ kg} \cdot \text{m}^2) + (28 \text{ kg} + 32 \text{ kg})(1.3 \text{ m})^2} = 12 \text{ rpm}$$

ASSESS Can this be right? The angular speed hardly changed! The answer is correct. The boy added rotational inertia but no angular momentum; if the girl had not jumped on, the boy would have caused the speed to drop (to 9.2 rpm; see Exercise 29). The girl brings addi-

tional angular momentum, enough to increase the speed a bit above its initial value.

What about energy? It isn't conserved. As the children leap on, frictional forces act to bring children and merry-go-round to rest with respect to one another. We end up with what's effectively a composite object with a single rotational speed, making this the rotational analog of an inelastic collision.

> ✓ **TIP** Angular Momentum in Straight-line Motion
>
> You don't have to be rotating to have angular momentum. The girl in this example was running in a straight line, yet she had nonzero angular momentum with respect to the merry-go-round's rotation axis. Problem 36 explores this point further.

In a popular demonstration, a student stands on a stationary turntable holding a wheel rotating about a vertical axis. The student flips the wheel upside down, and the turntable starts rotating. Figure 11.8 shows how angular momentum conservation explains this behavior. Once again, though, mechanical energy isn't conserved. In this case the student does work, exerting forces that result in torques on her body and the turntable. The end result is a greater rotational kinetic energy than was initially present.

GOT IT? 11.2 Suppose you step onto a nonrotating turntable like the one shown in Fig. 11.8, holding a nonrotating wheel with its axis vertical. (a) If you spin the wheel counterclockwise as viewed from above, which way will you rotate? (b) If you now turn the spinning wheel upside down, will your rotation rate increase, decrease, or remain unchanged? What about your direction of rotation?

11.5 Gyroscopes and Precession

Angular momentum—a vector quantity with direction as well as magnitude—is conserved in the absence of external torques. For symmetric objects, angular momentum has the same direction as the rotation axis, so the axis can't change direction unless an external torque acts. This is the principle behind the gyroscope—a spinning object whose rotation axis remains fixed in space. The faster a gyroscope spins, the larger its angular momentum and thus the harder it is to change its orientation. Gyroscopes are widely used for navigation, where their direction-holding capability provides an alternative to the magnetic compass. More sophisticated gyroscope systems guide missiles and submarines, stabilize cruise ships in heavy seas, and track the orientation of wireless computer mice. Space-based telescopes like the Hubble start and stop gyroscopic wheels oriented along three perpendicular axes; to conserve angular momentum, the entire telescope reorients itself to point toward a desired astronomical object. This approach avoids rocket exhaust that would foul the telescope's superb viewing, and ensures that there's no fuel to run out. Instead, solar-generated electricity operates the wheels' drive motors.

Precession

If an object does experience a net external torque, then, according to the rotational analog of Newton's law (Equation 11.5, $d\vec{L}/dt = \vec{\tau}$), its angular momentum must change. For rapidly rotating objects, the result is the surprising phenomenon of **precession**—a continual change in the direction of the rotation axis, which traces out a circle. You may have

The student stands on a stationary turntable holding a wheel that spins counterclockwise; the wheel's angular momentum points upward.

$\vec{L}_{total} = \vec{L}_{wheel}$

(a)

She flips the spinning wheel, reversing its angular momentum. The total angular momentum is conserved, so turntable and student (ts) must rotate the other way.

$\vec{L}_{ts}$

$\vec{L}_{total}$

$\vec{L}_{wheel}$

(b)

FIGURE 11.8 A demonstration of angular momentum conservation.

seen a toy gyroscope or top precess instead of simply falling over as you might expect. Figure 11.9 shows that precession comes about because the direction of the angular momentum change is the same as the direction of thc torque.

Precession on the atomic scale helps explain the medical imaging technique known as MRI (*magnetic resonance imaging*). Protons in the body's abundant hydrogen precess because of torque resulting from a strong magnetic field. The MRI imager detects signals emitted at the frequency of precession. By spatially varying the magnetic field, the device localizes the precessing protons and thus constructs high-resolution images of the body's interior.

On a much larger scale, Earth itself precesses. Because of its rotation, the planet bulges slightly at the equator. Solar gravity exerts a torque on the equatorial bulge, causing Earth's rotation axis to precess with a period of about 26,000 years (Fig. 11.10). The axis now points toward Polaris, which for that reason we call the North Star, but it won't always do so. This precession, in connection with deviations in Earth's orbit from a perfect circle, results in subtle climatic changes that are believed partly responsible for the onset of ice ages.

Change $\Delta\vec{L}$ is also into page, so the gyroscope precesses, its tip describing a circle.

$\vec{\tau}$ points into the page.

Gravity exerts a torque about the pivot; $\vec{\tau} = \vec{r} \times \vec{F}$ is into the page.

FIGURE 11.9 Why doesn't the spinning gyroscope fall over?

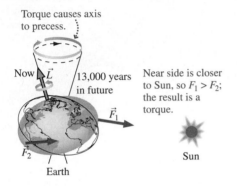

Torque causes axis to precess.

Now $\vec{L}$

13,000 years in future

$\vec{F}_1$

$\vec{F}_2$

Earth

Near side is closer to Sun, so $F_1 > F_2$; the result is a torque.

Sun

FIGURE 11.10 Earth's precession. The equatorial bulge is highly exaggerated.

GOT IT? 11.3 You push horizontally at right angles to the shaft of a spinning gyroscope, as shown in the figure. Does the shaft move (a) upward, (b) downward, (c) in the direction of your push, or (d) opposite the direction of your push?

APPLICATION Bicycling

The rotational analog of Newton's second law helps explain why bicycles don't tip over. The famous photo of Einstein cycling shows how. If the bicycle is perfectly vertical, the gravitational force exerts no torque. But if it tips to the rider's left, as in the photo, then there's a torque $\vec{\tau} = \vec{r} \times \vec{F}_g$ toward the rear. A stationary bicycle, with no angular momentum, would respond by tipping further left, rotating about a front-to-back axis and gaining angular momentum toward the rear. That's just as Newton requires: a change in angular momentum in the direction of the torque. But a moving bicycle already has angular momentum $\vec{L}$ of its rotating wheels; as the photo shows, it points generally to the rider's left. A rearward change in the angular momentum then requires just a slight turn of the front wheel to the left. The rider subconsciously makes that turn, at once satisfying Newton and helping to keep the bicycle stable.

The physics of cycling is a complicated subject, and the role of angular momentum described here is only one of several effects that contribute to bicycle stability.

Gravitational torque is toward back of bicycle, into page.

Wheel turns to left, changing angular momentum vector in direction of torque.

Big Picture

The big idea of this chapter is that rotational quantities can be described as vectors, with the vector direction at right angles to the plane in which the action—motion, acceleration, or effects associated with torque—is occurring. The direction is given by the right-hand rule. A new concept, angular momentum, is the rotational analog of linear momentum. The rotational analog of Newton's law equates the net torque on a system with the rate of change of its angular momentum. In the absence of a net torque, angular momentum is conserved.

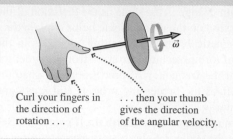

Curl your fingers in the direction of rotation . . .

. . . then your thumb gives the direction of the angular velocity.

Key Concepts and Equations

The **vector cross product** is a way of multiplying two vectors $\vec{A}$ and $\vec{B}$ to produce a third vector $\vec{C}$ of magnitude $C = AB \sin\theta$ and direction at right angles to the other two, as given by the right-hand rule. It's written as

$$\vec{C} = \vec{A} \times \vec{B}$$

Torque is defined as the cross product of the radius vector $\vec{r}$ from a given axis to the point where a force $\vec{F}$ is applied:

$$\vec{\tau} = \vec{r} \times \vec{F}$$

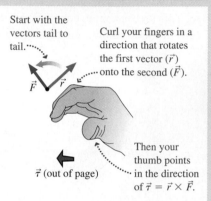

Start with the vectors tail to tail.

Curl your fingers in a direction that rotates the first vector ($\vec{r}$) onto the second ($\vec{F}$).

Then your thumb points in the direction of $\vec{\tau} = \vec{r} \times \vec{F}$.

$\vec{\tau}$ (out of page)

Angular momentum $\vec{L}$ is the rotational analog of linear momentum $\vec{p}$. It's always defined with respect to a particular axis. For a point particle at position $\vec{r}$ with respect to the axis, moving with linear momentum $\vec{p} = m\vec{v}$, the angular momentum is defined as

$$\vec{L} = \vec{r} \times \vec{p}$$

For a symmetric object with rotational inertia I rotating with angular velocity $\vec{\omega}$, the angular momentum becomes $\vec{L} = I\vec{\omega}$.

In terms of angular momentum, the rotational analog of Newton's law states that the rate of change of angular momentum is equal to the net external torque:

$$\frac{d\vec{L}}{dt} = \vec{\tau}$$

If the external torque on a system is zero, then its angular momentum cannot change.

Cases and Uses

Conservation of angular momentum explains the action of gyroscopes—spinning objects whose rotation axis remains fixed in the absence of a net external torque. If an external torque is applied, the rotation axis undergoes a circular motion known as **precession**. Precession occurs in systems ranging from subatomic particles to tops and gyroscopes and on to planets.

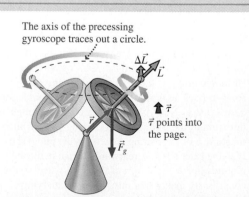

The axis of the precessing gyroscope traces out a circle.

$\vec{\tau}$ points into the page.

For Thought and Discussion

1. Does Earth's angular velocity vector point north or south?
2. Figure 11.11 shows four forces acting on a body. What are the directions of each of the associated torques about the point O? About the point P?

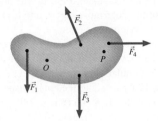

FIGURE 11.11 For Thought and Discussion 2

3. A satellite is small enough to be treated as a point particle. Can Earth's gravity exert a torque on the satellite about Earth's center?
4. Although it contains no parentheses, the expression $\vec{A} \times \vec{B} \cdot \vec{C}$ is unambiguous. Why?

5. The dot product of two vectors is equal to the magnitude of their cross product. What can you conclude about the angle between them?
6. Why does a tetherball move faster as it winds up its pole?
7. Why do helicopters have two rotors?
8. A group of polar bears are standing around the edge of a circular ice floe that's slowly rotating. If the bears all walk to the center, what will happen to the rotation rate?
9. Tornadoes in the northern hemisphere rotate counterclockwise as viewed from above. A far-fetched idea suggests that driving on the right side of the road may increase the frequency of tornadoes. Does this idea have *any* merit? Explain in terms of the angular momentum imparted to the air by passing cars.
10. Does a particle moving at constant speed in a straight line have angular momentum about a point on the line? About a point not on the line? In either case, is its angular momentum constant?
11. When you turn on a high-speed power tool such as a router, you feel the tool tending to twist in your hands. Why?

Exercises and Problems

Exercises

Section 11.1 Angular Velocity and Acceleration Vectors

12. A car is headed north at 70 km/h. Give the magnitude and direction of the angular velocity of its 62-cm-diameter wheels.
13. If the car of Exercise 12 makes a 90° left turn lasting 25 s, determine the average angular acceleration of the wheels.
14. A wheel is spinning at 45 rpm with its spin axis vertical. After 15 s, it's spinning at 60 rpm with its axis horizontal. Find (a) the magnitude of its average angular acceleration and (b) the angle the average angular acceleration vector makes with the horizontal.
15. A wheel is spinning about a horizontal axis, with angular speed 140 rad/s and with its angular velocity pointing east. Find the magnitude and direction of its angular velocity after an angular acceleration of 35 rad/s², pointing 68° west of north, is applied for 5.0 s.

Section 11.2 Torque and the Vector Cross Product

16. A rod is free to pivot about one end, and a force $\vec{F}$ is applied at the other end, as shown in Fig. 11.12. What are the magnitude and direction of the torque about the pivot point?

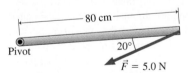

FIGURE 11.12 Exercise 16

17. A coordinate system lies with its x-y plane in the plane of this page and its z axis coming out of the page. A force $\vec{F}$ is applied at the point $x = 1$, $y = 1$. What is the direction of the torque about the origin if $\vec{F}$ points (a) in the x direction, (b) in the y direction, and (c) in the z direction?
18. A 12-N force is applied at the point $x = 3$ m, $y = 1$ m. Find the torque about the origin if the force points in (a) the x direction, (b) the y direction, and (c) the z direction.
19. A force $\vec{F} = 1.3\hat{\imath} + 2.7\hat{\jmath}$ N is applied at the point $x = 3.0$ m, $y = 0$ m. Find the torque about (a) the origin and (b) the point $x = -1.3$ m, $y = 2.4$ m.

Section 11.3 Angular Momentum

20. Express the units of angular momentum (a) using only the fundamental units kilogram, meter, and second; (b) in a form involving newtons; (c) in a form involving joules.

21. In the Olympic hammer throw, a contestant whirls a 7.3-kg steel ball on the end of a 1.2-m cable. If the contestant's arms reach an additional 90 cm from his axis of rotation and if the speed of the ball just prior to release is 27 m/s, what is the magnitude of its angular momentum?
22. A gymnast of rotational inertia 62 kg·m² is tumbling head over heels. If her angular momentum is 470 kg·m²/s, what is her angular speed?
23. A 640-g hoop 90 cm in diameter is rotating at 170 rpm about its central axis. What is its angular momentum?
24. A 7.4-cm-diameter baseball has a mass of 150 g and is spinning at 210 rad/s. What is the magnitude of its angular momentum? Treat the baseball as a solid sphere.

Section 11.4 Conservation of Angular Momentum

25. A potter's wheel, with rotational inertia 6.40 kg·m², is spinning freely at 19.0 rpm. The potter drops a 2.70-kg lump of clay onto the wheel, where it sticks a distance of 46.0 cm from the rotation axis. What is the subsequent angular speed of the wheel?
26. A 3.0-m-diameter merry-go-round with rotational inertia 120 kg·m² is spinning freely at 0.50 rev/s. Four 25-kg children sit suddenly on the edge of the merry-go-round. (a) Find the new angular speed, and (b) determine the total energy lost to friction between the children and the merry-go-round.
27. A uniform, spherical cloud of interstellar gas has mass 2.0×10^{30} kg and radius 1.0×10^{13} m, and is rotating with period 1.4×10^6 years. If the cloud collapses to form a star 7.0×10^8 m in radius, what will be the star's rotation period?
28. A skater's body has rotational inertia 4.2 kg·m² with his fists held to his chest and 5.7 kg·m² with arms outstretched. The skater is twirling at 3.0 rev/s while holding a 2.5-kg weight in each outstretched hand; the weights are 76 cm from his rotation axis. If he pulls his hands in to his chest, how fast will he be twirling?
29. Rework Example 11.3 to find the final angular speed in the case that only the boy jumps onto the merry-go-round.

Problems

30. You slip a wrench over a bolt. If you take the origin at the bolt, the other end of the wrench is at $x = 18$ cm, $y = 5.5$ cm. You apply a force $\vec{F} = 88\hat{\imath} - 23\hat{\jmath}$ to the end of the wrench. What is the torque on the bolt?

31. Vector $\vec{A}$ points 30° counterclockwise from the x axis. Vector $\vec{B}$ is twice as long as $\vec{A}$. Their product $\vec{A} \times \vec{B}$ has length A^2 and points in the negative z direction. What is the direction of vector $\vec{B}$?

32. Show that the cross product of two arbitrary vectors $\vec{A} = A_x\hat{\imath} + A_y\hat{\jmath} + A_z\hat{k}$ and $\vec{B} = B_x\hat{\imath} + B_y\hat{\jmath} + B_z\hat{k}$ can be written $\vec{A} \times \vec{B} = (A_yB_z - A_zB_y)\hat{\imath} + (A_zB_x - A_xB_z)\hat{\jmath} + (A_xB_y - A_yB_x)\hat{k}$.

33. Show that $\vec{A} \cdot (\vec{A} \times \vec{B}) = 0$ for any vectors $\vec{A}$ and $\vec{B}$.

34. Find the cross product of the vectors $\vec{A} = \hat{\imath} - 2\hat{\jmath} + 3\hat{k}$ and $\vec{B} = 2\hat{\imath} + 4\hat{\jmath} + 2\hat{k}$.

35. A weightlifter's barbell consists of two 25-kg masses on the ends of a 15-kg rod 1.6 m long. The weightlifter holds the rod at its center and spins it at 10 rpm about an axis perpendicular to the rod. What is the magnitude of the barbell's angular momentum?

36. A particle of mass m moves in a straight line at constant speed v. Show that its angular momentum about a point located a perpendicular distance b from its line of motion is mvb regardless of where the particle is on the line.

37. Two identical 1800-kg cars are traveling in opposite directions at 90 km/h. Each car's center of mass is 3.0 m from the center of the highway (Fig. 11.13). What are the magnitude and direction of the angular momentum of the system consisting of the two cars, about a point on the centerline of the highway?

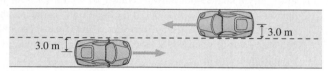

FIGURE 11.13 Problem 37

38. Figure 11.14 shows the dimensions of a 880-g wooden baseball bat whose rotational inertia about its center of mass is 0.048 kg·m². If the bat is swung so its far end moves at 50 m/s, find (a) its angular momentum about the pivot point P and (b) the constant torque applied about P to achieve this angular momentum in 0.25 s. *Hint:* Remember the parallel-axis theorem.

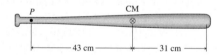

FIGURE 11.14 Problem 38

39. Engineers redesign a car's wheels with the goal of decreasing each wheel's angular momentum by 30% for a given linear speed for the car. Other design considerations require that the wheel diameter go from 38 cm to 35 cm. If the old wheel had a rotational inertia of 0.32 kg·m², what must be the new wheel's rotational inertia if the engineers are to achieve their goal?

40. A turntable of radius 25 cm and rotational inertia 0.0154 kg·m² is spinning freely at 22.0 rpm about its central axis, with a 19.5-g mouse on its outer edge. The mouse walks from the edge to the center. Find (a) the new rotation speed and (b) the work done by the mouse.

41. A 17-kg dog is standing on the edge of a stationary, frictionless turntable of rotational inertia 95 kg·m² and radius 1.81 m. The dog walks once around the turntable. What fraction of a full circle does the dog's motion make with respect to the ground?

42. A physics student is standing on an initially motionless, frictionless turntable with rotational inertia 0.31 kg·m². She's holding a wheel of rotational inertia 0.22 kg·m² spinning at 130 rpm about a vertical axis, as in Fig. 11.8. When she turns the wheel upside down, student and turntable begin rotating at 70 rpm. (a) What is the student's mass, considering her to be a cylinder 30 cm in diameter? (b) How much work did she do in turning the wheel

upside down? Neglect the distance between the axes of the turntable and wheel.

43. Eight 60-kg skaters join hands and skate down an ice rink at 4.6 m/s. Side by side, they form a line 12 m long. The skater at one end stops abruptly, and the line proceeds to rotate rigidly about that skater. Find (a) the angular speed, (b) the linear speed of the outermost skater, and (c) the force that must be exerted on the outermost skater.

44. The dot product of a pair of vectors is twice the magnitude of their cross product. What is the angle between the vectors?

45. A circular bird feeder 19 cm in radius has rotational inertia 0.12 kg·m². The feeder is suspended by a thin wire and is spinning slowly at 5.6 rpm. A 140-g bird lands on the rim of the feeder, coming in tangent to the rim at 1.1 m/s in a direction opposite the feeder's rotation. What is the rotation rate after the bird lands?

46. A force $\vec{F}$ applied at the point $x = 2.0$ m, $y = 0$ produces a torque $\vec{\tau} = 4.6\hat{k}$ N·m about the origin. If the x component of $\vec{F}$ is 3.1 N, what angle does it make with the x axis?

47. A turntable has rotational inertia 0.021 kg·m² and is rotating at 0.29 rad/s about a frictionless vertical axis. A wad of clay is tossed onto the turntable and sticks 15 cm from the rotation axis (Fig. 11.15). The clay hits with horizontal velocity component 1.3 m/s, at right angles to the turntable's radius, and in a direction that opposes the rotation. After the clay lands, the turntable has slowed to 0.085 rad/s. Find the mass of the clay.

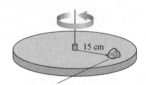

FIGURE 11.15 Problem 47

48. A uniform, solid, spherical asteroid with mass 1.2×10^{13} kg and radius 1.0 km is rotating with a period of 4.3 h. A meteoroid moving in the asteroid's equatorial plane crashes into the equator at 8.4 km/s. It hits at a 58° angle to the vertical and embeds itself at the surface. After the impact the asteroid's rotation period is 3.9 h. Find the meteoroid's mass.

49. About 99.9% of the solar system's total mass lies in the Sun. Using data from Appendix E, estimate what fraction of the solar system's angular momentum about its center is associated with the Sun. Where is most of the rest of the angular momentum?

50. An advanced civilization lives on a solid spherical planet of uniform density. Running out of room for their expanding population, the civilization's government calls an engineering firm specializing in planetary reconfiguration. Without adding any material or angular momentum, the engineers reshape the planet into a hollow shell whose thickness is one-fifth of its outer radius. Find the ratio of the new to the old (a) surface area and (b) length of day.

51. In Fig. 11.16 the lower disk, of mass 440 g and radius 3.5 cm, is rotating at 180 rpm on a frictionless shaft of negligible radius. The upper disk, of mass 270 g and radius 2.3 cm, is initially not rotating. It drops freely down the shaft onto the lower disk, and frictional forces act to bring the two disks to a common rotational speed. (a) What is that speed? (b) What fraction of the initial kinetic energy is lost to friction?

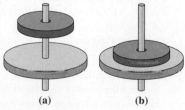

FIGURE 11.16 Problem 51

52. A massless spring of spring constant k is mounted on a turntable of rotational inertia I, as shown in Fig. 11.17. The turntable is on a frictionless vertical axle, though initially it's not rotating. The spring is compressed a distance Δx from its equilibrium, with a mass m placed against it. When the spring is released, the mass leaves the spring moving at right angles to a line through the center of the turntable, at a distance b from the center, and slides without friction across the table and off. Find expressions for (a) the linear speed of the mass and (b) the rotational speed of the turntable. *Hint:* What's conserved?

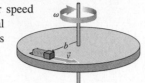

FIGURE 11.17 Problem 52

53. A solid ball of mass M and radius R is spinning with angular velocity ω_0 about a horizontal axis. It drops vertically onto a surface where the coefficient of kinetic friction with the ball is μ_k (Fig. 11.18). Find expressions for (a) the final angular velocity once it's achieved pure rolling motion and (b) the time it takes until it's in pure rolling motion.

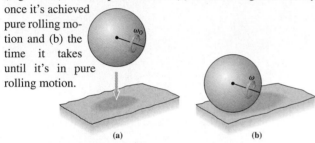

(a)   (b)

FIGURE 11.18 Problem 53

54. A system has total angular momentum $\vec{L}$ about an axis O. Show that the system's angular momentum about a parallel axis O' is given by $\vec{L}' = \vec{L} - \vec{h} \times \vec{P}$ where $\vec{P}$ is the system's linear momentum and $\vec{h}$ is the displacement vector from O to O' (see Fig. 11.19).

FIGURE 11.19 An object of arbitrary shape comprises a system whose angular momentum about axis O is $\vec{L}$. What's its angular momentum about axis O'? Also shown is the displacement $\vec{h}$ from O to O', and vectors from each axis to the ith mass element of the system (Problem 54).

55. A thin rod of length L and mass M is free to pivot about one end. (a) If it makes an angle θ with the horizontal, find the torque due to gravity about the pivot point. (You'll need to integrate the torques on the individual mass elements composing the rod.) (b) Find the value of θ such that when the rod is released from rest at this angle, the magnitude of the initial acceleration of the rod's tip is equal to g (the acceleration of gravity).

56. A 30-cm-diameter phonograph record is dropped onto a turntable being driven at $33\frac{1}{3}$ rpm. If the coefficient of friction between the record and turntable is 0.19, how far will the turntable rotate between the time when the record first contacts it and when the record is rotating at the full $33\frac{1}{3}$ rpm? Assume that the record is a homogeneous disk. *Hint:* You'll need to do an integral to calculate the torque.

57. Consider a rapidly spinning gyroscope whose axis is precessing uniformly in a horizontal circle, as shown in Figure 11.9. Show that the angular speed of the precession about the vertical axis through the center of the circle is mgr/L. *Hint:* You will need to relate ΔL to the radius of the circle described by the tip of L and the infinitesimal angular element around the vertical axis.

58. When a star like our Sun no longer has any hydrogen or helium "fuel" for thermonuclear reactions in its core, it can collapse and become a white dwarf star. Often the star will "blow off" its outer layers and lose some mass before it collapses into the rapidly spinning, dense white dwarf. Suppose a star with mass $1.0M_{Sun}$, with a radius of 6.96×10^8 m and rotating once every 25 days, becomes a white dwarf with a mass of $0.60M_{Sun}$ and a rotation period of 131 s. What is the radius of this white dwarf? (You may assume the progenitor star and the white dwarf star are both spherical.) Compare your answer with the radius of the Sun and the radius of Earth.

59. You have constructed a clever device to measure angular acceleration. It consists of a spinning wheel to which a torque is applied at right angles. A counter attached to the wheel gives you the angular speed of the wheel. Initially spinning at 5 rad/s, the wheel accelerates to 15 rad/s in 16.6 s. Your professor, who does research on computer storage devices, wants the acceleration to be between 0.75 and 1.0 rad/s^2. Does the device give these results?

60. Your professor, whom you now call Dr. Spinmaster, has other questions about his spinning disks. The disk in question is 35 cm in radius. In order to use an appropriate-sized motor, he wants to know the smallest force that needs to be applied at a point on the rim to provide a torque of 1.2 N·m about an axis through the center of the disk and perpendicular to it. A motor that can apply 4 N of force is available. Will it work?

61. The physics department has acquired a new anemometer for measuring wind speed (Fig. 11.20). It consists of four small cups each with mass 120 g mounted on the ends of essentially massless rods 32 cm long. To properly design the support, find the angular momentum of the assembly when it is rotating at 12 rev/s.

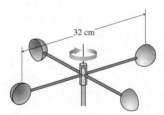

FIGURE 11.20 Problem 61

Answers to Chapter Questions

Answer to Chapter Opening Question
The rotation axis precesses—changes orientation—over a 26,000-year cycle. This alters the relation between sunlight intensity and seasons, triggering ice ages.

Answers to GOT IT? Questions
11.1 (a) $\vec{\tau}_3$; (b) $\vec{\tau}_6$; (c) $\vec{\tau}_1$; (d) $\vec{\tau}_4$

11.2 (a) You'll go clockwise to keep total angular momentum at zero.
(b) Total angular momentum remains unchanged at zero, so now you'll spin counterclockwise but at the same rate, assuming the wheel hasn't slowed.

11.3 (a)

Static Equilibrium

■ The Alamillo Bridge in Seville, Spain, is the work of architect Santiago Calatrava. What conditions must be met to ensure the stability of this dramatic structure?

▶ **To Learn**

By the end of this chapter you should be able to

■ Describe the two conditions necessary for a system to be in static equilibrium, neither accelerating nor rotating (12.1).

■ Calculate the forces and torques necessary to ensure static equilibrium (12.3).

■ Determine whether an equilibrium is stable or unstable (12.4).

◀ **To Know**

■ This chapter draws on Newton's second law as applied in Chapter 5, including the concept of force and the addition of force vectors to determine the net force (5.1–5.3).

■ This chapter is also based on the concepts of torque and the rotational analog of Newton's second law (10.2, 10.3), and the idea of torque as a vector (11.2).

■ In one sense the material here is a simplified, special case of what you've learned before, in that we allow neither acceleration nor rotation and therefore have zero on the right-hand side of Newton's law and its rotational analog.

Architect Santiago Calatrava envisioned the boldly improbable bridge shown above. But it took engineers to make sure that the bridge would be stable in the face of what looks like an obvious tendency to topple over to the left. The key to the engineers' success is **static equilibrium**—the condition in which a structure or system experiences neither a net force nor a net torque. Engineers use the principles of static equilibrium to design buildings, bridges, and aircraft. Scientists apply equilibrium principles at scales from molecular to astrophysical. Here we explore the conditions that the laws of physics require for static equilibrium.

12.1 Conditions for Equilibrium

A body is in **equilibrium** when the net external force and torque on it are both zero. In the special case when the body is also at rest, it's in **static equilibrium**. Systems in static equilibrium include not only engineered structures but also trees, molecules, and even your bones and muscles when you're at rest.

We can write the conditions for static equilibrium mathematically by setting the sums of all the external forces and torques both to zero:

$$\sum \vec{F}_i = \vec{0} \tag{12.1}$$

and

$$\sum \vec{\tau}_i = \sum (\vec{r}_i \times \vec{F}_i) = \vec{0} \tag{12.2}$$

Here the subscripts i label the forces $\vec{F}$ acting on an object, the positions $\vec{r}$ of the force-application points, and the associated torques $\vec{\tau}$.

In Chapters 10 and 11, we noted that torque depends on the choice of a rotation axis. Actually, the issue is not so much an axis but a single point—the point of origin of the vectors $\vec{r}$ that enter the expression $\vec{\tau} = \vec{r} \times \vec{F}$. In this chapter, where we have objects in equilibrium so they aren't rotating, we'll talk of this "pivot point" rather than a rotation axis. So the torque $\vec{\tau} = \vec{r} \times \vec{F}$ depends on the choice of pivot point. Then there seems to be an ambiguity in Equation 12.2, since we haven't specified a pivot point.

For an object to be in equilibrium it can't rotate about *any* point, so Equation 12.2 must hold no matter what point we choose. Must we then check every possible point? Fortunately, no. If the first equilibrium condition holds—that is, if the net force on an object is zero—and if the net torque about *some* point is zero, then the net torque about *any other* point is also zero. Problem 55 leads you through the proof of this statement.

In solving equilibrium problems, we're thus free to choose any convenient point about which to evaluate the torques. An appropriate choice is often the application point of one of the forces; then $\vec{r} = \vec{0}$ for that force, and the associated torque $\vec{r} \times \vec{F}$ is zero. This leaves Equation 12.2 with one term fewer than it would otherwise have.

EXAMPLE 12.1   **Choosing the Pivot: A Drawbridge**

The raised span of the drawbridge shown in Fig. 12.1a has its 11,000-kg mass distributed uniformly over its 14-m length. Find the magnitude of the tension in the supporting cable.

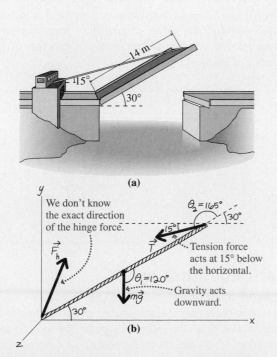

(a)

(b)

FIGURE 12.1 (a) A drawbridge. (b) Our sketch showing forces supporting the bridge.

INTERPRET Because the drawbridge is at rest, it's in static equilibrium.

DEVELOP Here we'll demonstrate how a sensible choice of the pivot point can make solving static-equilibrium problems easier. Figure 12.1b is a simplified diagram of the bridge, showing the three forces acting on it. These forces must satisfy both Equations 12.1 and 12.2, but we aren't asked about the hinge force $\vec{F}_h$, so it makes sense to choose the pivot at the hinge. We can then focus on Equation 12.2, $\sum \vec{\tau}_i = \vec{0}$, in which the only torques are due to gravity and tension. Gravity acts at the center of mass, half the bridge length L from the pivot (we'll prove this shortly). Therefore it exerts a torque $\tau_g = -(L/2)mg \sin\theta_1$, where θ_1 is the angle between the gravitational force and a vector from the pivot. This torque is into the page, or in the negative z direction—hence the negative sign. Similarly the tension force, applied at the full length L, exerts a torque $\tau_T = LT \sin\theta_2$. Equation 12.2 then becomes

$$-\frac{L}{2}mg \sin\theta_1 + LT \sin\theta_2 = 0$$

EVALUATE We solve for the tension T:

$$T = \frac{mg \sin\theta_1}{2 \sin\theta_2} = \frac{(11{,}000 \text{ kg})(9.8 \text{ m/s}^2)(\sin 120°)}{(2)(\sin 165°)} = 180 \text{ kN}$$

ASSESS This tension force is considerably larger than the approximately 110-kN weight of the bridge because the tension acts at a small angle to produce a torque that balances the torque due to gravity.

The point of this example is to show how a wise choice of the pivot point can eliminate a lot of work—in this case, allowing us to solve the problem using only Equation 12.2. If we had chosen a different pivot, then the force $\vec{F}_h$ would have appeared in the torque equation, and we would have had to eliminate it using the force equation, Equation 12.1 (see Exercise 15).   ∎

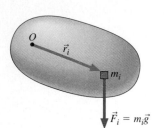

FIGURE 12.2 The gravitational force on the mass element m_i produces a torque about point O.

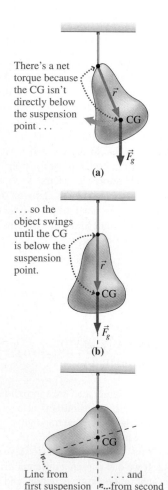

There's a net torque because the CG isn't directly below the suspension point . . .

(a)

. . . so the object swings until the CG is below the suspension point.

(b)

Line from first suspension point . . .

. . . and . . . from second point.

(c)

FIGURE 12.3 Finding the center of gravity.

GOT IT? 12.1 The figure shows three pairs of forces acting on an object. Which pair, acting as the *only* forces on the object, results in static equilibrium? Explain why the others don't.

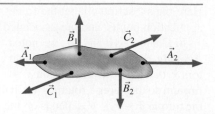

12.2 Center of Gravity

In Fig. 12.1b we drew the gravitational force acting at the center of mass of the bridge. That seems sensible, but is it correct? After all, gravity acts on all parts of an object. How do we know that the resulting torque is equivalent to the torque due to a single force acting at the center of mass? To see that it is, consider the gravitational forces on all parts of an object of mass M. The vector sum of those forces is $M\vec{g}$, but what about the torques? Figure 12.2 shows the ingredients we need to calculate the torque $\vec{\tau} = \vec{r} \times \vec{F}$ associated with one mass element; summing gives the total torque:

$$\vec{\tau} = \sum \vec{r}_i \times \vec{F}_i = \sum \vec{r}_i \times m_i \vec{g} = \left(\sum m_i \vec{r}_i\right) \times \vec{g}$$

We can rewrite this equation by multiplying the right-hand side by M/M, with M the total mass:

$$\vec{\tau} = \left(\frac{\sum m_i \vec{r}_i}{M}\right) \times M\vec{g}$$

The term in parentheses is the position of the center of mass (Section 9.1), and the right-hand term is the total weight. Therefore, the net torque on the body due to gravity is that of the gravitational force $M\vec{g}$ acting at the center of mass. In general, the point at which the gravitational force seems to act is called the **center of gravity**. We've just proven an important point: **The center of gravity coincides with the center of mass when the gravitational field is uniform.**

We can easily locate the center of gravity (CG) by suspending an object from a string attached to its edge. The torque about the suspension point is zero only if the center of gravity lies directly beneath that point (Fig. 12.3). In equilibrium, a vertical line from the suspension point must therefore pass through the center of gravity. If we now change the suspension point, we can determine another line that passes through the center of gravity. These two lines intersect at the center of gravity (Fig. 12.3c).

GOT IT? 12.2 The dancer in the figure is balanced; that is, she's in static equilibrium. Which of the three lettered points could be her center of gravity?

12.3 Examples of Static Equilibrium

It's frequently the case that all the forces acting on a system lie in a plane, so Equation 12.1—the statement that there's no net force in static equilibrium—becomes two equations for the two force components in that plane. And with all the forces in a plane, the torques are all at right angles to that plane, so Equation 12.2—the statement that there's no net torque—becomes a single equation. We'll restrict ourselves to such cases in which the conditions for static equilibrium reduce to three scalar equations. Sometimes, as in Example 12.1, the torque equation alone will give what we're looking for, but usually that's not the case.

Solving static-equilibrium problems is much like solving Newton's law problems; after all, the equations for static equilibrium are Newton's law and its rotational analog, both with acceleration set to zero. Here we adapt our Newton's law strategy from Chapter 4 to problems of static equilibrium. The examples that follow illustrate the use of this strategy.

PROBLEM SOLVING STRATEGY 12.1 **Static-Equilibrium Problems**

INTERPRET Interpret the problem to be sure it's about static equilibrium, and identify the object that you want to keep in equilibrium. Next, identify all the forces acting on the object.

DEVELOP Draw a diagram showing the forces acting on your object. Since you've got torques to calculate, it's important to show *where* each force is applied. So don't represent your object as a single dot but show it semirealistically with the force-application points. This is a static-equilibrium problem, so Equations 12.1, $\sum \vec{F}_i = \vec{0}$, and 12.2, $\sum \vec{\tau}_i = \vec{0}$, apply. Develop your solution by choosing a coordinate system that will help resolve the force vectors into components *and* choose its origin at an appropriate pivot point—usually the application point of one of the forces. In some problems the unknown is itself a force; in that case, draw a force vector that you think is appropriate and let the algebra take care of the signs and angles.

EVALUATE At this point the physics is done, and you're ready to evaluate your answer. Begin by writing the two components of Equation 12.1 in your coordinate system. Then evaluate the torques about your chosen origin, and write Equation 12.2 as a single scalar equation showing that the torques sum to zero. Now you've got three equations, and you're ready to solve. Since there are three equations, there will be three unknowns even if you're asked for only one final answer. You can use the equations to eliminate the unknowns you don't want.

ASSESS Assess your solution to see whether it makes sense. Are the numbers reasonable? Do the directions of forces and torques make sense in the context of static equilibrium? What happens in special cases—for example, when a force or mass goes to zero or gets very large, or for special values of angles among the various vectors?

EXAMPLE 12.2 **Static Equilibrium: Ladder Safety**

A ladder of mass m and length L is leaning against a wall, as shown in Fig. 12.4a. The wall is frictionless, and the coefficient of static friction between ladder and floor is μ. Find an expression for the minimum angle ϕ at which the ladder can lean without slipping.

INTERPRET This problem is about static equilibrium, and the ladder is the object we want to keep in equilibrium. We identify four forces acting on the ladder: gravity, normal forces from the floor and wall, and static friction from the floor.

DEVELOP Figure 12.4b shows the four forces and the unknown angle ϕ. We'll get the minimum angle when static friction is greatest: $f_s = \mu n_1$. Since we're dealing with static equilibrium, Equations 12.1 and 12.2 apply. In a horizontal/vertical coordinate system, Equation 12.1 has the two components:

Force, x: $\qquad \mu n_1 - n_2 = 0$

Force, y: $\qquad n_1 - mg = 0$

Now for the torques: If we choose the bottom of the ladder as the pivot, we eliminate two forces from the torque equation. That leaves only the gravitational torque and the torque due to the wall's normal force; both involve the unknown angle ϕ. The gravitational torque is into the page, or the negative z direction, so it's given by $\tau_g = -(L/2)mg \sin(90° - \phi) = -(L/2)mg \cos\phi$. The torque due to the wall is out of the page: $\tau_w = Ln_2 \sin(180° - \phi) = Ln_2 \sin\phi$. We used two trig identities here: $\sin(90° - \phi) = \cos\phi$ and $\sin(180° - \phi) = \sin\phi$. Then Equation 12.2 becomes

Torque: $\qquad Ln_2 \sin\phi - \dfrac{L}{2}mg \cos\phi = 0$

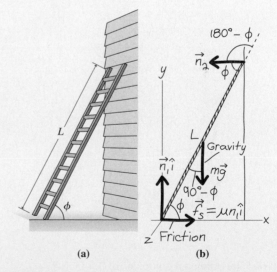

FIGURE 12.4 (a) At what angle will the ladder slip? (b) Our sketch.

EVALUATE We have three unknowns: n_1, n_2, and ϕ. The y component of the force equation gives $n_1 = mg$, showing that the floor supports the ladder's weight. Using this result in the x component of the force equation gives $n_2 = \mu mg$. Then the torque equation becomes

cont'd.

$\mu mgL \sin\phi - (L/2)mg\cos\phi = 0$. The term mgL cancels, giving $\mu\sin\phi - \frac{1}{2}\cos\phi = 0$. We solve for the unknown angle ϕ by forming its tangent:

$$\tan\phi = \frac{\sin\phi}{\cos\phi} = \frac{1}{2\mu}$$

ASSESS Make sense? The larger the frictional coefficient, the more horizontal force holding the ladder in place, and the smaller the angle

at which it can safely lean. On the other hand, a very small frictional coefficient makes for a very large tangent—meaning the angle approaches 90°. With no friction, you could stand the ladder only if it were strictly vertical. A word of caution: We worked this example with no one on the ladder. With the extra weight of a person, especially near the top, the minimum safe angle will be a lot larger. Problem 32 explores this situation.

EXAMPLE 12.3 **Static Equilibrium: In the Body**

Figure 12.5a shows a human arm holding a pumpkin, with masses and distances marked. Find the magnitudes of the biceps tension and the contact force at the elbow joint.

INTERPRET This problem is about static equilibrium, with the arm/pumpkin being the object in equilibrium. We identify four forces: the weights of the arm and the pumpkin, the biceps tension, and the contact force at the elbow.

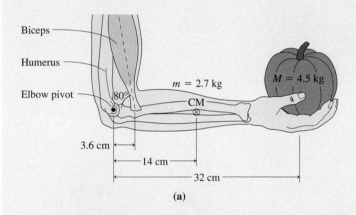

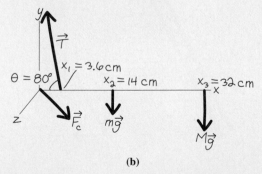

(b)

FIGURE 12.5 (a) Holding a pumpkin. (b) Our sketch.

DEVELOP Figure 12.5b shows the four forces, including the elbow contact force $\vec{F}_c$, whose exact direction we don't know. We can read the horizontal and vertical components of Equation 12.1, the force balance equation, from the diagram:

Force, x: $\qquad F_{cx} - T\cos\theta = 0$

Force, y: $\qquad T\sin\theta - F_{cy} - mg - Mg = 0$

Choosing the elbow as the pivot eliminates the contact force from the torque equation, giving

Torque: $\qquad x_1 T\sin\theta - x_2 mg - x_3 Mg = 0$

where the x values are the coordinates of the three force-application points.

EVALUATE We begin by solving the torque equation for the biceps tension:

$$T = \frac{(x_2 m + x_3 M)g}{x_1 \sin\theta} = 500\text{ N}$$

where we used the values in Fig. 12.5 to evaluate the numerical answer. The force equations then give the components of the elbow contact force:

$$F_{cx} = T\cos\theta = 87\text{ N} \quad\text{and}\quad F_{cy} = T\sin\theta - (m+M)g = 420\text{ N}$$

The magnitude of the elbow contact force then becomes $F_c = \sqrt{87^2 + 420^2}\text{ N} = 430\text{ N}$.

ASSESS These answers may seem huge—both the biceps tension and the elbow contact force are roughly ten times the weight of the pumpkin, on the order of 100 pounds. But that's because the biceps muscle is attached so close to the elbow; given this small lever arm, it takes a large force to balance the torque from the weight of pumpkin and arm. This example shows that the human body routinely experiences forces substantially greater than the weights of objects it's lifting.

GOT IT? 12.3 The figure shows a person in static equilibrium leaning against a wall. Which of the following must be true: (a) There must be a frictional force at the wall but not necessarily at the floor. (b) There must be a frictional force at the floor but not necessarily at the wall. (c) There must be frictional forces at both floor and wall.

12.4 Stability

If a body is disturbed from equilibrium, it generally experiences nonzero torques or forces that cause it to accelerate. Figure 12.6 shows two very different possibilities for the subsequent motion of two cones initially in equilibrium. Tip the cone on the left slightly, and a torque develops that brings it quickly back to equilibrium. Tip the cone on the right, and over it goes. The torque arising from even a slight displacement swings the cone permanently away from its original equilibrium. The former situation is an example of **stable equilibrium**, the latter of **unstable equilibrium**. Nearly all the equilibria we encounter in nature are stable, since a body in unstable equilibrium won't remain so. The slightest disturbance will set it in motion, bringing it to a very different equilibrium state.

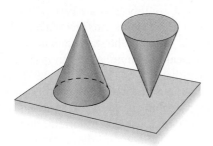

FIGURE 12.6 Stable (*left*) and unstable (*right*) equilibria.

APPLICATION   **Restoring the Statue of Liberty**

The Statue of Liberty, France's famous gift to the United States, was shipped to new York in 300 pieces, assembled, and dedicated in 1886. Liberty was the artistic work of French sculptor Frédéric-Auguste Bartholdi, who suggested that his creation should last as long as Egypt's pyramids. But after only 100 years, Liberty was ready for a major renovation. Corrosive air pollution had taken its toll, along with a chemical reaction between the statue's iron frame and its copper skin. And an assembly error had resulted in excessive torques on the statue's structural members.

Sculptor Bartholdi was no engineer, and without the work of French engineer Gustave Eiffel—designer of the famous tower—the statue could not have maintained itself in static equilibrium. Eiffel designed an inner skeleton of iron to provide the forces necessary to counteract the forces and torques associated with the weights of the statue's components and also with the wind. But Liberty's head and upper arm were mounted contrary to Eiffel's plans, probably as a result of a conscious aesthetic decision. The figure shows how the incorrect arm mounting—a two-foot offset and a correspondingly greater angle—resulted in excessive torques about the shoulder.

Liberty underwent extensive renovations during its centennial year. For historical integrity, renovators chose not to correct the original assembly error. Instead, they reinforced the support structure so it would withstand better the excess forces and torques.

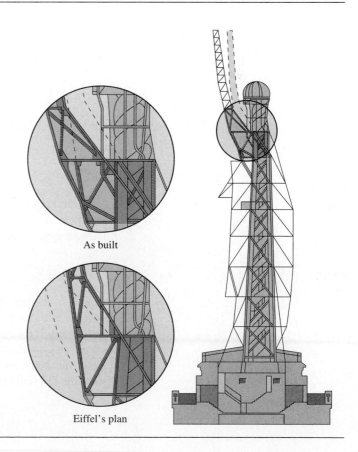

As built

Eiffel's plan

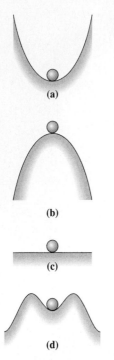

FIGURE 12.7 (a) Stable, (b) unstable, (c) neutrally stable, and (d) metastable equlilibria.

Figure 12.7 shows a ball in four different equilibrium situations. Clearly situation (a) is stable and (b) is unstable. Situation (c) is neither stable nor unstable; it's called **neutrally stable**. But what about situation (d)? For very small disturbances, the ball will return to its original state, so the equilibrium is stable. But for larger disturbances—large enough to push the ball over the highest points on the hill—it's unstable. Such an equilibrium is called **conditionally stable** or **metastable**.

A system disturbed from stable equilibrium may not return immediately. In Fig. 12.7a, for example, displacing the ball results in its rolling back and forth. Eventually friction dissipates its energy, and it comes to rest at equilibrium. Back-and-forth motion is common to many systems—from nuclei and atoms to skyscrapers and bridges—that are displaced from stable equilibrium. Such motion is the topic of the next chapter.

Stability is closely associated with potential energy. Because gravitational potential energy is directly proportional to height, the shapes of the hills and valleys in Fig. 12.7 are in fact potential-energy curves. In all cases of equilibrium, the ball is at a minimum or maximum of the potential-energy curve—at a place where the force (that is, the derivative of potential energy with respect to position) is zero. For the stable and metastable equilibria, the potential energy at equilibrium is a local minimum. A deviation from equilibrium requires that work be done against the force that tends to restore the ball to equilibrium. The unstable equilibrium, in contrast, occurs at a maximum in potential energy. Here, a deviation from equilibrium results in lower potential energy and in a force that accelerates the ball farther from equilibrium. For the neutrally stable equilibrium, there's no change whatever in potential energy as we move away from equilibrium; consequently the ball experiences no force. Figure 12.8 gives another example of equilibria in the context of potential energy.

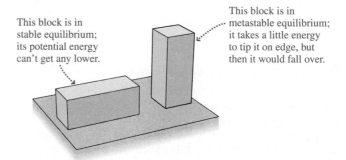

This block is in stable equilibrium; its potential energy can't get any lower.

This block is in metastable equilibrium; it takes a little energy to tip it on edge, but then it would fall over.

FIGURE 12.8 Identical blocks in stable and metastable equilibria.

We can sum up our understanding of equilibrium and potential energy in two simple mathematical statements. First, the force must be zero; that requires a local maximum or minimum in the potential energy:

$$\frac{dU}{dx} = 0 \quad \text{(equilibrium condition)} \tag{12.3}$$

where U is the potential energy of a system and x is a variable describing the system's configuration. For the simple systems we've been considering, x measures the position or orientation of an object, but for more complicated systems, it could be another quantity such as the system's volume or even its composition. For a stable equilibrium, we require a local minimum, so the potential-energy curve is concave upward. (See Tactics 12.1 to review the relevant calculus.) Mathematically,

$$\frac{d^2U}{dx^2} > 0 \quad \text{(stable equilibrium)} \tag{12.4}$$

This condition applies to metastable equilibria as well because they're *locally* stable. In contrast, unstable equilibrium occurs where the potential energy has a local maximum, or

$$\frac{d^2U}{dx^2} < 0 \qquad \text{(unstable equilibrium)} \qquad (12.5)$$

The intermediate case $d^2U/dx^2 = 0$ corresponds to neutral stability.

TACTICS 12.1 Finding Maxima and Minima

1 Begin by sketching a plot of the function, which will give a visual check for your numerical answers.

2 Next take the function's first derivative and set it to zero. As Fig. 12.7 suggests, a hill (maximum) or valley (minimum) is level right at its top or bottom. So by setting the first derivative to zero, you're requiring that its slope be zero and therefore requiring the function to be at a maximum or minimum.

3 Find the sign of the second derivative of the function at the points where you found the first derivative is zero. Your sketch should show this; where the curve is concave upward, as in Figs. 12.7a and c, the second derivative is positive and the point is a minimum. Where it's concave downward, as in Fig. 12.7b, d^2U/dt^2 is negative and you've got a maximum. If it wasn't obvious how to sketch the function, you can use calculus to determine the second derivative and then find its sign at the equilibrium points.

4 Check that the values you found for maxima and minima agree with your plot of the function.

EXAMPLE 12.4 Stability Analysis: Semiconductor Engineering

Physicists develop a new semiconductor device in which the potential energy of an electron is given by $U(x) = ax^2 - bx^4$, where x is the electron's position in nm, U is its potential energy in aJ (10^{-18} J), and constants a and b are 8 aJ/nm^2 and 1 aJ/nm^4, respectively. Find the equilibrium positions for the electron, and describe their stability.

INTERPRET This problem is about stability in the context of a given potential-energy function. We're interested in the electron, and we're asked to find the values of x where it's in equilibrium and then examine their stability.

DEVELOP The potential-energy curve gives us insight into this problem, so we've drawn it by plotting the function $U(x)$ in Fig. 12.9. Equation 12.3, $dU/dx = 0$, determines the equilibria, while Equations 12.4, $d^2U/dx^2 > 0$, and 12.5, $d^2U/dx^2 < 0$, determine the stability. Our plan is first to find the equilibrium positions using Equation 12.3 and then to examine their stability.

EVALUATE Equation 12.3 states that equilibria occur where the potential energy has a maximum or minimum—that is, where its derivative is zero. Taking the derivative of U and setting it to zero gives

$$0 = \frac{dU}{dx} = 2ax - 4bx^3 = 2x(a - 2bx^2)$$

This equation has solutions when $x=0$ and when $a=2bx^2$ or $x = \pm\sqrt{a/2b} = \pm 2$ nm. We could take second derivatives to evaluate the stability, but the situation is evident from our plot: $x=0$ lies at a local minimum of the potential-energy curve, so this equilibrium is conditionally stable. The other two equilibria, at maxima of U, are unstable.

ASSESS Do our numerical answers make sense? Yes: You can see that the potential-energy curve has zero slope at the points $x=-2$ nm, $x=0$, and $x=2$ nm, so we've found all the equilibria. Note that the equilibrium at $x=0$ is only metastable; given enough energy, an electron disturbed from this position could make it all the way over the peaks and never return to $x=0$. ∎

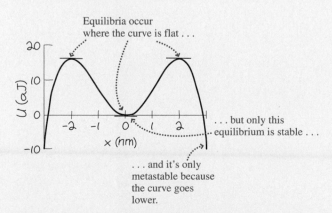

FIGURE 12.9 Our sketch of the potential-energy curve for Example 12.4.

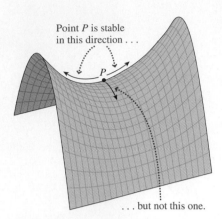

Point P is stable
in this direction . . .

. . . but not this one.

FIGURE 12.10 Equilibrium on a
saddle-shaped potential-energy curve.

Stability considerations apply to the overall arrangements of matter. A mixture of hydrogen and oxygen, for example, is in metastable equilibrium at room temperature. Lighting a match puts some atoms over the maxima in their potential-energy curves, at which point they rearrange into a state of lower potential energy—the state we call H_2O. Similarly, a uranium nucleus is at a local minimum of its potential-energy curve, and a little excess energy can result in its splitting into two smaller nuclei whose total potential energy is much lower. That transition from a less stable to a more stable equilibrium describes the basic physics of nuclear fission.

Potential-energy curves for complex structures like molecules or skyscrapers can't be described fully with one-dimensional graphs. If potential energy varies in different ways when the structure is altered in different directions, then in order to determine stability we need to consider all possible ways potential energy might vary. For example, a snowball sitting on a mountain pass—or any other system with a saddle shaped potential-energy curve—is stable against displacements in one direction but not another (Fig. 12.10). Stability analysis of complex physical systems, ranging from nuclei and molecules to bridges and buildings and machinery, and on to stars and galaxies, is an important part of contemporary work in engineering and science.

GOT IT? 12.4 Which of the labeled points in the figure are stable, metastable, unstable, or neutrally stable equilibria?

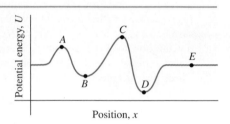

Big Picture

The big idea here is **static equilibrium**—the state in which a system at rest remains so because there's no net force to accelerate it and no net torque to start it rotating. An equilibrium is stable if a disturbance of the system results in its returning to the original equilibrium state.

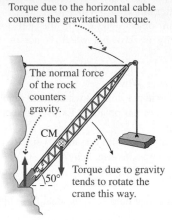

Torque due to the horizontal cable counters the gravitational torque.

The normal force of the rock counters gravity.

CM

Torque due to gravity tends to rotate the crane this way.

50°

Key Concepts and Equations

Static equilibrium requires that there be no net force and no net torque on a system; mathematically:

$$\sum \vec{F}_i = \vec{0}$$

and

$$\sum \vec{\tau}_i = \sum \vec{r}_i \times \vec{F}_i = \vec{0}$$

where the sums include all the forces applied to the system. Solving an equilibrium problem involves identifying all the forces $\vec{F}_i$ acting on the system, choosing an appropriate origin about which to evaluate the torques, and requiring that forces and torques sum to zero.

Equilibria occur where a system's potential energy $U(x)$ has a maximum or a minimum:

$$\frac{dU}{dx} = 0 \qquad \text{(equilibrium condition)}$$

$$\frac{d^2U}{dx^2} > 0 \qquad \text{(stable equilibrium)}$$

$$\frac{d^2U}{dx^2} < 0 \qquad \text{(unstable equilibrium)}$$

Stable equilibria occur at minima of U and unstable equilibria at maxima.

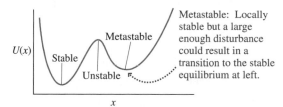

Metastable: Locally stable but a large enough disturbance could result in a transition to the stable equilibrium at left.

$U(x)$

Stable

Metastable

Unstable

x

Cases and Uses

The **center of gravity** of a system is the point where the force of gravity appears to act. When the gravitational field is uniform over the system, then the center of gravity coincides with the center of mass. This provides a handy way to locate the center of mass.

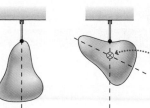

Suspend the object from any point; the CM lies somewhere directly below.

The same is true for any other point, so the CM is where the lines cross.

Four different types of equilibrium are **stable**, **unstable**, **neutrally stable**, and **metastable**.

The lowest point in a valley is stable.

The highest point on a hill is unstable.

A level surface is neutrally stable.

This point is metastable.

Note that the hill goes lower over here.

For Thought and Discussion

1. Give an example of an object on which the net force is zero, but which is not in static equilibrium.
2. Give an example of an object on which the net torque about the center of gravity is zero, but which is not in static equilibrium.
3. The best way to lift a heavy weight is to squat with your back vertical, rather than to lean over. Why?
4. Pregnant women often assume a posture in which their shoulders are held well back of their normal position. Explain in terms of torque and center of gravity.
5. When you carry a bucket of water with one hand, you often find it easier if you extend your opposite arm. Why?
6. Is a ladder more likely to slip when you stand near the top or the bottom? Explain.
7. How does a heavy keel on a boat help keep the boat from tipping over?

8. In addition to the wings, most airplanes have a smaller set of horizontal surfaces near the tail. Why? What does this suggest about the center of gravity of the airplane?
9. Does choosing a pivot point in an equilibrium problem mean that something is necessarily going to rotate about that point?
10. If you take the pivot point at the application point of a force in a static-equilibrium problem, that force doesn't enter the torque equation. Does that make the force irrelevant to the problem? Explain.
11. You're hanging a heavy picture on a wall, using wire attached to the top corners of the picture. Is the wire more likely to break if you run it tightly between the corners or if you give it some slack? Explain.
12. A short dog and a tall person are standing on a slope. If the angle of the slope is increased, which will fall over first? Why?
13. A stiltwalker is standing motionless on one stilt. What can you say about the location of the stiltwalker's center of mass?

Exercises and Problems

Exercises

Section 12.1 Conditions for Equilibrium

14. A body is subject to three forces: $\vec{F}_1 = 2\hat{i} + 2\hat{j}$ N, applied at the point $x = 2$ m, $y = 0$ m; $\vec{F}_2 = -2\hat{i} - 3\hat{j}$ N, applied at $x = -1$ m, $y = 0$; and $\vec{F}_3 = 1\hat{j}$ N, applied at $x = -7$ m, $y = 1$ m. (a) Show explicitly that the net force on the body is zero. (b) Show explicitly that the net torque about the origin is zero.
15. To show that the choice of pivot point doesn't matter, show that the torques in Exercise 14 sum to zero when evaluated about the points (3 m, 2 m) and $(-7$ m, 1 m$)$.
16. In Fig. 12.11 the forces shown all have the same magnitude F. For each of the cases shown, is it possible to place a third force so the three forces meet both conditions for static equilibrium? If so, specify the force and a suitable application point; if not, why not?

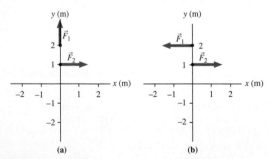

FIGURE 12.11 Exercise 16

17. Four forces act on a body, as shown in Fig. 12.12. Write the set of scalar equations that must hold for the body to be in equilibrium, evaluating the torques (a) about point O and (b) about point P.

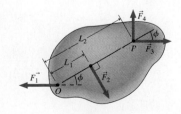

FIGURE 12.12 Exercise 17

Section 12.2 Center of Gravity

18. Figure 12.13a shows a thin, uniform square plate of mass m and side L. The plate is in a vertical plane. Find the magnitude of the gravitational torque on the plate about each of the three points shown.

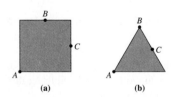

(a)   (b)

FIGURE 12.13 Exercises 18 and 19

19. Figure 12.13b shows a thin, uniform plate of mass m in the shape of an equilateral triangle of side L. The plate is in a vertical plane. Find the magnitude of the gravitational torque on the plate about each of the three points shown.
20. A 23-m-long log of irregular cross section is lying horizontally, supported by a wall at one end and a cable attached 4.0 m from the other end, as shown in Fig. 12.14. The log weighs 7.5×10^3 N, and the tension in the cable is 6.2×10^3 N. Where is the log's center of gravity?

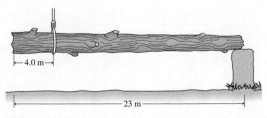

FIGURE 12.14 Exercise 20

Section 12.3 Examples of Static Equilibrium

21. A 60-kg uniform board 2.4 m long is supported by a pivot 80 cm from the left end and by a scale at the right end (Fig. 12.15). How far from the left end should a 40-kg child sit if the scale is to read zero?

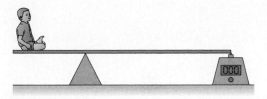

FIGURE 12.15 Exercises 21 and 22

22. Where should the child in Fig. 12.15 sit if the scale is to read (a) 100 N and (b) 300 N?

23. A 4.2-m-long beam is supported by a cable at its center. A 65-kg steelworker stands at one end of the beam. Where should a 190-kg bucket of concrete be suspended if the beam is to be in static equilibrium?

24. Figure 12.16 shows how a scale with a capacity of only 250 N can be used to weigh a heavier person. The board is 3.0 m long, has a mass of 3.4 kg, and is of uniform density. It is free to pivot about the end farthest from the scale. What is the weight of a person standing 1.2 m from the pivot end if the scale reads 210 N? Assume that the beam remains nearly horizontal.

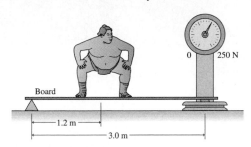

FIGURE 12.16 Exercise 24

Section 12.4 Stability

25. A portion of a roller-coaster track is described by $h = 0.94x - 0.010x^2$, where h and x are the height and horizontal position in meters. (a) Find a point where the roller-coaster car could be in static equilibrium on this track. (b) Is the equilibrium stable or unstable?

26. A particle's potential energy as a function of position is given by $U = 2x^3 - 2x^2 - 7x + 10$, with x in meters and U in joules. Find the positions of any stable and unstable equilibria.

Problems

27. Figure 12.17 shows a traffic signal, with the masses and positions of its various members indicated. The structure is mounted with two bolts, located symmetrically about the vertical member's centerline, as indicated. What tension force must the left-hand bolt be capable of withstanding?

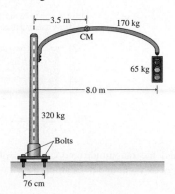

FIGURE 12.17 Problem 27

28. Figure 12.18a shows an outstretched arm with a mass of 4.2 kg. The arm is 56 cm long, and its center of gravity is 21 cm from the shoulder. The hand at the end of the arm holds a 6.0-kg mass. (a) What is the torque about the shoulder due to the weights of the arm and the 6.0-kg mass? (b) If the arm is held in equilibrium by the deltoid muscle, whose force on the arm acts 5.0° below the horizontal at a point 18 cm from the shoulder joint (Fig. 12.18b), what is the force exerted by the muscle?

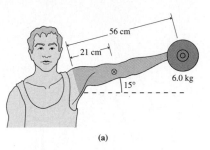

(a)

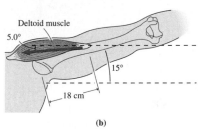

(b)

FIGURE 12.18 Problem 28

29. A uniform sphere of radius R is supported by a rope attached to a vertical wall, as shown in Fig. 12.19. The point where the rope is attached to the sphere is located so a continuation of the rope would intersect a horizontal line through the sphere's center a distance $R/2$ beyond the center, as shown. What is the smallest possible value for the coefficient of friction between wall and sphere?

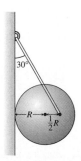

FIGURE 12.19 Problem 29

30. A garden cart loaded with firewood is being pushed horizontally when it encounters a step 8.0 cm high, as shown in Fig. 12.20. The mass of the cart and its load is 55 kg, and the cart is balanced so that its center of mass is directly over the axle. The wheel diameter is 60 cm. What is the minimum horizontal force that will get the cart up the step?

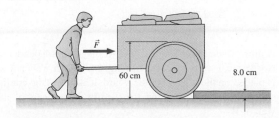

FIGURE 12.20 Problem 30

31. Figure 12.21 shows the foot and lower leg of a person standing on the ball of one foot. Three forces act on the foot to maintain this equilibrium: the tension force $\vec{T}$ in the Achilles tendon, contact force $\vec{F}_c$ at the ankle joint, and the normal force $\vec{n}$ of the ground that supports the person's weight. The person's mass is 70 kg, and the force-application points are as indicated in Fig. 12.21. Find the magnitude of (a) the tension in the Achilles tendon and (b) the contact force at the ankle joint.

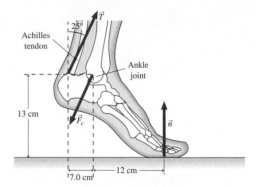

FIGURE 12.21 Problem 31

32. A uniform 5.0-kg ladder is leaning against a frictionless vertical wall, with which it makes a 15° angle. The coefficient of friction between ladder and ground is 0.26. Can a 65-kg person climb to the top of the ladder without it slipping? If not, how high can the person climb? If so, how massive a person would make the ladder slip?

33. The boom in the crane of Fig. 12.22 is free to pivot about point P and is supported by the cable that joins halfway along its 18-m total length. The cable passes over a pulley and is anchored at the back of the crane. The boom has mass 1700 kg distributed uniformly along its length, and the mass hanging from the end of the boom is 2200 kg. The boom makes a 50° angle with the horizontal. What is the tension in the cable that supports the boom?

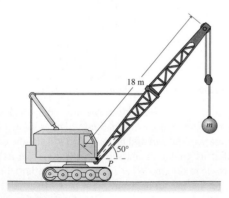

FIGURE 12.22 Problem 33

34. A uniform board of length L and weight W is suspended between two vertical walls by ropes of length $L/2$ each. When a weight w is placed on the left end of the board, it assumes the configuration shown in Fig. 12.23. Find the weight w in terms of the board weight W.

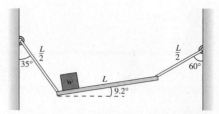

FIGURE 12.23 Problem 34

35. Figure 12.24 shows a 1250-kg car that has slipped over the edge of an embankment. People are trying to hold the car in place by pulling on a horizontal rope, as shown. The car's bottom is pivoted on the edge of the embankment, and its center of mass lies farther back, as shown. If the car makes a 34° angle with the horizontal, what force must the people apply to hold it in place?

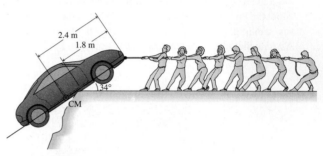

FIGURE 12.24 Problem 35

36. Repeat Example 12.2, now assuming that the coefficient of friction at the floor is μ_1 and at the wall is μ_2. Show that the minimum angle at which the board will not slip is now given by

$$\phi = \tan^{-1}\left(\frac{1 - \mu_1\mu_2}{2\mu_1}\right)$$

37. Figure 12.25 shows a 66-kg sign hung centered from a uniform rod of mass 8.2 kg and length 2.3 m. At one end the rod is attached to the wall by a pivot; at the other end it's supported by a cable that can withstand a maximum tension of 800 N. What is the minimum height h above the pivot for anchoring the cable to the wall?

FIGURE 12.25 Problem 37

38. Find the magnitude and direction of the force exerted by the drawbridge hinge in Example 12.1.

39. Climbers attempting to cross a stream place a 340-kg log against a vertical, frictionless ice cliff on the opposite side (Fig. 12.26). The log is inclined at 27°, and its center of gravity is one-third of the way along its 6.3-m length. If the coefficient of friction between the left end of the log and the ground is 0.92, what is the maximum mass for a climber and pack to cross without the log slipping?

FIGURE 12.26 Problem 39

40. A crane in a marble quarry is mounted on the rock walls of the quarry and is supporting a 2500-kg slab of marble as shown in Fig. 12.27. The center of mass of the 830-kg boom is located one-third of the way from the pivot end of its 15-m length, as shown. Find the tension in the horizontal cable that supports the boom.

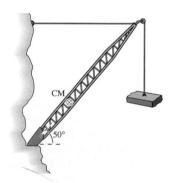

FIGURE 12.27 Problem 40

41. A uniform rectangular block is twice as long as it is wide. Letting θ be the angle that the long dimension makes with the horizontal, determine the angular positions of any static equilibria, and comment on their stability.

42. The potential energy as a function of position for a certain particle is given by

$$U(x) = U_0\left(\frac{x^3}{x_0^3} + a\frac{x^2}{x_0^2} + 4\frac{x}{x_0}\right)$$

where U_0, x_0, and a are constants. For what values of a will there be two static equilibria? Comment on the stability of these equilibria.

43. A cubical block rests on an inclined board with two sides parallel to the direction of the incline. The coefficient of static friction between block and board is 0.95. If the inclination angle of the board is increased, will the block first slide or first tip over?

44. A 160-kg highway sign of uniform density is 2.3 m wide and 1.4 m high. At one side it is secured to a pole with a single bolt, mounted a distance d from the top of the sign. The only other place where the sign contacts the pole is at its bottom corner. If the bolt can sustain a horizontal tension of 2100 N, what is the maximum permissible value for the distance d?

45. A 5.0-m-long ladder has mass 9.5 kg and is leaning against a frictionless wall, making a 66° angle with the horizontal. If the coefficient of friction between the ladder and ground is 0.42, what is the mass of the heaviest person who can safely ascend to the top of the ladder? (The center of mass of the ladder is at its center.)

46. To what vertical height on the ladder in Problem 45 could a 95-kg person reach before the ladder starts to slip?

47. A uniform, solid cube of mass m and side s is in stable equilibrium when sitting on a level tabletop. How much energy is required to bring it to an unstable equilibrium where it's resting on its corner?

48. An isosceles triangular block of mass m and height h is in stable equilibrium, resting on its base on a horizontal surface. How much energy does it take to bring it to unstable equilibrium, resting on its apex?

49. A uniform ladder of mass m is leaning against a frictionless vertical wall with which it makes an angle θ. The coefficient of static friction at the floor is μ. Find an expression for the maximum mass of a person who is able to climb to the top of the ladder without its slipping. Use your result to show that *anyone* can climb to the top if $\mu \geq \tan\theta$ but that *no one* can if $\mu < \frac{1}{2}\tan\theta$.

50. A 2.0-m-long rod has a density described by $\lambda = a + bx$, where λ is the density in kilograms per meter of length, $a = 1.0$ kg/m, $b = 1.0$ kg/m², and x is the distance in meters from the left end of the rod. The rod rests horizontally with each end supported by a scale. What do the two scales read?

51. What horizontal force applied at its highest point is necessary to keep a wheel of mass M from rolling down a slope inclined at angle θ to the horizontal?

52. A rectangular block twice as high as it is wide is resting on a board. The coefficient of static friction between board and incline is 0.63. If the board is tilted as shown in Fig. 12.28, will the block first tip over or first begin sliding?

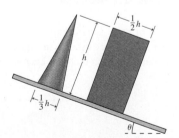

FIGURE 12.28 Problems 52, 53, and 54

53. What condition on the coefficient of friction in Problem 52 will cause the block to slide before it tips?

54. A uniform solid cone of height h and base diameter $\frac{1}{3}h$ is placed on the board of Fig. 12.28. The coefficient of static friction between the cone and incline is 0.63. As the slope of the board is increased, will the cone first tip over or begin sliding? *Hint:* Begin with an integration to find the center of mass.

55. In this problem you prove the statement in Section 12.1 that the choice of pivot point does not matter when applying the conditions for static equilibrium. Figure 12.29 shows an object on which the net force is assumed to be zero. Also, the net torque about the point O is zero. You're to show that the net torque about any other point P is also zero. To do so, write the net torque about P as $\vec{\tau}_P = \sum \vec{r}_{P_i} \times \vec{F}_i$, where the vectors $\vec{r}_P$ are from P to the force-application points, and the index i labels the different forces. Note in Fig. 12.29 that $\vec{r}_{P_i} = \vec{r}_{O_i} + \vec{R}$, where $\vec{R}$ is a vector from P to O. Use this result in your expression for $\vec{\tau}_P$ and apply the distributive law to get two separate sums. Use the assumptions that $\vec{F}_{\text{net}} = \vec{0}$ and $\vec{\tau}_O = \vec{0}$ to argue that both terms are zero. This completes the proof.

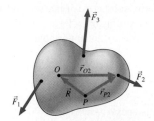

FIGURE 12.29 Problem 55

56. Three identical books of length L are stacked over the edge of a table as shown in Fig. 12.30. The top book overhangs the middle one by $\frac{1}{2}L$, so it just barely avoids falling. The middle book overhangs the bottom one by $\frac{1}{4}L$. How much of the bottom book can be allowed to overhang the edge of the table without the books falling?

θ,

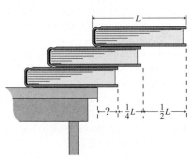

FIGURE 12.30 Problem 56

57. A uniform pole of mass M is at rest on an incline of angle   secured by a horizontal rope as shown in Fig. 12.31. What is the minimum coefficient of friction that will keep the pole from slipping?

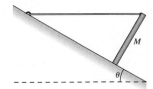

FIGURE 12.31 Problems 57 and 58

58. For what angle does the situation of Problem 57 require the greatest coefficient of friction?

59. Figure 12.32 shows a popular system for mounting bookshelves. An aluminum bracket is mounted on a vertical aluminum support by small tabs inserted into vertical slots. If each bracket in a shelf system supports 32 kg of books, with the center of gravity 12 cm out from the vertical support, what is the horizontal component of

the force exerted on the upper of the two bracket tabs? Assume contact between the bracket and support occurs only at the upper tab and at the bottom of the bracket, 4.5 cm below the upper tab.

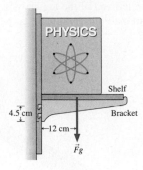

FIGURE 12.32 Problem 59

60. A 4.2-kg plant hangs from the bracket shown in Figure 12.33. The bracket has a mass of 0.85 kg, and its center of mass lies 9.0 cm from the wall. A single screw holds the bracket to the wall, as shown. Find the horizontal tension force in the screw. *Hint:* Imagine that the bracket is slightly loose and pivoting about its bottom end. Assume the wall is frictionless.

FIGURE 12.33 Problem 60

61. Figure 12.34 shows a wheel on a slope with inclination angle $\theta = 20°$, where the coefficient of friction is adequate to prevent the wheel from slipping; however, it might still roll. The wheel is a uniform disk of mass 1.5 kg, and it is weighted at one point on the rim with an additional 0.95-kg mass m. Find the angle ϕ shown in the figure such that the wheel will be in static equilibrium.

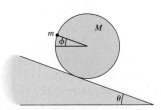

FIGURE 12.34 Problems 61 and 62

62. The wheel in Figure 12.34 has mass M and is weighted with an additional mass m as shown. The slope angle is θ. Show that static equilibrium is possible only if $m > \dfrac{M \sin \theta}{1 - \sin \theta}$.

63. An interstellar spacecraft from an advanced civilization is hovering above Earth, as shown in Figure 12.35. The ship consists of two pods of mass m separated by a rigid shaft of negligible mass that is one Earth radius (R_E) long. Find (a) the magnitude and direction of the net gravitational force on the ship and (b) the net torque about the center of mass. (c) Show that the ship's center of gravity is displaced approximately $0.083R_E$ from its center of mass.

FIGURE 12.35 Problem 63

64. You have been called on to testify in a product liability lawsuit. An infant sitting in a portable seat that is supported by the edge of a table fell to the floor (Fig. 12.36). The manufacturer claims the child was too heavy for the seat, and the parents claim the seat was defective. Testing showed that the seat can withstand the weight of a child when the force at A does not exceed 96.2 N and the force at B does not exceed 229 N. The seat's mass is 2.0 kg; the child's mass is 10 kg, and the center of mass of child and seat is 16 cm from the table edge. In whose favor should the judge rule?

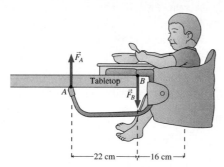

FIGURE 12.36 Problem 64

65. You walk into your professor's office to find him staring at the weather radar for the area where his cabin is located. Very heavy snow is forecast and he's concerned about the roof. Figure 12.37 shows the house with its high "cathedral" ceiling. He estimates that after the snowfall, each diagonal roof rafter will support 170 kg of snow and building materials. The horizontal tie beam near the roof peak can withstand a force of 7500 N. Neglect any horizontal component of force due to the vertical walls below the roof. Ignore the widths of the various structural members, treating contact forces as though they were concentrated at the roof peak and at the outside edge of the rafter/wall junction. Will the tie beam hold? Is it in compression or tension?

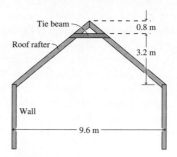

FIGURE 12.37 Problem 65

66. A friend asks for your help with the rigging for her sailboat. A uniform boom of mass 350 kg is attached to a vertical mast by a pivot, and its far end is supported by a cable, as shown in Fig. 12.38. The maximum angle ϕ the cable will make with the mast is 24°. Will a cable break if it can withstand a tension of 4000 N?

FIGURE 12.38 Problem 66

Answers to Chapter Questions

Answer to Chapter Opening Question
Both the net force and the net torque on all parts of the bridge must be zero.

Answers to GOT IT? Questions
12.1 Pair C; pair A produces nonzero net force, and pair B produces nonzero net torque.

12.2 B; it's located directly over the point of contact with the floor, ensuring there's no gravitational torque.

12.3 (b); a frictional force at the floor is necessary to balance the normal force from the wall.

12.4 D: stable; B: metastable; A and C: unstable; E: neutrally stable.

Mechanics

The big idea of Part One is Newton's realization that forces—pushes and pulls—don't cause motion but instead cause *change* in motion. Newton's second law quantifies this idea. With momentum $\vec{p} = m\vec{v}$ as Newton's measure of "quantity of motion," the second law equates the net force on an object to the rate of change of its momentum: $\vec{F} = d\vec{p}/dt$ or, for constant mass, $\vec{F} = m\vec{a}$. The second law encompasses the first, the law of inertia: In the absence of a net force, an object continues in uniform motion, unchanging in speed or direction—a state that includes the special case of being at rest. Newton's third law rounds out the picture, providing a fully consistent description of motion with its statement that forces come in pairs: If object A exerts a force on B, then B exerts a force of equal magnitude but opposite direction back on A.

From the concept of force and Newton's laws follow the essential ideas of work and energy, including kinetic and potential energy and the conservation of mechanical energy in the absence of nonconservative forces like friction. One important force is gravity, which Newton described through his law of universal gravitation and applied to explain the motions of the planets. Application of Newton's laws to systems of objects gives us the concept of center of mass and lets us describe the interactions of colliding objects. Finally, Newton's laws explain circular and rotational motion, the latter through the analogy between force and torque. That, in turn, gives us the conditions needed for static equilibrium—the state in which an object at rest remains so, subject neither to a net force nor to a net torque.

Newton's Laws

Newton's 1st law: Force causes a change in motion.
Newton's 2nd law: $\vec{F} = d\vec{p}/dt$ or, for constant mass, $\vec{F} = m\vec{a}$
Newton's 3rd law: $\vec{F}_{AB} = -\vec{F}_{BA}$

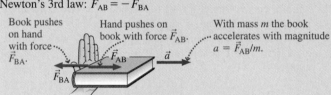

Book pushes on hand with force $\vec{F}_{BA}$.

Hand pushes on book with force $\vec{F}_{AB}$.

With mass m the book accelerates with magnitude $a = \vec{F}_{AB}/m$.

Work and energy

Work: $W = \vec{F} \cdot \Delta\vec{r}$ or, for a varying force, $W = \int \vec{F} \cdot d\vec{r}$

Work-energy theorem: $\Delta K = W$ with kinetic energy $K = \frac{1}{2}mv^2$

For conservative forces, work is stored as potential energy U. Then $K + U = $ constant.

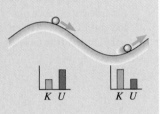

Universal gravitation

$$F = \frac{Gm_1 m_2}{r^2}$$

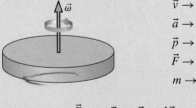

Conservation of momentum in a system

Initial state   Final state

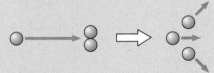

Initial momentum   =   Final momentum

$$\vec{P}_i = \Sigma m_i \vec{v}_i = m_1 \vec{v}_{1i} \quad \Rightarrow \quad \vec{P}_f = \Sigma m_f \vec{v}_f = m_1 \vec{v}_{1f} + m_2 \vec{v}_{2f} + m_3 \vec{v}_{3f}$$

Rotational motion

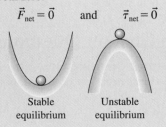

$$\vec{v} \rightarrow \vec{\omega}$$
$$\vec{a} \rightarrow \vec{\alpha}$$
$$\vec{p} \rightarrow \vec{L}$$
$$\vec{F} \rightarrow \vec{\tau}$$
$$m \rightarrow I$$

$$\vec{F} = m\vec{a} \rightarrow \vec{\tau} = I\vec{\alpha}$$
$$K = \frac{1}{2}mv^2 \rightarrow K = \frac{1}{2}I\omega^2$$

Static equilibrium

A system is in static equilibrium when the net force and the net torque on the system are both zero:

$$\vec{F}_{net} = \vec{0} \quad \text{and} \quad \vec{\tau}_{net} = \vec{0}$$

Stable equilibrium   Unstable equilibrium

Part One challenge problem

A solid ball of radius R is set spinning with angular speed ω about a horizontal axis. The ball is then lowered vertically with negligible speed until it just touches a horizontal surface and is released (see figure). If the coefficient of kinetic friction between the ball and the surface is μ, find (a) the linear speed of the ball once it achieves pure rolling motion, (b) the distance it travels before its motion is pure rolling, and (c) the fraction of the ball's initial rotational kinetic energy that's been lost to friction.

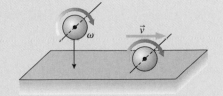

PART TWO

OVERVIEW

Oscillations, Waves, and Fluids

■ High-speed photo shows complex fluid behavior and spreading circular waves on water.

A tsunami crashes on shore, dissipating energy that has traveled across thousands of kilometers of open ocean. Near the epicenter of the earthquake that spawned the tsunami, a skyscraper sways in response, but suffers no damage thanks to a carefully engineered system that counters quake-induced vibrations. An electric guitar sounds loud during a rock concert, the sound waves following the vibrations of the guitar strings. Inside your watch, a tiny quartz crystal vibrates 32,768 times each second to keep near-perfect time. A radar-equipped police officer waits around the next turn in the highway ready to ticket your speeding car, while astrophysicists use the same principle to measure the expansion of the universe. A rafting party enters a narrow gorge, getting a wild ride as the river's speed increases. A plane cruises far overhead, supported by the force of air on its wings. All these examples involve the collective motion of many particles. In the next three chapters, we first explore the repetitive motion called oscillation and then show how oscillations in many-particle systems lead to wave motion. Finally, we apply the laws of motion to reveal the sometimes surprising and beautiful behavior of fluids like air and water.

203

13 Oscillatory Motion

■ Dancers from the Bandaloop Project perform on vertical surfaces, executing graceful slow-motion jumps. What determines the duration of these jumps?

▶ To Learn
By the end of this chapter you should be able to
- Describe the relations between the period and frequency of oscillatory motion and between ordinary frequency and angular frequency (13.1, 13.2).
- State what simple harmonic motion is, and explain why it occurs universally in the physical world (13.2).
- Describe simple harmonic motion quantitatively (13.3, 13.4).
- Explain damped harmonic motion and the phenomenon of resonance (13.6, 13.7).

◀ To Know
- The material in this chapter draws on Newton's second law as applied in Chapter 5 and its rotational analog as introduced in Chapter 10 (5.1–5.3, 10.3).
- You should be familiar with the behavior of springs (4.6) and with the potential energy of a spring (7.2).
- You should have a solid understanding of the sine and cosine functions and be able to take derivatives of both these functions (Appendix A).

D isplace a system from stable equilibrium, and forces or torques tend to restore the equilibrium. But, like the ball in Fig. 13.1, the system often overshoots its equilibrium and goes into **oscillatory motion** back and forth about equilibrium. Absent friction, this oscillation would continue forever; in reality, the system eventually settles into equilibrium.

Here the ball is in stable equilibrium.

Disturb the ball, and it oscillates about its equilibrium position.

FIGURE 13.1 Disturbing a system results in oscillatory motion.

Oscillatory motion occurs throughout the physical world. A uranium nucleus oscillates before it fissions. Water molecules oscillate to heat the food in a microwave oven. Carbon dioxide molecules in the atmosphere oscillate, absorbing energy and thus contributing to global warming. A watch—whether an old-fashioned mechanical one or a modern quartz timepiece—is a carefully engineered oscillating system. Buildings and bridges undergo oscillatory motion, sometimes with disastrous results. Even stars oscillate. And waves—from sound to ocean waves to seismic waves in the solid Earth—ultimately involve oscillatory motion.

Oscillatory motion is universal because systems in stable equilibrium naturally tend to return toward equilibrium no matter how they're displaced. And it's not just the qualitative phenomenon of oscillation that's universal: Remarkably, the mathematical description of oscillatory motion is the same for systems ranging from atoms and molecules through cars and bridges and on to stars and galaxies.

13.1 Describing Oscillatory Motion

Figure 13.2 shows two quantities that characterize oscillatory motion: **Amplitude** is the maximum displacement from equilibrium, and **period** is the time it takes for the motion to repeat itself. Another way to express the time aspect is **frequency**, or number of oscillation cycles per unit time. Frequency f and period T are complementary ways of conveying the same information, and mathematically they're inverses:

$$f = \frac{1}{T} \tag{13.1}$$

The unit of frequency is the **hertz** (Hz), named after the German Heinrich Hertz (1857–1894), who was the first to produce and detect radio waves. One hertz is equal to one oscillation cycle per second.

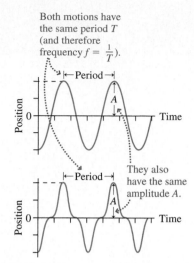

FIGURE 13.2 Position-time graphs for two oscillatory motions with the same amplitude A and period T (and therefore frequency).

EXAMPLE 13.1 **Amplitude, Period, Frequency:**
An Oscillatory Distraction

Tired of homework, a student holds one end of a flexible plastic ruler against a desk and idly strikes the other end, setting it into oscillation (Fig. 13.3). The student notes that 28 complete cycles occur in 10 s and that the end of the ruler moves a total distance of 8.0 cm. What are the amplitude, period, and frequency of this oscillatory motion?

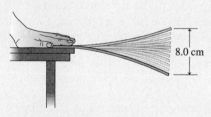

FIGURE 13.3 A ruler undergoing oscillatory motion.

INTERPRET We've got a case of oscillatory motion, and we're asked to describe it quantitatively in terms of amplitude, period, and frequency.

DEVELOP We can work from the definitions of these quantities: Amplitude is the maximum displacement from equilibrium, period is the time to complete a full oscillation, and frequency is the inverse of the period (Equation 13.1).

EVALUATE The ruler moves a total of 8.0 cm from one extreme to the other. Since the motion takes it to both sides of its equilibrium position, the amplitude is 4.0 cm. With 28 cycles in 10 s, the time per cycle, or the period, is

$$T = \frac{10\ \text{s}}{28} = 0.36\ \text{s}$$

The frequency is the inverse of the period: $f = 1/T = 1/0.36\ \text{s} = 2.8\ \text{Hz}$. We can also get this directly: 28 cycles/10 s = 2.8 Hz.

ASSESS Make sense? With a period that's less than 1 s, the frequency must be more than 1 cycle per second or 1 Hz. By the way, our definition of amplitude as the maximum displacement from equilibrium led to our 4.0-cm amplitude; the full 8.0 cm between extreme positions is called the **peak-to-peak amplitude**. ∎

Amplitude and frequency don't provide all the details of oscillatory motion, since two quite different motions can have the same frequency and amplitude (Fig. 13.2). The differences reflect the restoring forces that return systems to equilibrium. Remarkably, though, restoring forces in many physical systems have the same mathematical form—a form we encountered before, when we introduced the force of an ideal spring in Chapter 4.

13.2 Simple Harmonic Motion

In many systems, the restoring force that develops when the system is displaced from equilibrium increases approximately in direct proportion to the displacement—meaning that if you displace the system twice as far from equilibrium, the force tending to restore equilibrium becomes twice as great. In the rest of this chapter, we therefore consider the case where the restoring force is directly proportional to the displacement. This is an approximation for most real systems, but often a very good approximation, especially for small displacements from equilibrium.

The type of motion that results from a restoring force proportional to displacement is called **simple harmonic motion** (SHM). Mathematically, we describe such a force by writing

$$F = -kx \qquad \text{(restoring force in SHM)} \tag{13.2}$$

where F is the force, x the displacement, and k a constant of proportionality between them. The minus sign in Equation 13.2 indicates a *restoring* force: If the object is displaced in one direction, the force is in the *opposite* direction, so it tends to restore the equilibrium.

We've seen Equation 13.2 before: It's the force exerted by an ideal spring of spring constant k. So a system consisting of a mass attached to a spring undergoes simple harmonic motion (Fig. 13.4). Many other systems—including atoms and molecules—can be modeled as miniature mass-spring systems.

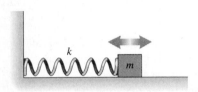

FIGURE 13.4 A mass attached to a spring undergoes simple harmonic motion.

How does a body in simple harmonic motion actually move? We can find out by applying Newton's second law, $F = ma$, to the mass-spring system of Fig. 13.4. Here the force on the mass m is $-kx$, so Newton's law becomes $-kx = ma$, where we take the x axis along the direction of motion, with $x = 0$ at the equilibrium position. Now, the acceleration a is the second derivative of position, so we can write our Newton's law equation as

$$m\frac{d^2x}{dt^2} = -kx \qquad \text{(Newton's 2}^{\text{nd}}\text{ law for SHM)} \tag{13.3}$$

The solution to this equation is the position x as a function of time. What sort of function might it be? We expect periodic motion, so let's try periodic functions like sine and cosine. Suppose we pull the mass in Fig. 13.4 to the right and, at time $t = 0$, release it. Since it starts with a nonzero displacement, cosine is the appropriate function (recall that $\cos(0) = 1$, and $\sin(0) = 0$). We don't know the amplitude or frequency, so we'll try a form that has two unknown constants:

$$x(t) = A\cos\omega t \tag{13.4}$$

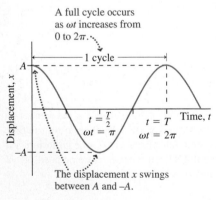

A full cycle occurs as ωt increases from 0 to 2π.

The displacement x swings between A and $-A$.

FIGURE 13.5 The function $A\cos\omega t$.

Because the cosine function itself varies between $+1$ and -1, A in Equation 13.4 is the amplitude—the greatest displacement from equilibrium (Fig. 13.5). What about ω? The cosine function undergoes a full cycle as its argument increases by 2π radians, or $360°$, as shown in Fig. 13.5. In Equation 13.4, the argument of the cosine is ωt. Since the time for a full cycle is the period T, the argument ωt must go from 0 to 2π as the time t goes from 0 to T. So we have $\omega T = 2\pi$, or

$$T = \frac{2\pi}{\omega} \tag{13.5}$$

The frequency of the motion is then

$$f = \frac{1}{T} = \frac{\omega}{2\pi} \tag{13.6}$$

Equation 13.6 shows that ω is a measure of the frequency, although it differs from the frequency f by the factor 2π. The quantity ω is called the **angular frequency**, and its units are radians per second or, since radians are dimensionless, simply inverse seconds (s^{-1}).

✓ **TIP** Why Radians?
Here, as in Chapter 10, we use the angular quantity ω because it provides the simplest mathematical description of the motion. In fact, the relationship between angular frequency and frequency in hertz is the same as Chapter 10's relationship between angular speed in radians per second and in revolutions per second. We'll explore this similarity further in Section 13.4.

Writing the displacement x in the form 13.4 doesn't guarantee that we have a solution; we still need to see whether this form satisfies Equation 13.3. With $x(t)$ given by Equation 13.4, its first derivative is

$$\frac{dx}{dt} = \frac{d}{dt}(A\cos\omega t) = -A\omega\sin\omega t$$

where we've used the chain rule for differentiation (see Appendix A). Then the second derivative is

$$\frac{d^2x}{dt^2} = \frac{d}{dt}\left(\frac{dx}{dt}\right) = \frac{d}{dt}(-A\omega\sin\omega t) = -A\omega^2\cos\omega t$$

We can now try out our assumed solution for x (Equation 13.4) and its second derivative in Equation 13.3. Substituting $x(t)$ and d^2x/dt^2 in the appropriate places gives

$$m(-A\omega^2\cos\omega t) \overset{?}{=} -k(A\cos\omega t)$$

where the ? indicates that we're still trying to find out whether this is indeed an equality. If it is, the equality must hold *for all values of time t*. Why? Because Newton's law holds at all times, and we derived our questionable equality from Newton's law. Fortunately, the time-dependent term $\cos\omega t$ appears on both sides of the equation, so we can cancel it. Also, the amplitude A and the minus sign cancel from the equation, leaving only $m\omega^2 = k$, or

$$\omega = \sqrt{\frac{k}{m}} \qquad \text{(angular frequency, simple harmonic motion)} \qquad (13.7a)$$

Thus, Equation 13.4 *is* a solution of Equation 13.3, *provided* the angular frequency ω is given by Equation 13.7a.

Frequency and Period in Simple Harmonic Motion

We can recast Equation 13.7a in terms of the more familiar frequency f and period T using Equation 13.6, $f = \omega/2\pi$. This gives

$$f = \frac{\omega}{2\pi} = \frac{1}{2\pi}\sqrt{\frac{k}{m}} \quad \text{and} \quad T = \frac{1}{f} = 2\pi\sqrt{\frac{m}{k}} \qquad (13.7b, c)$$

Do these relationships make sense? If we increase the mass m, it becomes harder to accelerate and we expect slower oscillations. This is reflected in Equations 13.7a and b, where m appears in the denominator. Increasing k, on the other hand, makes the spring stiffer and therefore results in greater force. That increases the oscillation frequency—as shown by the presence of k in the numerators of Equations 13.7a and b.

Physical systems display a wide range of m and k values and a correspondingly large range of oscillation frequencies. A molecule, with its small mass and its "springiness" provided by electric forces, may oscillate at 10^{14} Hz or more. A massive skyscraper, in contrast, typically oscillates at about 0.1 Hz.

Amplitude in Simple Harmonic Motion

The amplitude A canceled from our equations, so our analysis works for *any* value of A. This means that the oscillation frequency doesn't depend on amplitude. Independence of frequency and amplitude is a feature of simple harmonic motion, and arises because the restoring force is *directly proportional* to the displacement. When the restoring force does not have the simple form $F = -kx$, then frequency *does* depend on amplitude and the analysis of oscillatory motion becomes much more complicated. In many systems the relation $F = -kx$ breaks down if the displacement x gets too big; for this reason, simple harmonic motion usually occurs only for small oscillation amplitudes.

Phase

Equation 13.4 isn't the only solution to Equation 13.3; you can readily show that $x = A \sin \omega t$ works just as well. We chose the cosine because we took time $t = 0$ at the point of maximum displacement. Had we set $t = 0$ as the mass passed through its equilibrium point, sine would have been the appropriate function. More generally we can take the zero of time at some arbitrary point in the oscillation cycle. Then, as Fig. 13.6 shows, we can represent the motion by the form

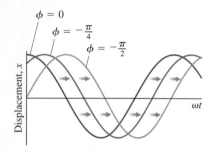

FIGURE 13.6 A negative phase constant shifts the curve to the right.

$$x(t) = A\cos(\omega t + \phi) \quad \text{(simple harmonic motion)} \tag{13.8}$$

where the **phase constant** ϕ has the effect of shifting the cosine curve to the left (for $\phi > 0$) or right ($\phi < 0$) but doesn't affect the frequency or amplitude.

Velocity and Acceleration in Simple Harmonic Motion

Equation 13.4 (or, more generally, Equation 13.8) gives the position of an object in simple harmonic motion as a function of time, so its first derivative must be the object's velocity:

$$v(t) = \frac{dx}{dt} = \frac{d}{dt}(A\cos\omega t) = -\omega A \sin\omega t \tag{13.9}$$

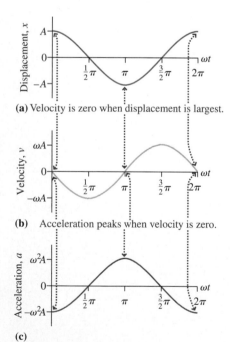

(a) Velocity is zero when displacement is largest.

(b) Acceleration peaks when velocity is zero.

(c)

FIGURE 13.7 Displacement, velocity, and acceleration in simple harmonic motion.

Because the maximum value of the sine function is 1, this expression shows that the maximum velocity is ωA. This makes sense because a higher-frequency oscillation requires that the object traverse the distance A in a shorter time—so it must move faster. Equation 13.9 shows that the velocity $v(t)$ is a sine function when the displacement $x(t)$ is a cosine. Thus velocity is a maximum when displacement is zero, and vice versa; mathematically, we express this by saying that displacement and velocity differ in phase by $\frac{\pi}{2}$ radians or 90°. Does this make sense? Sure, because at the extremes of its motion, the object is instantaneously at rest as it reverses direction: maximum displacement, zero speed. And when it passes through its equilibrium position, the object is going fastest. Figures 13.7a and b show graphically the relationship between displacement and velocity in simple harmonic motion.

Just as velocity is the derivative of position, so acceleration is the derivative of velocity, or the second derivative of position:

$$a(t) = \frac{dv}{dt} = \frac{d}{dt}(-\omega A \sin\omega t) = -\omega^2 A\cos\omega t \tag{13.10}$$

Thus the maximum acceleration is $\omega^2 A$. Since acceleration is a cosine function if velocity is a sine, each reaches its maximum value when the other is zero (Fig. 13.7b, c).

GOT IT? 13.1 Two identical mass-spring systems are displaced different amounts from equilibrium and then released at different times. Of the amplitudes, frequencies, periods, and phase constants of the subsequent motions, which are the same for both systems and which are different?

EXAMPLE 13.2   Simple Harmonic Motion: A Tuned Mass Damper

The tuned mass damper in New York's Citicorp Tower (see Application, below) consists of a 373-Mg concrete block that completes one oscillation in 6.80 s. The oscillation amplitude in a high wind is 110 cm. Determine the spring constant and the maximum speed and acceleration of the block.

INTERPRET This is a problem involving simple harmonic motion, with the concrete block and spring making up the oscillating system. We're given the period, mass, and amplitude.

DEVELOP Equation 13.7c, $T = 2\pi\sqrt{m/k}$, gives us the spring constant. Equations 13.9 and 13.10 show that the maximum speed and acceleration are $v_{max} = \omega A$ and $a_{max} = \omega^2 A$, and we can get the angular frequency ω from the period using Equation 13.5: $\omega = 2\pi/T$.

EVALUATE First we solve Equation 13.7c for the spring constant:

$$k = \frac{4\pi^2 m}{T^2} = \frac{(4\pi^2)(3.73 \times 10^5 \text{ kg})}{(6.80 \text{ s})^2} = 3.18 \times 10^5 \text{ N/m}$$

The angular frequency is $\omega = 2\pi/T = 0.924 \text{ s}^{-1}$. Then we have $v_{max} = \omega A = (1.10 \text{ m})(0.924 \text{ s}^{-1}) = 1.02 \text{ m/s}$ and $a_{max} = \omega^2 A = 0.939 \text{ m/s}^2$.

ASSESS The large spring constant and relatively low velocity and acceleration make sense given the huge mass involved. Note that we had to convert the mass, given as 373 Mg (373×10^6 g), to kilograms before evaluating.   ∎

13.3 Applications of Simple Harmonic Motion

Simple harmonic motion occurs in any system where the tendency to return to equilibrium increases in direct proportion to the displacement from equilibrium. Analysis of such systems is like that of the mass-spring system we just considered but may involve different physical quantities.

The Vertical Mass-Spring System

A mass hanging vertically from a spring is subject to gravity as well as the spring force (Fig. 13.8). In equilibrium the spring stretches enough for its force to balance gravity: $mg - kx_1 = 0$, where x_1 is the new equilibrium position. Stretching the spring an additional amount Δx increases the spring force by $k\,\Delta x$, and this increased force tends to restore the equilibrium. So once again we have a restoring force that's directly proportional to displacement. And here, with the same spring constant k and mass m, our previous analysis still applies and we get simple harmonic motion with frequency $\omega = \sqrt{k/m}$. Thus gravity changes only the equilibrium position and doesn't affect the frequency.

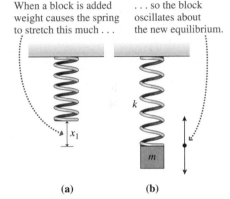

When a block is added weight causes the spring to stretch this much . . .   . . . so the block oscillates about the new equilibrium.

(a)   **(b)**

FIGURE 13.8 A vertical mass-spring system oscillates about a new equilibrium position x_1, with the same frequency $\omega = \sqrt{k/m}$.

APPLICATION   Swaying Skyscrapers

Skyscrapers are tall, thin, flexible structures. High winds and earthquakes can set them oscillating, much like the ruler of Example 13.1. Wind-driven oscillations are uncomfortable to occupants of a building's upper floors, and earthquake-induced oscillations can be downright destructive.

Modern skyscrapers use so-called tuned mass dampers to counteract building oscillations. These devices are essentially large mass-spring systems mounted high in the building. They're engineered to oscillate with the same frequency as the building (hence the term "tuned") but 180° out of phase, thus reducing the amplitude of the building's own oscillation. The result is increased comfort for the building's occupants and improved safety for buildings in earthquake-prone regions. Tuned mass dampers also find applications in tall smokestacks, airport control towers, power-plant cooling towers, bridges, and even ski lifts. By suppressing vibrations, tuned mass dampers enable architects and engineers to design structures that don't need as much intrinsic stiffness, so they can be lighter and less expensive. The photo shows the world's largest tuned mass damper and the building that houses it, Taiwan's Taipei 101 skyscraper. The damper helps the building survive earthquakes and typhoons. Example 13.2 explores another tuned mass damper.

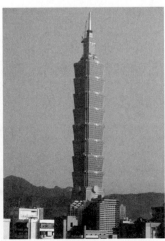

FIGURE 13.9 A torsional oscillator.

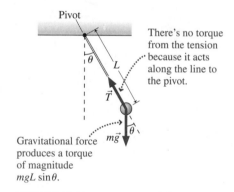

There's no torque from the tension because it acts along the line to the pivot.

Gravitational force produces a torque of magnitude $mgL\sin\theta$.

FIGURE 13.10 Forces on a pendulum.

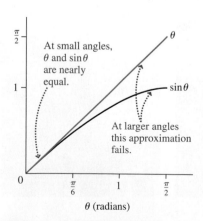

At small angles, θ and $\sin\theta$ are nearly equal.

At larger angles this approximation fails.

θ (radians)

FIGURE 13.11 For θ much less than 1 radian, $\sin\theta$ and θ are nearly equal.

The Torsional Oscillator

Figure 13.9 shows a disk suspended from a wire. Rotate the disk slightly, and a torque develops in the wire. Let go, and the disk oscillates by rotating back and forth. This is a **torsional oscillator**, and it's best described using the language of rotational motion. The **angular displacement** θ, **restoring torque** τ, and **torsional constant** κ relate the torque and displacement: $\tau = -\kappa\theta$, where again the minus sign indicates that the torque is opposite the displacement, tending to restore the system to equilibrium. The rotational analog of Newton's law, $\tau = I\alpha$, describes the system's behavior; here the rotational inertia I plays the role of mass. But the angular acceleration α is the second derivative of the angular position, so Newton's law becomes

$$I\frac{d^2\theta}{dt^2} = -\kappa\theta \tag{13.11}$$

This is identical to Equation 13.3 for the linear oscillator, with I replacing m, θ replacing x, and κ replacing k. So we can immediately write $\theta(t) = A\cos\omega t$ for the angular displacement and, in analogy with Equation 13.7a,

$$\omega = \sqrt{\frac{\kappa}{I}} \tag{13.12}$$

for the angular frequency.

Torsional oscillators constitute the timekeeping mechanism in mechanical watches, and they can provide accurate measures of rotational inertia.

The Pendulum

A **simple pendulum** consists of a point mass suspended from a massless string. Real systems approximate this ideal when a suspended object's size is negligible compared with the suspension length and its mass is much greater than that of the suspension. The dancers in the chapter's opening photo are essentially simple pendulums, as is the pendulum in a grandfather clock. Figure 13.10 shows a pendulum of mass m and length L displaced slightly from equilibrium. The gravitational force exerts a torque given by $\tau = -mgL\sin\theta$, where the minus sign indicates that the torque tends to rotate the pendulum back toward equilibrium. The rotational analog of Newton's law, $\tau = I\alpha$, then becomes

$$I\frac{d^2\theta}{dt^2} = -mgL\sin\theta$$

where we've written the angular acceleration as the second derivative of the angular displacement. This looks like Equation 13.11 for the torsional oscillator—but not quite, since the torque involves $\sin\theta$ rather than θ itself. Thus the restoring torque is not *directly* proportional to the angular displacement, and the motion is therefore *not* simple harmonic.

If, however, the amplitude of the motion is small, then it *approximates* simple harmonic motion. Figure 13.11 shows that for small angles, $\sin\theta$ and θ are essentially equal. For a small-amplitude pendulum we can therefore replace $\sin\theta$ with θ to get

$$I\frac{d^2\theta}{dt^2} = -mgL\theta$$

This is essentially Equation 13.11, with mgL playing the role of κ. So the small-amplitude pendulum undergoes simple harmonic motion, with its angular frequency given by Equation 13.12 with $\kappa = mgL$:

$$\omega = \sqrt{\frac{mgL}{I}} \tag{13.13}$$

For a *simple* pendulum, the rotational inertia I is that of a point mass m a distance L from the rotation axis, or $I = mL^2$, as we found in Chapter 10. Then we have

$$\omega = \sqrt{\frac{mgL}{mL^2}} = \sqrt{\frac{g}{L}} \quad \text{(simple pendulum)} \tag{13.14}$$

or, from Equation 13.5,

$$T = \frac{2\pi}{\omega} = 2\pi\sqrt{\frac{L}{g}} \quad \text{(simple pendulum)} \tag{13.15}$$

These equations show that the frequency and period of a simple pendulum are independent of its mass, depending only on length and gravitational acceleration.

EXAMPLE 13.3 **A Pendulum: Rescuing Tarzan**

Tarzan stands on a branch as a leopard threatens. Fortunately, Jane is on a nearby branch of the same height, holding a 25-m-long vine attached directly above the point midway between her and Tarzan. She grasps the vine and steps off with negligible velocity. How soon does she reach Tarzan?

INTERPRET This is a problem about a pendulum, which we identify as consisting of Jane and the vine. The period of the pendulum is the time for a full swing back and forth, so the answer we're after—the time to reach Tarzan—is half the period.

DEVELOP We sketched the situation in Fig. 13.12. Equation 13.15, $T = 2\pi\sqrt{L/g}$, determines the period, so we can use this equation to find the half-period.

EVALUATE Equation 13.15 gives

$$\frac{1}{2}T = \left(\frac{1}{2}\right)(2\pi)\sqrt{\frac{L}{g}} = (\pi)\sqrt{\frac{25\text{ m}}{9.8\text{ m/s}^2}} = 5.0\text{ s}$$

FIGURE 13.12 Our sketch for Example 13.3. Vine length is not to scale.

ASSESS This seems a reasonable answer for a problem involving human-scale objects and many meters of vine. One caution: Jane's rescue will be successful only if the vine is strong enough—not only to support her weight but also to provide the acceleration that keeps her moving in a circular arc. You can explore that issue in Problem 54. ∎

GOT IT? 13.2 What happens to the period of a pendulum if (a) its mass is doubled; (b) it's moved to a planet whose gravitational acceleration is one-fourth that of Earth; and (c) its length is quadrupled?

The Physical Pendulum

A **physical pendulum** is an object of arbitrary shape that's free to swing (Fig. 13.13). It differs from a simple pendulum in that mass may be distributed over its entire length. Physical pendulums are everywhere: Examples include the legs of humans and other animals (see Example 13.4), a skier on a chair lift, a boxer's punching bag, a frying pan hanging from a rack, and a crane lifting any object of significant extent. In our analysis of the simple pendulum, we used the fact that mass was concentrated at the bottom only in the final step, when we wrote mL^2 for the rotational inertia. Our analysis before that step therefore applies to the physical pendulum as well.

In particular, a physical pendulum displaced slightly from equilibrium undergoes simple harmonic motion with frequency given by Equation 13.13. But how are we to interpret the length L in that equation? Because gravity—which provides the restoring torque for *any* pendulum—acts on an object's center of gravity, L must be the distance from the pivot to the center of gravity, as marked in Fig. 13.13.

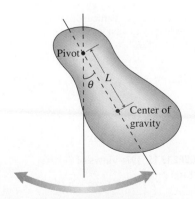

FIGURE 13.13 A physical pendulum.

EXAMPLE 13.4   A Physical Pendulum: Walking

When walking, the leg not in contact with the ground swings forward, acting like a physical pendulum. Approximating the leg as a uniform rod, find the period of this pendulum motion for a leg of length 90 cm.

INTERPRET This problem is about a physical pendulum, here identified as a uniform rod approximating the leg.

DEVELOP Figure 13.14 is our drawing, showing the leg as a rod pivoting at the hip. The center of mass of a uniform rod is at its center, so the effective length L is half the leg's length, or 45 cm. Equation 13.13, $\omega = \sqrt{mgL/I}$, determines the angular frequency, from which we can get the period using Equation 13.5, $T = 2\pi/\omega$. We also need the rotational inertia; from Table 10.2, that's $I = \frac{1}{3}M(2L)^2$, where we use $2L$ because Table 10.2's expression involves the *full* length of the rod.

EVALUATE Putting this all together, we evaluate to get the answer:

$$T = \frac{2\pi}{\omega} = 2\pi\sqrt{\frac{I}{mgL}} = 2\pi\sqrt{\frac{\frac{1}{3}m(2L)^2}{mgL}} = 2\pi\sqrt{\frac{4L}{3g}}$$

Using $L = 0.45$ m gives $T = 1.6$ s.

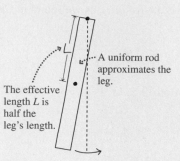

FIGURE 13.14 A human leg treated as a physical pendulum pivoting at the hip.

ASSESS The leg swings forward to complete a full stride in half a period, or 0.8 s. This seems a reasonable value for the pace in walking. ∎

13.4 Circular and Harmonic Motion

Look down on the solar system, and you see Earth in circular motion about the Sun (Fig. 13.15a). But look in from the plane of Earth's orbit, and Earth appears to be moving back and forth (Fig. 13.15b). Figure 13.16 shows that this apparent back-and-forth motion is a single component of the actual circular motion, and that this component describes a sinusoidal function of time. Specifically, the position vector $\vec{r}$ for Earth or any other object in circular motion makes an angle that increases linearly with time: $\theta = \omega t$, where we measure θ with respect to the x axis and take $t = 0$ when the object is on the x axis. Then the two components $x = r\cos\theta$ and $y = r\sin\theta$ of the object's position become

$$x(t) = r\cos\omega t \quad \text{and} \quad y(t) = r\sin\omega t$$

These are the equations for two different simple harmonic motions, one in the x direction and the other in the y direction. Because one is a cosine and the other is a sine, they're out of phase by $\frac{\pi}{2}$ or 90°.

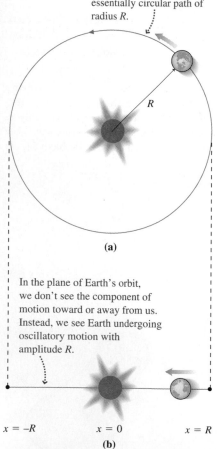

FIGURE 13.15 Two views of Earth's orbital motion.

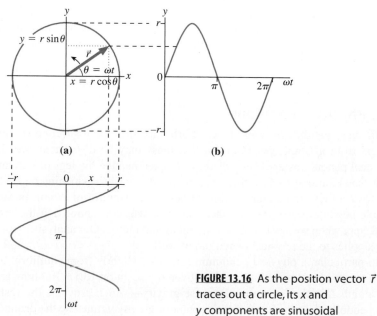

FIGURE 13.16 As the position vector $\vec{r}$ traces out a circle, its x and y components are sinusoidal functions of time.

So we can think of circular motion as resulting from perpendicular simple harmonic motions, with the same amplitude and frequency but 90° out of phase. This should help you to understand why we use the term "angular frequency" for simple harmonic motion even though no angle is involved. The argument ωt in the description of simple harmonic motion is the same as the physical angle θ in the corresponding circular motion. The time for one cycle of simple harmonic motion is the same as the time for one revolution in the circular motion, so the values of T and therefore ω are exactly the same.

You can verify that two mutually perpendicular simple harmonic motions of the same amplitude and frequency sum vectorially to give circular motion (see Problem 52). If the amplitudes or frequencies aren't the same, then interesting complex motions occur, as shown in Fig. 13.17.

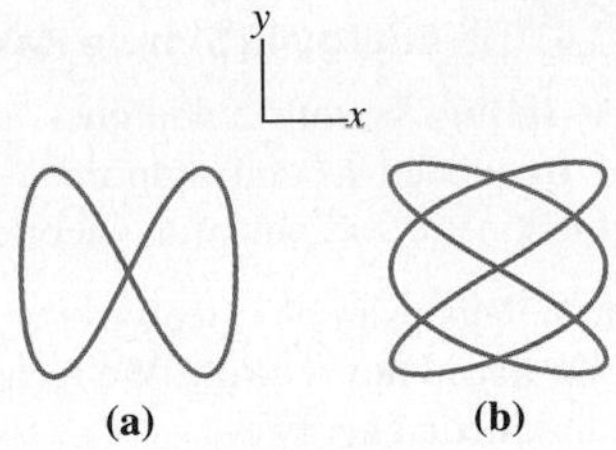

GOT IT? 13.3 Figure 13.17 shows the paths traced out in the horizontal plane by two pendulums swinging with different frequencies in two perpendicular directions. What is the ratio of x-direction frequency to y-direction frequency for the two paths shown?

FIGURE 13.17 Complex paths resulting from different frequencies in different directions. Can you determine the frequency ratios?

13.5 Energy in Simple Harmonic Motion

Displace a mass-spring system from equilibrium, and you do work as you build up potential energy in the spring. Release the mass, and it accelerates toward equilibrium, gaining kinetic energy at the expense of potential energy. It passes through its equilibrium position with maximum kinetic energy and no potential energy, then slows and builds potential energy as it compresses the spring. If there's no energy loss, this process continues indefinitely. In oscillatory motion, energy is continuously transferred back and forth between its kinetic and potential forms (Fig. 13.18).

For a mass-spring system, the potential energy is given by Equation 7.4: $U = \frac{1}{2}kx^2$, where x is the displacement from equilibrium. Meanwhile, the kinetic energy is $K = \frac{1}{2}mv^2$. We can illustrate explicitly the interchange of kinetic and potential energy in simple harmonic motion by using x from Equation 13.4 and v from Equation 13.9 in the expressions for potential and kinetic energy. Then we have

$$U = \tfrac{1}{2}kx^2 = \tfrac{1}{2}k(A\cos\omega t)^2 = \tfrac{1}{2}kA^2\cos^2\omega t$$

and

$$K = \tfrac{1}{2}mv^2 = \tfrac{1}{2}m(-\omega A\sin\omega t)^2 = \tfrac{1}{2}m\omega^2A^2\sin^2\omega t = \tfrac{1}{2}kA^2\sin^2\omega t$$

where we used $\omega^2 = k/m$. Both energy expressions have the same maximum value—$\frac{1}{2}kA^2$—equal to the initial potential energy of the stretched spring. But the potential energy is a maximum when the kinetic energy is zero, and vice versa. What about the total energy? It's

$$E = U + K = \tfrac{1}{2}kA^2\cos^2\omega t + \tfrac{1}{2}kA^2\sin^2\omega t = \tfrac{1}{2}kA^2$$

where we used $\sin^2\omega t + \cos^2\omega t = 1$.

Our result is a statement of the conservation of mechanical energy—the principle we introduced in Chapter 7—applied to a simple harmonic oscillator. Although the kinetic and potential energies K and U both vary with time, their sum—the total energy E—does not (Fig. 13.19).

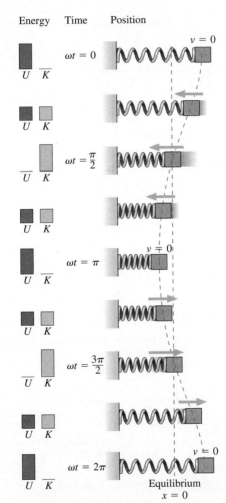

FIGURE 13.18 Kinetic and potential energy in simple harmonic motion.

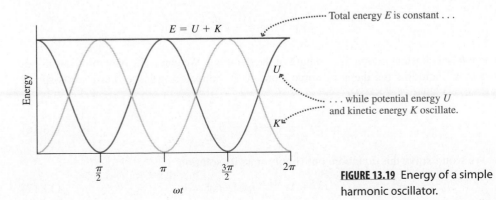

FIGURE 13.19 Energy of a simple harmonic oscillator.

EXAMPLE 13.5 **Energy in Simple Harmonic Motion**

A mass-spring system undergoes simple harmonic motion with angular frequency ω and amplitude A. Find its speed at the point where the kinetic and potential energies are equal.

INTERPRET This example involves the concept of energy conservation in simple harmonic motion. We're asked to find a speed, which is related to kinetic energy.

DEVELOP When the kinetic energy equals the potential energy, each must be half the total energy. What is that total? The speed is at its maximum, $v_{\text{max}} = \omega A$ from Equation 13.9, when the energy is all kinetic. Thus the total energy is $E = \frac{1}{2}mv_{\text{max}}^2 = \frac{1}{2}m\omega^2 A^2$. The speed v

we're after occurs when the kinetic energy has half this value, or $K = \frac{1}{2}mv^2 = \frac{1}{2}\left(\frac{1}{2}m\omega^2 A^2\right) = \frac{1}{4}m\omega^2 A^2$.

EVALUATE Solving for v gives our answer:

$$v = \frac{\omega A}{\sqrt{2}}$$

ASSESS Make sense? Yes. The speed at this point must obviously be less than the maximum speed, since half the energy is tied up as potential energy in the spring. And because kinetic energy depends on the *square* of the speed, it's lower not by a factor of 2 but by $\sqrt{2}$. ∎

Potential-Energy Curves and Simple Harmonic Motion

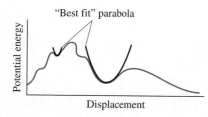

FIGURE 13.20 Near their minima, potential-energy curves often approximate parabolas; the result is simple harmonic motion.

We arrived at the expression $U = \frac{1}{2}kx^2$ for the potential energy of a spring by integrating the spring force, $-kx$, over distance. Since every simple harmonic oscillator has a restoring force or torque proportional to displacement, integration always results in a potential energy proportional to the *square* of the displacement—that is, in a parabolic potential-energy curve. Conversely, any system with a parabolic potential-energy curve exhibits simple harmonic motion. The simplest mathematical approximation to a smooth curve is a parabola, and for that reason potential-energy curves for complex systems often approximate parabolas near their stable equilibrium points (Fig. 13.20). Small disturbances from these equilibria therefore result in simple harmonic motion, and that's why simple harmonic motion is so common throughout the physical world.

GOT IT? 13.4 Two different mass-spring systems are oscillating with the same amplitude and frequency. If one has twice as much total energy as the other, how do (a) their masses and (b) their spring constants compare? (c) What about their maximum speeds?

13.6 Damped Harmonic Motion

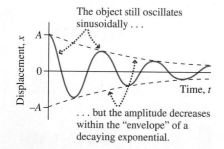

FIGURE 13.21 Weakly damped motion.

In real oscillating systems, forces such as friction or air resistance normally dissipate the oscillation energy. This energy loss causes the oscillation amplitude to decrease, and the motion is said to be **damped**.

If dissipation is sufficiently weak that only a small fraction of the system's energy is lost in each oscillation cycle, then we expect that the system should behave essentially as in the undamped case, except for a gradual decrease in amplitude (Fig. 13.21).

In many systems the damping force is approximately proportional to the velocity and in the opposite direction:

$$F_{\text{d}} = -bv = -b\frac{dx}{dt}$$

where b is a constant giving the strength of the damping. We can write Newton's law as before, now including the damping force along with the restoring force. For a mass-spring system, we have

$$m\frac{d^2x}{dt^2} = -kx - b\frac{dx}{dt} \tag{13.16}$$

We won't solve this equation, but simply state its solution:

$$x(t) = Ae^{-bt/2m}\cos(\omega t + \phi) \tag{13.17}$$

This equation describes sinusoidal motion whose amplitude decreases exponentially with time. How fast depends on the damping constant b and mass m: When $t = 2m/b$, the amplitude has dropped to $1/e$ of its original value. When the damping is so weak that only a small fraction of the total energy is lost in each cycle, the frequency ω in Equation 13.17 is essentially equal to the undamped frequency $\sqrt{k/m}$. But with stronger damping, the damping force slows the motion, and the frequency becomes lower. As long as oscillation occurs, the motion is said to be **underdamped** (Fig. 13.22a). For sufficiently strong damping, though, the effect of the damping force is as great as that of the spring force. Under this condition, called **critical damping**, the system returns to its equilibrium state without undergoing any oscillations (Fig. 13.22b). If the damping is made still stronger, the system becomes *overdamped*. The damping force now dominates, so the system returns more slowly to equilibrium (Fig. 13.22c).

Many physical systems, from atoms to the human leg, can be modeled as damped oscillators. Engineers often design systems with specific amounts of damping. Automobile shock absorbers, for example, coordinate with the springs to give critical damping. This results in rapid return to equilibrium while absorbing the energy imparted by road bumps.

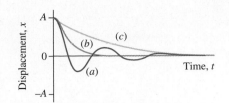

FIGURE 13.22 (a) Underdamped, (b) critically damped, and (c) overdamped oscillations.

EXAMPLE 13.6 **Damped Simple Harmonic Motion: Bad Shocks**

A car's suspension acts like a mass-spring system with $m = 1200$ kg and $k = 58$ kN/m. Its worn-out shock absorbers provide a damping constant $b = 230$ kg/s. After the car hits a pothole, how many oscillations will it make before the amplitude drops to half its initial value?

INTERPRET We interpret this problem as being about damped simple harmonic motion, and we identify the car as the oscillating system.

DEVELOP Our plan is to find out how long it takes the amplitude to decrease by half and then find the number of oscillation cycles in this time. Equation 13.17, $x(t) = Ae^{-bt/2m}\cos(\omega t + \phi)$, describes the motion, with the factor $e^{-bt/2m}$ giving the decrease in amplitude. At $t = 0$ this factor is 1, so we want to know when it's equal to one-half: $e^{-bt/2m} = \frac{1}{2}$.

EVALUATE Taking the natural logarithms of both sides gives $bt/2m = \ln 2$, where we used the facts that $\ln(x)$ and e^x are inverse functions and $\ln(1/x) = -\ln(x)$. Then

$$t = \frac{2m}{b}\ln 2 = \frac{(2)(1200\ \text{kg})}{230\ \text{kg/s}}\ln 2 = 7.23\ \text{s}$$

is the time for the amplitude to drop to half its original value. For weak damping, the period is very close to the undamped period, which is

$$T = 2\pi\sqrt{\frac{m}{k}} = 2\pi\sqrt{\frac{1200\ \text{kg}}{58 \times 10^3\ \text{N/m}}} = 0.904\ \text{s}$$

Then the number of cycles during the 7.23 s it takes the amplitude to drop in half is

$$\frac{7.23\ \text{s}}{0.904\ \text{s}} = 8$$

ASSESS That the number of oscillations is much greater than 1 tells us that the damping is weak, justifying our use of the undamped period. It also tells us that those are really bad shocks! ∎

13.7 Driven Oscillations and Resonance

Pushing a child on a swing, you can build up a large amplitude by giving a relatively small push once each oscillation cycle. If your pushing were not in step with the swing's natural oscillatory motion, then the same force would have little effect.

When an external force acts on an oscillatory system, we say that the system is **driven**. Consider a mass-spring system, which you might drive as suggested in Fig. 13.23. Suppose the driving force is given by $F_0\cos\omega_\text{d}t$, where ω_d is called the **driving frequency**. Then Newton's law is

$$m\frac{d^2x}{dt^2} = -kx - b\frac{dx}{dt} + F_0\cos\omega_\text{d}t \qquad (13.18)$$

where the first term on the right-hand side is the restoring force, the second the damping force, and the third the driving force. Since the system is being driven at the frequency ω_d, we expect it to undergo oscillatory motion at this frequency. So we guess that the solution to Equation 13.18 might have the form

$$x = A\cos(\omega_\text{d}t + \phi)$$

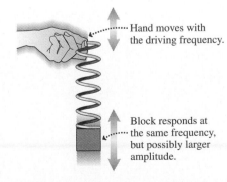

FIGURE 13.23 Driving a mass-spring system results in a large amplitude if the driving frequency is near the natural frequency $\sqrt{k/m}$.

Substituting this expression and its derivatives into Equation 13.18 (see Problem 75) shows that the equation is satisfied if

$$A(\omega) = \frac{F_0}{m\sqrt{(\omega_d^2 - \omega_0^2)^2 + b^2\omega_d^2/m^2}} \qquad (13.19)$$

where ω_0 is the undamped **natural frequency** $\sqrt{k/m}$, as distinguished from the driving frequency ω_d.

Figure 13.24 shows **resonance curves**—plots of Equation 13.19 as a function of driving frequency—for several values of the damping constant. As long as the system is underdamped, the curve has a maximum at some nonzero frequency, and for weak damping, that maximum occurs at very nearly the natural frequency. The weaker the damping, the more sharply peaked is the resonance curve. Thus, in weakly damped systems, it's possible to build up large-amplitude oscillations with relatively small driving forces—a phenomenon known as **resonance**.

Most physical systems, from molecules to cars, and loudspeakers to buildings and bridges, exhibit one or more natural modes of oscillation. If these oscillations are weakly damped, then the buildup of large-amplitude oscillations through resonance can cause serious problems—sometimes even destroying the system (Fig. 13.25). Engineers designing complex structures spend a lot of their time exploring all possible oscillation modes and taking steps to avoid resonance. In an earthquake-prone area, for example, a building's natural frequencies would be designed to avoid the frequency of typical earthquake motions. A loudspeaker should be engineered so its natural frequency isn't in the range of sound it's intended to reproduce. Damping systems such as the shock absorbers of Example 13.6 or the tuned mass damper of Example 13.2 help limit resonant oscillations in cases where natural frequencies aren't easily altered.

Resonance is also important in microscopic systems. The resonant behavior of electrons in a special tube called a magnetron produces the microwaves that cook food in a microwave oven; the same resonant process heats ionized gases in some experiments designed to harness fusion energy. Carbon dioxide in Earth's atmosphere absorbs infrared radiation because CO_2 molecules—acting like miniature mass-spring systems—resonate at some of the frequencies of infrared radiation. The result is the greenhouse effect, which now threatens Earth with significant climatic change. The process called nuclear magnetic resonance (NMR) uses the resonant behavior of protons to probe the structure of matter and is the basis of magnetic resonance imaging (MRI) used in medicine. In NMR, the resonance involves the natural precession frequency of the protons due to magnetic torques; we described the classic model of this process in Chapter 11.

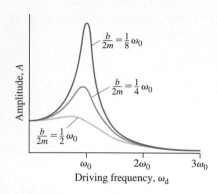

FIGURE 13.24 Resonance curves for several damping strengths; ω_0 is the undamped natural frequency $\sqrt{k/m}$.

FIGURE 13.25 Collapse of the Tacoma Narrows Bridge—only four months after its opening in 1940—followed the resonant growth of large-amplitude oscillations.

Big Picture

The big idea here is **simple harmonic motion** (SHM), oscillatory motion that is ubiquitous and that occurs whenever a disturbance from equilibrium results in a restoring force or torque that is directly proportional to the displacement. Position in SHM is a sinusoidal function of time:

$$x(t) = A\cos\omega t$$

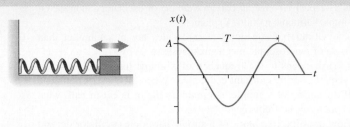

Key Concepts and Equations

Period T is the time to complete one oscillation cycle; its inverse is **frequency**, or number of oscillations per unit time:

$$f = \frac{1}{T}$$

Another measure of frequency is **angular frequency** ω, given by

$$\omega = 2\pi f = \frac{2\pi}{T}$$

Angular frequency can be understood in terms of the close relationship between circular motion and simple harmonic motion.

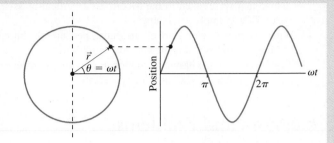

Absent friction and other dissipative forces, energy in SHM is conserved, although it's transformed back and forth between kinetic and potential forms:

$$E = \tfrac{1}{2}mv^2 + \tfrac{1}{2}kx^2 = \text{constant}$$

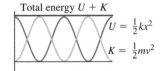

Total energy $U + K$

$U = \tfrac{1}{2}kx^2$

$K = \tfrac{1}{2}mv^2$

When dissipative forces act, the motion is **damped**. For small dissipative forces the oscillation amplitude decreases over time:

$$x(t) = Ae^{-bt/2m}\cos(\omega t + \phi)$$

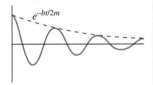

If a system is driven at a frequency near its natural oscillation frequency ω_0, then large-amplitude oscillations can build; this is **resonance**. The amplitude A depends on the driving force F_0, the driving frequency ω_d, the natural frequency $\omega_0 = \sqrt{k/m}$, and the damping constant b:

$$A(\omega) = \frac{F_0}{m\sqrt{(\omega_d^2 - \omega_0^2)^2 + b^2\omega_d^2/m^2}}$$

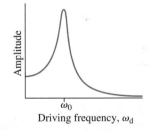

Driving frequency, ω_d

Cases and Uses

$$\omega = \sqrt{\frac{k}{m}}$$

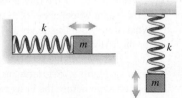

In systems involving rotational oscillations, the analogous relations involves the torsional constant and rotational inertia:

$$\omega = \sqrt{\frac{\kappa}{I}}$$

A special case is the **pendulum**, for which (with small-amplitude oscillations)

$$\omega = \sqrt{\frac{mgL}{I}}$$

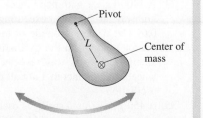

In the case of a **simple pendulum**, the angular frequency reduces to

$$\omega = \sqrt{\frac{g}{L}}$$

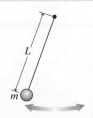

For Thought and Discussion

1. Is a vertically bouncing ball an example of oscillatory motion? Of simple harmonic motion? Explain.
2. The vibration frequencies of molecules are much higher than those of macroscopic mechanical systems. Why?
3. What happens to the frequency of a simple harmonic oscillator when the spring constant is doubled? When the mass is doubled?
4. If the spring of a simple harmonic oscillator is cut in half, what happens to the frequency?
5. How does the frequency of a simple harmonic oscillator depend on its amplitude?
6. How would the frequency of a horizontal mass-spring system change if it were taken to the Moon? Of a vertical mass-spring system? Of a simple pendulum?
7. When is the acceleration of an undamped simple harmonic oscillator zero? When is the velocity zero?
8. Explain how simple harmonic motion might be used to determine the mass of objects in an orbiting spacecraft.
9. One pendulum consists of a solid rod of mass m and length L, and another consists of a compact ball of the same mass m on the end of a massless string of the same length L. Which has the greater period? Why?
10. The x and y components of motion of a body are both simple harmonic with the same frequency and amplitude. What shape is the path of the body if the component motions are (a) in phase, (b) $\pi/2$ out of phase, and (c) $\pi/4$ out of phase?
11. Why is critical damping desirable in a car's suspension system?
12. Explain why the frequency of a damped system is lower than that of the equivalent undamped system.
13. Opera singers have been known to break glasses with their voices. How?
14. What will happen to the period of a mass-spring system if it is placed in a jetliner accelerating down the runway? What will happen to the period of a pendulum in the same situation?
15. How can a system have more than one resonant frequency?

Exercises and Problems

Exercises

Section 13.1 Describing Oscillatory Motion

16. A doctor counts 77 heartbeats in 1 minute. What are the period and frequency of the heart's oscillations?
17. A violin string playing the note A oscillates at 440 Hz. What is the period of its oscillation?
18. The vibration frequency of a hydrogen chloride molecule is 8.66×10^{13} Hz. How long does it take the molecule to complete one oscillation?
19. Write expressions for simple harmonic motion (a) with amplitude 10 cm, frequency 5.0 Hz, and maximum displacement at $t = 0$; and (b) with amplitude 2.5 cm, angular frequency 5.0 s^{-1}, and maximum velocity at $t = 0$.
20. The top of a skyscraper sways back and forth, completing 9 oscillation cycles in 1 minute. Find the period and frequency of the motion.

Section 13.2 Simple Harmonic Motion

21. A 200-g mass is attached to a spring of constant $k = 5.6$ N/m and set into oscillation with amplitude $A = 25$ cm. Determine (a) the frequency in hertz, (b) the period, (c) the maximum velocity, and (d) the maximum force in the spring.
22. An astronaut in an orbiting spacecraft is "weighed" by being strapped to a spring of constant $k = 400$ N/m and set into simple harmonic motion. If the oscillation period is 2.5 s, what is the astronaut's mass?
23. A simple model for an automobile suspension consists of a mass attached to a spring. If the mass is 1900 kg and the spring constant is 26 kN/m, with what frequency and period will the car undergo simple harmonic motion?
24. The quartz crystal in a digital quartz watch executes simple harmonic motion at 32,768 Hz. (This is 2^{15} Hz, chosen so 15 divisions by 2 give a signal at 1.00000 Hz.) If each face of the crystal undergoes a maximum displacement of 100 nm, find the maximum velocity and acceleration of the crystal faces.
25. A 50-g mass is attached to a spring and undergoes simple harmonic motion. Its maximum acceleration is 15 m/s^2 and its maximum speed is 3.5 m/s. Determine (a) the angular frequency, (b) the spring constant, and (c) the amplitude of the motion.
26. A particle undergoes simple harmonic motion with amplitude 25 cm and maximum speed 4.8 m/s. Find (a) the angular frequency, (b) the period, and (c) the maximum acceleration.
27. A particle undergoes simple harmonic motion with maximum speed 1.4 m/s and maximum acceleration 3.1 m/s^2. Find (a) the angular frequency, (b) the period, and (c) the amplitude of the motion.

Section 13.3 Applications of Simple Harmonic Motion

28. How long should you make a simple pendulum so its period is (a) 200 ms, (b) 5.0 s, and (c) 2.0 min?
29. At the heart of a grandfather clock is a simple pendulum 1.45 m long; the clock ticks each time the pendulum reaches its maximum displacement in either direction. What is the time interval between ticks?
30. A 640-g hollow ball 21 cm in diameter is suspended by a wire and is undergoing torsional oscillations at a frequency of 0.78 Hz. What is the torsional constant of the wire?
31. A meter stick is suspended from one end and set swinging. What is the period of the resulting oscillations, assuming they have small amplitude?

Section 13.4 Circular and Harmonic Motion

32. A wheel rotates at the rate of 600 rpm. Viewed from the edge, a point on the wheel appears to undergo simple harmonic motion. What are (a) the frequency in Hz and (b) the angular frequency for this SHM?
33. The x and y components of motion of a body are harmonic with frequency ratio 1.75:1. How many oscillations must each component undergo before the body returns to its initial position?

Section 13.5 Energy in Simple Harmonic Motion

34. A 1400-kg car with poor shock absorbers is bouncing down the highway at 20 m/s, executing vertical harmonic motion at 0.67 Hz. If the amplitude of the oscillations is 18 cm, what is the total energy in the oscillations? What fraction of the car's kinetic energy is this? Neglect the rotational energy of the wheels and the fact that not all the car's mass participates in the oscillation.
35. A 450-g mass on a spring is oscillating at 1.2 Hz. The total energy of the oscillation is 0.51 J. What is the amplitude?

36. A torsional oscillator of rotational inertia 1.6 kg·m² and torsional constant 3.4 N·m/rad has a total energy of 4.7 J. What are its maximum angular displacement and maximum angular speed?

Sections 13.6 and 13.7 Damped Harmonic Motion and Resonance

37. The vibration of a piano string can be described by an equation analogous to Equation 13.17. If the quantity analogous to $b/2m$ in that equation has the value 2.8 s⁻¹, how long will it take the vibration amplitude to drop to half its original value?

38. A mass-spring system has $b/m = \omega_0/5$, where b is the damping constant and ω_0 the natural frequency. How does its amplitude when driven at frequencies 10% above and below ω_0 compare with its amplitude at ω_0?

39. A car's front suspension has a natural frequency of 0.45 Hz. The car's front shock absorbers are worn out, so they no longer provide critical damping. The car is driving on a bumpy road with bumps 40 m apart. At a certain speed, the driver notices that the car begins to shake violently. What speed?

Problems

40. A simple model of a carbon dioxide (CO_2) molecule consists of three mass points (the atoms) connected by two springs (electric forces), as suggested in Fig. 13.26. One way this system can oscillate is if the carbon atom stays fixed and the two oxygens move symmetrically on either side of it. If the frequency of this oscillation is 4.0×10^{13} Hz, what is the effective spring constant? The mass of an oxygen atom is 16 u.

FIGURE 13.26 Problem 40

41. Two identical mass-spring systems consist of 430-g masses on springs of constant $k = 2.2$ N/m. Both are displaced from equilibrium, and the first is released at time $t = 0$. How much later should the second be released so the two oscillations differ in phase by $\pi/2$?

42. A mass m slides along a frictionless horizontal surface at speed v_0. It strikes a spring of constant k attached to a rigid wall, as shown in Fig. 13.27. After a completely elastic encounter with the spring, the mass heads back in the direction it came from. In terms of k, m, and v_0, determine (a) how long the mass is in contact with the spring and (b) the maximum compression of the spring.

FIGURE 13.27 Problem 42

43. Show by substitution that $x(t) = A \sin \omega t$ is a solution to Equation 13.3.

44. A physics student, bored by a lecture on simple harmonic motion, idly picks up his pencil (mass 9.2 g, length 17 cm) by the tip with his frictionless fingers, and allows it to swing back and forth with small amplitude. If the pencil completes 6279 full cycles during the lecture, how long does the lecture last?

45. A pendulum of length L is mounted in a rocket. What is its period if the rocket is (a) at rest on its launch pad; (b) accelerating upward with acceleration $a = \frac{1}{2}g$; (c) accelerating downward with acceleration $a = \frac{1}{2}g$; and (d) in free fall?

46. A 340-g mass is attached to a vertical spring and lowered slowly until it rests at a new equilibrium position, which is 30 cm below the spring's original equilibrium. The system is then set into simple harmonic motion. What is the period of the motion?

47. A mass is attached to a vertical spring, which then goes into oscillation. At the high point of the oscillation, the spring is in the original unstretched equilibrium position it had before the mass was attached; the low point is 5.8 cm below this. What is the period of oscillation?

48. Derive the period of a simple pendulum by considering the horizontal displacement x and the force acting on the bob, rather than the angular displacement and torque.

49. A solid disk of radius R is suspended from a spring of linear spring constant k and torsional constant κ, as shown in Fig. 13.28. In terms of k and κ, what value of R will give the same period for the vertical and torsional oscillations of this system?

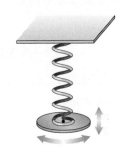

FIGURE 13.28 Problem 49

50. A thin steel beam 8.0 m long is suspended from a crane and is undergoing torsional oscillations. Two 75-kg steelworkers leap onto opposite ends of the beam, as shown in Fig. 13.29. If the frequency of torsional oscillations diminishes by 20%, what is the mass of the beam?

FIGURE 13.29 Problem 50

51. A cyclist turns her bicycle upside down to tinker with it. After she gets it upside down, she notices the front wheel executing a slow, small-amplitude, back-and-forth rotational motion with a period of 12 s. Considering the wheel to be a thin ring of mass 600 g and radius 30 cm, whose only irregularity is the presence of the tire valve stem, determine the mass of the valve stem.

52. An object undergoes simple harmonic motion in two mutually perpendicular directions, its position given by $\vec{r} = A \sin \omega t\, \hat{\imath} + A \cos \omega t\, \hat{\jmath}$. (a) Show that the object remains a fixed distance from the origin (i.e., that its path is circular), and find that distance. (b) Find an expression for the object's velocity. (c) Show that the speed remains constant, and find its value. (d) What is the angular speed of the object in its circular path?

53. A pendulum consists of a 320-g solid ball 15.0 cm in diameter, suspended by an essentially massless string 80.0 cm long. Calculate the period of this pendulum, treating it first as a simple pendulum and then as a physical pendulum. How much error is introduced by the simple pendulum approximation? *Hint:* Remember the parallel-axis theorem.

54. If Jane and Tarzan are initially 8.0 m apart in Fig. 13.12, and Jane's mass is 60 kg, what is the maximum tension in the vine, and at what point does it occur?

55. A thin, uniform hoop of mass M and radius R is suspended from a thin horizontal rod and set oscillating with small amplitude, as shown in Fig. 13.30. Show that the period of the oscillations is $2\pi\sqrt{2R/g}$. *Hint:* You may find the parallel-axis theorem useful.

FIGURE 13.30 Problem 55

56. A mass m is mounted between two springs of constants k_1 and k_2, as shown in Fig. 13.31. Show that the angular frequency of oscillation is given by $\omega = \sqrt{(k_1 + k_2)/m}$.

FIGURE 13.31 Problem 56

57. The equation for an ellipse is

$$\frac{x^2}{a^2} + \frac{y^2}{b^2} = 1$$

Show that two-dimensional simple harmonic motion whose two components have different amplitudes and are $\pi/2$ out of phase gives rise to elliptical motion. How are a and b related to the amplitudes?

58. Show that the potential energy of a simple pendulum is proportional to the square of the angular displacement in the small-amplitude limit.

59. The total energy of a mass-spring system is the sum of its kinetic and potential energy: $E = \frac{1}{2}mv^2 + \frac{1}{2}kx^2$. Assuming E remains constant, differentiate both sides of this expression with respect to time and show that Equation 13.3 results. *Hint:* Remember that $v = dx/dt$.

60. A solid cylinder of mass M and radius R is mounted on an axle through its center. The axle is attached to a horizontal spring of constant k, and the cylinder rolls back and forth without slipping (Fig. 13.32). Write the statement of energy conservation for this system, and differentiate it to obtain an equation analogous to Equation 13.3 (see Problem 59). Comparing your result with Equation 13.3, determine the angular frequency of the motion.

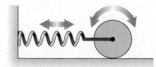

FIGURE 13.32 Problem 60

61. A mass m is free to slide on a frictionless track whose height y as a function of horizontal position x is given by $y = ax^2$, where a is a constant with the units of inverse length. The mass is given an initial displacement from the bottom of the track and then released. Find an expression for the period of the resulting motion.

62. A 250-g mass is mounted on a spring of constant $k = 3.3$ N/m. The damping constant for this system is $b = 8.4 \times 10^{-3}$ kg/s. How many oscillations will the system undergo during the time the amplitude decays to $1/e$ of its original value?

63. A harmonic oscillator is underdamped provided that the damping constant b is less than $\sqrt{2}m\omega_0$, where ω_0 is the natural frequency of undamped motion. Show that for an underdamped oscillator, Equation 13.19 has a maximum at a driving frequency less than ω_0.

64. A massless spring of spring constant $k = 74$ N/m is hanging from the ceiling. A 490-g mass is hooked onto the unstretched spring and allowed to drop. Find (a) the amplitude and (b) the period of the resulting motion.

65. A meter stick is suspended from a frictionless rod inserted through a small hole at the 25-cm mark. Find the period of small-amplitude oscillations about the stick's equilibrium position.

66. Two balls each of unknown mass m are mounted on opposite ends of a 1.5-m-long rod of mass 850 g. The system is suspended from a wire attached to the center of the rod and set into torsional oscillations. If the torsional constant of the wire is 0.63 N·m/rad and the period of the oscillations is 5.6 s, what is the unknown mass m?

67. Two mass-spring systems with the same mass are undergoing oscillatory motion with the same amplitudes. System 1 has twice the frequency of system 2. How do (a) their energies and (b) their maximum accelerations compare?

68. Two mass-spring systems have the same mass and the same total energy. The amplitude of system 1 is twice that of system 2. How do (a) their frequencies and (b) their maximum accelerations compare?

69. A 500-g mass is suspended from a thread 45 cm long that can sustain a maximum tension of 6.0 N before breaking. What is the maximum allowable amplitude for pendulum motion of this system?

70. A 500-g block on a frictionless, horizontal surface is attached to a rather limp spring of constant $k = 8.7$ N/m. A second block rests on the first, and the whole system executes simple harmonic motion with a period of 1.8 s. When the amplitude of the motion is increased to 35 cm, the upper block just begins to slip. What is the coefficient of static friction between the blocks?

71. Repeat Problem 61 for a small solid ball of mass M and radius R that rolls without slipping on the parabolic track.

72. The magnitude of the gravitational acceleration *inside* Earth is given approximately by $g(r) = g_0(r/R_E)$, where g_0 is the surface value, r is the distance from Earth's center, and R_E is Earth's radius; the acceleration is directed toward Earth's center. Suppose a narrow hole were drilled straight through the center of Earth and out the other side. Neglecting air resistance, show that an object dropped into this hole executes simple harmonic motion, and find an expression for the period. Evaluate and compare with the period of a satellite in a circular orbit not far above Earth's surface.

73. A 1.2-kg block rests on a frictionless surface and is attached to a horizontal spring of constant $k = 23$ N/m (Fig. 13.33). The block is oscillating with amplitude 10 cm and with phase constant $\phi = -\pi/2$. A block of mass 0.80 kg is moving from the right at 1.7 m/s. It strikes the first block when the latter is at the rightmost point in its oscillation. The collision is completely inelastic, and the two blocks stick together. Determine the frequency, amplitude, and phase constant (relative to the *original* $t = 0$) of the resulting motion.

FIGURE 13.33 Problem 73

74. A disk of radius R is suspended from a pivot somewhere between its center and edge (Fig. 13.34). For what pivot point will the period of this physical pendulum be a minimum?

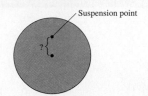

Suspension point

FIGURE 13.34 Problem 74

75. Show by direct substitution that $x = A\cos(\omega_d t + \phi)$ satisfies Equation 13.18 with A given by Equation 13.19.
76. Suppose that the potential energy, $U(x)$, for a particle of mass m, has a minimum at $x = a$. Find an expression for the angular frequency of the simple harmonic motion for small oscillations about $x = a$ in terms of the potential energy and its derivatives.
77. Show that $x(t) = a\cos\omega t - b\sin\omega t$ represents simple harmonic motion, as in Equation 13.8, with $A = \sqrt{a^2 + b^2}$ and $\phi = \tan^{-1}(b/a)$.
78. You've come up with a clever means of determining the mass of birds. As shown in Fig. 13.35, a bird feeder consists of a disk 50 cm in diameter suspended by a wire at the center. Two birds sit on opposite sides and the feeder goes into torsional oscillation at 2.6 Hz. The torsional constant of the wire is 5.00 N·m/rad. Assuming each bird has the same mass, what is it?

340 g

50 cm

FIGURE 13.35 Problem 78

79. While waiting for your plane to take off, you suspend your keys from a thread and set the resulting pendulum oscillating. It completes exactly 90 cycles in 1 minute. You repeat the experiment as the plane accelerates down the runway, and now find the pendulum completes exactly 91 cycles in 1 minute. Find the plane's acceleration.
80. You have devised a simple means of measuring the rotational inertia of a child on a swing. A child twirls around in a swing, twisting the ropes as shown in Figure 13.36. As a result the child and swing rise slightly, with the rise h in cm equal to the square of the number of full turns. When the child stops twisting, the swing begins torsional oscillation. You measure the period at 6.91 s. The mass of the child and swing is 20 kg. What is the rotational inertia?

h

FIGURE 13.36 Problem 80

Answers to Chapter Questions

Answer to Chapter Opening Question
The dancers undergo pendulum motion, whose period is determined entirely by their rope lengths and the acceleration of gravity.

Answers to GOT IT? Questions
13.1 Frequencies and periods are the same; amplitudes and phase constants are different because of the different initial displacements and times of release, respectively.
13.2 (a) No change; (b) doubles; (c) doubles.
13.3 (a) 1:2; (b) 3:2.
13.4 The more energetic oscillator has (a) twice the mass and (b) twice the spring constant. (c) Their maximum speeds are equal.

14 Wave Motion

■ Ocean waves travel thousands of kilometers across the open sea before breaking on shore. How much water moves with the waves?

Humans and other animals communicate using sound waves. Light and related waves enable us to visualize our surroundings and provide virtually all our information about the universe beyond Earth. Our cell phones keep us connected via radio waves. Physicians probe our bodies with ultrasound waves. Radio waves connect our wireless laptops to the Internet and cook the food in our microwave ovens. Earthquakes trigger waves in the solid Earth and may generate dangerous ocean tsunamis. **Wave motion** is an essential feature of our physical environment.

All these examples involve a disturbance that moves or **propagates** through space. The disturbance carries energy, but not matter. Air doesn't move from your mouth to a listener's ear, but sound energy does. Water doesn't move across the open ocean, but wave energy does. So, **a wave is a traveling disturbance that transports energy but not matter.**

14.1 Waves and Their Properties

In this chapter we'll deal with **mechanical waves**, which are disturbances of some material **medium**, such as air, water, a violin string, or Earth's interior. Visible and infrared light waves, radio waves, ultraviolet and X rays, in contrast, are **electromagnetic waves**. They share many properties with mechanical waves, but they don't require a material medium. We'll treat electromagnetic waves in Chapters 29–32.

Mechanical waves occur when a disturbance in one part of a medium is communicated to adjacent parts. Figure 14.1 shows a multiple mass-spring system that serves as a model for many types of mechanical waves. Disturb one mass, and it goes into simple harmonic motion. But because the masses are connected, that motion is communicated to the adjacent mass. As a result, both the disturbance and its associated energy propagate along the mass-spring system, disturbing successive masses as they go.

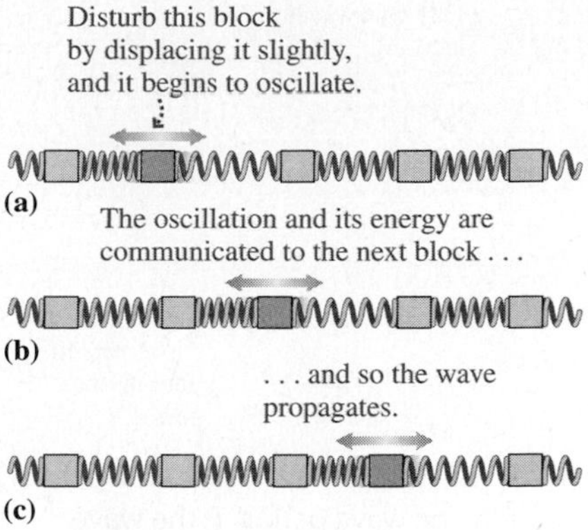

✓ **TIP** Wave Motions

A wave moves energy from place to place, but not matter. However, that doesn't mean that the matter making up the wave medium doesn't move. It does, undergoing localized oscillatory motion as the wave passes. But once the wave is gone, the disturbed matter returns to its equilibrium state. Don't confuse this localized motion of the medium with the motion of the wave itself. Both occur, but only the latter carries energy from one place to another.

Disturb this block by displacing it slightly, and it begins to oscillate.

(a) The oscillation and its energy are communicated to the next block . . .

(b) . . . and so the wave propagates.

(c)

FIGURE 14.1 Wave propagation in a mass-spring system.

Longitudinal and Transverse Waves

In Fig. 14.1, we disturbed the system by displacing one block so its subsequent oscillations were back and forth along the structure—in the same direction as the wave propagation. The result is a **longitudinal wave**. Sound is a longitudinal wave, as we'll see in Section 14.4. We could equally well displace a mass at right angles, as in Fig. 14.2. Then we get a **transverse wave**, whose disturbance is at right angles to the wave propagation. Some waves include both longitudinal and transverse motions, as shown for a water wave in Fig. 14.3.

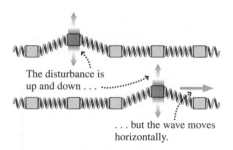

The disturbance is up and down . . .

. . . but the wave moves horizontally.

FIGURE 14.2 A transverse wave.

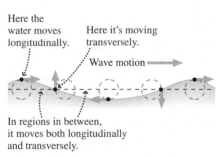

Here the water moves longitudinally.

Here it's moving transversely.

Wave motion ⟶

In regions in between, it moves both longitudinally and transversely.

FIGURE 14.3 A water wave has both longitudinal and transverse components.

Wave Amplitude

The maximum value of a wave's disturbance is the wave **amplitude**. For a water wave, amplitude is the maximum height above the undisturbed water level; for a sound wave, it's the maximum excess air pressure; for the waves of Figs. 14.1 and 14.2, it's the maximum displacement of a mass.

Wave Shape

Wave disturbances come in many shapes, called **waveforms** (Fig. 14.4). An isolated disturbance is a **pulse**, which occurs when the medium is disturbed only briefly. A **continuous wave** results from an ongoing periodic disturbance. Intermediate between these extremes is a **wave train**, resulting from a periodic disturbance lasting a finite time.

(a)

(b)

(c)

FIGURE 14.4 (a) A pulse, (b) a continuous wave, and (c) a wave train.

Wavelength, Period, and Frequency

A continuous wave repeats in both space and time. The **wavelength** λ is the *distance* over which the wave pattern repeats (Fig. 14.5). The wave **period** T is the *time* for one complete oscillation. The **frequency** f, or number of wave cycles per unit time, is the inverse of the period.

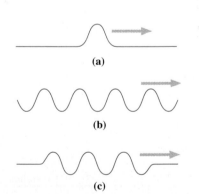

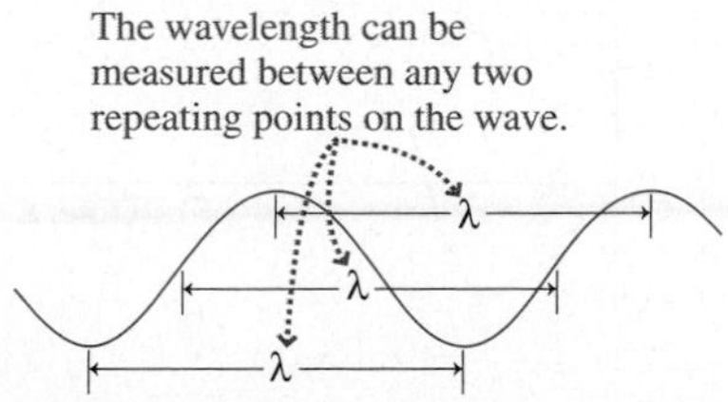

The wavelength can be measured between any two repeating points on the wave.

FIGURE 14.5 The wavelength λ is the distance over which the wave pattern repeats.

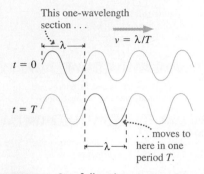

FIGURE 14.6 One full cycle passes a given point in one wave period T; the wave speed is therefore $v = \lambda/T$.

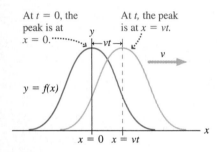

FIGURE 14.7 The wave pulse moves a distance vt in time t, but its shape stays the same.

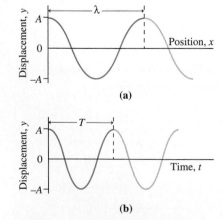

FIGURE 14.8 A sinusoidal wave (a) as a function of position at fixed time $t = 0$ and (b) as a function of time at fixed position $x = 0$.

Wave Speed

A wave travels at a specific speed through its medium. The speed of sound in air is about 340 m/s. Small ripples on water move at about 20 cm/s, while earthquake waves travel at several kilometers per second. The physical properties of the medium ultimately determine the wave speed, as we'll see in Section 14.3.

Wave speed, wavelength, and period are related. In one wave period, a fixed observer sees one complete wavelength go by (Fig. 14.6). Thus, the wave moves one wavelength in one period, so its speed is

$$v = \frac{\lambda}{T} = \lambda f \qquad \text{(wave speed)} \tag{14.1}$$

where the second equality follows because period and frequency are inverses.

GOT IT? 14.1 A boat bobs up and down on a water wave, moving a vertical distance of 2 m in 1 s. A wave crest moves a horizontal distance of 10 m in 2 s. Is the wave speed (a) 2 m/s or (b) 5 m/s? Explain.

14.2 Wave Math

Figure 14.7 shows "snapshots" of a wave pulse at time $t = 0$ and at some later time t. Initially the wave disturbance y is some function of position: $y = f(x)$. Later the pulse has moved to the right a distance vt, but its shape, described by the function f, is the same. We can represent this displaced pulse by replacing x with $x - vt$ as the argument of the function f. Then x has to be larger—by the amount vt—to give the same value of f as it did before. For example, this particular pulse peaks when the argument of f is zero. Initially, that occurred when x was zero. Replacing x by $x - vt$ ensures that the argument becomes zero when $x = vt$, putting the peak at this new position. As time increases, so does vt and therefore the value of x corresponding to the peak. Thus $f(x - vt)$ correctly represents the moving pulse.

Although we considered a single pulse, this argument applies to *any* function $f(x)$, including continuous waves: Replace the argument x with $x - vt$, and the function $f(x - vt)$ describes a wave moving in the positive x direction with speed v. You can convince yourself that a function of the form $f(x + vt)$ describes a wave moving in the negative x direction.

A particularly important case is a **simple harmonic wave**, for which a "snapshot" at time $t = 0$ shows a sinusoidal function. We'll choose coordinates so that $x = 0$ is at a maximum of the wave, making the function a cosine (Fig. 14.8a). Then $y(x, 0) = A \cos kx$, where A is the amplitude and k is a constant, called the **wave number**. We can find k because we know that the wave repeats in one wavelength λ. Since the period of the cosine function is 2π, we therefore want kx to be 2π when x equals λ. Then $k\lambda = 2\pi$, or

$$k = \frac{2\pi}{\lambda} \qquad \text{(wave number)} \tag{14.2}$$

To describe a wave moving with speed v, we replace x in the expression $A \cos kx$ with $x - vt$, giving $y(x, t) = A \cos[k(x - vt)]$. If we now sit at the point $x = 0$, we'll see an oscillation described by $y(0, t) = A \cos(-kvt) = A \cos(kvt)$, where the last step follows because $\cos(-x) = \cos x$. But we found that $k = 2\pi/\lambda$, and Equation 14.1 shows that $v = \lambda/T$, so the argument of the cosine function becomes $kvt = (2\pi/\lambda)(\lambda/T)t = 2\pi t/T$.

In Chapter 13, we introduced the **angular frequency** $\omega = 2\pi/T$ in describing simple harmonic motion; here the same quantity arises in describing wave motion. And no wonder: At a fixed point in space, the wave medium undergoes simple harmonic motion with angular frequency $\omega = 2\pi/T$ (Fig. 14.8b). Putting this all together, we can write a traveling sinusoidal wave in the simple form

$$y(x, t) = A \cos(kx \pm \omega t) \qquad \text{(sinusoidal wave)} \tag{14.3}$$

where we've written $\pm$ so we can describe a wave going in the positive x direction ($-$ sign) or the negative x direction ($+$ sign). The argument of the cosine is called the wave's **phase**. Note that k and ω are related to the more familiar wavelength λ and period T in the same way: $k = 2\pi/\lambda$ and $\omega = 2\pi/T$. Just as ω is a measure of frequency—oscillation cycles per unit *time*, with an extra factor of 2π—so is k a measure of **spatial frequency**—oscillation cycles per unit *distance*, again with that factor of 2π to make the math simpler. The relations between k, λ and ω, T allow us to rewrite the wave speed of Equation 14.1 in terms of k and ω:

$$v = \frac{\lambda}{T} = \frac{2\pi/k}{2\pi/\omega} = \frac{\omega}{k} \tag{14.4}$$

EXAMPLE 14.1   **Describing a Wave: Surfing**

A surfer paddles out beyond the breaking surf to where the waves are sinusoidal in shape, with crests 14 m apart. The surfer bobs a vertical distance 3.6 m from trough to crest, a process that takes 1.5 s. Find the wave speed, and describe the wave using Equation 14.3.

INTERPRET This is a problem about a simple harmonic wave—that is, a wave with sinusoidal shape.

DEVELOP We'll take $x = 0$ at the location of a wave crest when $t = 0$, so Equation 14.3, $y(x, t) = A\cos(kx \pm \omega t)$, applies. Let's take the positive x direction toward shore, so we'll use the minus sign in Equation 14.3. In Fig. 14.9a we sketched a "snapshot" of the wave, showing the spatial information we're given. Figure 14.9b shows the temporal information.

EVALUATE The 1.5-s trough-to-crest time in Fig. 14.9b is half the full crest-to-crest period T, so $T = 3.0$ s. The crest-to-crest distance in Fig. 14.9a is the wavelength λ, so $\lambda = 14$ m. Then Equation 14.1 gives

$$v = \frac{\lambda}{T} = \frac{14 \text{ m}}{3.0 \text{ s}} = 4.7 \text{ m/s}$$

To describe the wave with Equation 14.3 we need the amplitude A, wave number k, and angular frequency ω. The amplitude is half the crest-to-trough displacement, or $A = 1.8$ m, as shown in Fig. 14.9a. The wave number k and angular frequency ω then follow from λ and T: $k = 2\pi/\lambda = 0.449$ m^{-1} and $\omega = 2\pi/T = 2.09$ s^{-1}. Then the wave description is

$$y(x, t) = 1.8\cos(0.449x - 2.09t)$$

with y and x in meters and t in seconds.

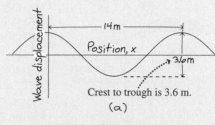

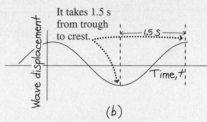

FIGURE 14.9 Our sketch of displacement versus (a) position and (b) time.

ASSESS As a check on our answer, let's see whether our values of ω and k satisfy Equation 14.4: $v = \omega/k = 2.09$ s$^{-1}/0.449$ m$^{-1} = 4.7$ m/s. Thus the pairs λ, T and ω, k are equivalent ways to describe the same wave. ∎

GOT IT? 14.2 The figure shows two waves propagating with the same speed. Which has the greater (a) amplitude, (b) wavelength, (c) period, (d) wave number, (e) frequency (f or ω)?

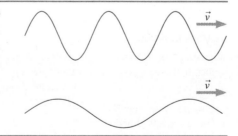

14.3 Waves on a String

The speed of a wave is determined by the properties of the wave medium. Scientists and engineers calculate wave speeds by applying the laws of physics—a challenging task in all but the simplest cases. Here we tackle the case of transverse waves on a stretched string. Our results are directly applicable to musical instrument strings, suspension bridge cables, and other elongated structures.

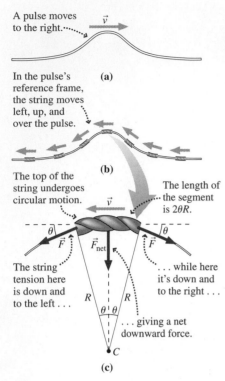

A pulse moves to the right. $\vec{v}$

In the pulse's reference frame, the string moves left, up, and over the pulse.

(a)

The top of the string undergoes circular motion. $\vec{v}$

The length of the segment is $2\theta R$.

(b)

The string tension here is down and to the left . . .

. . . while here it's down and to the right . . .

$\vec{F}$ $\vec{F}_{net}$ $\vec{F}$

. . . giving a net downward force.

C

(c)

FIGURE 14.10 A wave pulse moving on a string.

Our string has mass per unit length μ in kilograms per meter, and it's stretched to a tension force F. Consider a wave pulse propagating to the right, as shown in Fig. 14.10a. We'll use Newton's law to analyze the string's motion and determine the speed of the pulse. It's easiest to do this in a frame of reference moving with the pulse; in that frame, the entire string moves generally to the *left* with the pulse speed v. At the pulse location, however, the string's motion deviates from horizontal as it rides up and down over the pulse (Fig. 14.10b).

Whatever the pulse shape, a small section at the top forms a circular arc of some radius R, as shown in Fig. 14.10c. Then the string right at the top of the pulse undergoes circular motion with speed v and radius R; if its mass is m, Newton's law requires that a force of magnitude mv^2/R act toward the center of curvature to keep the string on its circular path. This force is provided by the difference in the direction of the string tension between the two ends of the section; as Fig. 14.10c shows, the tension at each end contributes a downward component $F\sin\theta$. Then the net force on the segment has magnitude $2F\sin\theta$ and points toward the center of curvature.

Now we make an additional assumption: that the disturbance of the string is small, in the sense that the string remains almost horizontal even at the pulse. Then the angle θ is small, and we can apply the approximation $\sin\theta \approx \theta$. Therefore the net force on the string section becomes approximately $2F\theta$. Furthermore, the small-disturbance approximation means that the tension doesn't vary significantly from its undisturbed value, so the F in this expression is essentially the same F we're using to characterize the tension throughout the string. Finally, our curved string section forms a circular arc whose length, from Fig. 14.10c, is $2\theta R$. Multiplying by the mass per unit length μ gives its mass: $m = 2\theta R\mu$. Now we can apply Newton's law, equating the net force $2F\theta$ to the mass times acceleration:

$$2F\theta = \frac{mv^2}{R} = \frac{2\theta R\mu v^2}{R} = 2\theta\mu v^2$$

Solving for the wave speed v then gives

$$v = \sqrt{\frac{F}{\mu}} \qquad (14.5)$$

Does this make sense? The greater the tension F, the greater the acceleration of the disturbed string, and the more rapidly the wave should propagate. The string's inertia, on the other hand, limits the string's acceleration, and therefore a greater mass per unit length should slow the wave propagation. Equation 14.5, with F in the numerator and μ in the denominator, reflects both these trends.

We've made no assumptions here other than to assume that the disturbance is small. Therefore Equation 14.5 applies to small-amplitude pulses, continuous waves, and wave trains of any shape.

EXAMPLE 14.2   **Wave Speed and Tension Force: Rock Climbing**

A 43-m-long rope of mass 5.0 kg joins two climbers. One climber strikes the rope, and 1.4 s later the second climber feels the effect. What's the rope tension?

INTERPRET We're asked for the rope tension. Although wave speed isn't mentioned explicitly, we just learned to relate wave speed and rope tension. Striking the rope produces a wave, which the second climber feels. We're given the time it takes that wave to propagate along the rope.

DEVELOP Equation 14.5, $v = \sqrt{F/\mu}$, gives the relations among rope tension, mass per unit length, and wave speed. Our plan is to solve for the rope tension, but first we need to find μ and v from the given information.

EVALUATE We're given the rope's mass m and length L, so its mass per unit length is $\mu = m/L$. We're given the time t for the wave to travel the rope length L, so the wave speed is $v = L/t$. Solving Equation 14.5 for F then gives

$$F = \mu v^2 = \left(\frac{m}{L}\right)\left(\frac{L}{t}\right)^2 = \frac{mL}{t^2} = \frac{(5.0\text{ kg})(43\text{ m})}{(1.4\text{ s})^2} = 110\text{ N}$$

ASSESS Is this number reasonable? A typical adult weights around 700 N, so the rope is supporting only a small fraction of the lower climber's weight—a reasonable situation.   ■

Wave Power

Waves carry energy. For a wave on a string, the vertical component of the tension force does work that transfers energy along the string. Figure 14.11 shows that the vertical force on the string at the left side of the pulse is approximately $-F\theta$. As we showed in Chapter 6, power—the rate of doing work—is the product of force and velocity, so the power here is $P = -F\theta u$, where u is the vertical velocity of the string—*not* the wave speed. For a simple harmonic wave, the string velocity is the rate of change of its position $y(x, t) = A\cos(kx - \omega t)$:

$$u = \frac{dy}{dt} = A\omega \sin(kx - \omega t)$$

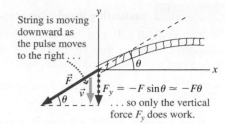

String is moving downward as the pulse moves to the right . . .

$F_y = -F\sin\theta \simeq -F\theta$

. . . so only the vertical force F_y does work.

FIGURE 14.11 The vertical force component does work on the string; for small θ, $\sin\theta \simeq \theta$, so $F_y \simeq F\theta$.

where we used the chain rule, differentiating cosine to $-$sine and then multiplying by the derivative, $-\omega$, of the cosine's argument $kx - \omega t$. As Fig. 14.11 shows, the tangent of the angle θ is the slope, dy/dx, of the string. For small angles, $\tan\theta \simeq \theta$, so $\theta \simeq dy/dx = -kA\sin(kx - \omega t)$. Putting these results for u and θ in our expression for power gives $P = -F\theta u = F\omega kA^2 \sin^2(kx - \omega t)$. The sine term shows that the power fluctuates in space and time. Usually we're interested in the *average* power, $\overline{P} = \frac{1}{2}F\omega kA^2$, which follows because the average value of $\sin^2$ is $\frac{1}{2}$ (Fig. 14.12). We can give this a more physical meaning if we use Equations 14.4 and 14.5 to write $k = \omega/v$ and $F = \mu v^2$, with v the wave speed. Then we have

$$\overline{P} = \tfrac{1}{2}\mu\omega^2 A^2 v \qquad (14.6)$$

This equation gives the sensible result that wave power is directly proportional to the speed v at which energy moves along the wave.

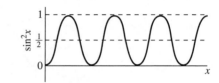

FIGURE 14.12 The function $\sin^2 x$ swings symmetrically between 0 and 1, so its average value is $\frac{1}{2}$.

Wave Intensity

Total power is useful in describing waves confined to narrow structures like strings for mechanical waves or optical fibers for electromagnetic waves. But for waves in three-dimensional media, like sound in air, it makes more sense to talk about the **intensity**, or the rate at which the wave carries energy across a unit area perpendicular to the wave propagation. Intensity is thus power per unit area, measured in watts per square meter (W/m^2).

Wavefronts are surfaces on which the wave phase is constant—for example, wave crests. A **plane wave** is one whose wavefronts are planes. Since the wave doesn't spread out, its intensity remains constant (Fig. 14.13a). But as waves propagate from a localized source, they spread out and the intensity drops. **Spherical waves** originate from point sources and their spherical wavefronts spread in all directions. Since the area of a sphere is $4\pi r^2$, the intensity of a spherical wave decreases as the inverse square of the distance from its source:

$$I = \frac{P}{A} = \frac{P}{4\pi r^2} \qquad \text{(spherical wave)} \qquad (14.7)$$

Note that energy isn't lost here; rather, the same energy is spread over ever-larger areas as the wave propagates (Fig. 14.13b). Table 14.1 lists some typical wave intensities.

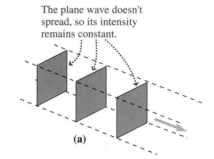

The plane wave doesn't spread, so its intensity remains constant.

(a)

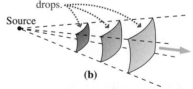

The spherical wave spreads over ever-larger areas, so its intensity drops.

Source

(b)

FIGURE 14.13 (a) Plane and (b) spherical waves.

TABLE 14.1 Wave Intensities

Wave	Intensity, W/m²
Sound, 4 m from loud rock band	1
Sound, jet aircraft at 50 m	10
Sound, whisper at 1 m	10^{-10}
Light, sunlight at Earth's orbit	1368
Light, sunlight at Jupiter's orbit	50
Light, 1 m from typical camera flash	4000
Light, at target of laser fusion experiment	10^{18}
TV signal, 5 km from 50-kW transmitter	1.6×10^{-4}
Microwaves, inside microwave oven	6000
Earthquake wave, 5 km from Richter 7.0 quake	4×10^4

EXAMPLE 14.3 **Evaluating Wave Intensity: A Reading Light**

Your book is 1.9 m from a 75-watt light bulb, and the light is barely adequate for reading. How far from a 40-W bulb would the book have to be to get the same intensity at the page?

INTERPRET This is a problem about wave intensity, and we identify the light bulbs as sources of spherical waves.

DEVELOP Equation 14.7, $I = P/(4\pi r^2)$, gives the intensity. We want both bulbs to produce the same intensity, so we have $I = P_{75}/(4\pi r_{75}^2) = P_{40}/(4\pi r_{40}^2)$.

EVALUATE We then solve for the unknown distance r_{40}:

$$r_{40} = r_{75}\sqrt{\frac{P_{40}}{P_{75}}} = (1.9 \text{ m})\sqrt{\frac{40 \text{ W}}{75 \text{ W}}} = 1.4 \text{ m}$$

ASSESS Make sense? Although the 40-W bulb has only about half the power output, the decrease in distance is not as great as you might expect because the intensity depends on the inverse *square* of the distance. ∎

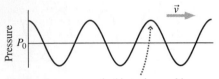

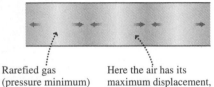

Molecules converge in this region, making the pressure a maximum. Since the molecules come from both directions, the net displacement at the center of the region is zero.

Rarefied gas (pressure minimum) occurs as air molecules move away from this region.

Here the air has its maximum displacement, but the pressure is unchanged from its equilibrium value.

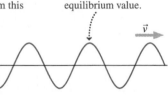

FIGURE 14.14 A sound wave consists of alternating regions of compression (higher density and pressure) and rarefaction (lower density and pressure) propagating through the air.

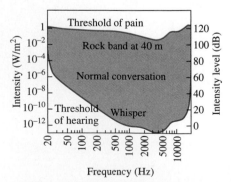

FIGURE 14.15 The human ear responds to sound whose intensity and frequency lie within the shaded region.

GOT IT? 14.3 Two identical stars are different distances from Earth, and the intensity of the light from the more distant star as received at Earth is only 1% that of the closer star. Is the more distant star (a) twice as far away, (b) 100 times as far away, (c) 10 times as far away, or (d) $\sqrt{10}$ times as far away?

14.4 Sound Waves

Sound waves are longitudinal mechanical waves that propagate through gases, liquids, and solids. Most familiar to us is sound in air. Here the wave disturbance comprises a small change in air pressure and density accompanied by a back-and-forth motion of the air (Fig. 14.14). The speed of sound in air and other gases depends on the background pressure P (force per unit area) and density ρ (mass per unit volume):

$$v = \sqrt{\frac{\gamma P}{\rho}} \tag{14.8}$$

where γ is a constant characteristic of the gas. For air and other diatomic gases, γ is $\frac{7}{5}$; for monatomic gases like helium, it's $\frac{5}{3}$. Sound propagates faster in liquids and solids because they're less compressible.

Sound and the Human Ear

The human ear responds to a wide range of sound intensities and frequencies, as shown in Fig. 14.15. Audible frequencies range from around 20 Hz to 20 kHz, although the upper limit drops with age. Figure 14.15 shows that the minimum intensity for audible sound increases at high and low frequencies; that's the reason for the "loudness" switch on your stereo system, which boosts lows and highs to make the sound richer at low volumes. Dolphins, bats, and other creatures can hear much higher frequencies than we humans; bats locate their prey with sound waves at frequencies approaching 100 kHz. Medical ultrasound frequencies extend to tens of MHz.

Decibels

Figure 14.15 shows that the human ear responds to an extremely broad range of sound intensities, covering some 12 orders of magnitude; that's why Fig. 14.15 has a logarithmic scale. We therefore quantify sound levels using a logarithmic unit called the **decibel** (dB). The **sound intensity level** β in decibels is defined by

$$\beta = 10 \log\left(\frac{I}{I_0}\right) \tag{14.9}$$

where I is the intensity in W/m² and $I_0 = 10^{-12}$ W/m² is a reference level chosen as the approximate threshold of hearing at 1 kHz. Since the logarithm of 10 is 1, an increase of 10 dB corresponds to a factor-of-10 increase in the intensity I. Similarly, as you can show in Problem 63, a 3-dB increase is approximately a doubling in intensity. Your ears, however, don't respond linearly, and for intensity levels above about 40 dB, you perceive a 10-dB increase as making the sound roughly twice as loud.

EXAMPLE 14.4 **Decibels: Turn Down the TV!**

Your sister is watching TV, and the sound blasts your ears at 75 dB. You yell to her to turn down the volume, and she lowers the intensity level to 60 dB. By what factor has the power dropped?

INTERPRET This problem is about the relation between power and sound intensity level as measured in decibels.

DEVELOP Equation 14.9, $\beta = 10\log(I/I_0)$, relates the decibel level to the intensity, or power per unit area. At a fixed distance, the sound intensity is proportional to the power from the TV speaker, so in this example we can replace I by P in Equation 14.9.

EVALUATE Call the original 75-dB level β_1; then Equation 14.9 reads $\beta_1 = 10\log(P_1/P_0) = 10\log P_1 - 10\log P_0$, where P_1 is the corresponding power and P_0 is the reference-level power (whose value we aren't going to need). Here we used the identity $\log(a/b) = \log a - \log b$. At the turned-down power P_2, the equation reads $\beta_2 = 10\log P_2 - 10\log P_0$. Subtracting our two equations gives

$$\beta_2 - \beta_1 = 10\log P_2 - 10\log P_1 = 10\log\left(\frac{P_2}{P_1}\right)$$

Therefore $\log(P_2/P_1) = (\beta_2 - \beta_1)/10 = (60 - 75)/10 = -1.5$. The answer we want is the ratio P_2/P_1, and because logarithms and exponentials are inverses, we can get it by exponentiating both sides. Since we're working with base-10 logarithms, we use 10 as the base, getting $P_2/P_1 = 10^{-1.5} = 0.032$.

ASSESS Although we worked this problem formally using Equation 14.9, you can often work with decibels in your head. Here the intensity level has dropped by 15 dB, corresponding to 1.5 orders of magnitude in actual intensity. So the intensity—and therefore the power the TV emits—has dropped by a factor of $10^{-1.5}$, or $1/(10\sqrt{10})$. Since $\sqrt{10}$ is about 3, that's a factor of about 1/30. Because you perceive each 10-dB change as a factor of about 2 in loudness, the reduced volume will sound somewhere between one-fourth and one-half as loud as before. ∎

14.5 Interference

Figure 14.16 shows two wave trains approaching from opposite directions. Where they meet, experiment shows that the net displacement is the sum of the individual displacements. This is true for most waves, at least when the amplitude isn't too large. Waves whose displacements simply add are said to obey the **superposition principle**.

At the point shown in Fig. 14.16b, the wave crests coincide and so do the troughs. The resulting wave is, momentarily, twice as big. This is **constructive interference**—two waves superposing to produce a larger wave displacement. A little later, in Fig. 14.16c, the two waves cancel; this is **destructive interference**. Wave interference occurs throughout physics, from mechanical waves to light and even with the quantum-mechanical waves that describe matter at the atomic scale. Here we take a quick look at wave interference; we'll consider the interference of light waves in more detail in Chapter 32.

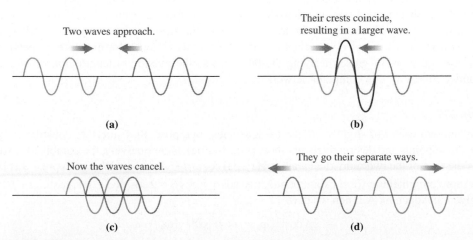

FIGURE 14.16 Wave superposition showing (b) constructive interference and (c) destructive interference.

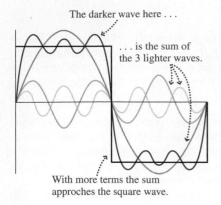

FIGURE 14.17 A square wave built up as a sum of simple harmonic waves. In this case the sum has the form $y(t) = A\sin(\omega t) + \frac{1}{3}A\sin(3\omega t) + \frac{1}{5}A\sin(5\omega t) + \cdots$. Only the first three terms are shown.

Fourier Analysis

The superposition principle lets us build complex wave shapes by superposing simpler ones. The French mathematician Jean Baptiste Joseph Fourier (1768–1830) showed that *any* periodic wave can be written as a sum of simple harmonic waves, a process now known as **Fourier analysis**. Figure 14.17 shows a square wave—important, for example, as the "clock" signal that sets the speed of your computer—represented as a superposition of individual sine waves. Fourier analysis has applications ranging from music to structural engineering to communications, because it helps us understand how a complex wave behaves if we know how its harmonic components behave. The mix of Fourier components in the waveform from a musical instrument determines the exact sound we hear and accounts for the different sounds from different instruments even when they're playing the same note (Fig. 14.18).

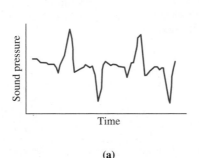
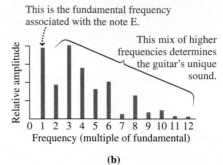

FIGURE 14.18 (a) An electric guitar plays the note E, producing a complex waveform. (b) Fourier analysis shows the relative strengths of the individual sine waves whose sum produces the waveform.

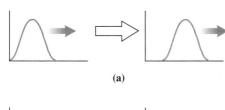

(a)

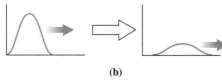

(b)

FIGURE 14.19 (a) A wave pulse in a nondispersive medium holds its shape as it propagates. (b) In a dispersive medium, the pulse shape changes.

Dispersion

When wave speed is independent of wavelength, the simple harmonic components making up a complex waveform travel at the same speed. As a result, the waveform maintains its shape. But for some media, wave speed depends on wavelength. Then, individual harmonic waves travel at different speeds, and a complex waveform changes shape as it moves. This phenomenon is called **dispersion** and is illustrated in Fig. 14.19. Waves on the surface of deep water, for example, have speed given by

$$v = \sqrt{\frac{\lambda g}{2\pi}} \tag{14.10}$$

where λ is the wavelength and g the acceleration of gravity. Because v depends on λ, the waves are dispersive. Long-wavelength waves from a storm at sea have the highest speeds and therefore reach shore well in advance of both the storm and the shorter-wavelength waves. Dispersion is also important in communications systems; for example, dispersion of the square wave pulses carrying digital data sets the maximum lengths for wires and optical fibers used in computer networks.

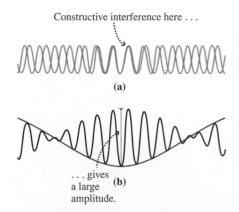

FIGURE 14.20 The origin of beats.

Beats

When two waves of slightly different frequencies superpose, they interfere constructively at some points and destructively at others (Fig. 14.20a). Quantitatively, the combined wave is the sum of the two individual waves: $y(t) = A\cos\omega_1 t + A\cos\omega_2 t$. We can express this in a more enlightening form using the identity $\cos\alpha + \cos\beta = 2\cos\left[\frac{1}{2}(\alpha - \beta)\right]\cos\left[\frac{1}{2}(\alpha + \beta)\right]$ given in Appendix A. Then we have

$$y(t) = 2A\cos\left[\tfrac{1}{2}(\omega_1 - \omega_2)t\right]\cos\left[\tfrac{1}{2}(\omega_1 + \omega_2)t\right]$$

The second cosine factor represents a sinusoidal oscillation at the average of the two individual frequencies. The first term oscillates at a lower frequency—half the difference of the individual frequencies. If we think of the entire term $2A\cos\left[\frac{1}{2}(\omega_1 - \omega_2)t\right]$ as the "amplitude" of the higher-frequency oscillation, then this amplitude itself varies with time, as Fig. 14.20b shows. Note that there are *two* amplitude peaks for each cycle of the slow oscillation, so the frequency with which the amplitude varies is simply $\omega_1 - \omega_2$.

For sound waves, interference of two nearly equal frequencies produces intensity variations called **beats**; the closer the two frequencies, the longer the period between beats. Pilots, for example, synchronize airplane engines by reducing the beat frequency toward zero; musicians use the same trick to tune instruments. Beating of electromagnetic waves forms the basis for some very sensitive measurements.

Interference in Two Dimensions

Waves propagating in two and three dimensions exhibit a rich variety of interference phenomena. Figure 14.21 shows one of the simplest and most important examples—the interference of waves from two point sources oscillating at the same frequency. Points on a perpendicular line midway between the sources are equidistant from both sources, and therefore waves arrive at this line in phase. Thus, they interfere constructively, producing a large amplitude. Some distance away, the waves arrive exactly half a period out of phase. They therefore interfere destructively, producing a **nodal line** where the wave amplitude is very small. Since waves travel half a wavelength in half a period, the nodal line occurs where the distances to the two sources differ by half a wavelength. Additional nodal lines occur where those distances differ by $1\frac{1}{2}$ wavelengths, $2\frac{1}{2}$ wavelengths, and so forth. In practice, two-source interference is observable only when the source separation is comparable to the wavelength. If it's much larger, then the regions of constructive and destructive interference are so close that they blur together.

Two-source interference also results when plane waves pass through two closely spaced apertures that act as sources of circular or spherical wavefronts. Such two-slit interference experiments are important in optics and modern physics, and are of historical interest because they were first used to demonstrate the wave nature of light.

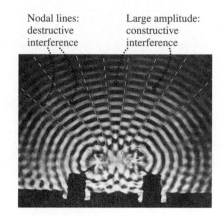

Nodal lines: destructive interference

Large amplitude: constructive interference

FIGURE 14.21 Water waves from two sources interfere to produce regions of low and high amplitude.

EXAMPLE 14.5 **Wave Interference in Two Dimensions: Calm Water**

Ocean waves pass through two small openings, 20 m apart, in a breakwater. You're in a boat 75 m from the breakwater and initially midway between the openings, but the water is pretty rough. You row 33 m parallel to the breakwater and, for the first time, find yourself in relatively calm water. What's the wavelength of the waves?

INTERPRET This is a problem about wave interference. The water is rough at your initial location because constructive interference produces large-amplitude waves. You find calm water at the first nodal line, where destructive interference reduces the wave amplitude.

DEVELOP We sketched the situation in Fig. 14.22. We've seen that the first nodal line occurs when the path lengths from two sources differ by half a wavelength. So our plan is to calculate the wavelength by applying this fact to the distances AP and BP.

EVALUATE Applying the Pythagorean theorem gives

$$AP = \sqrt{(75\text{ m})^2 + (43\text{ m})^2} = 86.5\text{ m}$$
$$BP = \sqrt{(75\text{ m})^2 + (23\text{ m})^2} = 78.4\text{ m}$$

The wavelength is twice the difference between these lengths, so

$$\lambda = 2(AP - BP) = 2(86.5\text{ m} - 78.4\text{ m}) = 16\text{ m}$$

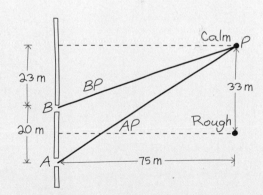

FIGURE 14.22 Calm water at P implies that paths AP and BP differ by half a wavelength.

ASSESS We expect two-source interference to be obvious when the source spacing is comparable to the wavelength. Here the 20-m spacing is indeed comparable to the 16-m wavelength, so our answer makes sense. ∎

GOT IT? 14.4 Light shines through two small holes into a dark room, and a screen is mounted opposite the holes. The hole spacing is comparable to the wavelength of the light. Looking at the screen, will you see (a) two bright spots opposite the two holes or (b) a pattern of light and dark patches? Explain.

14.6 Reflection and Refraction

You shout in a mountain valley and hear echoes. You look in a mirror and see your reflection. A metal screen reflects microwaves to keep them in your oven. A physician's ultrasound probes your body, reflecting off internal structures. A bat uses reflected sound to home in on its prey. All these are examples of wave **reflection**.

You can see that wave reflection *must* occur when a wave hits a medium in which it can't propagate; otherwise, where would the wave energy go? Figures 14.23 and 14.24 detail the reflection process for waves on a stretched string, in the two cases where the string end is clamped at a rigid wall and free to move up and down. In the first case, the wave amplitude must remain zero at the end, so the incident and reflected pulses interfere destructively and the reflected wave is therefore inverted. In the second case, the displacement is a maximum at the free end, and the reflected wave is not inverted.

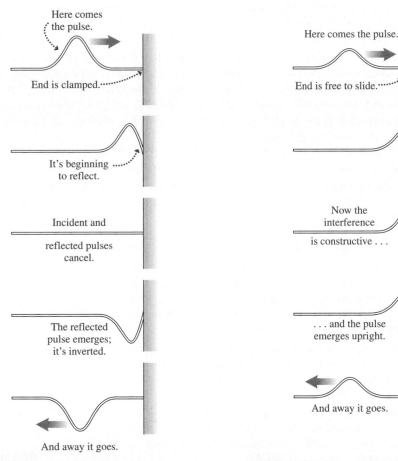

FIGURE 14.23 Reflection of a wave pulse at the rigidly clamped end of string.

FIGURE 14.24 Reflection of a wave pulse at a free end.

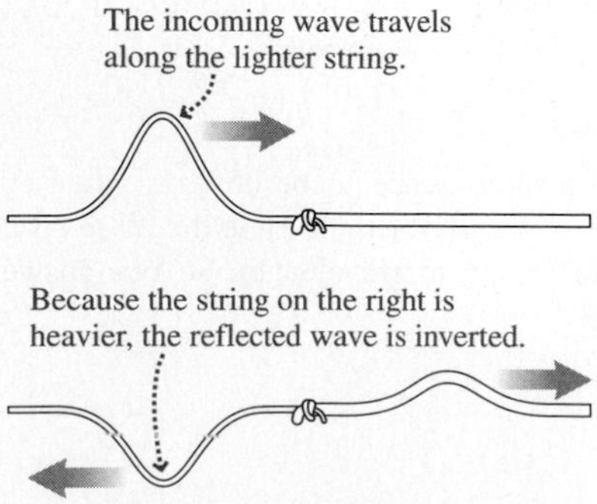

The incoming wave travels along the lighter string.

Because the string on the right is heavier, the reflected wave is inverted.

FIGURE 14.25 Partial reflection occurs at the junction between two strings.

Between the extremes of a rigid wall and a perfectly free end lies the case of one string connected to another with a different mass per unit length. In this case, some wave energy is transmitted to the second string and some is reflected back along the first (Fig. 14.25).

The phenomenon of partial reflection and transmission at a junction of strings has its analog in the behavior of all sorts of waves at interfaces between different media. For example, shallow-water waves are partially reflected if the water depth changes abruptly.

Light incident on even the clearest glass undergoes partial reflection because of the difference in the light-transmitting capabilities of air and glass (Fig. 14.26). Partial reflection of ultrasound waves at the interfaces of body tissues with different densities makes ultrasound a valuable medical diagnostic.

When waves strike an interface between two media at an oblique angle and are capable of propagating in the second medium, the phenomenon of **refraction** occurs. In refraction, the direction of wave propagation changes because of a difference in wave speed between the two media (Fig. 14.27). We'll discuss the mathematics of refraction in Chapter 30.

FIGURE 14.26 Even though glass is transparent, partial reflection occurs at the interface between air and glass.

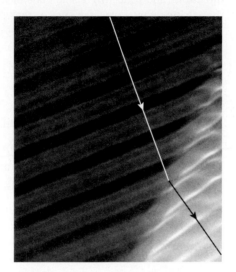

FIGURE 14.27 Waves in shallow water refract at the interface between two different water depths.

APPLICATION   **Probing the Earth**

Waves propagating and reflecting inside the Earth help geologists deduce the planet's interior structure. That's because Earth's interior supports two types of waves. Longitudinal waves, also called P waves, propagate in both solids and liquids. Transverse, or S waves, propagate only in solids. Earthquakes generate S waves that propagate throughout the solid Earth. But as the figure suggests, they can't get through the liquid outer core, so they leave a "shadow" where seismographs don't record any S-wave activity. This effect is our clearest evidence that Earth has a liquid core.

P waves, however, do propagate through the liquid core. But they undergo partial reflections farther in—evidence for an abrupt change in core density. Careful analysis shows that wave speeds in the inner core are consistent with its being solid—giving our planet the solid–liquid–solid structure suggested in the figure.

Studies of Earth's large-scale structure generally use earthquake waves, although inner-core evidence also comes from underground nuclear explosions. At a smaller scale, explosive charges or machines that "thump" the ground produce waves whose reflections from rock layers down to a few kilometers depth help reveal oil and gas deposits.

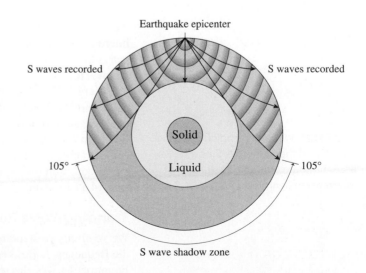

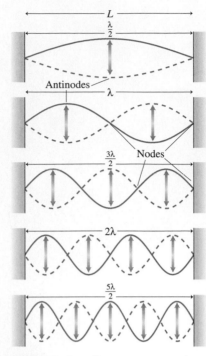

FIGURE 14.28 Standing waves on a string clamped at both ends; shown are the fundamental and four overtones.

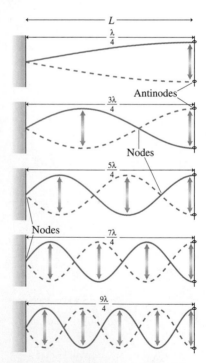

FIGURE 14.29 When one end of the string is fixed and the other free, the string can accommodate only an odd number of quarter-wavelengths.

14.7 Standing Waves

Imagine a string clamped tightly at both ends. Waves propagate back and forth by reflecting at the ends. But because the ends are clamped, the wave displacement at the ends must always be zero. Only certain waves can satisfy this requirement; as Fig. 14.28 suggests, they're waves for which an integer number of half-wavelengths just fits on the length L of the string.

The waves in Fig. 14.28 are **standing waves**, so called because they essentially stand still, confined to the length of the string. At each point the string executes simple harmonic motion perpendicular to its undisturbed state. We can describe standing waves mathematically as arising from the superposition of two waves propagating in opposite directions and reflecting at the ends of the string. If we take the x axis to coincide with the string, then we can write the string displacements in two such waves as $y_1(x, t) = A\cos(kx - \omega t)$ for the wave propagating in the $+x$ direction (recall Equation 14.3) and $y_2(x, t) = -A\cos(kx + \omega t)$ for the wave propagating in the $-x$ direction. (The minus sign in y_2 accounts for the phase change that occurs on reflection.) Their superposition is then

$$y(x, t) = y_1 + y_2 = A[\cos(kx - \omega t) - \cos(kx + \omega t)]$$

Appendix A lists a trig identity for the difference of two cosines:

$$\cos\alpha - \cos\beta = -2\sin[\tfrac{1}{2}(\alpha + \beta)]\sin[\tfrac{1}{2}(\alpha - \beta)]$$

Applying this identity with $\alpha = kx - \omega t$ and $\beta = kx + \omega t$ gives

$$y(x, t) = 2A\sin kx \sin \omega t \qquad (14.11)$$

Equation 14.11 is the mathematical description of a standing wave, and it affirms our qualitative description that each point on the string simply oscillates up and down. Pick any point—that is, any fixed value of x—and Equation 14.11 does indeed describe simple harmonic motion in the y direction, through the factor $\sin\omega t$. The amplitude of that motion depends on the point x you've chosen, and is given by the factor that multiplies $\sin\omega t$—namely, $2A\sin kx$.

Because the string is clamped at both ends, the amplitude at the ends must be zero. Our amplitude factor $2A\sin kx$ does give $y = 0$ in Equation 14.11 at $x = 0$, but what about at $x = L$? Here we'll get zero only if $\sin kL = 0$—and that requires kL to be a multiple of π. So we must have $kL = m\pi$, where m is any integer. But the wave number k is related to the wavelength λ by $k = 2\pi/\lambda$. Our condition $kL = m\pi$ can then be written

$$L = \frac{m\lambda}{2}, \quad m = 1, 2, 3, \dots \qquad (14.12)$$

This is just the condition we already guessed from Fig. 14.28—namely, that the string length L be an integer number of half-wavelengths.

Given a particular string length L, Equation 14.12 limits the allowed standing waves on the string to a discrete set of wavelengths. Those allowed waves are called **modes** or **harmonics**, and the integer m is the **mode number**. The $m = 1$ mode is the **fundamental** and is the longest-wavelength standing wave that can exist on the string. The higher modes are **overtones**.

Figure 14.28 shows that there are points where the string doesn't move at all. These are called **nodes**. Points where the amplitude of the wave displacement is a maximum, in contrast, are **antinodes**.

When a string is clamped rigidly at one end but is free at the other, its clamped end is a node but its free end is an antinode. Figure 14.29 shows that the string length must then be an odd multiple of a quarter-wavelength—a result that you can also get from Equation 14.11 by requiring $\sin kL = 1$ to give maximum amplitude at $x = L$.

Standing-Wave Resonance

We've discussed standing waves in terms of constraints on the wavelength λ rather than on the frequency f. But because waves on a string have a fixed speed v, and because $f\lambda = v$, Equation 14.12's discrete set of allowed wavelengths corresponds to a set of discrete frequencies. The lowest allowed frequency, the fundamental, corresponds to the longest wavelength; the overtones have higher frequencies.

Because a stretched string can oscillate in any of its allowed frequencies, the resonant behavior that we discussed in Chapter 13 can occur close to any of those frequencies. Buildings and other structures, in analogy with our simple string, support a variety of standing-wave modes. For example, a skyscraper is like the string of Fig. 14.29, with its base clamped to Earth but its top free to swing. Engineers must be sure to identify all possible modes of structures they design in order to avoid harmful resonances. The disastrous oscillations of the Tacoma Narrows Bridge shown in Fig. 13.25 are actually torsional standing waves.

Other Standing Waves

Standing waves are common phenomena. Water waves in confined spaces exhibit standing waves, and entire lakes can develop very slow oscillations corresponding to low-mode-number standing waves. Standing electromagnetic waves occur inside closed metal cavities; in microwave ovens the nodes of the standing-wave pattern would result in "cold" spots were not either the food or the source of microwaves kept in motion. Standing sound waves in the Sun help astrophysicists probe the solar interior. And even atomic structure can be understood in terms of standing waves associated with electrons.

Musical Instruments

Our analysis of standing waves on strings applies directly to stringed musical instruments such as violins, guitars, and pianos. Standing-wave vibrations in the instrument strings are communicated to the air as sound waves, usually through the intermediary of a sounding box or electronic amplifiers. For instruments in the violin family, the body of the instrument itself undergoes standing-wave vibrations, excited by the vibration of the string, that establish each individual instrument's peculiar sound quality (Fig. 14.30). Similarly, the stretched membranes of drums exhibit a variety of standing-wave patterns representing the allowed modes on these two-dimensional surfaces.

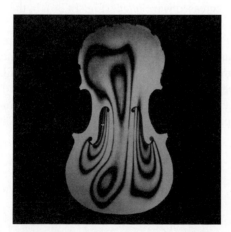

FIGURE 14.30 Standing waves on a violin, imaged using holographic interference of laser light waves.

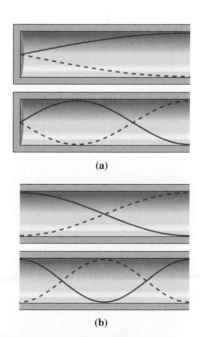

(a)

(b)

FIGURE 14.31 Standing waves in wind instruments: (a) open at one end and (b) open at both ends.

Wind instruments generate standing sound waves in air columns, as suggested in Fig. 14.31. These must be open at one end to allow sound to escape; in many instruments the column is effectively open at both ends. An open end has its pressure fixed at atmospheric pressure; it is therefore a pressure node and thus, from Fig. 14.14, a displacement antinode. As a result, an instrument open at one end supports odd-integer multiples of a quarter-wavelength (Fig. 14.31a), in analogy with Fig. 14.29. An instrument open at both ends, on the other hand, supports integer multiples of a half-wavelength (Fig. 14.31b).

EXAMPLE 14.6 **Standing-Wave Modes: The Double Bassoon**

The double bassoon is the lowest-pitched instrument in a normal orchestra. The instrument is "folded" to achieve an effective air column 5.5 m long, and it acts like a pipe open at both ends. What is the frequency of the double bassoon's fundamental note? Assume the sound speed is 343 m/s.

INTERPRET This is a problem about standing-wave modes in a hollow pipe open at both ends.

DEVELOP Figure 14.31b applies to a pipe that's open at both ends. So our sketch of the fundamental mode in Fig. 14.32 looks like the upper of the two pictures in Fig. 14.31b. We can find the wavelength and then use Equation 14.1, $v = \lambda f$, to get the frequency.

EVALUATE The wavelength is twice the instrument's 5.5-m length, or 11 m. Then Equation 14.1 gives

$$f = \frac{v}{\lambda} = \frac{343 \text{ m/s}}{11 \text{ m}} = 31 \text{ Hz}$$

ASSESS This frequency is the note B_0, which lies near the low-frequency limit of the human ear. Like most wind instruments, the bassoon has a number of holes that, when uncovered, alter the positions of the antinodes and therefore change the pitch.

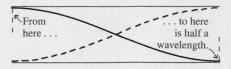

FIGURE 14.32 Sketch for Example 14.6.

GOT IT? 14.5 A string 1 m long is clamped tightly at one end and is free to slide up and down at the other. Which of the following are possible wavelengths for standing waves on this string: $\frac{4}{5}$m, 1 m, $\frac{4}{3}$m, $\frac{3}{2}$m, 2 m, 3 m, 4 m, 5 m, 6 m, 7 m, 8 m?

14.8 The Doppler Effect and Shock Waves

The speed v of a wave is its speed relative to the medium through which it propagates. A point source at rest in the medium radiates waves uniformly in all directions (Fig. 14.33). But when the source moves, wave crests bunch up in the direction toward which the source is moving, resulting in a decreased wavelength (Fig. 14.34). In the opposite direction, wave crests spread out and the wavelength increases.

The wave speed is determined by the properties of the medium, so it doesn't change with source motion. Thus the equation $v = \lambda f$ still holds. This means that an observer in front of the moving source, where λ is smaller, experiences a higher wave frequency as more wave crests pass per unit time. Similarly, an observer behind the source experiences a lower frequency. This change in wavelength and frequency from a moving source is the **Doppler effect** or **Doppler shift**, after the Austrian physicist Christian Johann Doppler (1803–1853).

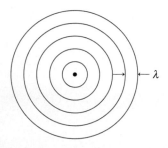

FIGURE 14.33 Circular waves from a source at rest with respect to the medium.

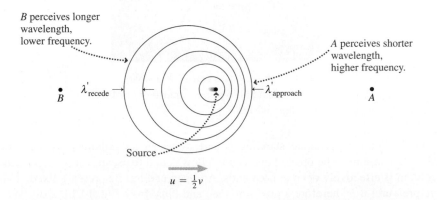

B perceives longer wavelength, lower frequency.

A perceives shorter wavelength, higher frequency.

λ'_{recede}

$\lambda'_{\text{approach}}$

Source

$u = \frac{1}{2}v$

FIGURE 14.34 Origin of the Doppler effect, shown for a source moving with half the wave speed.

To analyze the Doppler effect, let λ be the wavelength measured when the source is stationary, and λ' the wavelength when the source is moving at speed u through a medium where the wave speed is v. At the source, the time between wave crests is the wave period T, and a wave crest moves one wavelength λ in this time. But during the same time T, the moving source covers a distance uT, after which it emits the next wave crest. So the distance between wave crests, as seen by an observer in front of the moving source, is $\lambda' = \lambda - uT$. Writing $T = \lambda/v$, we get

$$\lambda' = \lambda - u\frac{\lambda}{v} = \lambda\left(1 - \frac{u}{v}\right) \qquad \text{(source approaching)} \qquad (14.13a)$$

The situation is similar in the direction opposite the source motion, except that now the wavelength *increases* by the amount $\lambda u/v$, giving

$$\lambda' = \lambda\left(1 + \frac{u}{v}\right) \qquad \text{(source receding)} \qquad (14.13b)$$

We can recast these expressions in terms of frequency using the relations $\lambda = v/f$ and $\lambda' = v/f'$, where f' is the frequency of waves from the moving source as measured by an observer at rest in the medium. Substituting these relations in our expressions for λ' and then solving for f' gives

$$f' = \frac{f}{1 \pm u/v} \qquad \text{(Doppler shift, moving source)} \qquad (14.14)$$

for the Doppler-shifted frequency, where the $+$ and $-$ signs correspond to receding and approaching sources, respectively.

You've probably experienced the Doppler effect for sound when standing near a highway. A loud truck approaches with a high-pitched sound "aaaaaaaaaaa." As it passes, the pitch drops abruptly: "aaaaaaaaaeiooooooooooo," and stays low as the truck recedes. Practical uses of the Doppler effect are numerous. The Doppler shift in reflected ultrasound measures blood flow and fetal heartbeat. Police radar uses the Doppler shift of high-frequency radio waves reflected from moving cars. The Doppler shift of starlight reveals stellar motions, and Doppler-shifted light from distant galaxies is evidence that our entire universe is expanding.

EXAMPLE 14.7 **Doppler Effect: The Wrong Note**

A car speeds down the highway with its stereo blasting. An observer with perfect pitch is standing by the roadside and, as the car approaches, notices that a musical note that should be G ($f = 392\ \text{Hz}$) sounds like A (440 Hz). How fast is the car moving?

INTERPRET This problem is about the Doppler effect in sound from a moving source.

DEVELOP Equation 14.14, $f' = f/(1 \pm u/v)$, relates the original and shifted frequencies to the source speed u, so our plan is to solve this equation for u. We'll use the minus sign because the source is approaching. We'll also need the sound speed v, which Example 14.6 gave as 343 m/s.

EVALUATE Solving Equation 14.14 for u gives

$$u = v\left(1 - \frac{f}{f'}\right) = (343\ \text{m/s})\left(1 - \frac{392\ \text{Hz}}{440\ \text{Hz}}\right) = 37.4\ \text{m/s}$$

ASSESS Our answer—some 134 km/h or 84 mi/h—seems reasonable for a speeding car, though not a particularly safe speed! And it's a little more than 10% of the sound speed, consistent with the roughly 10% change in the sound frequency. ∎

Moving Observers

A Doppler shift in frequency, but not wavelength, also occurs when a moving observer approaches a stationary source—meaning a source at rest with respect to the wave medium. An observer moving toward a stationary source passes wave crests more often than would happen if the observer were at rest, and thus measures a shorter wave period and therefore a higher frequency. The result, as you can show in Problem 76, is a shifted frequency given by

$$f' = f\left(1 \pm \frac{u}{v}\right) \qquad \text{(Doppler shift, moving observer)} \qquad (14.15)$$

with the positive sign for an observer approaching the source and the negative sign for an observer receding. For observer velocities u small compared with the wave speed v, Equations 14.14 and 14.15 give essentially the same results.

Waves from a stationary source that reflect from a moving object undergo a Doppler shift *twice*. First, because the frequency as received at the reflecting object is shifted, according to Equation 14.15, due to the object's motion relative to the source. Then a stationary observer sees the reflected waves as coming from a moving source, so there's another shift, this time given by Equation 14.14. Police radar and other Doppler-based speed measurements make use of this double Doppler shift that occurs on reflection.

The Doppler Effect for Light

Although light and other electromagnetic waves do not require a material medium, they, too, are subject to the Doppler shift. Both Doppler formulas we derived here apply to electromagnetic waves, but only as approximations when the relative speed between source and observer is much lower than the speed of light.

The Doppler shift for electromagnetic waves is the same whether it's the source that moves or the observer. This reflects a profound fact at the root of Einstein's relativity: that "stationary" and "moving" are meaningful only as relative terms. Electromagnetic waves, unlike mechanical waves, do not require a medium—and therefore terms such as "stationary source" and "moving observer" are meaningless. All that matters is the relative motion between source and observer. We'll explore this point further in Chapter 33.

Shock Waves

Equation 14.13a suggests that wavelength goes to zero if a source approaches at exactly the wave speed. This happens because wave crests can't get away from the source, so they pile up just ahead of it to form a large-amplitude wave called a **shock wave** (Fig. 14.35). When the source moves faster than the wave speed, waves pile up on a cone whose half-angle is given by $\sin\theta = v/u$, as shown. The ratio u/v is called the **Mach number,** and the cone angle is the **Mach angle**.

Shock waves occur in a wide variety of physical situations. Sonic booms are shock waves from supersonic aircraft. The bow wave of a boat is a shock wave on the water surface. On a much larger scale, a huge shock wave forms in space as the solar wind—a high-speed flow of particles from the Sun—encounters Earth's magnetic field.

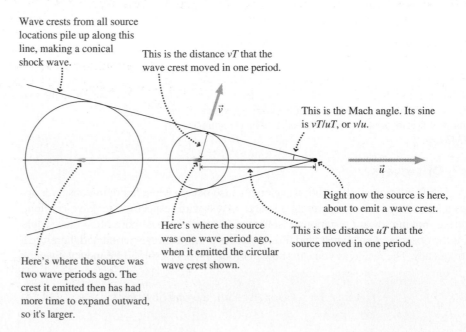

Wave crests from all source locations pile up along this line, making a conical shock wave.

This is the distance vT that the wave crest moved in one period.

This is the Mach angle. Its sine is vT/uT, or v/u.

Right now the source is here, about to emit a wave crest.

This is the distance uT that the source moved in one period.

Here's where the source was one wave period ago, when it emitted the circular wave crest shown.

Here's where the source was two wave periods ago. The crest it emitted then has had more time to expand outward, so it's larger.

FIGURE 14.35 Shock waves form when the source speed u exceeds the wave speed v.

Big Picture

Waves are the big idea here. A wave is a propagating disturbance that carries energy but not matter. Waves are characterized by their amplitude, wavelength, and speed. They can be **longitudinal** or **transverse**.

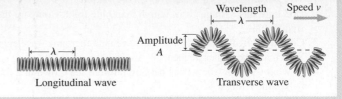

Longitudinal wave

Transverse wave

Key Concepts and Equations

Wave **period** is the time for one complete wave cycle. Period and frequency are inverses, and wavelength λ, period T or frequency f, and wave speed v are all related:

$$v = \frac{\lambda}{T} = \lambda f$$

A **simple harmonic wave** is sinusoidal in shape. The wave disturbance is a function of position and time and is most simply described in terms of its **wave number** k and **angular frequency** ω:

$$y(x, t) = A\cos(kx - \omega t)$$

They're related to wavelength and period by

$$k = \frac{2\pi}{\lambda} \quad \text{and} \quad \omega = \frac{2\pi}{T}.$$

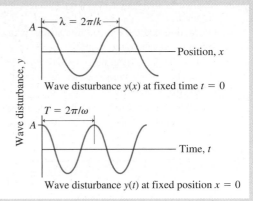

Wave disturbance $y(x)$ at fixed time $t = 0$

Wave disturbance $y(t)$ at fixed position $x = 0$

Wave **intensity** is the power per unit area carried by the wave: $I = P/A$. For a spherical wave that spreads in all directions from a localized source, intensity decreases as the inverse square of the distance from the source: $I = P/(4\pi r^2)$.

Cases and Uses

Wave speed is a characteristic of the medium.

Transverse waves on strings: $v = \sqrt{\dfrac{F}{\mu}}$

Longitudinal sound waves in a gas: $v = \sqrt{\dfrac{\gamma P}{\rho}}$, about 343 m/s in air under standard conditions

Surface waves in deep water: $v = \sqrt{\dfrac{\lambda g}{2\pi}}$

Standing waves on strings

Clamped at both ends, string length is an integer multiple of a half-wavelength: $L = m\lambda/2$

$m = 2$; $L = \lambda$ shown

Nodes

Clamped at one end, string length is an odd-integer multiple of a quarter-wavelength

Nodes

$L = \frac{3}{4}\lambda$ shown

The **Doppler effect** is a frequency and/or wavelength shift due to the motion u of an observer or source relative to the medium with wave speed v.

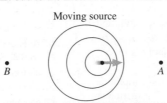

Moving source

B A

Moving source: $f' = \dfrac{f}{(1 \pm u/v)}$, $+$ for receding, $-$ for approaching; λ also changes

Moving observer: $f' = f(1 \pm u/v)$, $+$ for approaching, $-$ for receding; no change in λ

For Thought and Discussion

1. What distinguishes a wave from an oscillation?
2. Red light has a longer wavelength than blue light. Compare their frequencies.
3. Consider a light wave and a sound wave with the same wavelength. Which has the higher frequency?
4. In what sense is "the wave" passing through the crowd at a football game really a wave?
5. Must a wave be either transverse or longitudinal? Explain.
6. As a wave propagates on a stretched string, the string moves back and forth sideways. Is the string speed related to the wave speed? Explain.
7. If you doubled the tension in a string, what would happen to the speed of waves on the string?
8. A heavy cable is hanging vertically, its bottom end free. How will the speed of transverse waves near the top and bottom of the cable compare? Why?
9. The intensity of light from a localized source decreases as the inverse square of the distance from the source. Does this mean that the light loses energy as it propagates?
10. The maximum frequency the human ear can detect is about 20 kHz. If you walk into a room in which two sources are emitting sound waves at 100 kHz and 102 kHz, will you hear anything? Explain.
11. If you double the pressure of a gas while keeping its density the same, what happens to the sound speed?
12. Water is about a thousand times more dense than air, yet the speed of sound in water is greater than in air. How is this possible?
13. If you place a perfectly clear piece of glass in perfectly clear water, you can still see the glass. Why?
14. When a wave source moves relative to the medium, a stationary observer measures changes in both wavelength and frequency. But when the observer moves and the source is stationary, only the frequency changes. Why the difference?
15. Why does a boat easily produce a shock wave on the water surface, but it takes very high-speed aircraft to produce sonic booms?

Exercises and Problems

Exercises

Section 14.1 Waves and Their Properties

16. Ocean waves with 18-m wavelength travel at 5.3 m/s. What is the time interval between wave crests passing under a boat moored at a fixed location?
17. Ripples in a shallow puddle are propagating at 34 cm/s. If the wave frequency is 5.2 Hz, what are (a) the period and (b) the wavelength?
18. An 88.7-MHz FM radio wave propagates at the speed of light. What is its wavelength?
19. Calculate the wavelengths of (a) a 1.0-MHz AM radio wave, (b) a channel 9 TV signal (190 MHz), (c) a police radar (10 GHz), (d) infrared radiation from a hot stove $(4.0 \times 10^{13}$ Hz), (e) green light $(6.0 \times 10^{14}$ Hz), and (f) 1.0×10^{18} Hz X rays. All are electromagnetic waves that propagate at 3.0×10^{8} m/s.
20. A seismograph located 1200 km from an earthquake detects waves from the quake 5.0 min after the quake occurs. The seismograph oscillates in step with the waves, at a frequency of 3.1 Hz. Find the wavelength of the waves.

Section 14.2 Wave Math

21. Ultrasound used in a particular medical imager has frequency 4.8 MHz and wavelength 0.31 mm. Find (a) the angular frequency, (b) the wave number, and (c) the wave speed.
22. An ocean wave has period 4.1 s and wavelength 10.8 m. Find (a) its wave number and (b) its angular frequency.
23. Find (a) the amplitude, (b) the wavelength, (c) the period, and (d) the speed of a wave whose displacement is given by $y = 1.3 \cos(0.69x + 31t)$, where x and y are in cm and t is in seconds. (e) In which direction is the wave propagating?
24. A simple harmonic wave of wavelength 16 cm and amplitude 2.5 cm is propagating along a string in the negative x direction at 35 cm/s. Find (a) the angular frequency and (b) the wave number. (c) Write a mathematical expression describing the displacement y of this wave (in centimeters) as a function of position and time. Assume the displacement at $x = 0$ is a maximum when $t = 0$.

Section 14.3 Waves on a String

25. The main cables supporting New York's George Washington Bridge have a mass per unit length of 4100 kg/m and are under tension of 250 MN. At what speed would a transverse wave propagate on these cables?
26. A transverse wave 1.2 cm in amplitude is propagating on a string; the wave frequency is 44 Hz. The string is under 21-N tension and has mass per unit length of 15 g/m. Determine the wave speed.
27. A transverse wave with 3.0-cm amplitude and 75-cm wavelength is propagating on a stretched spring whose mass per unit length is 170 g/m. If the wave speed is 6.7 m/s, find the spring tension.
28. A rope is stretched between supports 12 m apart; its tension is 35 N. If one end of the rope is tweaked, the resulting disturbance reaches the other end 0.45 s later. What is the total mass of the rope?
29. A rope with 280 g of mass per meter is under 550-N tension. A wave with frequency 3.3 Hz and amplitude 6.1 cm is propagating on the rope. What is the average power carried by the wave?

Section 14.4 Sound Waves

30. Show that the quantity $\sqrt{P/\rho}$ from Equation 14.8 has the units of speed.
31. Find the sound speed in air under standard conditions with pressure 101 kN/m^2 and density 1.20 kg/m^3.
32. Timers in sprint races start their watches when they see smoke from the starting gun, not when they hear the sound. Why? How much error would be introduced by timing a 100-m race from the sound of the shot?
33. The factor γ for nitrogen dioxide (NO_2) is 1.29. Find the sound speed in NO_2 at a pressure of 4.8×10^4 N/m^2 and density 0.35 kg/m^3.
34. A gas with density 1.0 kg/m^3 and pressure 8.0×10^4 N/m^2 has sound speed 365 m/s. Are the gas molecules monatomic or diatomic?
35. Divers in an underwater habitat breathe a special mixture of oxygen and neon, with pressure 6.2×10^5 N/m^2 and density 4.5 kg/m^3. The effective γ value for the mixture is 1.61. Find the frequency in this mixture for a 50-cm-wavelength sound wave, and compare with its frequency in air under normal conditions.

Section 14.5 Interference

36. You're in an airplane whose two engines are running at 560 rpm and 570 rpm. How often do you hear the sound intensity increase as a result of wave interference?

37. What is the wavelength of the ocean waves in Example 14.5 if the calm water you encounter at 33 m is the *second* calm region on your voyage from the center line?

Section 14.7 Standing Waves

38. A 2.0-m-long string is clamped at both ends. (a) What is the longest-wavelength standing wave that can exist on this string? (b) If the wave speed is 56 m/s, what is the lowest standing-wave frequency?
39. When a stretched string is clamped at both ends, its fundamental standing-wave frequency is 140 Hz. (a) What is the next higher frequency? (b) If the same string, with the same tension, is now clamped at one end and free at the other, what is the fundamental frequency? (c) What is the next higher frequency in case (b)?
40. A string is clamped at both ends and tensioned until its fundamental vibration frequency is 85 Hz. If the string is then held rigidly at its midpoint, what is the lowest frequency at which it will vibrate?
41. Estimate the fundamental frequency of the human vocal tract by assuming it to be a cylinder 15 cm long that is closed at one end.

Section 14.8 The Doppler Effect and Shock Waves

42. A car horn emits 380-Hz sound. If the car moves at 17 m/s with its horn blasting, what frequency will a person standing in front of the car hear?
43. The stationary siren on a firehouse is blaring at 85 Hz. What is the frequency perceived by a firefighter racing toward the station at 120 km/h?
44. A fire truck's siren at rest wails at 1400 Hz; standing by the roadside as the truck approaches, you hear it at 1600 Hz. How fast is the truck going?
45. Red light emitted by hydrogen atoms at rest in the laboratory has wavelength 656 nm. Light emitted in the same process on a distant galaxy is received at Earth with wavelength 708 nm. Describe the galaxy's motion relative to Earth.

Problems

46. Figure 14.36 shows a simple harmonic wave at time $t = 0$ and later at $t = 2.6$ s. Write a mathematical description of this wave.

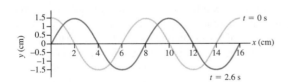

FIGURE 14.36 Problem 46

47. Transverse waves propagate at 18 m/s on a string whose tension is 14 N. What will be the wave speed if the tension is increased to 40 N?
48. A uniform cable hangs vertically under its own weight. Show that the speed of waves on the cable is given by $v = \sqrt{yg}$, where y is the distance from the bottom of the cable.
49. Figure 14.37 shows a wave train consisting of two cycles of a sine wave propagating along a string. Obtain an expression for the total energy in this wave train, in terms of the string tension F, the wave amplitude A, and the wavelength λ.

FIGURE 14.37 Problem 49

50. A loudspeaker emits energy at the rate of 50 W, spread in all directions. What is the intensity of sound 18 m from the speaker?
51. The light intensity 3.3 m from a light bulb is 0.73 W/m². What is the power output of the bulb, assuming it radiates equally in all directions?
52. Light emerges from a 5.0-mW laser in a beam 1.0 mm in diameter. The beam shines on a wall, producing a spot 3.6 cm in diameter. What are the beam intensities (a) at the laser and (b) at the wall?
53. Two waves have the same angular frequency ω, wave number k, and amplitude A, but they differ in phase: $y_1 = A\cos(kx - \omega t)$ and $y_2 = A\cos(kx - \omega t + \phi)$. Show that their superposition is also a simple harmonic wave, and determine its amplitude A_s as a function of the phase difference ϕ.
54. A wave on a taut wire is described by the equation $y = 1.5\sin(0.10x - 560t)$, where x and y are in cm and t is in seconds. If the wire tension is 28 N, what are (a) the amplitude, (b) the wavelength, (c) the period, (d) the wave speed, and (e) the power carried by the wave?
55. A spring of mass m and spring constant k has an unstretched length L_0. Find an expression for the speed of transverse waves on this spring when it has been stretched to a length L.
56. When a 340-g spring is stretched to a total length of 40 cm, it supports transverse waves propagating at 4.5 m/s. When it's stretched to 60 cm, the waves propagate at 12 m/s. Find (a) the unstretched length of the spring and (b) its spring constant.
57. At a point 15 m from a source of spherical sound waves, you measure a sound intensity of 750 mW/m². How far do you need to walk, directly away from the source, until the intensity is 270 mW/m²?
58. Figure 14.38 shows two observers 20 m apart on a line that connects them and a spherical light source. If the observer nearer the source measures a light intensity 50% greater than the other observer, how far is the nearer observer from the source?

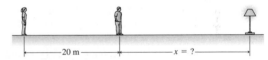

FIGURE 14.38 Problem 58

59. An ideal spring is stretched to a total length L_1. When that length is doubled, the speed of transverse waves on the spring triples. Find an expression for the unstretched length of the spring.
60. Show that the time it takes a wave to propagate up the cable in Problem 48 is $t = 2\sqrt{L/g}$, where L is the cable length.
61. You see an airplane straight overhead at an altitude of 5.2 km. Sound from the plane, however, seems to be coming from a point back along the plane's path at a 35° angle to the vertical (Fig. 14.39). What is the plane's speed, assuming an average 330-m/s sound speed?

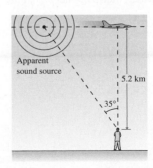

FIGURE 14.39 Problem 61

62. What are the intensity and pressure amplitudes in sound waves with intensity levels of (a) 65 dB and (b) −5 dB?

63. Show that a doubling of sound intensity corresponds to very nearly a 3-dB increase in the decibel level.

64. Sound intensity from a localized source decreases as the inverse square of the distance, according to Equation 14.7. If the distance from the source doubles, what happens to (a) the intensity and (b) the decibel level?

65. At a distance 2.0 m from a localized sound source you measure the intensity level as 75 dB. How far away must you be for the perceived loudness to drop in half (i.e., to an intensity level of 65 dB)?

66. The A-string (440 Hz) on a piano is 38.9 cm long and is clamped tightly at both ends. If the string is under 667-N tension, what is its mass?

67. Show that the standing-wave condition of Equation 14.12 is equivalent to the requirement that the time it takes a wave to make a round trip from one end of the medium to the other and back be an integer multiple of the wave period.

68. What length is necessary for an organ pipe to produce a 22-Hz tone (a) if the pipe is closed at one end and (b) if it's open at both ends?

69. You're standing by the roadside as a truck approaches, and you measure the dominant frequency in the truck noise at 1100 Hz. As the truck passes, the frequency drops to 950 Hz. What is the truck's speed?

70. You're between two loudspeakers emitting 180-Hz tones. How fast would you have to walk or run to perceive a beat frequency of 1.5 Hz between the two?

71. A supersonic plane flies directly over you at 2.2 times the sound speed. You hear its sonic boom 19 s later. What is the plane's altitude, assuming a constant 340 m/s sound speed?

72. A 1.5-m-long pipe has one end open. Among its possible standing-wave frequencies is 225 Hz; the next higher frequency is 375 Hz. Find (a) the fundamental frequency and (b) the sound speed.

73. A wave source recedes from you at 8.2 m/s, and the wavelength you measure is 20% greater than what you would measure if the source were at rest. What is the wave speed?

74. Obstetricians use ultrasound to monitor fetal heartbeat. If 5.0-MHz ultrasound reflects off the moving heart wall with a 100-Hz frequency shift, what is the speed of the heart wall? *Hint:* You have *two* shifts to consider.

75. What is the frequency shift of a 70-GHz police radar signal when it reflects off a car moving at 120 km/h? (Radar waves travel at the speed of light.) *Hint:* See Problem 74.

76. You move at speed u toward a wave source that's stationary with respect to the medium in which waves of wavelength λ propagate with speed v. Your speed relative to the wave crests is therefore $v + u$. Show that for you the time between wave crests is $T' = \lambda/(v + u)$, and from this show that you perceive a frequency given by Equation 14.15, with the + sign.

77. Suppose you're operating a machine with two motors. If the two motors are not operating at the same rpm, you will hear a periodic variation in the intensity of the sound—that is, beats. Suppose the tachometer that indicates the rotation rate of one motor is malfunctioning. The operational one shows 3600 rpm. You hear a beat every 30 s. You are certain the other motor is running fast. What is its rotational speed in rpm?

78. Perhaps you know of the sousaphone, a tubalike brass instrument named for its inventor, John Philip Sousa. Now your physics professor has designed a new instrument called the worfaphone. (This instrument will be used by the producers of a new *Star Trek* movie in a scene featuring Klingon Opera.) It consists of a series of long metal cylinders each of length x and diameter $x/10$. A cylinder is closed at one end and open at the other. A string is stretched across the open end. The string has mass per unit length μ and tension T. The string must be put under enough tension so that the fundamental frequency of the sound waves in the cylinder equals the third harmonic (i.e., second overtone) of the waves in the vibrating string. Your professor asks you to derive an equation for the tension of the string in terms of the other parameters of the instrument and the speed of sound in air, v.

79. Having grown tired of hearing loud music from passing cars at 3 AM and the boisterous yelling of bar patrons at closing, a professor flees the city and purchases a home in a quiet suburban neighborhood. He is, however, concerned about noise from freeway traffic. Using a sound-level meter, he stands at an overpass over the interstate highway. The location is 2.6 km from his home. The overpass is about 10 m from the passing cars. He measures the sound level at 80 dB. What will the sound level be at the new home assuming there is no decrease in the sound intensity due to absorption by trees and other objects? Compare with that of a quiet whisper, about 20 dB.

80. Your little sister builds two tree houses and stretches a rope between them for sending messages to the other tree house. She hangs a mass of 1.4 kg on one end of the rope that passes over a pulley. The other end is tied to the other tree house. When she plucks the rope, a wave propagates at 18 m/s. This is too slow, and she desires to increase the wave speed to 30 m/s. She asks you, "What mass should I use?"

Answers to Chapter Questions

Answer to Chapter Opening Question

None. The waves transport energy, but not matter.

Answers to GOT IT? Questions

14.1 (b) 5 m/s, because that's the speed of the wave crest. 2 m/s is the speed of the localized disturbance, not the wave speed.

14.2 (a) Upper wave; (b) lower; (c) lower; (d) upper; (e) upper (both f and ω).

14.3 (c).

14.4 (b), because of wave interference analogous to that shown in Fig. 14.21.

14.5 $\frac{4}{5}$ m, $\frac{4}{3}$ m, 4 m—one-fourth of each value fits into 1 m an odd number of times.

15 Fluid Motion

■ Why is only the "tip of the iceberg" above water?

A tornado whirls across a darkened sky. You fly, your plane supported by air pressure on its wings. Gas from a giant star forms a cosmic whirlpool before plunging into a black hole. The force of your foot on your car's brake pedal is amplified by fluid in the brake cylinders. Your own body is sustained by air moving into and out of your lungs, and by the flow of blood throughout your tissues. All these examples involve fluid motion.

Fluid is matter that flows under the influence of external forces. Fluids include both liquids and gases. The intermolecular forces are weaker in fluids than in solids, and as a result the molecules move around readily. In a liquid, those forces are strong enough to keep the molecules in close contact, while in a gas they're almost negligible and the molecules are usually widely spaced. Mobility of the individual molecules means that a fluid spreads out to take the shape of its container.

15.1 Density and Pressure

If we could observe a fluid on the molecular scale, we would find large numbers of molecules in continuous motion, colliding frequently with each other and with the walls of their containers. This molecular behavior is governed by the laws of mechanics, and in principle we could study fluids by applying those laws to all the individual molecules. But even a drop of water contains about 10^{21} molecules; to calculate the motions of all those molecules would take the fastest computers many times the age of the universe!

▶ **To Learn**
By the end of this chapter you should be able to
■ Explain the relation between pressure and force (15.1).
■ Calculate pressure as a function of depth in liquids (15.2).
■ Explain why some objects float and others sink, and determine quantitatively the position of floating objects and the apparent weight of submerged objects (15.3).
■ Express the conservation of mass and energy for fluids through the continuity equation and Bernoulli's equation, and use the two to solve problems in fluid dynamics (15.4, 15.5).

◀ **To Know**
■ Force (Chapter 4) and energy (Chapter 6) are ultimately the concepts behind fluid motion.
■ Kinetic energy and gravitational potential energy are both important in describing fluid flow (6.3, 7.2).

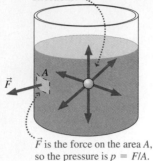

The fluid exerts pressure internally as well as on the container. The internal pressure is the same in all directions. ⋯

$\vec{F}$ is the force on the area A, so the pressure is $p = F/A$.

FIGURE 15.1 Pressure, the force per unit area, is exerted equally in all directions.

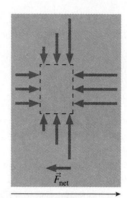

$\vec{F}_{net}$

Increasing pressure

(a)

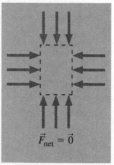

$\vec{F}_{net} = \vec{0}$

Constant pressure

(b)

FIGURE 15.2 If pressure varies with position, then there's a net force on a volume of fluid.

Because the number of molecules is so large, we approximate a fluid by considering it to be continuous rather than composed of discrete particles. In this approximation, valid for fluid samples large compared with the distance between molecules, we describe the fluid by specifying macroscopic properties such as density and pressure.

Density

Density (symbol ρ, Greek rho) measures the mass per unit volume; its SI units are kg/m^3. The density of water is normally about 1000 kg/m^3; that of air about a factor of 1000 smaller. Because their molecules are essentially in contact, liquids are **incompressible**, meaning that their densities remain nearly constant. Gases, in contrast, are **compressible**: With relatively large intermolecular distances, their densities change readily.

Pressure

Pressure measures the normal force per unit area exerted by a fluid (Fig. 15.1):

$$p = \frac{F}{A} \quad \text{(pressure)} \tag{15.1}$$

The SI pressure unit is N/m^2, given the name **pascal** (Pa) after the French mathematician, scientist, and philosopher Blaise Pascal (1623–1662). Another commonly used pressure unit is the **atmosphere** (atm), defined as Earth's normal atmospheric pressure at sea level and equal to 101.3 kPa (14.7 pounds per square inch, or psi).

Pressure is a scalar quantity; at a given point in a fluid, pressure is exerted equally in all directions (Fig. 15.1), so it makes no sense to associate a direction with it. This property explains an aspect of pressure that you may find puzzling. Although the atmosphere bears down on your body with a pressure of 14.7 pounds on every square inch, you certainly don't feel that burden. That's because the force arising from this pressure is everywhere perpendicular to your body, and your body fluids respond by compressing until they're at the same pressure. If you've had your ears "pop" in a fast elevator or airplane, or when diving underwater, you know the pain that can develop when the pressure on your body is temporarily imbalanced.

15.2 Hydrostatic Equilibrium

For a fluid to remain at rest, the net force everywhere in the fluid must be zero; this condition is **hydrostatic equilibrium**. In the absence of any external forces, hydrostatic equilibrium requires that the pressure be constant throughout the fluid; otherwise, pressure differences would result in a net force, and the fluid would move in response. As Fig. 15.2 suggests, it's pressure *difference*, rather than pressure itself, that gives rise to forces within fluids.

Hydrostatic Equilibrium with Gravity

Hydrostatic equilibrium in the presence of gravity requires a pressure force to counteract the gravitational force. Since forces arise only from pressure differences, the fluid pressure must therefore vary with depth.

Figure 15.3 shows the forces on a fluid element of area A, thickness dh, and mass dm. A gravitational force acts downward on this fluid element; for it to be in equilibrium there must therefore be an upward pressure force—and that requires a greater pressure on the lower side. Suppose the pressures at the top and bottom are p and $p + dp$, respectively. Since pressure is force per unit area, the net pressure force is $dF_{press} = (p + dp)A - pA = A\, dp$. The gravitational force is $dF_g = -g\, dm$, where the minus sign designates the downward direction. But the mass dm is the density times the volume, so $dF_g = -g\,dm = -g\rho A\,dh$. Hydrostatic equilibrium requires that these forces sum to zero: $A\,dp - g\rho A\,dh = 0$, or

$$\frac{dp}{dh} = \rho g \quad \text{(hydrostatic equilibrium)} \tag{15.2}$$

This equation shows that dp/dh—the variation in pressure with depth h—is positive, confirming that pressure increases with depth. For a liquid, which is essentially incompressible, ρ is constant and Equation 15.2 shows that pressure increases linearly with depth:

$$p = p_0 + \rho g h \tag{15.3}$$

where p_0 is the pressure at the liquid surface.

Equation 15.2 applies to any fluid in a uniform gravitational field; Equation 15.3 follows from Equation 15.2 for the special case of a liquid. It's also possible to integrate Equation 15.2 to find the pressure in a gas that's subject to the gravitational force. Because the gas density isn't constant, this is a little more involved mathematically. Problem 70 explores the variation of pressure with height in Earth's atmosphere.

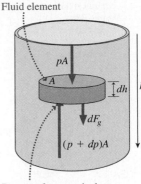

Fluid element

Pressure force on the bottom must be greater in order to balance gravity.

FIGURE 15.3 Forces on a fluid element in hydrostatic equilibrium.

EXAMPLE 15.1   **Calculating Pressure: Ocean Depths**

(a) At what water depth is the pressure twice atmospheric pressure? (b) What is the pressure at the bottom of the 11-km-deep Marianas Trench, the deepest point in the ocean? Take atmospheric pressure as 100 kPa and the density of water as 1000 kg/m³.

INTERPRET This problem is about hydrostatic equilibrium, with water the fluid.

DEVELOP We determine that Equation 15.3, $p = p_0 + \rho g h$, applies, with p_0 equal to the atmospheric pressure at the water surface. Then at twice atmospheric pressure, $p = 2p_0$, and we can solve for h to answer part (a). Because pressure increases linearly with depth, we can extrapolate our result for part (a) to find the answer to part (b).

EVALUATE Solving our equation for the depth h and substituting the given numbers in, we find for part (a):

$$h = \frac{p - p_0}{\rho g} = \frac{2.0 \times 10^5 \text{ Pa} - 1.0 \times 10^5 \text{ Pa}}{(1000 \text{ kg/m}^3)(9.8 \text{ m/s}^2)} = 10 \text{ m}$$

For part (b) we note that the pressure continues to increase by 100 kPa for every 10 m of depth. In the Marianas Trench, 11×10^3 m deep, the pressure increase is then

$$(11 \times 10^3 \text{ m})(100 \text{ kPa}/10 \text{ m}) = 110 \text{ MPa}$$

ASSESS This is over a thousand times atmospheric pressure, or more than 8 tons per square inch! Creatures living at these depths are in pressure equilibrium with their surroundings. To bring them to the surface for study, scientists must maintain their natural pressure or they'll explode. A similar plight awaits scuba divers who hold their breath while ascending; air in the lungs expands, bursting the alveoli. ∎

Measuring Pressure

Figure 15.4 shows a **barometer**, in which air pressure acts on the open pool of mercury, pushing the liquid into the evacuated tube. Since $p_0 = 0$ in the vacuum at the top of the tube, Equation 15.3 becomes simply $p = \rho g h$, showing that the height h of the mercury is directly proportional to atmospheric pressure p. Standard atmospheric pressure of 101.3 kPa supports a mercury column 760 mm or 29.92 in. high. Pressure varies slightly with meteorological conditions, and weather forecasters regularly report atmospheric pressure in millimeters or inches of mercury. Mercury's high density makes for a reasonable-sized barometer. Example 15.1 shows that a water-filled barometer would need to be 10 m long!

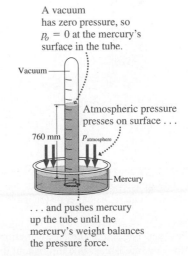

A vacuum has zero pressure, so $p_0 = 0$ at the mercury's surface in the tube.

Vacuum

Atmospheric pressure presses on surface . . .

760 mm $p_{\text{atmosphere}}$

Mercury

. . . and pushes mercury up the tube until the mercury's weight balances the pressure force.

FIGURE 15.4 A mercury barometer.

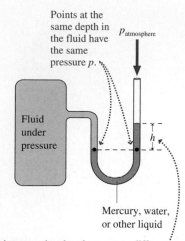

Points at the same depth in the fluid have the same pressure p.

$p_{\text{atmosphere}}$

Fluid under pressure

h

Mercury, water, or other liquid

h is proportional to the pressure difference between fluid and atmosphere.

FIGURE 15.5 A manometer used to measure the pressure difference between a closed container and the atmosphere.

A **manometer** is a U-shaped tube filled with liquid and used to measure pressure differences. A pressure difference between the two ends results in a height difference h between the liquid surfaces (Fig. 15.5). Equation 15.3 shows that h is directly proportional to the pressure difference.

Barometers and manometers are the classic pressure-measuring instruments, and understanding them will help you grasp the meaning of pressure. But pressure-measuring devices today are usually electronic, using the pressure force to alter electrical properties and produce an electrical signal proportional to pressure.

The term **gauge pressure** describes the excess pressure above atmospheric. Inflation instructions for tires and sports equipment specify gauge pressure. A tire inflated to 200 kPa (about 30 psi) has an absolute pressure of about 300 kPa because of the additional 100-kPa atmospheric pressure.

Pascal's Law

Equation 15.3 shows that an increase in surface pressure p_0 results in the same pressure increase throughout the fluid. More generally, a pressure increase anywhere is felt throughout the fluid—a fact known as **Pascal's law**. Pascal applied this principle in his invention of the hydraulic press. Today hydraulic systems, based on Pascal's law, control machinery ranging from automobile brakes to aircraft wings, bulldozers, cranes, and robots.

EXAMPLE 15.2 **Applying Pascal's Law: A Hydraulic Lift**

In the hydraulic lift of Fig. 15.6, a large piston supports a car; the total mass of car and piston is 3200 kg. What force must be applied to the smaller piston to support the car?

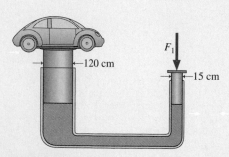

F_1

120 cm

15 cm

FIGURE 15.6 A hydraulic lift.

INTERPRET We interpret this as a problem involving Pascal's law. Whatever pressure results from the force on the smaller piston is transmitted through the fluid to the larger piston and thus supports the car.

DEVELOP We're given a drawing. Having determined that Pascal's law applies, and neglecting pressure variations with depth, we conclude that the pressure is the same throughout the system. Our plan, then, is

to write expressions involving the pressures at both pistons and use the fact that they're equal to solve for the unknown force. We'll use the fact that the pressure on a piston is the applied force divided by the piston's area.

EVALUATE The small piston exerts a pressure $p = F_1/A_1 = F_1/\pi R_1^2$, where F_1 is the unknown force. The pressure at the large piston is the same and produces a force $F_2 = pA_2$. This force supports the weight mg of piston and car; therefore we have

$$mg = pA_2 = p\pi R_2^2 = \frac{F_1}{\pi R_1^2}\pi R_2^2 = F_1\left(\frac{R_2}{R_1}\right)^2$$

Solving for F_1 gives our answer:

$$F_1 = mg\left(\frac{R_1}{R_2}\right)^2 = (3200 \text{ kg})(9.8 \text{ m/s}^2)\left(\frac{15 \text{ cm}}{120 \text{ cm}}\right)^2 = 490 \text{ N}$$

We used the diameters from Fig. 15.3, rather than the radii, because their ratio is the same.

ASSESS How can a 490-N force—about 100 lb—support the car? Through the constant fluid pressure, this smaller force is effectively multiplied by the ratio of the piston areas. If we lifted the car farther, the work done in moving the small piston—the product of the force and the distance moved—would be equal to the work done on the large piston. Since the force on the large piston is greater, the distance moved is smaller, and energy is conserved. ∎

15.3 Archimedes' Principle and Buoyancy

Why do some objects float while others sink? Figure 15.7a shows the upward pressure force on an arbitrary fluid volume balancing the downward gravitational force. Now imagine replacing the fluid volume with a solid object of identical shape (Fig. 15.7b). The remaining fluid hasn't changed, so it continues to exert an upward force on the object—a force whose magnitude equals the weight of the *original fluid volume*. This force is the **buoyancy force**, and in giving its magnitude we've stated **Archimedes' principle**: The buoyancy force on an object is equal to the weight of the fluid displaced by the object.

If the submerged object weighs more than the displaced fluid, then the gravitational force exceeds the buoyancy force and the object sinks. If the object weighs less than the displaced fluid, buoyancy is greater and the object rises. Therefore, an object floats or sinks depending on whether its average density is greater or less than that of the fluid. In between is the case of **neutral buoyancy**, when an object's average density is the same as that of the fluid.

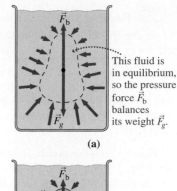

This fluid is in equilibrium, so the pressure force $\vec{F}_b$ balances its weight $\vec{F}_g$.

(a)

Replace the fluid with a solid object, and the pressure force doesn't change. But the weight may.

(b)

FIGURE 15.7 The buoyancy force $\vec{F}_b$ arises because pressure increases with depth.

EXAMPLE 15.3 **Finding the Buoyancy Force: Working Underwater**

You're setting up a raft in a swimming area, and you need to move a 60-kg concrete block on the lake bottom. What's the apparent weight of the block as you lift it underwater? The density of concrete is 2200 kg/m³.

INTERPRET We interpret this as a problem about buoyancy; the concrete will seem to weigh less underwater because of the upward buoyancy force. We identify the apparent weight as the force you'll need to apply to lift the block off the lake bottom.

DEVELOP Figure 15.8 is our sketch, showing gravity and the buoyancy force on the block; you'll need to apply a force equal but opposite to their sum. Archimedes' principle applies, giving a buoyancy force

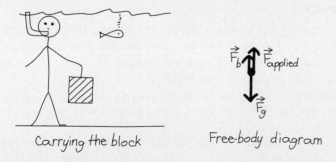

Carrying the block   Free-body diagram

FIGURE 15.8 What's the apparent weight of the concrete block?

equal to the weight of water that occupies the same volume as the concrete block. So our plan is to find that force and compare it with the gravitational force on the block.

EVALUATE The concrete block's mass is m_c, so its weight is the gravitational force $F_g = m_c g$. Its volume is $V_c = m_c/\rho_c$, which also equals the volume of the displaced water: $V_w = V_c = m_c/\rho_c$. Archimedes' principle says that the weight of this displaced water is the magnitude of the buoyancy force, so $F_b = m_w g = V_w \rho_w g = m_c g(\rho_w/\rho_c)$. Then the upward buoyancy force and the downward gravitational force sum to give a downward force of magnitude:

$$F_g - F_b = m_c g - m_c g\left(\frac{\rho_w}{\rho_c}\right) = m_c g\left(1 - \frac{\rho_w}{\rho_c}\right)$$

$$= (60\text{ kg})(9.8\text{ m/s}^2)\left(1 - \frac{1}{2.2}\right) = 320\text{ N}.$$

You have to apply an upward force of equal magnitude to lift the block off the bottom.

ASSESS This is about 70 lb—a lot more manageable than the block's weight mg of nearly 600 N or about 130 lb in air. Knowing the apparent weight of a submerged object would let us turn this problem around to determine its density. Archimedes purportedly used his principle in this way to find the density of the king's crown, and thus verify that it was indeed made of gold. ∎

Floating Objects

Archimedes' principle still holds for a floating object. But now the buoyancy force must balance the object's weight—which will happen if the fluid displaced by the submerged part of the object has a weight equal to that of the object. This condition determines how high in the water the object floats, as the next example illustrates.

EXAMPLE 15.4 **Floating Objects: The Tip of the Iceberg**

The average density of a typical arctic iceberg is 0.86 that of seawater. What fraction of an iceberg's volume is submerged?

INTERPRET We interpret this problem also as being about buoyancy, but now we have a floating object with buoyancy balancing gravity. Only the submerged portion contributes to the buoyancy force, so the condition of force balance will enable us to find how much of the iceberg is submerged.

DEVELOP Figure 15.9 is our sketch, showing gravitational and buoyancy forces of equal magnitude. Archimedes' principle applies here, and states that the buoyancy force is equal to the weight of water displaced by the submerged portion of the iceberg. So our plan is to find the gravitational and buoyancy forces, and then equate their magnitudes to get the submerged volume. Since we're looking for volume, we'll write any masses as products of density and volume.

EVALUATE The iceberg's weight is $w_{ice} = m_{ice}g = \rho_{ice}V_{ice}g$, where V_{ice} is the volume of the *entire* iceberg. Only the submerged portion displaces water, so the volume of displaced water is V_{sub}, and the weight of the displaced water is therefore $w_{water} = m_{water}g = \rho_{water}V_{sub}g$. By Archimedes' principle, w_{water} is equal in magnitude to the buoyancy

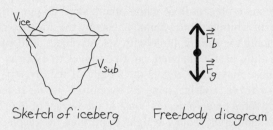

FIGURE 15.9 How much of the iceberg is submerged?

force, which balances gravity when the iceberg is in equilibrium. Equating the two gives $\rho_{water}V_{sub}g = \rho_{ice}V_{ice}g$, which we solve to get

$$\frac{V_{sub}}{V_{ice}} = \frac{\rho_{ice}}{\rho_{water}} = 0.86$$

ASSESS Our result means that 86% of the iceberg's volume is under water, leaving only 14% showing. Tip of the iceberg, indeed! Note that the volume ratio is just the density ratio ρ_{ice}/ρ_{water}, showing that the closer an object's density is to that of water, the lower it floats. ∎

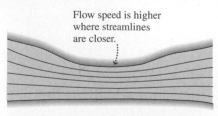

(a)

With CB above CM, torque tends to right the boat.

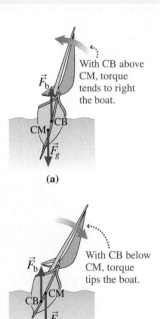

(b)

With CB below CM, torque tips the boat.

FIGURE 15.10 A boat's stability requires the center of buoyancy (CB) to be above the center of mass (CM).

Flow speed is higher where streamlines are closer.

FIGURE 15.11 Streamlines represent flow velocity in a river.

GOT IT? 15.1 Arctic sea ice is melting as a result of global warming. Does this process contribute to a rise in sea level? Explain.

Center of Buoyancy

The buoyancy force acts not at the center of mass of a floating object, but at the center of mass of the water that would be there if the object weren't. This point is called the **center of buoyancy**, and for an object to float in stable equilibrium, the center of buoyancy must lie above the center of mass. Otherwise, a net torque results that tends to tip the object. The stability of watercraft depends critically on this condition (Fig. 15.10).

15.4 Fluid Dynamics

We now turn our attention to moving fluids, which are described by the flow velocity at each point in the fluid and at each instant of time. We can describe flow velocity by drawing continuous lines called **streamlines** that are everywhere tangent to the local flow direction (Fig. 15.11). Their spacing is a measure of flow speed, with closely spaced streamlines indicating regions of higher speed. Small particles introduced into moving fluids follow streamlines and therefore give a visual indication of the flow velocity pattern.

In **steady flow**, the pattern of fluid motion remains the same at each point, even though individual fluid elements are in continuous motion. A river in steady flow always looks the same, even though you're not seeing the same water each time you look. At a given point, the water velocity is always the same. **Unsteady flow**, in contrast, involves fluid motion that changes with time. The blood in your arteries is in unsteady flow; with each contraction of the heart ventricles, the pressure rises and the flow velocity increases. We'll restrict our quantitative description of fluid motion to steady flow.

Like all other motion in classical physics, fluid motion is governed by Newton's laws. It's possible to write Newton's second law in a form that involves explicitly the fluid velocity as a function of position and time. But the resulting equation is difficult to solve in any but the simplest cases. Instead of applying Newton's law directly, we'll approach fluid dynamics using energy conservation.

GOT IT? 15.2 The photo shows smoke particles tracing streamlines in a test of a car's aerodynamic properties. Is the flow speed greater (a) over the top or (b) at the back?

Conservation of Mass: The Continuity Equation

In mechanics we had no trouble keeping track of the individual objects. But a fluid is continuous and deformable, so it's not easy to follow an individual fluid element as it moves. Yet fluid is conserved; as it moves, new fluid is neither created nor destroyed.

Consider a steady fluid flow represented by streamlines, as shown in Fig. 15.12a. We shaded a **flow tube**—a small tubelike region bounded on its sides by streamlines and on its ends by areas at right angles to the flow. The flow tube has a sufficiently small cross section that fluid velocity and other properties don't vary significantly over any cross section; however, fluid properties may vary along the flow tube. Although our flow tube has no physical boundaries, it nevertheless acts like a pipe because fluid flows *along*, not across, the streamlines. In steady flow, the rate at which fluid enters the tube at its left end must equal the rate at which it exits at the right.

Figure 15.12b shows a small fluid element just about to enter the flow tube, a process that will take some time Δt. Suppose the fluid is moving at speed v_1; since it takes time Δt to cross the tube end, its length is $v_1 \Delta t$. With cross-sectional area A_1, length $v_1 \Delta t$, and density ρ_1, the mass of the entering fluid is $m = \rho_1 A_1 v_1 \Delta t$.

Another fluid element is shown just about to leave the tube. Suppose it has the *same* mass m as the entering fluid element. Then it must exit the tube in the *same* time Δt in order to keep the total mass in the tube constant. Its mass can be written as $m = \rho_2 A_2 v_2 \Delta t$.

Equating our two expressions for m shows that $\rho_1 v_1 A_1 = \rho_2 v_2 A_2$. Since the endpoints of the tube are arbitrary, we conclude that the quantity $\rho v A$ must have the same value anywhere along the flow tube:

$$\rho v A = \text{constant along a flow tube} \quad \left(\begin{array}{c} \text{continuity equation,} \\ \text{any fluid} \end{array} \right) \quad (15.4)$$

Equation 15.4 is the **continuity equation**, which expresses the conservation of mass in steady fluid flow. The units of $\rho v A$ here are $(\text{kg/m}^3)(\text{m/s})(\text{m}^2)$, or simply kg/s. This quantity is therefore the **mass flow rate** or mass of fluid per unit time passing through the flow tube. Equation 15.4 says that the mass flow rate is constant in steady flow.

For a liquid, the density ρ is constant, and the continuity equation becomes simply

$$v A = \text{constant along a flow tube} \quad \left(\begin{array}{c} \text{continuity equation,} \\ \text{liquid} \end{array} \right) \quad (15.5)$$

Now the constant quantity is just $v A$, with units of $(\text{m/s})(\text{m}^2)$, or m³/s. This is the **volume flow rate**. In a fluid of constant density, constancy of mass flow rate also implies constancy of volume flow rate because a given mass of fluid doesn't change volume. The continuity equation makes obvious physical sense. Where the liquid has a large cross-sectional area, it flows slowly to transport a given volume of fluid per unit time. But in a constricted area, it must flow faster to carry the same volume. With a gas, obeying Equation 15.4 but not necessarily 15.5, the situation is slightly more ambiguous because density variations also play a role. For flow speeds below the speed of sound in a gas, it turns out that smaller area implies a higher flow speed just as for a liquid. But when the gas flow speed exceeds the sound speed, density changes become so great that flow speed actually decreases with smaller area.

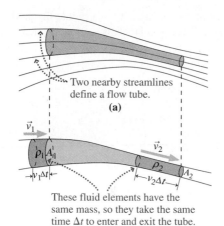

Two nearby streamlines define a flow tube.
(a)

These fluid elements have the same mass, so they take the same time Δt to enter and exit the tube.
(b)

FIGURE 15.12 In steady flow, fluid enters and leaves a flow tube at the same rate.

EXAMPLE 15.5 **Using the Continuity Equation: Ausable Chasm**

The Ausable River in upstate New York is about 40 m wide. Under typical early summer conditions, it's 2.2 m deep and flows at 4.5 m/s. Just before it reaches Lake Champlain, the river enters Ausable Chasm, a deep gorge only 3.7 m wide. If the flow rate in the gorge is 6.0 m/s, how deep is the river at this point? Assume a rectangular cross section with uniform flow speed.

INTERPRET The concept behind this problem is mass conservation, embodied in the continuity equation for a liquid, Equation 15.5. Since the flow is uniform over the river's cross section, we can treat the entire river as a single flow tube.

DEVELOP Equation 15.5 says that the product vA is constant. For the river's rectangular cross section, the area A is the product of width w

and depth d. Then Equation 15.5 becomes $v_1 w_1 d_1 = v_2 w_2 d_2$, where the subscripts indicate values upstream and in the gorge. Our plan is to solve for the depth d_2 in the gorge.

EVALUATE Solving gives

$$d_2 = \frac{v_1 w_1 d_1}{v_2 w_2} = \frac{(4.5 \text{ m/s})(40 \text{ m})(2.2 \text{ m})}{(6.0 \text{ m/s})(3.7 \text{ m})} = 18 \text{ m}$$

ASSESS This is about 60 feet, quite a depth for a small river! But conservation of mass requires it. In the gorge, the river is much narrower but its flow speed is only a little higher, so it's got to be a lot deeper. ∎

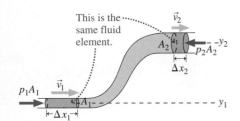

FIGURE 15.13 A flow tube showing the same fluid element entering and leaving. The work done by pressure and gravitational forces equals the change in kinetic energy of the fluid element.

Conservation of Energy: Bernoulli's Equation

We now turn to conservation of fluid energy. Figure 15.13 shows the same fluid element as it enters and again as it leaves a flow tube. If it enters with speed v_1 and leaves with speed v_2, the change in its kinetic energy is

$$\Delta K = \tfrac{1}{2} m (v_2^2 - v_1^2)$$

The work-energy theorem (Equation 6.14) equates this change to the net work done on the fluid element. As the element enters the tube, it's subject to a pressure force $p_1 A_1$ from the fluid to its left. This external force acts over the length Δx_1 of the fluid element as it enters, so it does work $W_1 = p_1 A_1 \Delta x_1$. Similarly, as it leaves the tube, the fluid element experiences a force $p_2 A_2$ from the fluid to its right. Because this force is opposite the flow direction, it does negative work $W_2 = -p_2 A_2 \Delta x_2$. External forces from adjacent flow tubes act at right angles to the flow, so they do no work. Finally, the fluid element rises a distance $y_2 - y_1$ as it traverses the tube; therefore gravity does negative work $W_g = -mg(y_2 - y_1)$. Summing these three contributions and applying the work-energy theorem, we have $W_1 + W_2 + W_g = \Delta K$, or $p_1 A_1 \Delta x_1 - p_2 A_2 \Delta x_2 - mg(y_2 - y_1) = \tfrac{1}{2} m(v_2^2 - v_1^2)$. The quantities $A_1 \Delta x_1$ and $A_2 \Delta x_2$ are the volumes of the fluid element as it enters and leaves the flow, respectively. If we restrict ourselves to incompressible fluids, then those volumes are equal. Dividing through by this common volume $V = A \Delta x$ and noting that $m/V = \rho$, we get for our equation $p_1 + \tfrac{1}{2} \rho v_1^2 + \rho g y_1 = p_2 + \tfrac{1}{2} \rho v_2^2 + \rho g y_2$, or

$$p + \tfrac{1}{2} \rho v^2 + \rho g y = \text{constant along a flow tube} \quad \text{(Bernoulli's equation)} \quad (15.6)$$

This is called **Bernoulli's equation,** after the Swiss mathematician Daniel Bernoulli (1700–1782).

What do the terms in Bernoulli's equation mean? The quantity $\tfrac{1}{2} \rho v^2$ looks like the kinetic energy $\tfrac{1}{2} mv^2$, except it has mass per unit volume ρ instead of mass m. It's therefore the kinetic energy per unit volume, or kinetic-energy density. Similarly, $\rho g y$ is the gravitational potential energy per unit volume. Pressure p, too, has the units of energy density and represents internal energy of the fluid. Bernoulli's equation therefore says that the total energy per unit volume of fluid is conserved as the fluid moves.

Bernoulli's equation in the form 15.6 applies to incompressible fluids. It neglects fluid friction, also called *viscosity*, that may dissipate fluid kinetic energy. It also neglects energy transfers associated with machinery such as turbines or pumps that may extract or add to the fluid's energy. Engineers often include those effects in Bernoulli's equation.

15.5 Applications of Fluid Dynamics

The laws of mass and energy conservation that we just derived for fluids allow us to analyze a wide variety of natural and technological phenomena. We'll usually need both the continuity equation and Bernoulli's equation, considering the values of the appropriate constant quantities at two points in a fluid flow. As you study the examples and applications that follow, remember that they're ultimately based in the same Newtonian principles we've been using to describe mechanical systems.

PROBLEM SOLVING STRATEGY 15.1   **Fluid Dynamics**

The continuity equation and Bernoulli's equation are the keys to solving problems in fluid dynamics. Here's a strategy that will help you focus these two equations on a problem.

INTERPRET The form of Bernoulli's equation we derived applies only to incompressible fluids. So be sure you're dealing either with a liquid or with a gas flowing at speeds well below its sound speed.

DEVELOP

- Identify a flow tube. This may be a physical pipe or other structure, or a mathematical tube bounded by streamlines.
- Draw a sketch of the situation, showing the flow tube.
- Determine the point where you're interested in solving for some aspect of the flow, and another point where you know the quantities that go into the continuity equation and Bernoulli's equation. Note those quantities that you know at each point. Mark the two points on your sketch.
- Write the the continuity equation and Bernoulli's equation, with the known quantities forming the terms on one side and the other side containing your unknown(s).

EVALUATE Evaluate by solving your equations for the unknown quantity or quantities. Often this will involve solving the continuity equation first and then using the result in Bernoulli's equation.

ASSESS Ask whether your result makes sense. Does flow speed increase at a constriction? Does pressure go up when flow speed drops, or vice versa? Are there any limitations that apply, or insights to be gained?

EXAMPLE 15.6   **Bernoulli's Equation: Draining a Tank**

A large, open tank is filled to height h with liquid of density ρ. Find the speed of liquid emerging from a small hole at the base of the tank.

INTERPRET We're dealing with a flow of water, an incompressible liquid. So we can apply our problem-solving strategy for fluid dynamics.

DEVELOP We take the tank to be a rather oddly shaped flow tube, and Fig. 15.14 is our sketch. We're interested in the water's velocity at the hole, so the hole is one of the points we'll use in the fluid equations. Since the hole is open to the atmosphere, the pressure at the hole is atmospheric pressure p_a. The top surface is also open to the atmosphere, so here the pressure is also p_a. Now, because the hole is very small in relation to the tank, the water level drops only slowly. Therefore we can make the approximation $v = 0$ at the top—and thus we know both p and v at the top. Although we didn't write a formal equation here, that approximation follows from the continuity equation because the ratio of hole to top surface area is so small. We also need the potential-energy terms in Bernoulli's equation. If we take $y = 0$ at the hole, then those terms are zero at the hole and $\rho g h$ at the top. Then Bernoulli's equation, $p + \frac{1}{2}\rho v^2 + \rho g y = \text{constant}$, becomes

$$p_a + \rho g h = p_a + \tfrac{1}{2}\rho v_{\text{hole}}^2$$

where the terms on the left are at the top surface and those on the right are at the hole. We've taken care of the continuity equation through our assumption of negligible flow speed at the top.

EVALUATE Atmospheric pressure cancels, and we solve for the unknown flow velocity at the hole:

$$v_{\text{hole}} = \sqrt{2gh}$$

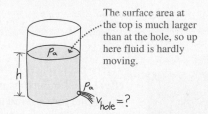

The surface area at the top is much larger than at the hole, so up here fluid is hardly moving.

FIGURE 15.14 How fast does the liquid emerge from the tank?

ASSESS This is the same result we would get by dropping an object from a height h—and for the same reason: conservation of energy. Draining a gram of water from the hole is energetically equivalent to removing a gram of water from the top and dropping it. Just as the speed of a falling object is independent of its mass, so the speed of the liquid is independent of its density. As the liquid drains, of course, the height decreases and so does the flow rate. That's a calculus challenge you can try in Problem 69.

✓**TIP** Reasonable Approximations

Making reasonable approximations is often important in solving realistic problems. Look for opportunities to approximate a physical quantity, especially when other terms appear more significant. But always be sure that your approximations are reasonable. In this example, we reasoned that the fluid's speed at the top of the tank was negligible because it's proportional to the ratio of the hole to the top surface area, a very small value.

■

Venturi Flows and the Bernoulli Effect

A constriction in a pipe carrying incompressible fluid requires that the flow speed increase in order to maintain constant mass flow. Such a constriction is a **venturi**. Because of the increased speed, Bernoulli's equation requires the pressure to be lower in the venturi. The next example shows how this effect provides a measure of fluid flow.

EXAMPLE 15.7 **Measuring Flow Speed: A Venturi Flowmeter**

An incompressible fluid of density ρ flows through a horizontal pipe of cross-sectional area A_1. The pipe has a venturi constriction of area A_2, and a gauge measures the pressure difference Δp between the unconstricted pipe and the venturi. Find an expression for the flow speed in the unconstricted pipe.

INTERPRET This is a problem about incompressible fluid flow, so our strategy applies.

DEVELOP For a flow tube, we choose a section of pipe that includes the venturi. Figure 15.15 is a sketch showing some streamlines through this tube. We're interested in the flow velocity in the unconstricted pipe, so any point outside the venturi will do. The other point should

be in the venturi. The continuity equation then reads $v_1 A_1 = v_2 A_2$, where the subscript 1 refers to the unconstricted pipe and 2 to the venturi. The pipe is horizontal, so the potential-energy term $\rho g h$ in Bernoulli's equation is the same on both sides, and it drops out. Bernoulli's equation then reads

$$p_1 + \tfrac{1}{2}\rho v_1^2 = p_2 + \tfrac{1}{2}\rho v_2^2$$

EVALUATE We can eliminate the velocity v_2 by solving the continuity equation: $v_2 = (A_1/A_2)v_1 = bv_1$, where we defined b as the ratio of the larger to smaller area. Using this result in Bernoulli's equation gives $p_1 + \tfrac{1}{2}\rho v_1^2 = p_2 + \tfrac{1}{2}\rho b^2 v_1^2$. In terms of the pressure difference $\Delta p = p_1 - p_2$, this becomes $\Delta p = \tfrac{1}{2}\rho b^2 v_1^2 - \tfrac{1}{2}\rho v_1^2 = \tfrac{1}{2}\rho v_1^2(b^2 - 1)$. We then solve for v_1 to get our answer:

$$v_1 = \sqrt{\frac{2\,\Delta p}{\rho(b^2 - 1)}}$$

ASSESS Make sense? The pressure difference results from the change in speed; no flow, no pressure difference. So it's reasonable that v increases with Δp. But a given pressure difference Δp is easier to get with a larger area ratio b, so flow speed depends inversely on b. Finally, the greater inertia of a denser fluid means a given pressure difference produces less acceleration, implying a lower initial speed; that's why ρ appears in the denominator.

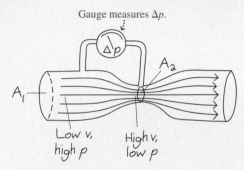

Gauge measures Δp.

Low v, high p High v, low p

FIGURE 15.15 Our sketch of a venturi flowmeter.

High v, low p

Low v, high p

FIGURE 15.16 A ping-pong ball supported by downward-flowing air. High-velocity flow is inside the narrow part of the funnel.

The occurrence of lower pressure with higher flow speeds, and vice versa—the **Bernoulli effect**—has numerous manifestations. The dirt around a prairie dog's hole is mounded up in a way that forces wind to accelerate over the hole, resulting in lower pressure above the hole. Biologists speculate that prairie dogs have evolved this design to provide natural ventilation. The Bernoulli effect can be strikingly counterintuitive. Figure 15.16 shows a ping-pong ball suspended by *downward* airflow in an inverted funnel. Rapid divergence of the flow results in lower speed and therefore higher pressure below the ball.

GOT IT? 15.3 A large tank is filled with liquid to the level h_1 shown in the figure. It drains through a small pipe whose diameter varies; emerging from each section of pipe are vertical tubes open to the atmosphere. Although the picture shows the same liquid level in each pipe, they really won't be the same. Rank order the levels h_1 through h_4.

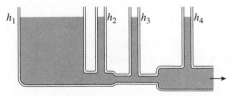

h_1 h_2 h_3 h_4

Flight and Lift

Airplanes, helicopters, and birds fly using forces resulting from their dynamic interaction with the air. Hydrofoil boats, water skis, and high-performance sailboards have analogous interactions with water. Projectiles such as baseballs and missiles, though not supported by the air, have their trajectories substantially modified by aerodynamic forces.

One of the simplest examples of aerodynamic **lift** is the helicopter. Its whirling blades are tilted so they force air downward as they move, just like a giant fan (Fig. 15.17). By Newton's third law, the air exerts an upward force on the blades, ultimately supporting the helicopter. An airplane wing works in the same way, except that it moves forward in a straight line instead of describing a circle. Wings are shaped to maximize the downward deflection of the air even with the wing horizontal, but in principle even a flat board would function as a wing if it were tilted to the oncoming air. Figure 15.18 shows the airflow around a wing. Note how the flow, initially horizontal, leaves the wing moving downward—a clear indication that the wing has exerted a downward force on the air. The third law requires a corresponding upward force, and that's what supports the plane.

Baseball's "curve ball" provides another example of aerodynamic lift. Figure 15.19a is a top view of the airflow around a baseball that's not spinning; the flow is symmetric and the air isn't deflected. But if the ball spins as shown in Fig. 15.19b, air is dragged around the ball and deflected. A corresponding third-law force then acts on the ball, curving its path.

Bernoulli's equation is frequently invoked to explain lift forces. It's true, as Figs. 15.18 and 15.19b suggest, that flow speeds are higher, and therefore—according to Bernoulli's equation—pressures are lower on top of a wing or spinning ball. Forces associated with that pressure difference provide the lift, so Bernoulli can help explain what's going on. But those pressure differences are manifestations of a simpler underlying phenomenon—namely, the paired forces of Newton's third law.

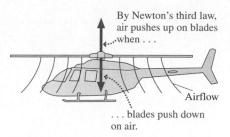

FIGURE 15.17 Newton's third law explains the helicopter's flight.

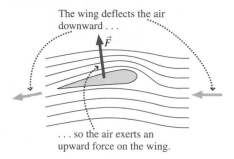

FIGURE 15.18 Flow past a wing.

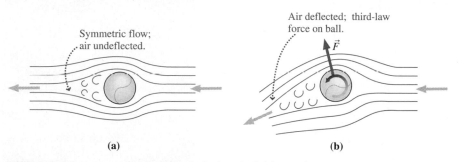

FIGURE 15.19 Top views of airflow around a baseball: (a) no spin; (b) spinning, thus a curve ball.

APPLICATION **Wind Energy**

Wind turbines extract kinetic energy from moving air. In a wind with speed v, Bernoulli's equation shows that the air has kinetic-energy density $\frac{1}{2}\rho v^2$. A chunk of air that passes through a wind turbine in time Δt has length $v\,\Delta t$ and volume $vA\,\Delta t$, where A is the area swept out by the blades. The kinetic energy in this volume is the energy density times the volume: $\Delta K = \left(\frac{1}{2}\rho v^2\right)\left(vA\,\Delta t\right) = \frac{1}{2}\rho v^3 A\,\Delta t$. Dividing by $A\,\Delta t$ gives the energy per time per unit area—that is, the power per unit area available from the wind:

$$\text{wind power per unit area} = \tfrac{1}{2}\rho v^3$$

Unfortunately, we can't extract *all* this energy because then the air would come to a complete stop behind the turbine, halting the flow. A careful analysis shows that the maximum rate for wind-energy extraction is $\frac{8}{27}\rho v^3$, about 59% of the wind's energy. Given air's density of 1.2 kg/m³, this means a 10-m/s wind amounts to some 350 W/m². The factor v^3 shows that the available power increases rapidly at higher speeds. The best practical wind turbines can achieve about 80% of the theoretical maximum. Wind is the fastest-growing component of the world's energy supply, and in some European countries it provides as much as 20% of the electrical energy.

15.6 Viscosity and Turbulence

Moving fluid interacts with the surfaces it contacts, resulting in a kind of fluid friction called **viscosity**. Viscosity also results from the transfer of momentum among adjacent layers within a fluid. Viscosity is especially important right near fluid boundaries because viscous forces bring the fluid to a complete stop at the boundary (Fig. 15.20). This boundary effect produces drag forces on objects moving through fluids—but it's the same drag at the surfaces of airplane and ship propellers that exerts a force on the fluid. Without viscosity, propellers would spin uselessly and planes and ships would go nowhere.

Viscosity depends on fluid properties and dimensions. Honey is more viscous than water, but at the tiny scales of a human capillary or a bacterium wiggling its flagella for propulsion, water too can be extremely viscous. Viscosity is also important in stabilizing flows that would otherwise become **turbulent**, or chaotically unsteady. Turbulence results from the growth of waves that gain energy at the expense of the flow, turning a smooth flow into a chaotic mess (Fig. 15.21). Turbulence is still not fully understood and presents ongoing challenges to scientists and engineers.

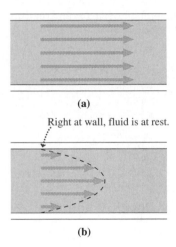

(a)

Right at wall, fluid is at rest.

(b)

FIGURE 15.20 Velocity profiles in (a) inviscid and (b) viscous flow.

FIGURE 15.21 Smooth flow becomes turbulent.

Big Picture

Fluid is matter that readily deforms and flows under the influence of forces. Pressure, density, and flow velocity characterize fluids. Liquids and slowly moving gases are **incompressible**, meaning their density doesn't change significantly. A fluid that isn't moving is in **hydrostatic equilibrium**. In the presence of gravity, equilibrium requires that fluid pressure increase with depth.

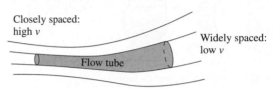

Increasing
p

A solid maintains its shape.   A liquid takes the shape of its container.   A gas fills a closed container.

Key Concepts and Equations

Pressure is the force per unit area: $p = F/A$. The pressure in a fluid exerts itself equally in all directions.

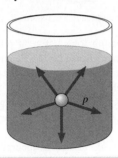

Streamlines represent a moving fluid.

Closely spaced: high *v*

Flow tube

Widely spaced: low *v*

The **continuity equation** describes the conservation of mass along a flow tube:

$$\rho v A = \text{constant (any fluid)}$$

$$v A = \text{constant (incompressible fluid)}$$

Bernoulli's equation describes the conservation of energy:

$$p + \tfrac{1}{2}\rho v^2 + \rho g h = \text{constant} \quad \text{(incompressible fluid, neglecting viscosity)}$$

Viscosity, or fluid friction, is especially important when fluids interact with solid objects.

Cases and Uses

Archimedes' principle states that the **buoyancy force** $\vec{F}_b$ due to pressure on an object is equal to the weight of the displaced fluid. For an object less dense than a fluid, the buoyancy force exceeds gravity and the object floats; otherwise, it sinks or is in neutral buoyancy.

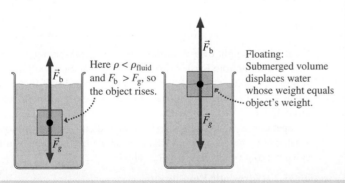

Here $\rho < \rho_{\text{fluid}}$ and $F_b > F_g$, so the object rises.

Floating: Submerged volume displaces water whose weight equals object's weight.

Bernoulli's principle helps explain lift forces, although ultimately these are based in Newton's third law.

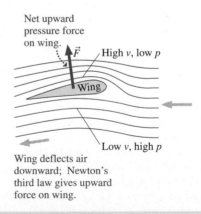

Net upward pressure force on wing.
$\vec{F}$
High *v*, low *p*
Wing
Low *v*, high *p*
Wing deflects air downward; Newton's third law gives upward force on wing.

For Thought and Discussion

1. Why do your ears "pop" when you drive up a mountain?
2. The cabins of commercial jet aircraft are usually pressurized to the pressure of the atmosphere at about 2 km above sea level. Why don't you feel the lower pressure on your entire body?
3. Water pressure at the bottom of the ocean arises from the weight of the overlying water. Does this mean that the water exerts pressure only in the downward direction? Explain.
4. The three containers in Fig. 15.22 are filled to the same level and are open to the atmosphere. How do the pressures at the bottoms of the three containers compare?

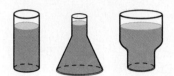

FIGURE 15.22 For Thought and Discussion 4

5. Why is it easier to float in the ocean than in fresh water?
6. Figure 15.23 shows a cork suspended from the bottom of a sealed container of water. The container is on a turntable rotating about a vertical axis, as shown. Explain the position of the cork.

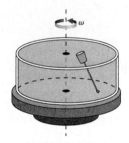

FIGURE 15.23 For Thought and Discussion 6

7. Meteorologists in the United States usually report barometer readings in inches. What are they talking about?
8. A mountain stream, frothy with entrained air bubbles, presents a serious hazard to hikers who fall into it, for they may sink in the stream where they would float in calm water. Why?
9. Why are dams thicker at the bottom than at the top?
10. It's not possible to breathe through a snorkel from a depth greater than a meter or so (Fig. 15.24). Why not?

FIGURE 15.24 For Thought and Discussion 10

11. A helium-filled balloon stops rising long before it reaches the "top" of the atmosphere, but a cork released from the bottom of a lake rises all the way to the surface of the water. Explain the difference between these two behaviors.
12. A barge filled with steel beams overturns in a lake, spilling its cargo. Does the water level in the lake rise, fall, or remain the same?
13. Under what conditions can a gas be treated as incompressible?
14. Why do airplanes take off into the wind?
15. Is the flow speed behind a wind turbine greater or less than the flow speed in front? Is the pressure behind the turbine higher or lower than in front? Is there a violation of Bernoulli's equation here? Explain.

Exercises and Problems

Exercises

Section 15.1 Density and Pressure

16. The density of molasses is 1600 kg/m³. Find the mass of the molasses in a 0.75-L jar.
17. The density of atomic nuclei is about 10^{17} kg/m³, while the density of water is 10^3 kg/m³. Roughly what fraction of the volume of water is *not* empty space?
18. Compressed air with mass 8.8 kg is stored in a gas cylinder with a volume of 0.050 m³. (a) What is the density of the compressed air? (b) How large a volume would the same gas occupy at typical atmospheric density of 1.2 kg/m³?
19. The pressure unit **torr** is defined as the pressure that will support a column of mercury 1 mm high. Meteorologists often give barometric pressure in **inches of mercury**, defined analogously. Express each of these in SI units. The density of mercury is 1.36×10^4 kg/m³.
20. Measurement of small pressure differences—for example, between the interior of a chimney and the ambient atmosphere—is often given in **inches of water**, where 1 in. of water is the pressure that will support a 1-in.-high water column. Express this in SI units.

21. What is the weight of a column of air with cross-sectional area 1 m² extending from Earth's surface to the top of the atmosphere?
22. A 4300-kg circus elephant balances on one foot. If the foot is a circle 30 cm in diameter, what pressure does it exert on the ground?
23. A paper clip is made from wire 1.5 mm in diameter. You unbend a paper clip and push the end against the wall. What force must you exert to give a pressure of 120 atm?

Section 15.2 Hydrostatic Equilibrium

24. What is the density of a fluid whose pressure increases at the rate of 100 kPa for every 6.0 m of depth?
25. A research submarine can withstand an external pressure of 50 MPa when its internal pressure is 100 kPa. How deep can it dive?
26. Scuba equipment provides the diver with air at the same pressure as the surrounding water. But at pressures higher than about 1 MPa, the nitrogen in air becomes dangerously narcotic. At what depth does nitrogen narcosis become a hazard?
27. A vertical tube open at the top contains 5.0 cm of oil (density 0.82 g/cm³) floating on 5.0 cm of water. Find the *gauge* pressure at the bottom of the tube.

28. A child attempts to drink water through a 100-cm-long straw but finds that the water rises only 75 cm. By how much has the child reduced the pressure in her mouth below atmospheric pressure?

29. Barometric pressure in the eye of a hurricane is 0.91 atm (27.2 in. of mercury). How does the level of the ocean surface under the eye compare with the level under a distant fair-weather region where the pressure is 1.0 atm?

Section 15.3 Archimedes' Principle and Buoyancy

30. On land, the most massive concrete block you can carry is 25 kg. How massive a block could you carry underwater, if the density of concrete is 2200 kg/m³?

31. A 5.4-g jewel has an apparent weight of 32 mN when submerged in water. Could the jewel be diamond (density 3.51 g/cm³)?

32. The density of Styrofoam is 160 kg/m³. What percent error is introduced by weighing a Styrofoam block in air, which exerts an upward buoyancy force, rather than in vacuum? The density of air is 1.2 kg/m³.

33. A steel drum has volume 0.23 m³ and mass 16 kg. Will it float in water when filled with (a) water or (b) gasoline (density 860 kg/m³)? Neglect the thickness of the steel.

Sections 15.4 and 15.5 Fluid Dynamics and Applications

34. A 2.5-cm-diameter pipe is full of water flowing at 1.8 m/s. If the pipe narrows to 2.0-cm diameter, what is the flow speed in the narrow section?

35. A fluid is flowing steadily, roughly from left to right. At left it flows rapidly; it then slows down, and finally speeds up again. Its final speed at right is not as great as its initial speed at left. Sketch a streamline pattern that could represent this flow.

36. Show that pressure has the units of energy density.

37. A typical mass flow rate for the Mississippi River is 1.8×10^7 kg/s. Find (a) the volume flow rate and (b) the flow speed in a region where the river is 2.0 km wide and an average of 6.1 m deep.

38. A fire hose 10 cm in diameter delivers water at the rate of 15 kg/s. The hose terminates in a nozzle 2.5 cm in diameter. What are the flow speeds (a) in the hose and (b) in the nozzle?

39. A typical human aorta, or main artery from the heart, is 1.8 cm in diameter and carries blood at a speed of 35 cm/s. What will be the flow speed around a clot that reduces the flow area by 80%?

Problems

40. When a couple with a total mass of 120 kg lies on a water bed, the pressure in the bed increases by 4700 Pa. What surface area of the two bodies is in contact with the bed?

41. A fully loaded Volvo station wagon has a mass of 1950 kg. If each of its four tires is inflated to a gauge pressure of 230 kPa, what is the total tire area in contact with the road?

42. An airplane's emergency escape window measures 50 cm by 90 cm. The interior pressure is 0.75 atm, and the plane is at an altitude where atmospheric pressure is 0.25 atm. Is there any danger that a passenger could open the window? Answer by calculating the force needed to pull the window straight inward.

43. A vertical tube 1.0 cm in diameter and open at the top contains 5.0 g of oil (density 0.82 g/cm³) floating on 5.0 g of water. Find the *gauge* pressure (a) at the oil-water interface and (b) at the bottom.

44. A 1500-m-wide dam holds back a lake 95 m deep. What force does the water exert on the dam? *Hint:* You'll need to consider the force $d\vec{F}$ on a small area dA, and integrate.

45. A U-shaped tube open at both ends contains water and a quantity of oil occupying a 2.0-cm length of the tube, as shown in Fig. 15.25. If the oil's density is 0.82 times that of water, what is the height difference h?

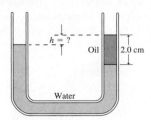

FIGURE 15.25 Problem 45

46. A hydraulic cylinder moving a robotic arm has a diameter of 5.0 cm and can exert a maximum force of 5.6 kN. (a) What pressure must the hydraulic lines be capable of withstanding? (b) The cylinder at the other end of the hydraulic system has a 1.0-cm diameter. What force must be applied to it to get the maximum force out of the cylinder driving the arm?

47. A garage lift has a 45-cm-diameter piston supporting the load. Compressed air with a maximum pressure of 500 kPa is applied to a small piston at the other end of the hydraulic system. What is the maximum mass the lift can support?

48. Archimedes purportedly used his principle to verify that the king's crown was pure gold by weighing the crown while it was submerged in water. Suppose the crown's actual weight was 25.0 N. What would be its apparent weight if it were made of (a) pure gold and (b) 75% gold and 25% silver, by volume? The densities of gold, silver, and water are 19.3 g/cm³, 10.5 g/cm³, and 1.00 g/cm³, respectively.

49. A partially full beer bottle with interior diameter 52 mm is floating upright in water. A drinker takes a swig and replaces the bottle in the water, where it now floats 28 mm higher than before. How much beer did the drinker drink?

50. A glass beaker measures 10 cm high by 4.0 cm in diameter. Empty, it floats in water with one-third of its height submerged. How many 15-g rocks can be placed in the beaker before it sinks?

51. A typical supertanker has mass 2.0×10^6 kg and carries twice that much oil. If 9.0 m of the ship is submerged when it's empty, what is the minimum water depth needed for it to navigate when full? Assume the sides of the ship are vertical.

52. A balloon contains gas of density ρ_g and is to lift a mass M, including the balloon but not the gas. Show that the minimum mass of gas required is

$$m = \frac{M\rho_g}{\rho_a - \rho_g}$$

where ρ_a is the atmospheric density.

53. (a) How much helium (density 0.18 kg/m³) is needed to lift a balloon carrying two people in a basket, if the total mass of people, basket, and balloon (but not gas) is 280 kg? (b) Repeat for a hot-air balloon whose air density is 10% less than that of the surrounding atmosphere.

54. A 55-kg swimmer climbs onto a Styrofoam block whose density is 160 kg/m³. If the water level comes right to the top of the Styrofoam, what is the block's volume?

55. If the blood pressure in the unobstructed artery of Exercise 39 is 16 kPa gauge (about 120 mm of mercury, in the unit commonly reported by doctors), what will it be at the clot? The density of blood is 1060 kg/m³. *Note:* Too low a pressure can actually collapse the artery, momentarily cutting off blood flow.

56. In Fig. 15.26 a horizontal pipe of cross-sectional area A is joined to a lower pipe of cross-sectional area $\frac{1}{2}A$. The entire pipe is full of liquid with density ρ, and the left end is at atmospheric pressure p_a. A small open tube extends upward from the lower pipe. Find the height h_2 of liquid in the small tube (a) when the right end of the lower pipe is closed, so the liquid is in hydrostatic equilibrium, and (b) when the liquid flows with speed v in the upper pipe.

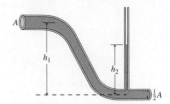

FIGURE 15.26 Problem 56

57. The water in a garden hose is at a gauge pressure of 140 kPa and is moving at negligible speed. The hose terminates in a sprinkler consisting of many small holes. What is the maximum height reached by the water emerging from the holes?

58. The venturi flowmeter shown in Fig. 15.27 is used to measure the flow rate of water in a solar collector system. The flowmeter is inserted in a pipe with diameter 1.9 cm; at the venturi of the flowmeter the diameter is reduced to 0.64 cm. The manometer tube contains oil with density 0.82 times that of water. If the difference in oil levels on the two sides of the manometer tube is 1.4 cm, what is the volume flow rate?

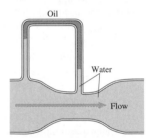

FIGURE 15.27 Problem 58

59. A 1.0-cm-diameter venturi flowmeter is inserted in a 2.0-cm-diameter pipe carrying water (density 1000 kg/m³). What are (a) the flow speed in the pipe and (b) the volume flow rate if the pressure difference between venturi and unconstricted pipe is 17 kPa?

60. A spherical rubber balloon with mass 0.85 g and diameter 30 cm is filled with helium (density 0.18 kg/m³). How many 1.0-g paper clips can you hang from the balloon before it loses its buoyancy?

61. Water at a pressure of 230 kPa is flowing at 1.5 m/s through a pipe, when it encounters an obstruction where the pressure drops by 5%. What fraction of the pipe's area is obstructed?

62. A venturi flowmeter in an oil pipeline has a radius half that of the pipe. The flow speed in the unconstricted flow is 1.9 m/s. If the pressure difference between the unconstricted flow and the venturi is 16 kPa, what is the density of the oil?

63. A drinking straw 20 cm long and 3.0 mm in diameter stands vertically in a cup of juice 8.0 cm in diameter. A section of straw 6.5 cm long extends above the juice. A child sucks on the straw, and the level of juice in the glass begins dropping at 0.20 cm/s. (a) By how much does the pressure in the child's mouth differ from atmospheric pressure? (b) What is the greatest height from which the child could drink, assuming this same mouth pressure?

64. Water emerges from a faucet of diameter d_0 in steady, near-vertical flow with speed v_0. Show that the diameter of the falling water column is given by $d = d_0[v_0^2/(v_0^2 + 2gh)]^{1/4}$, where h is the distance below the faucet (Fig. 15.28).

FIGURE 15.28 Problem 64

65. How massive an object can be supported by a 5.0-cm-diameter suction cup mounted on a vertical wall, if the coefficient of friction between cup and wall is 0.72? Assume normal atmospheric pressure.

66. Figure 15.29 shows a simplified diagram of a Pitot tube, used for measuring aircraft speeds. The tube is mounted on the underside of the aircraft wing with opening A at right angles to the flow and opening B pointing into the flow. The gauge prevents airflow through the tube. Use Bernoulli's equation to show that the air speed relative to the wing is given by $v = \sqrt{2\,\Delta p/\rho}$, where Δp is the pressure difference between the tubes and ρ is the density of air. *Hint:* The flow must be stopped at B, but continues past A with its normal speed.

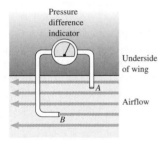

FIGURE 15.29 Problem 66

67. A wind turbine has a blade diameter of 65 m. What is the theoretical maximum power output of this turbine in winds of (a) 10 m/s and (b) 15 m/s? (c) If the machine actually achieves 50% of its theoretical maximum output, how many such turbines would be needed to displace a 1-GW nuclear power plant, assuming an average wind speed of 12 m/s?

68. A pencil is weighted so it floats vertically with length L submerged. It's pushed vertically downward without being totally submerged, then released. Show that it undergoes simple harmonic motion with period $T = 2\pi\sqrt{L/g}$.

69. A can of height h and cross-sectional area A_0 is initially full of water. A small hole of area $A_1 \ll A_0$ is cut in the bottom of the can. Find an expression for the time it takes all the water to drain from the can. *Hint:* Call the water depth y, use the continuity equation to relate dy/dt to the outflow speed at the hole, and then integrate.

70. Density and pressure in Earth's atmosphere are proportional: $\rho = p/h_0 g$, where h_0 is a constant with the approximate value 8.2 km and g is the acceleration of gravity. (a) Integrate Equation 15.2 for this case to show that atmospheric pressure as a function of height h above the surface is given by $p = p_0 e^{-h/h_0}$, where p_0 is the surface pressure. (b) At what height will the pressure have dropped to half its surface value?

71. (a) Use the result of Problem 70 to express Earth's atmospheric density as a function of height (this is straightforward). (b) Use the result of part (a) to find the height below which half of Earth's atmospheric mass lies (this will require integration).

72. A circular pan of liquid (density ρ) is centered on a horizontal turntable rotating with angular speed ω. Its axis coincides with the rotation axis, as shown in Fig. 15.30. Atmospheric pressure is p_a. Find expressions for (a) the pressure at the bottom of the pan and (b) the height of the liquid surface as functions of the distance r from the axis, given that the height at the center is h_0.

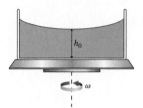

FIGURE 15.30 Problem 72

73. The density of a solid sphere of radius $R = 10$ cm and mass $M = 15$ kg varies with distance from the center according to the equation $\rho = \rho_0 e^{r/R}$. Determine the constant ρ_0.

74. The difference in air pressure between the inside and outside of a ball is very nearly a constant Δp. Show by direct integration that the pressure force on one hemisphere is $\pi R^2 \Delta p$ where R is the ball's radius.

75. What is the torque that the water exerts about the bottom edge of the dam in Problem 44?

76. One vertical wall of an above-ground swimming pool is in the shape of a regular trapezoid, with one base 10 m long on level ground and the other 20 m long a height of 3 m above it. If the pool is filled to the top with water, what is the net fluid force on the wall? (*Hint:* consider both the force exerted by the water on one side of the wall, and the force exerted by the atmosphere on the other side.)

77. You're a private investigator assisting a large food manufacturer in tracking down counterfeit salad dressing. The genuine dressing is one part (by volume) vinegar (density 1.0 g/cm³) to three parts olive oil (density 0.92 g/cm³). The counterfeit dressing would be diluted with water (density 1.0 g/cm³) to increase its density. You measure the density and find it to be 0.97 g/cm³. Has the dressing been altered?

78. You're attempting to convince your pre-med roommate that physics is relevant to medicine. You're studying fluids, so you begin talking about how the heart is like a big pump. "In fact," you exclaim, "your lower blood pressure number is all physics." To prove your point, you calculate the pressure of a column of blood (1060 kg/m³) 1 m high (the distance your heart is above your feet) and convert the result to mm of Hg. What value do you find?

79. A plumber comes to your ancient apartment building where you have a part-time job as a caretaker. He is checking the hot-water heating system. He notes the pressure is 18 psi. He asks you, "How high is the building?" "Three stories, each about 11 feet high," you reply. "OK, about 33 feet," he says, pausing to do some calculations in his head. "The pressure is fine," he states. On what basis did he come to that conclusion? Use the weight density of water in the English system—namely, 62.2 lb/ft³.

80. A friend is taking an architecture class and wants to design a ship. The ship's hull has a V-shaped cross section, as shown in Fig. 15.31. The ship has total length L, measured from bow to stern, and keel-to-deck height h_0. When empty, the distance from the water line to the keel is h_1. When fully loaded, the ship's draft approaches h_0. In order to be as realistic as possible in designing the ship, your friend asks you the maximum load the ship can carry in terms of h_0, h_1, L, and θ. Ignore any volume of the above-deck portion because it doesn't hold any cargo.

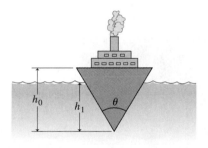

FIGURE 15.31 Problem 80

Answers to Chapter Questions

Answer to Chapter Opening Question
Because the density of ice is only slightly less than that of water.

Answers to GOT IT? Questions
15.1 No. When floating, the ice displaces a mass of water equal in volume to its submerged portion—which is less than the ice's total volume by the ice-to-water density ratio. When the ice melts, it shrinks by the same factor, so it continues to displace the same volume. Melting land ice, in contrast, does contribute to sea-level rise, as does the thermal expansion of seawater.

15.2 (a), over the top, where the streamlines are closer together.

15.3 $h_1 > h_4 > h_2 > h_3$, reflecting higher pressure with lower flow speed.

Oscillations, Waves and Fluids

Part Two has extended Newtonian mechanics to systems that undergo oscillatory motion and wave motion, or that involve the motion of fluids. Behind these more complex motions are the fundamental concepts of force, mass, and energy and their roles in characterizing motion.

Oscillatory motion

Oscillatory motion describes the back-and-forth motion of a system disturbed from a stable equilibrium.

When the force or torque tending to restore equilibrium is directly proportional to the displacement, the result is simple harmonic motion.

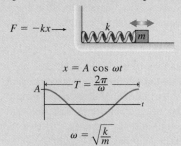

$$F = -kx$$

$$x = A \cos \omega t$$

$$T = \frac{2\pi}{\omega}$$

$$\omega = \sqrt{\frac{k}{m}}$$

Wave motion

A wave is a propagating disturbance that carries energy but not matter.

Simple harmonic waves are sinusoidal:

$$y(x, t) = A \cos(kx - \omega t)$$

Angular frequency $\omega = 2\pi f$

Wave number $k = \dfrac{2\pi}{\lambda}$

Wave period $T = \dfrac{1}{f}$

Wave speed $v = \dfrac{\omega}{k} = \dfrac{\lambda}{T} = f\lambda$

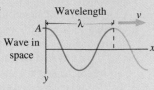

Wave in space

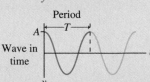

Wave in time

Interference

When waves overlap, the result is interference, which is constructive when the waves reinforce and destructive when they tend to cancel.

Nodal lines: destructive interference   Large amplitude: constructive interference

Standing waves

Standing waves occur when the medium has limited extent. Only certain wavelengths and frequencies are allowed, depending on the medium's length:

Two of the allowed standing waves on a string fixed at both ends

This wavelength isn't allowed

Fluid statics

Fluids in hydrostatic equilibrium exhibit a depth-dependent pressure that results in an upward buoyancy force $\vec{F}_b$.

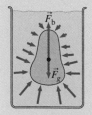

Archimedes's principle states that the buoyancy force is equal to the weight of the displaced fluid.

Fluid dynamics

Moving fluids obey conservation of mass and, in the absence of fluid friction (viscosity), they also conserve energy.

The continuity equation and Bernoulli's equation express these conservation laws. Both equations hold along a flow tube:

Continuity: $\rho v A = $ constant

Bernoulli: $\dfrac{p}{\rho} + \dfrac{1}{2}v^2 + gh = $ constant

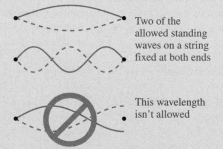

Closely spaced: high v

Widely spaced: low v

Flow tube

Part 2 challenge problem

A cylindrical log of total mass M and uniform diameter d has an uneven mass distribution that causes it to float in a vertical position, as shown in the figure. (a) Find an expression for the length L of the submerged portion of the log when it's floating in equilibrium, in terms of M, d, and the water density ρ. (b) If the log is displaced vertically from its equilibrium position and released, it will undergo simple harmonic motion. Find an expression for the period of this motion, neglecting viscosity and other frictional effects.

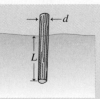

Thermodynamics

Humanity consumes energy at the prodigious rate of some 10^{13} watts. Nearly all that energy comes from the combustion of fossil fuels—a process governed by the laws of thermodynamics. Engines that extract mechanical energy from the heat of burning fuels propel our cars, trucks, and airplanes, and produce most of our electricity. Despite the efforts of the cleverest engineers, the laws of thermodynamics set fundamental limitations on our ability to convert heat to mechanical energy efficiently. The energy and environmental challenges humanity faces today are often grounded in the facts of thermodynamics.

Many natural systems, too, are fundamentally thermodynamic. Without the Sun's energy, radiated across a hundred million miles of empty space, Earth would be a lifeless, frozen rock. Heat flows throughout Earth, its oceans, and its atmosphere govern processes ranging from continental drift to ocean currents to weather and climate. Concern over human-induced climate change is rooted in thermodynamic properties of the atmosphere as they affect thermal energy flows. On a grander scale, thermodynamic principles govern much of the energy that flows throughout the universe.

Thermodynamics—the study of heat and its connection to the all-important concept of energy—is the subject of the next four chapters.

■ This huge steam turbine converts the energy of high-pressure steam to mechanical energy and then electricity. Systems like this one produce nearly all the world's electrical energy, and their operation and efficiency are governed by the laws of thermodynamics.

16

Temperature and Heat

■ How does this photo reveal heat loss from the house? And how can you tell that the car was recently driven?

Your own body gives you a good sense of "hot" and "cold." Questions about heat and temperature are ultimately about energy, and these concepts are crucial to an understanding of the energy flows that drive natural systems like Earth's climate and technologies such as engines, power plants, and refrigerators.

Properties like mass and kinetic energy apply equally to microscopic atoms and molecules and to cars and planets. But other properties, including temperature and pressure, apply only to macroscopic systems. It makes no sense to talk about the temperature or pressure of a single air molecule. **Thermodynamics** is the branch of physics that deals with these macroscopic properties. Ultimately, the thermodynamic behavior of matter follows from the motions of its constituent particles in response to the laws of mechanics. **Statistical mechanics** relates the macroscopic description of matter to the underlying microscopic processes. Historically, thermodynamics developed before the atomic theory of matter was fully established. The subsequent explanation of thermodynamics through statistical mechanics—the mechanics of atoms and molecules—was a triumph for physics.

16.1 Heat, Temperature, and Thermodynamic Equilibrium

Take a bottle of soda from the refrigerator, and eventually it reaches room temperature. At that point the soda and the room are in **thermodynamic equilibrium**, a state in which their macroscopic properties are no longer changing. To check for thermodynamic equilibrium we can consider any macroscopic property—length, volume, pressure, electrical resistance, whatever. If any macroscopic property changes when two systems are placed together, then they weren't originally in thermodynamic equilibrium. When changes cease, the systems have reached equilibrium.

The phrase "placed together" here has a definite meaning, stated more precisely as "placed in thermal contact." Two systems are in **thermal contact** if heating one of them results in macroscopic changes in the other. If that doesn't readily happen—for example, with a Styrofoam cup of coffee and its surroundings—then the systems are **thermally insulated**.

We're now ready to define temperature: **Two systems have the same temperature if they are in thermodynamic equilibrium**. Consider two systems A and C in thermal contact with a third system B but not with each other (Fig. 16.1a). Even though they're not in direct contact, A and C have the same temperature; that is, if you place A and C in thermal contact (Fig. 16.1b), no further changes occur. This fact—that two systems in equilibrium with a third system are therefore in equilibrium with each other—is so fundamental that it's called the **zeroth law of thermodynamics**.

A **thermometer** is a system with a conveniently observed macroscopic property that changes with temperature. It could be the length of a mercury column, gas pressure, electrical resistance, or the bending of a bimetal strip in a dial thermometer. Let the thermometer come to equilibrium with some system, and its temperature-dependent physical property provides a measure of temperature. The zeroth law assures consistency, in that two systems for which the thermometer gives the same reading must have the same temperature.

Gas Thermometers and the Kelvin Scale

Figure 16.2 shows a **constant-volume gas thermometer**, whose temperature indication is the pressure of gas held at constant volume. We define zero temperature as that temperature at which the gas pressure would become zero. A second known point is the so-called **triple point** of water—the unique temperature at which solid, liquid, and gaseous water can coexist in equilibrium (more on this in the next chapter). The SI temperature unit, the **kelvin** (symbol K; *not* "degrees kelvin" or °K), is defined by setting the triple point at 273.16 K. Other temperatures then follow from a linear relationship as shown in Fig. 16.3.

Since a gas can't have negative pressure, the zero of the kelvin scale is an absolute lower limit on temperature and is called **absolute zero**. We'll explore the meaning of absolute zero further in Chapter 19.

Gas thermometers are useful because they work over a wide temperature range. More important, as we'll see in the next chapter, all gases behave in essentially the same way in the limit of low pressure, so gas thermometry provides a reproducible temperature standard.

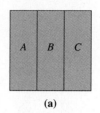

Systems A and C are each in thermodynamic equilibrium with B.

If A and C are placed in thermal contact, their macroscopic properties don't change—showing that they're already in equilibrium.

(a)   **(b)**

FIGURE 16.1 The zeroth law of thermodynamics.

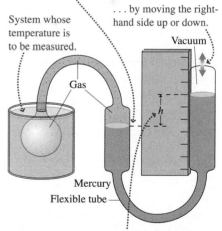

The mercury level in the left-hand side of the tube is maintained constant at this level . . .

System whose temperature is to be measured.

. . . by moving the right-hand side up or down.

Vacuum

Gas

h

Mercury

Flexible tube

The height difference h between the two mercury levels is a measure of the gas pressure and therefore of the temperature.

FIGURE 16.2 A constant-volume gas thermometer.

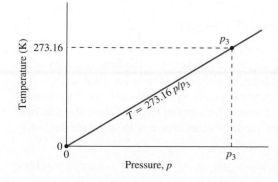

$T = 273.16 \, p/p_3$

FIGURE 16.3 The Kelvin temperature scale defined using a gas thermometer; p_3 is the gas pressure at the triple point of water.

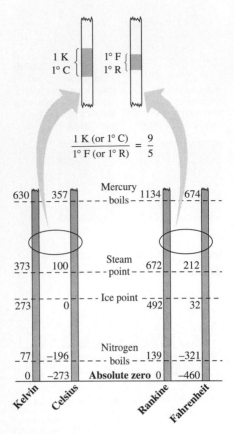

FIGURE 16.4 Relationships among four temperature scales.

Temperature Scales

Other temperature scales include Celsius (°C), Fahrenheit (°F), and Rankine (°R) (Fig. 16.4). One Celsius degree represents the same temperature difference as one kelvin, but the zero of the Celsius scale occurs at 273.15 K, so

$$T_C = T - 273.15 \qquad (16.1)$$

where T is the temperature in kelvins. On the Celsius scale the melting point of ice at standard atmospheric pressure is exactly 0°C, while the boiling point is 100°C. The triple point of water occurs at 0.01°C, which accounts for the 273.15 difference between the kelvin and Celsius scales. Equation 16.1 shows that absolute zero occurs at −273.15°C.

The Fahrenheit and Rankine scales, from the British unit system, are used primarily in the United States. Fahrenheit has water melting at 32°F and boiling at 212°F, so the relation between Fahrenheit and Celsius temperatures is

$$T_F = \tfrac{9}{5}T_C + 32 \qquad (16.2)$$

A Rankine degree is the same size as a Fahrenheit degree, but the zero of the Rankine scale is at absolute zero (Fig. 16.4). Engineers in the United States often use Rankine.

Heat and Temperature

A match will burn your finger, but it doesn't provide much heat. This example shows our intuitive sense of temperature and heat: Heat measures an *amount* of "something," whereas temperature is the *intensity* of that "something."

Scientists once considered heat to be a material fluid, called **caloric**, that flowed from hot bodies to colder ones. But in the late 1700s, the American-born scientist Benjamin Thompson observed essentially limitless amounts of heat being produced in the boring of cannon, and he concluded that heat could not be a conserved fluid. Instead, Thompson suggested, heat was associated with mechanical work done by the boring tool. In the next half-century, a series of experiments confirmed the association between heat and energy. These culminated in the work of the British physicist James Joule (1818–1889), who quantified the relation between heat and energy. In so doing, Joule brought thermal phenomena under the powerful conservation-of-energy principle. In recognition of this major synthesis in physics, the SI energy unit bears Joule's name.

We rarely make statements about the amount of "heat" in an object; we're more concerned that the temperature be appropriate. Rather, we think of heat as something that gets transferred from one object to another, causing a temperature change. The scientific definition reflects this sense of heat as energy in transit: **Heat is energy being transferred from one object to another because of a temperature difference alone**. Strictly speaking, **heat** refers only to energy in transit. Once heat has been transferred, we say that the **internal energy** of the object has increased, not that it contains more heat. This distinction reflects the fact that processes other than heating—such as transfer of mechanical or electrical energy—can also change an object's temperature.

16.2 Heat Capacity and Specific Heat

Experimentally, we find that the heat ΔQ transferred to an object and the resulting temperature change ΔT are directly proportional: $\Delta Q = C\,\Delta T$, where C is the **heat capacity** of the object. Since heat is a measure of energy transfer, the units of heat capacity are J/K. The heat capacity C applies to a specific object and depends on its mass and on the substance from which it's made. We characterize different substances in terms of their **specific heat** c, or heat capacity per unit mass. The heat capacity of an object is then the product of its mass and specific heat, so we can write

$$\Delta Q = mc\,\Delta T \qquad (16.3)$$

The SI units of specific heat are J/kg·K. Table 16.1 lists specific heats of common materials.

TABLE 16.1 Specific Heats of Some Common Materials*

Substance	Specific Heat, c	
	SI Units: J/kg·K	cal/g·°C, kcal/kg·°C, or Btu/lb·°F
Aluminum	900	0.215
Concrete	880	0.24
Copper	386	0.0923
Iron	447	0.107
Glass	753	0.18
Mercury	140	0.033
Steel	502	0.12
Stone (granite)	840	0.20
Water:		
Liquid	4184	1.00
Ice, −10°C	2050	0.49
Wood	1400	0.33

*Temperature range 0°C to 100°C except as noted.

Scientists first studied thermodynamic phenomena before they knew the relation between heat and energy, and they used other units for heat. The **calorie** (cal) was defined as the heat needed to raise the temperature of 1 g of water from 14.5°C to 15.5°C; consequently, the specific heat of water is 1 cal/g·°C. Several different definitions of the calorie exist today, based on different methods for establishing the heat-energy equivalence. In this book we use the so-called thermochemical calorie, defined as exactly 4.184 J. The "calorie" used in describing the energy content of foods is actually a kilocalorie. In the British system, still widely used in engineering in the United States, the unit of heat is the **British thermal unit** (Btu). One Btu is the amount of heat needed to raise the temperature of 1 lb of water from 63°F to 64°F, and is equal to 1054 J.

EXAMPLE 16.1   **Specific Heat: Waiting to Shower**

Your whole family has showered before you, dropping the temperature in the water heater to 18°C. If the heater holds 150 kg of water, how much energy will it take to bring it up to 50°C? If the energy is supplied by a 5.0-kW electric heating element, how long will that take?

INTERPRET Here we're interested in the energy it takes to raise the water temperature, so we interpret this problem as involving specific heat. For the second part, we're given the heater's power output and asked for the time, so we need to recall (Chapter 6) that power is energy per time.

DEVELOP Equation 16.3, $\Delta Q = mc\,\Delta T$, relates energy and temperature change via specific heat, so our plan is to calculate the required energy from this equation. We'll then use the relation between power and energy to find the time.

EVALUATE Equation 16.3 gives the energy:

$$\Delta Q = mc\,\Delta T = (150\ \text{kg})(4184\ \text{J/kg·K})(50°C - 18°C) = 20\ \text{MJ}$$

where we found the specific heat of water in Table 16.1. The heating element supplies energy at the rate of 5.0 kW or 5.0×10^3 J/s. At that rate the time needed to supply 20 MJ is

$$\Delta t = \frac{2.0 \times 10^7\ \text{J}}{5.0 \times 10^3\ \text{J/s}} = 4000\ \text{s}$$

or a little over an hour.

ASSESS That's a long time to wait, but it's not an unreasonable answer!

✓**TIP** Is That C or K?
It doesn't matter when we're talking about temperature *differences*. That's why we could mix units, multiplying the specific heat in J/kg·K by the difference of Celsius temperatures.

Heat capacity and specific heat vary slightly with temperature, and they also depend on whether an object's pressure or its volume changes as it's heated. For solids and liquids, which don't expand much, that distinction is not very important. But it makes a big difference whether a gas is confined or allowed to expand when heated. Consequently, gases have two different specific heats, depending on whether volume or pressure is constant. We'll deal with that issue in Chapter 18, where we explore the thermodynamic behavior of gases.

The Equilibrium Temperature

When objects at different temperatures are in thermal contact, heat flows from the hotter object to the cooler one until they reach thermodynamic equilibrium. If the objects are thermally insulated from their surroundings, then all the energy leaving the hotter object ends up in the cooler one. Mathematically, this statement reads

$$m_1 c_1 \,\Delta T_1 + m_2 c_2 \,\Delta T_2 = 0 \tag{16.4}$$

For the hotter object, ΔT is negative, so the two terms in Equation 16.4 have opposite signs. One term represents the outflow of heat from the hotter object, the other inflow into the cooler one.

GOT IT? 16.1 A hot rock with mass 250 g is dropped into an equal mass of cool water. Which temperature changes more, that of the rock or the water? Explain.

EXAMPLE 16.2 **Finding the Equilibrium Temperature: Cooling Down**

An aluminum frying pan of mass 1.5 kg is at 180°C, when it's plunged into a sink containing 8.0 kg of water at 20°C. Assuming that none of the water boils and that no heat is lost to the surroundings, find the equilibrium temperature of the water and pan.

INTERPRET Here we have two objects, initially at different temperatures, that come to thermal equilibrium. So this is a problem about the equilibrium temperature, with the system of interest comprising the pan and the water.

DEVELOP Equation 16.4, $m_1 c_1 \,\Delta T_1 + m_2 c_2 \,\Delta T_2 = 0$, applies. However, we're asked for the common equilibrium temperature T, so we write the temperature differences ΔT in terms of T and the initial temperatures T_p and T_w of pan and water. Equation 16.4 then becomes $m_p c_p (T - T_p) + m_w c_w (T - T_w) = 0$.

EVALUATE We now solve for the equilibrium temperature T:

$$T = \frac{m_p c_p T_p + m_w c_w T_w}{m_p c_p + m_w c_w}$$

Using the given values of m_p, T_p, m_w, and T_w, and taking c_p and c_w from Table 16.1, we find that $T = 26°C$.

ASSESS The water has much greater mass and higher specific heat, so it makes sense that its 6°C temperature change is a lot less than the 154°C drop in the pan's temperature.

16.3 Heat Transfer

How is heat transferred? Engineers need to know so they can design heating and cooling systems. Scientists need to know so they can anticipate temperature changes, as in global warming. Here we'll consider three common heat-transfer mechanisms: conduction, convection, and radiation. In some situations, a single mechanism dominates; in other cases, we may need to take all three into account.

Conduction

Conduction is heat transfer through direct physical contact. It occurs as molecules in a hotter region collide with and transfer energy to those in an adjacent cooler region. **Thermal conductivity** (symbol k; SI unit W/m·K) characterizes this process. Common materials exhibit a broad range of thermal conductivities, from about 400 W/m·K for copper—a good conductor—to 0.029 W/m·K for Styrofoam, a good thermal insulator. Table 16.2 lists some thermal conductivities; they're given in both SI and British units

TABLE 16.2 Thermal Conductivities*

Material	Thermal Conductivity, k	
	SI Units: W/m· K	British Units: Btu· in./h ·ft²· °F
Air	0.026	0.18
Aluminum	237	1644
Concrete (varies with mix)	1	7
Copper	401	2780
Fiberglass	0.042	0.29
Glass	0.7–0.9	5–6
Goose down	0.043	0.30
Helium	0.14	0.97
Iron	80.4	558
Steel	46	319
Styrofoam	0.029	0.20
Water	0.61	4.2
Wood (pine)	0.11	0.78

*Temperature range 0°C to 100°C.

because the latter are widely used in heat-loss calculations for buildings. The k values in Table 16.2 reflect physical properties of the materials. Metals, for example, are good thermal conductors because they contain free electrons that move quickly. Insulators like fiberglass and Styrofoam owe their insulating properties to a physical structure that traps small volumes of air or other gas.

Figure 16.5 shows a slab of thickness Δx and area A. One side is at temperature T and the other at $T + \Delta T$. The temperature difference ΔT drives a conductive heat flow through the slab; not surprisingly, we find that the heat flow is proportional to the temperature difference, the slab area, and the thermal conductivity k. The thicker the slab, on the other hand, the more resistance to heat flow, so the flow depends inversely on thickness. Therefore

$$H = -kA\frac{\Delta T}{\Delta x} \quad \text{(conductive heat flow)} \quad (16.5)$$

where $H = dQ/dt$ is the rate of heat flow in watts, and where the minus sign shows that the flow is opposite the direction of increasing temperature—that is, from hotter to cooler.

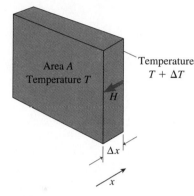

FIGURE 16.5 Heat flows from the hotter to the cooler face of the slab.

EXAMPLE 16.3 **Conduction: Warming a Lake**

A lake with a flat bottom and steep sides has surface area 1.5 km² and is 8.0 m deep. On a summer day, the surface water is at a temperature of 30°C and the bottom water is at 4.0°C. What is the rate of heat conduction through the lake? Assume that the temperature declines uniformly from surface to bottom.

INTERPRET This is a problem about heat conduction.

DEVELOP Our sketch, Fig. 16.6, shows that we can treat the lake like the slab shown in Fig. 16.5, provided we neglect heat flow out the sides of the lake. Then Equation 16.5, $H = -kA(\Delta T/\Delta x)$, will give the heat-flow rate.

EVALUATE Substituting numerical values, including water's thermal conductivity from Table 16.2, we get

$$H = -kA\frac{\Delta T}{\Delta x}$$
$$= -(0.61 \text{ W/m} \cdot \text{K})(1.5 \times 10^6 \text{ m}^2)\frac{30°C - 4.0°C}{8.0 \text{ m}} = -3.0 \text{ MW}$$

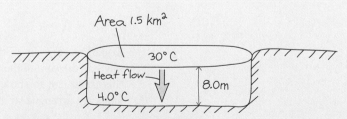

FIGURE 16.6 Our sketch for Example 16.3.

ASSESS This is a significant flow of energy, but with direct sunlight averaging about 1 kW on every square meter, the lake's 1.5-km² surface area absorbs plenty of solar energy, and that's what maintains the temperature difference that drives the conductive heat flow. Figure 16.5 shows x increasing in the direction of increasing temperature, so negative sign in our answer indicates that the flow is downward. ∎

If H weren't the same through both slabs, energy would accumulate at the interface.

Area A
Temperature T_1

H

T_3

T_2

R_2

R_1

Δx_2

Δx_1

FIGURE 16.7 A composite slab.

Equation 16.5 is strictly correct only when the temperature varies uniformly from one surface to the other. That's the case when two surfaces at different temperatures have the same area. With other geometries—as in the insulation surrounding a cylindrical pipe—we need to write $\Delta T/\Delta x$ as the derivative dT/dx and integrate to find the heat flow. Problems 72 and 78 explore this situation.

Often heat flows through several different materials. A building wall, for example, may contain wood, plaster, and fiberglass insulation. Figure 16.7 shows such a composite structure, with temperature T_1 on one side and T_3 on the other. The heat-flow rate H must be the same through both slabs so energy doesn't accumulate at the interface between the two. Then Equation 16.5 gives

$$H = -k_1 A \frac{T_2 - T_1}{\Delta x_1} = -k_2 A \frac{T_3 - T_2}{\Delta x_2}$$

where k_1 and k_2 are the thermal conductivities of the two materials, and T_2 is the temperature at the interface. We can express the heat-flow rate in terms of the surface temperatures T_1 and T_3 alone if we define the **thermal resistance** R of each slab:

$$R = \frac{\Delta x}{kA} \tag{16.6}$$

The SI units of R are K/W. Unlike the thermal conductivity k, which is a property of a *material*, R is a property of a *particular piece* of material, reflecting both its conductivity and its geometry. In terms of thermal resistance, our heat-flow equation becomes

$$H = -\frac{T_2 - T_1}{R_1} = -\frac{T_3 - T_2}{R_2}$$

so $R_1 H = T_1 - T_2$ and $R_2 H = T_2 - T_3$. Adding these two equations gives

$$(R_1 + R_2)H = T_1 - T_2 + T_2 - T_3 = T_1 - T_3$$

or

$$H = \frac{T_1 - T_3}{R_1 + R_2} \tag{16.7}$$

Equation 16.7 shows that the composite slab acts like a single slab whose thermal resistance is the sum of the resistances of the two slabs that compose it. We could easily extend this treatment to show that the thermal resistances of three or more slabs add when the slabs are arranged so the same heat flows through all of them.

GOT IT? 16.2 The figure shows three slabs with the same thickness but different thermal conductivities: k, $3k$, and $2k$; the left side is hotter, as shown. Rank order the three temperature differences ΔT.

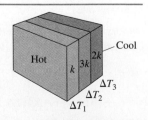

Insulating properties of building materials are described by the $\mathcal{R}$-**factor**, which is the thermal resistance for a slab of unit area:

$$\mathcal{R} = RA = \frac{\Delta x}{k} \qquad (16.8)$$

The SI units of $\mathcal{R}$ are $\text{m}^2\cdot\text{K/W}$, and that's how you'll find it listed if you buy insulation in Europe or other SI-based regions. In the United States, $\mathcal{R}$ is in $\text{ft}^2\cdot{}^{\circ}\text{F}\cdot\text{h/Btu}$, although the units are almost never stated. This means that $\mathcal{R}$-19 fiberglass insulation loses $\frac{1}{19}$ Btu per hour for each square foot of insulation for each degree Fahrenheit temperature difference across the insulation (Fig. 16.8).

FIGURE 16.8 Each square foot of this $\mathcal{R}$-19 fiberglass insulation loses $\frac{1}{19}$ Btu per hour for every °F of temperature difference ΔT.

EXAMPLE 16.4 **Calculating Heat Loss: The Cost of Oil**

Figure 16.9 shows a house whose walls consist of plaster $(\mathcal{R}=0.17)$, $\mathcal{R}$-11 fiberglass insulation, plywood $(\mathcal{R}=0.65)$, and cedar shingles $(\mathcal{R}=0.55)$. The roof has the same construction except it uses $\mathcal{R}$-30 fiberglass insulation. The average outdoor temperature in winter is 20°F, and the house is maintained at 70°F. The house's oil furnace produces 100,000 Btu for every gallon of oil, and oil costs $2.20 per gallon. How much does it cost to heat the house for a month?

INTERPRET Although the problem asks for the monthly cost of oil, this isn't economics! We interpret this as a problem about heat loss and identify the walls and roof as systems for which we need to know the heat flow. This is a rare case of a problem stated in English units.

DEVELOP We're given the drawing in Fig. 16.9. We have the $\mathcal{R}$-factors; in English units, their inverses give the heat-loss rate on a square-foot basis. So our plan is to find the square footage of the walls and roof separately, calculate the total heat-loss rate, and then find the amount and cost of oil to compensate for a month's heat loss.

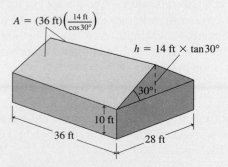

$$A = (36\text{ ft})\left(\frac{14\text{ ft}}{\cos 30^{\circ}}\right)$$

$$h = 14\text{ ft} \times \tan 30^{\circ}$$

FIGURE 16.9 House for Example 16.4.

EVALUATE The $\mathcal{R}$-factors for the wall materials sum to give $\mathcal{R}_{\text{wall}} = 12.4$; similarly, $\mathcal{R}_{\text{roof}} = 31.4$. The perimeter of the house measures $2 \times 28\text{ ft} + 2 \times 36\text{ ft} = 128\text{ ft}$, so the 10-ft vertical walls have area 1280 ft². There are also the triangular gables. Since there are two of them, each with area $\frac{1}{2}bh$, they give another bh or $(28\text{ ft})(14\text{ ft} \tan 30^{\circ}) = 226\text{ ft}^2$, so $A_{\text{wall}} = 1506\text{ ft}^2$. These $\mathcal{R}$-12.4 walls lose 1/12.4 Btu/h/ft²/°F. With 1506 ft² and a temperature difference of 50°F, the total heat-loss rate through the walls is

$$H_{\text{wall}} = \left(\tfrac{1}{12.4}\text{ Btu/h/ft}^2/{}^{\circ}\text{F}\right)(1506\text{ ft}^2)(50^{\circ}\text{F}) = 6073\text{ Btu/h}$$

The area of the pitched roof is larger than that of a flat roof by the factor $1/\cos 30^{\circ}$, so the heat-loss rate through the roof is

$$H_{\text{roof}} = \left(\tfrac{1}{31.4}\text{ Btu/h/ft}^2/{}^{\circ}\text{F}\right)\frac{(36\text{ ft})(28\text{ ft})}{\cos 30^{\circ}}(50^{\circ}\text{F}) = 1853\text{ Btu/h}$$

The total heat-loss rate is then 7926 Btu/h. In a month, this results in a heat loss of $Q = (7926\text{ Btu/h})(30\text{ days/month})(24\text{ h/day}) = 5.7\text{ MBtu}$.

Now for the oil: With 10^5 Btu (0.1 MBtu) per gallon, we'll burn 57 gallons per month to produce that 5.7 MBtu. At $2.20/gal, that will cost $126.

ASSESS If you've paid for heat in a northern climate, you know that this figure is, if anything, low. That's because we neglected heat losses through windows, doors, and the floor, as well as cold-air infiltration. On the other hand, we also left out any solar energy gained through the windows on sunny days. Problem 69 provides a more realistic look at this house. ∎

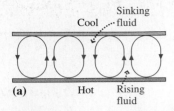

(b)

FIGURE 16.10 (a) Convection between two plates at different temperatures. (b) Top view of convection cells in a laboratory experiment. Fluid rises at the centers and sinks at the edges of the hexagonal cells.

Convection

Convection is heat transfer by fluid motion. It occurs as heated fluid becomes less dense and therefore rises. Figure 16.10a shows two plates at different temperatures, with fluid between them. Fluid heated by the lower plate rises and transfers heat to the upper plate. The cooled fluid sinks, and the process repeats. The pattern of rising and sinking fluid often acquires a striking regularity, as shown in Fig. 16.10b.

Convection is important in many technological and natural environments. When you heat water on a stove, convection carries heat through the water. Houses usually rely on convection from heat sources near floor level to circulate warm air throughout a room. Insulating materials trap air and thereby inhibit convection that would otherwise cause excessive heat loss. Convection associated with solar heating of Earth's surface drives the vast air movements that establish our overall climate. Violent convection, as in thunderstorms, is associated with localized temperature differences. On a much longer time scale, convection in Earth's mantle drives continental drift. Convection plays a crucial role in many astrophysical processes, including the generation of magnetic fields in stars and planets.

As with conduction, the convective heat-loss rate often is approximately proportional to the temperature difference. But the calculation of convective heat loss is complicated because of the associated fluid motion. The study of convection processes is an important research area in many fields of contemporary science and engineering.

Radiation

Turn a stove burner to "high" and it glows brightly; turn it to "low" and you can still sense its heat although it doesn't glow visibly. Either way, the burner loses energy by emitting electromagnetic waves, or **radiation**. The radiated power P increases rapidly with temperature, as described by the **Stefan-Boltzmann law**:

$$P = e\sigma AT^4 \quad \left(\begin{array}{c}\text{Stefan-Boltzmann law;} \\ \text{radiated power}\end{array}\right) \tag{16.9}$$

where A is the area of the emitting surface, T the temperature in kelvins, and σ the **Stefan-Boltzmann constant**, approximately 5.67×10^{-8} W/m²·K⁴. The quantity e is the **emissivity**, a number from 0 to 1 that measures the material's effectiveness in emitting radiation. For radiation of a given wavelength, a material is equally good at emitting and absorbing radiation. A perfect emitter has $e = 1$ and is also a perfect absorber. Such an object would appear black at room temperature and is therefore called a **blackbody**. A shiny object, in contrast, reflects most of the radiation that hits it and is therefore also a poor emitter. Wood stoves are usually painted black to increase their emissivity; Thermos bottles, on the other hand, have a shiny coating to reduce radiation.

Because of the strong T^4 dependence on temperature, radiation tends to be less important at low temperatures but dominant at high temperatures. And it's not just the amount of radiation that changes with temperature; as our stove burner example suggests, it's also the wavelength. Objects near room temperature, for example, emit mostly invisible infrared radiation; very hot objects like the Sun or a lightbulb filament emit more visible light. We'll take a quantitative look at this relation in Chapter 34.

GOT IT? 16.3 Name the dominant form of heat transfer from (a) a red-hot stove burner with nothing on it, (b) a burner in direct contact with a pan of water; and (c) the bottom to the top of the water in the pan once it's begun to boil.

EXAMPLE 16.5 **Calculating Radiation: The Sun's Temperature**

The Sun radiates energy at the rate $P = 3.9 \times 10^{26}$ W, and its radius is 7.0×10^8 m. Treating the Sun as a blackbody ($e = 1$), find its surface temperature.

INTERPRET This is a problem about the radiation from a hot object.

DEVELOP The Stefan-Boltzmann law, Equation 16.9, gives the radiated power in terms of the temperature, emissivity, and surface area: $P = e\sigma A T^4$. Our plan is to solve this equation for T. For the Sun, radiation comes from the entire spherical surface of area $4\pi R^2$, as our sketch shows (Fig. 16.11).

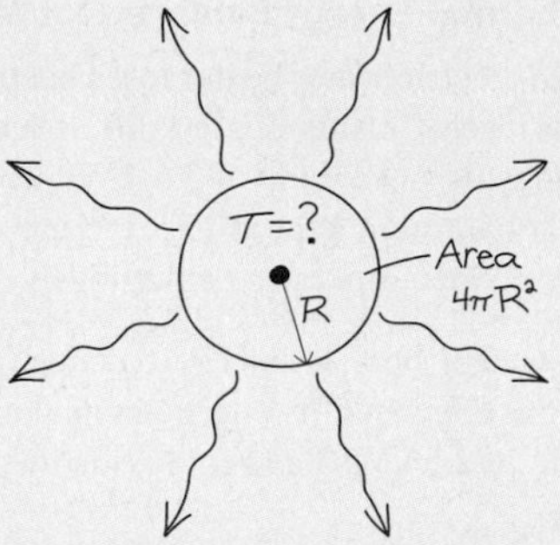

EVALUATE Using the Sun's spherical surface area and solving for T give

$$T = \left(\frac{P}{4\pi R^2 \sigma}\right)^{1/4}$$

$$= \left[\frac{3.9 \times 10^{26} \text{ W}}{4\pi (7.0 \times 10^8 \text{ m})^2 (5.7 \times 10^{-8} \text{ W/m}^2 \cdot \text{K}^4)}\right]^{1/4} = 5.8 \times 10^3 \text{ K}$$

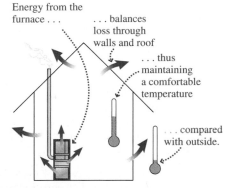

FIGURE 16.11 The Sun radiates from its spherical surface area $4\pi R^2$.

ASSESS Make sense? Yes: Our answer has the unit of temperature and agrees with observational measurements. ∎

16.4 Thermal-Energy Balance

You keep your house at a comfortable temperature in winter by balancing heat loss with energy from your heating system (Fig. 16.12). This state of **thermal-energy balance** occurs throughout science and engineering. Understanding thermal-energy balance enables engineers to specify a building's heat sources, and helps scientists predict Earth's future climate.

Engineered systems actively control the thermal-energy balance to achieve a desired temperature. But even without active control, systems with a fixed rate of energy input naturally tend toward energy balance. That's because all heat-loss mechanisms give increased loss with increasing temperature. If the rate of energy input to a system is greater than the loss rate, then the system gains energy and its temperature increases—and so, therefore, does the loss rate. Eventually the two come to balance at some fixed temperature. If the loss exceeds the gain, the system cools until again it's in balance. Problems involving thermal-energy balance are similar regardless of the energy-loss mechanism or whether the application is to a technological or a natural system.

Energy from the furnace . . . . . . balances loss through walls and roof . . . thus maintaining a comfortable temperature . . . compared with outside.

FIGURE 16.12 A house in thermal-energy balance.

PROBLEM SOLVING STRATEGY 16.1 **Thermal-Energy Balance**

INTERPRET Interpret the problem to be sure it deals with heat gains and losses. Identify the system of interest, the source(s) of energy input to the system, and the significant heat-loss mechanism(s).

DEVELOP Determine which equation(s) govern the heat loss; these will necessarily involve the system's temperature. Your plan is then to equate the rate of energy loss with the rate of energy input.

EVALUATE Write an equation that expresses equality of energy loss and input. Then evaluate by solving for the quantity the problem asks for—often the system's temperature.

ASSESS If your answer is a temperature, does it seem reasonable? Is the temperature of a heated system higher than that of its surroundings?

EXAMPLE 16.6 Thermal-Energy Balance: Hot Water

A poorly insulated electric water heater loses heat by conduction at the rate of 120 W for each Celsius degree difference between the water and its surroundings. It's heated by a 2.5-kW electric heating element and is located in a basement kept at 15°C. What's the water temperature if the heating element operates continuously?

INTERPRET The concept here is energy balance, and we identify the system of interest as the water. Its energy input comes from the heating element at the rate of 2.5 kW. The heat loss is by conduction.

DEVELOP Figure 16.13 is a sketch suggesting energy balance in the heater. We're given the conductive heat loss of 120 W/°C, meaning that the total heat-loss rate is $H = (120 \text{ W/°C})(\Delta T)$. We then equate the heat-loss rate to the energy-input rate: $(120 \text{ W/°C})(\Delta T) = 2.5 \text{ kW}$.

EVALUATE Solving for (ΔT) gives

$$\Delta T = \frac{2.5 \text{ kW}}{120 \text{ W/°C}} = 21°C$$

With the basement at 15°C, the water temperature is then 36°C.

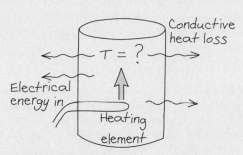

FIGURE 16.13 Balance between the heat supplied by the electric element and the conductive loss determines the water temperature.

ASSESS Is this answer reasonable? Not if you want a hot shower; our answer is 1°C below body temperature! But we're told the insulation is bad, so it's time for a new water heater! ∎

EXAMPLE 16.7 Thermal-Energy Balance: A Solar Greenhouse

A solar greenhouse has 300 ft² of opaque $\mathcal{R}$-30 walls, and 250 ft² of $\mathcal{R}$-1.8 double-pane glass that admits solar energy at the average rate of 40 Btu/h/ft². Find the greenhouse temperature on a day when the outdoor temperature is 15°F.

INTERPRET Again the concept is energy balance, now with the greenhouse as the system of interest. We're given $\mathcal{R}$-factors, suggesting that the energy loss is by conduction through walls and glazing. The energy input is sunlight.

DEVELOP As we saw in Example 16.4, the $\mathcal{R}$-factor determines a heat-loss rate that is related directly to area and temperature difference and inversely to the $\mathcal{R}$-factor. So we have

$$H_w = \frac{A_w \, \Delta T}{\mathcal{R}_w} = \left(\frac{300}{30}\right) \Delta T = (10 \text{ Btu/h/°F}) \, \Delta T$$

for the heat loss through the walls and

$$H_g = \frac{A_g \, \Delta T}{\mathcal{R}_g} = \left(\frac{250}{1.8}\right) \Delta T = (139 \text{ Btu/h/°F}) \, \Delta T$$

for the heat loss through the glass, giving a total heat loss $H = (149 \text{ Btu/h/°F}) \, \Delta T$. Meanwhile, the energy input through the entire 250-ft² of glass is $(40 \text{ Btu/h/ft}^2)(250 \text{ ft}^2) = 1.0 \times 10^4$ Btu/h. Our plan is to equate energy input and loss and then solve for ΔT.

EVALUATE Equating loss and gain gives

$$(149 \text{ Btu/h/°F}) \, \Delta T = 1.0 \times 10^4 \text{ Btu/h}.$$

We then solve for ΔT:

$$\Delta T = \frac{1.0 \times 10^4 \text{ Btu/h}}{149 \text{ Btu/h/°F}} = 67°F$$

So when it's 15°F outside, the greenhouse is at a tropical 82°F.

ASSESS This seems a reasonable greenhouse temperature. Our calculation assumes that solar input remains constant; in a real greenhouse the temperature would fluctuate as the Sun's angle changes and clouds pass over. We could minimize these fluctuations by giving the greenhouse a large heat capacity, perhaps by incorporating a massive concrete slab or concrete walls. ∎

APPLICATION **The Greenhouse Effect and Global Warming**

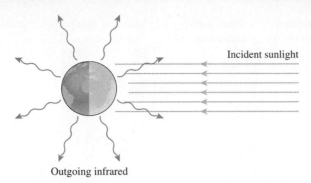

Earth's energy balance.

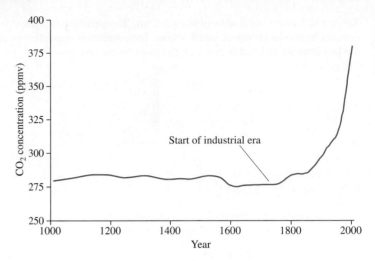

Increase in atmospheric CO_2 over the past 1000 years, in parts per million by volume (ppmv).

Earth absorbs energy from the Sun at a rate $S = 960$ W/m^2 averaged over the planet's cross-sectional area πR_E^2. It therefore warms and, in thermal-energy balance, radiates energy at the same rate. Earth is much cooler than the Sun, so this outgoing radiation is invisible infrared; furthermore, it's radiated from the planet's entire surface area, $4\pi R_E^2$. Assuming emissivity $e = 1$ in the infrared, energy balance using Equation 16.9 gives $\pi R_E^2 S = \sigma 4\pi R_E^2 T^4$. Solving yields $T = 255$ K $= -18°$C or 0°F. Is this reasonable? It's certainly in the right ballpark—not so hot as to boil the oceans or freeze the atmosphere. But 0°F seems a bit cold for a global average temperature. And it is: Earth's average temperature is around 15°C or 59°F. Why the discrepancy?

The answer lies with Earth's atmosphere. The dominant atmospheric gases, nitrogen and oxygen, are largely transparent to both incoming sunlight and outgoing infrared. But others—the so-called **greenhouse gases**, especially water vapor and carbon dioxide—let sunlight pass through but impede outgoing infrared. As a result, Earth's surface temperature has to be higher to get the same total radiation to space. This is the **natural greenhouse effect**, and it explains the 33°C temperature difference between our naïve calculation and Earth's actual surface temperature. Neighbor planets confirm this reasoning. Mars, with very little atmosphere, exhibits almost no greenhouse warming. Venus, whose atmosphere is 100 times denser than Earth's and largely CO_2, has a "runaway" greenhouse effect that keeps its surface hotter than an oven.

As the CO_2 graph shows, we humans have increased atmospheric carbon dioxide some 36% since the start of the industrial era, to levels the planet has not seen for millions of years. Combustion of fossil fuels is the dominant source of this CO_2, although processes like deforestation also contribute, as do other greenhouse gases such as methane. Basic physics then dictates that Earth's surface temperature should rise. How much and how fast depend on complex interactions among atmosphere, surface, oceans, and life, and on future greenhouse emissions. Nevertheless, a consensus among climate scientists

suggests that Earth warmed by some 0.6°C during the 20th century, mostly attributable to human activities (see the temperature graph). Further warming in the range of 1.5°C–6°C is projected by 2100. Although this may seem modest, the rate of increase is far greater than most natural climate change. And the increase is expected to be greater over land and at high latitudes. Even a few degrees' increase in the global average will result in significant climate change and a rise in sea level.

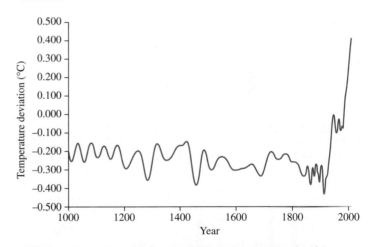

Global temperature over the past 1000 years, given as deviations from the average temperature for 1961–1990. Data before 1856 are a reconstruction based on tree rings and other proxies for temperature; data from 1856 on are from thermometer records.

Big Picture

The big ideas here are **temperature** and **heat**. **Temperature** is a property common to systems in **thermodynamic equilibrium**. Temperature is quantified in SI units using the **kelvin scale**, defined in terms of gas-based thermometers.

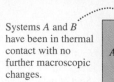

Systems A and B have been in thermal contact with no further macroscopic changes.

They've reached thermodynamic equilibrium and so have the same temperature.

Heat is energy in transit as a result of a temperature difference.

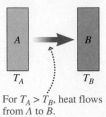

For $T_A > T_B$, heat flows from A to B.

Key Concepts and Equations

Heat capacity and **specific heat** quantify the energy ΔQ required to raise an object's temperature by ΔT:

$$\Delta Q = mc\ \Delta T$$

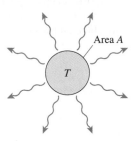

Mass m
Specific heat c
Temperature T

Add energy ΔQ

Temperature increase ΔT

$\Delta Q = mc\ \Delta T$

Three important heat-transfer mechanisms are:

Conduction

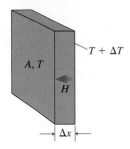

$T + \Delta T$

A, T

H

Δx

Convection

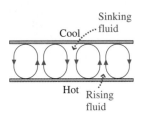

Cool   Sinking fluid

Hot   Rising fluid

Radiation

Area A

T

$$H = -kA\frac{\Delta T}{\Delta x} \quad \text{(conductive heat flow)}$$

$$P = e\sigma AT^4 \quad \left(\begin{array}{c}\text{Stefan-Boltzmann law;}\\ \text{radiated power}\end{array}\right)$$

Cases and Uses

Temperature scales include Kelvin (K), Celsius (°C), Fahrenheit (°F), and Rankine (°R).

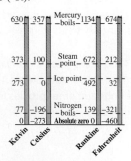

		Mercury boils		
630	357		1134	674
373	100	Steam point	672	212
273	0	Ice point	492	32
77	−196	Nitrogen boils	139	−321
0	−273	Absolute zero 0		−460

Kelvin   Celsius   Rankine   Fahrenheit

The Kelvin and Celsius scales are related by $T_C = T - 273.15$. The relation between Fahrenheit and Celsius scales is $T_F = \frac{9}{5}T_C + 32$.

Equilibrium temperature: Combining two systems at different temperatures results in a common equilibrium temperature given by $m_1 c_1\ \Delta T_1 + m_2 c_2\ \Delta T_2 = 0$.

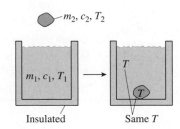

m_2, c_2, T_2

m_1, c_1, T_1

T

Insulated   Same T

Energy balance: A system experiencing both energy input and energy loss comes to energy balance at the temperature for which the energy-loss rate equals the rate of energy input.

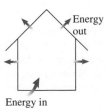

Energy out

Energy in

For Thought and Discussion

1. If system *A* is not in thermodynamic equilibrium with system *B*, and *B* is not in equilibrium with *C*, can you draw any conclusions about the temperatures of the three systems?
2. Does a thermometer measure its own temperature or the temperature of its surroundings? Explain.
3. Compare the relative sizes of the kelvin, the degree Celsius, the degree Fahrenheit, and the degree Rankine.
4. If you put a thermometer in direct sunlight, what do you measure: the air temperature, the temperature of the Sun, or some other temperature?
5. Why does the temperature in a stone building usually vary less than in a wooden building?
6. Why do large bodies of water exert a temperature-moderating effect on their surroundings?
7. A Thermos bottle consists of an evacuated, double-wall glass liner. The glass is coated with a thin layer of aluminum. How does a Thermos bottle work?
8. Stainless-steel cookware often has a layer of aluminum or copper embedded in the bottom. Why?
9. What method of energy transfer dominates in baking? In broiling?
10. After a calm, cold night, the temperature a few feet above ground often drops just as the Sun comes up. Explain in terms of convection.
11. Glass and fiberglass are made from the same material, yet have dramatically different thermal conductivities. Why?
12. To keep your hands warm while skiing, you should wear mittens instead of gloves. Why?
13. Since Earth is exposed to solar radiation, why doesn't Earth have the same temperature as the Sun?

Exercises and Problems

Exercises

Section 16.1 Heat, Temperature, and Thermodynamic Equilibrium

14. A Canadian meteorologist predicts an overnight low of $-15°C$. How would a U.S. meteorologist express that same prediction?
15. Normal room temperature is 68°F. What is this in Celsius?
16. The outdoor temperature rises by 10°C. What is that rise in Fahrenheit?
17. At what temperature do the Fahrenheit and Celsius scales coincide?
18. The normal boiling point of nitrogen is 77.3 K. Express this in Celsius and Fahrenheit.
19. A sick child's temperature reads 39.1 on a Celsius thermometer. What's the child's temperature on the Fahrenheit scale?

Section 16.2 Heat Capacity and Specific Heat

20. Find the heat capacity of a 55-tonne slab of concrete (1 tonne = 1000 kg).
21. Find the energy needed to raise a 2.0-kg chunk of aluminum by 18°C.
22. It takes 7.5 kJ to increase the temperature of a 1-kg block of material by 3.0°C. What's the material's specific heat?
23. The average human diet contains about 2000 kcal per day. If all this food energy is released rather than stored as fat, what is the approximate average power output of the human body?
24. Walking at 3 km/h requires an energy expenditure rate of about 200 W. How far would you have to walk to "burn off" a 300-kcal hamburger?
25. You bring a 350-g wrench into the house from your car. The house is 15°C warmer than the car, and it takes 2.52 kJ of energy to warm the wrench by this amount. Find (a) the heat capacity of the wrench and (b) the specific heat of the metal it's made from.
26. (a) How much heat does it take to bring a 3.4-kg iron skillet from 20°C to 130°C? (b) If the heat is supplied by a stove burner at the rate of 2.0 kW, how long will it take to heat the pan?

Section 16.3 Heat Transfer

27. Building heat loss in the United States is usually expressed in Btu/h. What is 1 Btu/h in SI units?
28. Find the heat-loss rate through a 1.0-m^2 slab of (a) wood and (b) Styrofoam, each 2.0 cm thick, if one surface is at 20°C and the other at 0°C.

29. The top of a steel wood stove measures 90 cm by 40 cm and is 0.45 cm thick. The fire maintains the inside surface of the stovetop at 310°C, while the outside surface is at 295°C. Find the rate of heat conduction through the stovetop.
30. How thick a concrete wall would be needed to give the same insulating value as $3\frac{1}{2}$ in. of fiberglass?
31. An 8.0 m by 12 m house is built on a concrete slab 23 cm thick. What is the heat-loss rate through the floor if the interior is at 20°C while the ground is at 10°C?
32. What is the $\mathcal{R}$-factor of a wall that loses 0.040 Btu each hour through each square foot for each °F temperature difference?
33. Compute the $\mathcal{R}$-factors for 1-in. thicknesses of air, concrete, fiberglass, glass, Styrofoam, and wood.
34. A horseshoe has a surface area of 50 cm^2, and a blacksmith heats it to a red-hot 810°C. At what rate does it emit energy by radiation?

Section 16.4 Thermal-Energy Balance

35. An oven loses energy at the rate of 14 W per °C temperature difference between its interior and the 20°C temperature of the kitchen. What average power must be supplied to maintain the oven at 180°C?
36. A home heating system can supply energy at the maximum rate of 40 kW. If the house loses energy at the rate of 8.0 kW per °C temperature difference between interior and exterior, what's the minimum outdoor temperature at which the heating system can maintain 20°C (68°F) indoors?
37. The filament of a 100-W lightbulb is at 3.0 kK. What's the surface area of the filament?
38. The solar input at Venus amounts to about 1.9 kW/m^2. Use an energy-balance calculation like that done for Earth in the greenhouse effect application to estimate Venus's surface temperature in the absence of a greenhouse effect, taking $e = 1$. The actual surface temperature is close to 500°C.

Problems

39. A constant-volume gas thermometer is filled with air whose pressure is 101 kPa at the normal melting point of ice. What would its pressure be at (a) the normal boiling point of water, (b) the normal boiling point of oxygen (90.2 K), and (c) the normal boiling point of mercury (630 K)?

40. A constant-volume gas thermometer is at 55-kPa pressure at the triple point of water. By how much does its pressure change for each kelvin temperature change?

41. In the gas thermometer of Fig. 16.14, the height h is 60.0 mm at the triple point of water. When the thermometer is immersed in boiling sulfur dioxide, the height drops to 57.8 mm. What is the boiling point of SO_2 in kelvins and in degrees Celsius?

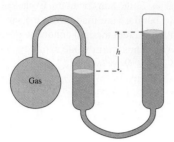

Gas

h

FIGURE 16.14 Problem 41

42. If your mass is 60 kg, what is the minimum number of calories you would "burn off" climbing a 1700-m-high mountain? *Note:* The actual metabolic energy used would be much greater.

43. Typical fats contain about 9 kcal per gram. If the energy in body fat could be utilized with 100% efficiency, how much mass could a runner lose in a 26.2-mile marathon? The runner's energy expenditure rate is 125 kcal/mile.

44. A circular lake 1.0 km in diameter is 10 m deep (Fig. 16.15). Solar energy is incident on the lake at an average rate of 200 W/m². If the lake absorbs all this energy and does not exchange heat with its surroundings, how long will it take to warm from 10°C to 20°C?

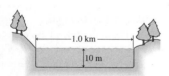

1.0 km

10 m

FIGURE 16.15 Problem 44

45. How much heat is required to raise an 800-g copper pan from 15°C to 90°C if (a) the pan is empty; (b) the pan contains 1.0 kg of water; and (c) the pan contains 4.0 kg of mercury?

46. Initially, 100 g of water and 100 g of another substance listed in Table 16.1 are at 20°C. Heat is then transferred to each substance at the same rate for 1.0 min. At the end of that time, the water is at 32°C and the other substance at 76°C. (a) What is the other substance? (b) What is the heating rate?

47. How long will it take a 625-W microwave oven to bring 250 mL of water from 10°C to the boiling point?

48. Two neighbors return from Florida to find their houses at a frigid 35°F. Each house has a furnace with heat output of 100,000 Btu/h. One house is made of stone and weighs 75 tons. The other is made of wood and weighs 15 tons. How long does it take each house to reach 65°F? Neglect heat loss, and assume the entire house mass reaches a uniform 65°F temperature.

49. A stove burner supplies heat at the rate of 1.0 kW, a microwave oven at 625 W. You can heat water in the microwave in a paper cup of negligible heat capacity, but the stove requires a pan whose heat capacity is 1.4 kJ/K. (a) How much water do you need before it becomes quicker to heat on the stovetop? (b) What will be the rate at which the temperature of this much water rises?

50. When a nuclear power plant's reactor is shut down, radioactive decay continues to produce heat at about 10% of the reactor's normal power level of 3.0 GW. In a major accident, a pipe breaks and all the reactor cooling water is lost. The reactor is immediately shut down, the break sealed, and 420 m³ of 20°C water injected into the reactor. If the water were not actively cooled, how long would it take to reach its normal boiling point?

51. A 1.2-kg iron tea kettle sits on a 2.0-kW stove burner. If it takes 5.4 min to bring the kettle and the water in it from 20°C to the boiling point, how much water is in the kettle?

52. Two cars collide head-on at 90 km/h. If all their kinetic energy ended up as heat, what would be the temperature increase of the wrecks? The specific heat of the cars is essentially that of iron.

53. A 1500-kg car moving at 40 km/h is brought to a sudden stop. If all the car's energy is dissipated in heating its four 5.0-kg steel brake disks, by how much do the disk temperatures increase?

54. A child complains that her cocoa is too hot. The cocoa is at 90°C. Her father pours 2 oz of milk at 3°C into the 6 oz of cocoa. Assuming milk and cocoa have the same specific heat as water, what is the new temperature of the cocoa?

55. A piece of copper at 300°C is dropped into 1.0 kg of water at 20°C. If the equilibrium temperature is 25°C, what is the mass of the copper?

56. A biology lab's walk-in cooler measures 3.0 m by 2.0 m by 2.3 m and is insulated with 8.0-cm-thick Styrofoam. If the surrounding building is at 20°C, at what average rate must the cooler's refrigeration unit remove heat in order to maintain 4.0°C in the cooler?

57. One end of an iron rod 40 cm long and 3.0 cm in diameter is in ice water, the other in boiling water (Fig. 16.16). The rod is well insulated so no heat is lost out the sides. What is the heat-flow rate along the rod?

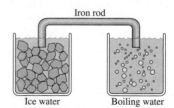

Iron rod

Ice water   Boiling water

FIGURE 16.16 Problem 57

58. (a) What is the $\mathcal{R}$-factor for a wall consisting of $\frac{1}{4}$-in. pine paneling, $\mathcal{R}$-11 fiberglass insulation, $\frac{3}{4}$-in. pine sheathing, and 2.0-mm aluminum siding? (b) What is the heat-loss rate through a 20 ft by 8 ft section of wall when the temperature difference across the wall is 55°F?

59. You're considering installing a 5 ft by 8 ft picture window in a north-facing wall that now has $\mathcal{R}$-19 insulation. The window has $\mathcal{R} = 2.1$ ft²·°F·h/Btu. If you install the window, how much more oil, at 100,000 Btu per gallon, will you have to burn in a winter month when the outdoor temperature averages 15°F and the indoor temperature is 68°F?

60. Repeat Problem 59 for a south-facing window where the average sunlight intensity is 180 W/m².

61. A house is insulated so its total heat loss is 370 W/°C. On a night when the outdoor temperature is 12°C, the owner throws a party and 40 people come. The average power output of the human body is 100 W. If there are no other heat sources in the house, what will be the house temperature during the party?

62. An electric stove burner has surface area 325 cm² and emissivity $e = 1.0$. The burner is at 900 K and the electric power input to the burner is 1500 W. If room temperature is 300 K, what fraction of the burner's heat loss is by radiation?

63. An electric current passes through a metal strip 0.50 cm by 5.0 cm by 0.10 mm, heating it at the rate of 50 W. The strip has emissivity $e = 1.0$ and its surroundings are at 300 K. What will be the temperature of the strip if it's enclosed in (a) a vacuum bottle transparent to all radiation and (b) an insulating box with thermal resistance $R = 8.0$ K/W that blocks all radiation?

64. The average human body produces heat at the rate of 100 W and has a total surface area of about 1.5 m². What is the coldest outdoor temperature in which a down sleeping bag with 4.0-cm loft (thickness) can be used without the body temperature dropping below 37°C? Consider only conductive heat loss.

65. A blacksmith heats a 1.1-kg iron horseshoe to 550°C, then plunges it into a bucket containing 15 kg of water at 20°C. What is the final temperature?

66. What is the power output of a microwave oven that can heat 430 g of water from 20°C to the boiling point in 5.0 min? Neglect the heat capacity of the container.

67. A cylindrical log 15 cm in diameter and 65 cm long is glowing red hot in a fireplace. If it's emitting radiation at the rate of 34 kW, what is its temperature? The log's emissivity is essentially 1.

68. A star whose surface temperature is 50 kK radiates 4.0×10^{27} W. If the star behaves like a blackbody, what is its radius?

69. Rework Example 16.4, now assuming that the house has ten single-glazed windows, each measuring 2.5 ft by 5.0 ft. Four of the windows are on the south and admit solar energy at the average rate of 30 Btu/h·ft². *All* the windows lose heat; their $\mathcal{R}$-factor is 0.90. (a) What is the total heating cost for the month? (b) How much is the solar gain worth?

70. A black wood stove with surface area 4.6 m² is made from cast iron 4.0 mm thick. The interior wall of the stove is at 650°C, while the exterior is at 647°C. (a) What is the rate of heat conduction through the stove wall? (b) What is the rate of heat loss by radiation from the stove? (c) Use the results of parts (a) and (b) to find how much heat the stove loses by a combination of conduction and convection in the surrounding air.

71. What should be the average temperature on Pluto, at its 5.9×10^{12}-m distance from the Sun? Treat Pluto as a blackbody.

72. In a cylindrical pipe where area isn't constant, Equation 16.5 takes the form $H = -kA(dT/dr)$. Use this equation to show that the heat-loss rate from a cylindrical pipe of radius R_1 and length L is

$$H = \frac{2\pi k L (T_1 - T_2)}{\ln(R_2/R_1)}$$

where the pipe is surrounded by insulation of outer radius R_2 and thermal conductivity k and where T_1 and T_2 are the temperatures at the pipe surface and at the outer surface of the insulation, respectively. *Hint:* Consider the heat flow through a thin section of pipe, with thickness dr, as shown in Fig. 16.17. Then integrate.

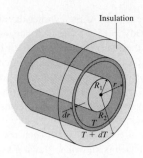

FIGURE 16.17 Problem 72

73. A friend of yours is writing a novel about the impact of a small asteroid on Earth. Such an event has been credited with causing a "nuclear winter" that may have killed the dinosaurs. Dust from the impact would darken the atmosphere and cause a reduction in the solar radiation received at the surface and thus lower the average surface temperature. Your friend would like to know the change in Earth's average temperature of 287 K if the impact caused a 10% reduction in solar intensity.

74. Your professor, who is building his own cabin, is examining a data sheet for some European-made insulation. According to the design specs for his cabin, he needs at least $\mathcal{R}$-19 insulation in the walls. The data sheet gives the $\mathcal{R}$-factor of the European insulation in SI units—namely, 3.5 m²·K/W. Will the European insulation work?

75. You're the public affairs manager for an electric power utility. A nuclear power plant has been refueled and needs to power up again. The news media want to know how long it will take before the reactor is online again. The reactor needs to heat 5.4×10^6 kg of cooling water from 10°C to 350°C (it's a pressurized-water reactor, so the water remains liquid). The reactor's thermal power output is 1.42 GW. Ignoring the heat capacity of the reactor vessel and plumbing, how long will that take?

76. A young relative gets a pet rabbit that will live in a backyard hutch. The hutch is enclosed and has a thermal resistance of 0.25 K/W. A rabbit-care book says the hutch should be kept at a temperature of 0°C or higher so the rabbit's water won't freeze. On a day when the outside temperature is $-15°C$, you put a 50-W heat lamp in the hutch. Will the bunny be able to take a drink of water? *Note:* Neglect the heat due to the animal's metabolism.

77. At low temperatures the specific heat of a solid is approximately proportional to the cube of the absolute temperature; for copper $c = 31(T/343 \text{ K})^3$ J/g·K. Integrate Equation 16.3 in differential form, $dQ = mc\,dT$, to find the heat required to bring a 40-g sample of copper from 10 K to 25 K.

78. Use the method outlined in Problem 72 to show that the steady heat flow rate in the direction of the axis of a truncated cone with conductivity k, faces of radii R_1 and R_2, and length L is $H = \pi k R_1 R_2 (T_1 - T_2)/L$, where T_1 and T_2 are the temperatures on the faces and insulation prevents any heat flow out the sides (see Figure 16.18).

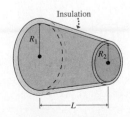

FIGURE 16.18 Problem 78

79. A house is at 20°C on a winter day when the outside temperature is a steady −15°C. The heat capacity of the house is 6.5 MJ/K and its thermal resistance is 6.67 mK/W. If the furnace suddenly fails (and there are no other sources of heat) how long will it take for the house temperature to reach the freezing point? (*Hint:* Combine the differential forms of Equation 16.3 for heat capacity and Equation 16.5, in terms of the thermal resistance, to show that the rate of temperature change is proportional to the temperature difference between the house and its surroundings. This relation is known as Newton's law of cooling.)

80. A 1000-W electric clothes iron has a surface area 300 cm^2 and emissivity $e = 0.97$. If the surface temperature of the iron is 500 K and its surroundings are at 300 K, (a) what is the *net* rate of energy transfer by radiation from the iron? (b) What must be the rate of energy transfer by conduction and convection?

Answers to Chapter Questions

Answer to Chapter Opening Question

The photo is taken in infrared light, and the amount of infrared radiation increases rapidly with increasing temperature. The car's wheels are glowing with infrared, a result of frictional heating when the brakes were recently applied.

Answers to GOT IT? Questions

16.1 The rock's temperature changes more because its specific heat is lower.

16.2 $\Delta T_2 < \Delta T_3 < \Delta T_1$; since H and Δx are the same for each slab, the product $k\,\Delta T$ must be constant, so a higher conductivity means a lower ΔT.

16.3 (a) Radiation; (b) conduction; (c) convection.

17 The Thermal Behavior of Matter

■ What unusual property of water is evident in this photo?

Matter responds to heating in several ways. It may get hotter or it may melt. It may change size, shape, or pressure. This chapter explores the thermal behavior of matter. We start with a simple gaseous state, whose behavior follows from Newtonian mechanics at the molecular level. We then move to liquids and solids, whose behavior is still grounded in the molecular properties of matter, but whose description is more empirical.

17.1 Gases

Gases are simple because their molecules are far apart and only rarely interact. That makes gas behavior and its physical explanation particularly straightforward. Developing that explanation will clarify the relation between macroscopic properties—such as temperature and pressure—and the underlying microscopic properties of gas molecules.

The Ideal-Gas Law

The macroscopic state of a gas in thermodynamic equilibrium is determined by its temperature, pressure, and volume. Moreover, it turns out that all gases exhibit, to a very good approximation, the same relation among these three quantities.

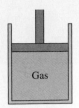

FIGURE 17.1 A piston-cylinder system.

A simple system for studying gas behavior consists of a gas-filled cylinder sealed by a movable piston (Fig. 17.1). This is not just a pedagogical abstraction: Practical devices including engines, pumps, and air compressors contain piston-cylinder systems, while lungs, balloons, gas bubbles, and many other natural systems are analogous to our piston-cylinder system.

If we maintain the system of Fig. 17.1 at constant temperature and move the piston to vary the gas volume, we find that the pressure varies inversely with the volume. If we increase the temperature while holding the volume fixed, the pressure rises in direct proportion to the temperature. If we double the amount of gas while holding temperature and volume constant, the pressure doubles. Putting all these results together, we can write

$$pV = NkT \quad \text{(ideal-gas law)} \tag{17.1}$$

with p, V, and T the pressure, volume, and temperature, respectively, and N the number of molecules in the gas. The constant $k = 1.38 \times 10^{-23}$ J/K is **Boltzmann's constant**, named for the Austrian physicist Ludwig Boltzmann (1844–1906), who was instrumental in developing the microscopic description of thermal phenomena. Equation 17.1 is the **ideal-gas law**. Most real gases obey this law to a very good approximation.

Because the number of molecules N in a typical gas sample is astronomically large, we often express the ideal-gas law in terms of the number of **moles** (mol) of gas molecules. One mole is an SI unit equal to Avogadro's number, $N_A = 6.022 \times 10^{23}$, of atoms or molecules. Formally, Avogadro's number is defined as the number of carbon-12 atoms in 12 grams of carbon-12.

If we have n moles of a gas, then $N = nN_A$ is the number of molecules, so the ideal-gas law becomes

$$pV = nN_A kT = nRT \tag{17.2}$$

where $R = N_A k = 8.314$ J/K·mol is called the **universal gas constant**.

EXAMPLE 17.1 **The Ideal Gas Law: STP**

What volume is occupied by 1.00 mol of an ideal gas at standard temperature and pressure (STP), where $T = 0°C$ and $p = 101.3$ kPa?

INTERPRET We're dealing with an ideal gas, and we're given the amount of gas, the temperature, and the pressure.

DEVELOP Because we're given the number of moles n (and not the number of molecules), we'll use the ideal-gas law in the form of Equation 17.2, $pV = nRT$, to find the volume.

EVALUATE Solving for V gives

$$V = \frac{nRT}{p} = \frac{(1.00 \text{ mol})(8.314 \text{ J/K} \cdot \text{mol})(273 \text{ K})}{1.01 \times 10^5 \text{ Pa}}$$
$$= 22.4 \times 10^{-3} \text{ m}^3 = 22.4 \text{ L}$$

where we expressed $T = 0°C$ as 273 K.

ASSESS This result may be familiar to you from earlier chemistry or physics courses: 1 mole of any ideal gas—no matter what its chemical composition—occupies 22.4 L at standard temperature and pressure. Recall that 1 liter (L) is 1000 cm^3 or 10^{-3} m^3. ∎

The ideal-gas law is remarkably simple. Neither its form nor the constants k and R depend on the substance making up the gas or on the mass of the gas molecules. Yet most real gases follow the ideal-gas law very closely over a wide range of pressures. This nearly ideal behavior is what gives gas thermometers their high precision over a wide temperature range.

Kinetic Theory of the Ideal Gas

Why do gases obey such a simple relation among temperature, pressure, and volume? Here we answer that question with an analysis based ultimately on Newtonian mechanics.

We start with some simplifying assumptions:

1. The gas consists of many identical molecules, each with mass m but negligible size and no internal structure. This assumption is approximately true for real gases when the distance between molecules is large compared with their size. This allows us to neglect intermolecular collisions, an assumption that simplifies our analysis but isn't crucial to the ideal gas.

2. The molecules don't exert action-at-a-distance forces on each other. Thus there's no intermolecular potential energy, and therefore the molecules have only kinetic energy. This assumption is fundamental to an ideal gas.

3. The molecules move in random directions with a distribution of speeds that's independent of direction.

4. Collisions with the container walls are elastic, conserving the molecules' energy and momentum. Here's where we tie our gas model to Newtonian mechanics.

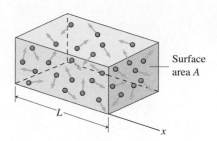

FIGURE 17.2 Gas molecules confined to a rectangular box.

Consider N molecules confined to a rectangular box with length L (Fig. 17.2). Each molecule that collides with a wall exerts a force. There are so many molecules that individual collisions aren't evident; instead the wall experiences an essentially constant average force. The gas pressure p is a measure of this force on a unit area. We're going to find an expression for p and show that it takes in the form of the ideal-gas law.

Figure 17.3 shows one molecule colliding with the right-hand wall. Since the collision is elastic, the y component of the molecule's velocity is unchanged, while the x component reverses sign. Thus the molecule undergoes a momentum change of magnitude $2mv_{xi}$, where i labels this particular molecule. After the molecule collides with the right-hand wall, nothing will change its x velocity until it hits the left-hand wall and its x velocity again reverses. So it will be back at the right-hand wall in the time $\Delta t_i = 2L/v_{xi}$ that it takes to go back and forth along the container.

Now each time our molecule collides with the right-hand wall, it delivers momentum $2mv_{xi}$ to the wall. Newton's second law says that force is the rate of change of momentum. So we can calculate the average force $\overline{F_i}$ due to one molecule by dividing the momentum delivered, $2mv_{xi}$, by the time, $2L/v_{xi}$ between collisions:

$$\overline{F_i} = \frac{2mv_{xi}}{2L/v_{xi}} = \frac{mv_{xi}^2}{L}$$

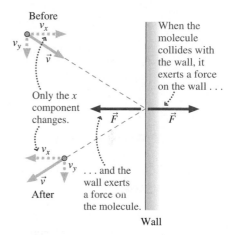

FIGURE 17.3 A molecule undergoes an elastic collision, reversing its x component and transferring momentum $2mv_x$ to the wall.

To get the total force on the wall, we sum over all N molecules with their different x velocities. Dividing by the wall area A then gives the pressure:

$$p = \frac{\overline{F}}{A} = \frac{\sum \overline{F_i}}{A} = \frac{\sum mv_{xi}^2/L}{A} = \frac{m\sum v_{xi}^2}{AL}$$

The last step follows because the box length L and molecular mass m are the same for all molecules, so they factor out of the sum. We can simplify by noting that the denominator AL is just the volume V. Let's also multiply by 1 in the form N/N, with N the number of molecules. Then we have

$$p = \frac{m\sum v_{xi}^2}{AL} = \frac{mN}{V}\frac{\sum v_{xi}^2}{N}$$

In the final expression here, the term $\sum v_{xi}^2/N$ is the average of the squares of all the x-velocity components of all the molecules; we designate this quantity $\overline{v_x^2}$. So the pressure becomes

$$p = \frac{mN}{V}\overline{v_x^2}$$

We still haven't used assumption 3—that the molecules move in random directions with speeds independent of direction. If we grab a molecule at random, that means we're just as likely to find it moving in the x direction, the y direction, the z direction, or any direction in between—and its speed, on average, won't depend on its direction of motion. So the average quantities $\overline{v_x^2}$, $\overline{v_y^2}$, and $\overline{v_z^2}$ must be equal. Since the three directions x, y, and z are perpendicular, the average of the molecular speeds squared is $\overline{v^2} = \overline{v_x^2} + \overline{v_y^2} + \overline{v_z^2}$. We've just argued that all three terms on the right are equal, so we can write $\overline{v^2} = 3\overline{v_x^2}$, or $\overline{v_x^2} = \frac{1}{3}\overline{v^2}$. Then our expression for pressure becomes

$$p = \frac{mN}{3V}\overline{v^2}$$

Multiplying through by V and by 1 in the form 2/2, we have

$$pV = \tfrac{2}{3} N\left(\tfrac{1}{2} m\overline{v^2}\right)$$

This looks a lot like the ideal-gas law (Equation 17.1), except that instead of kT we have $\tfrac{2}{3}\left(\tfrac{1}{2} m\overline{v^2}\right)$. Take a good look at the quantity in parentheses: You'll see that it's just the average kinetic energy of a gas molecule.

Think about what we've done here. We applied the fundamental laws of mechanics to an ideal gas and came up with an equation that looks like the experimentally verified ideal-gas law, except that it's expressed in terms of a microscopic quantity—molecular kinetic energy—rather than the macroscopic quantity temperature. Since our equation describes the behavior of an ideal gas, it *must be* the ideal-gas law. Comparing with the ideal-gas law in the form 17.1, we must therefore have

$$\tfrac{1}{2} m\overline{v^2} = \tfrac{3}{2} kT \qquad \text{(temperature and molecular energy)} \qquad (17.3)$$

Our derivation shows why, in terms of Newtonian mechanics, a gas obeying our four assumptions should obey the ideal-gas law. In Equation 17.3 we get an added bonus—a microscopic understanding of the meaning of temperature: **Temperature measures the average kinetic energy associated with random translational motion of the molecules.**

EXAMPLE 17.2 **Molecular Energy and Speed: An Air Molecule**

Find the average kinetic energy of a molecule in air at room temperature (20°C or 293 K), and determine the speed of a nitrogen molecule (N_2) with this energy.

INTERPRET This problem asks about the linkage between thermodynamic quantities and molecular energy. We just found that linkage: The temperature of a gas is a measure of the average kinetic energy of its molecules.

DEVELOP Equation 17.3, $\tfrac{1}{2} m\overline{v^2} = \tfrac{3}{2} kT$, quantifies the relation between temperature and molecular kinetic energy. Once we find the molecular kinetic energy, we'll need the molecular mass to determine the speed. We can get that using the atomic weight of nitrogen and the fact that an N_2 molecule contains two atoms.

EVALUATE We first evaluate the average molecular kinetic energy:
$$\overline{K} = \tfrac{1}{2} m\overline{v^2} = \tfrac{3}{2} kT = \tfrac{3}{2}(1.38 \times 10^{-23} \text{ J/K})(293 \text{ K}) = 6.07 \times 10^{-21} \text{ J}$$

We can solve for the corresponding speed if we know the molecular mass m. A nitrogen molecule consists of two atoms each with mass 14 u (see Appendix D), so its mass is

$$m = 2(14 \text{ u})(1.66 \times 10^{-27} \text{ kg/u}) = 4.65 \times 10^{-26} \text{ kg}$$

Since $\overline{K} = \tfrac{1}{2} m\overline{v^2}$, the speed corresponding to this kinetic energy is

$$v = \sqrt{\frac{2\overline{K}}{m}} = \sqrt{\frac{2(6.07 \times 10^{-21} \text{ J})}{4.65 \times 10^{-26} \text{ kg}}} = 511 \text{ m/s}$$

ASSESS Make sense? Not surprisingly, the answer is the same order of magnitude as the speed of sound ($\sim$340 m/s) in air at room temperature. At the microscopic level, the speed of the individual molecules limits the rate at which information can be transmitted by disturbances—sound waves—propagating through the gas. ∎

We call the speed calculated in Example 17.2 the **thermal speed**. In terms of temperature, Equation 17.3 shows

$$v_{\text{th}} = \sqrt{\frac{3kT}{m}} \qquad (17.4)$$

GOT IT? 17.1 If you double the kelvin temperature of a gas, what happens to the thermal speed of the gas molecules?

The Distribution of Molecular Speeds

The thermal speed v_{th} is a typical molecular speed, but it doesn't tell us much about the distribution of speeds. Are molecular speeds limited to a narrow band about v_{th}? Or are lots of molecules moving much faster or much slower?

In the 1860s, the Scottish physicist James Clerk Maxwell showed that elastic collisions among molecules result in a speed distribution that peaks near the thermal speed but may extend considerably higher. Figure 17.4 plots this **Maxwell-Boltzmann distribution** for two different temperatures. Note that increasing temperature results in a higher thermal speed, as expected, but that it also broadens the distribution so there are more molecules at lower and higher speeds. The high-speed "tail" of the distribution is especially important to chemists because high-energy molecules participate most readily in chemical reactions. The rapid extension of the high-energy tail with increasing temperature shows why reaction rates are strongly temperature sensitive, and explains why foods keep much longer with even modest refrigeration. High-energy molecules are also the first to evaporate from a liquid, leaving slower, cooler molecules behind and thus explaining evaporative cooling. Without this effect, Earth's atmosphere would be much drier and it would rain far less frequently.

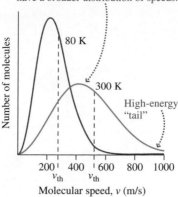

FIGURE 17.4 Maxwell-Boltzmann distribution of molecular speeds for nitrogen (N_2) at temperatures of 80 K and 300 K.

Real Gases

The ideal-gas law is a good approximation to the behavior of most real gases, but it's not perfect because our assumptions aren't entirely realistic. Two factors are especially important. First, real molecules take up space. This reduces the available volume, altering the ideal-gas law. Second, electrical effects that we'll explore in Chapter 20 result in a weak attractive force between nearby molecules. As they move apart, molecules do work against this **van der Waals force**, and their kinetic energy drops. Again, the effect is a deviation from ideal-gas behavior. That deviation is small for diffuse gases with widely spaced molecules, but at high density it becomes significant.

17.2 Phase Changes

Step out of a steamy shower, and you'll find the mirror fogged with water condensed on the cool glass. Climb a mountain in winter, and you'll be treated to the lovely spectacle of every branch and pine needle covered with a delicate coating of frost that's formed right from the air. Burn a rewritable CD or DVD, and you've stored information with a laser that melts tiny spots on the spinning disc. These examples involve **phase changes** between gas and liquid, gas and solid, and solid and liquid.

Heat and Phase Changes

Drop ice cubes into a drink and stir. What's the temperature of the drink? It's 0°C, and it stays at 0°C as long as any ice remains. The melting of a pure solid occurs at a fixed temperature. During the process, energy goes into breaking the molecular bonds that hold the material in its solid form. This increases the potential energy of the molecules but not their kinetic energy. Since temperature is a measure of molecular kinetic energy, that means the temperature doesn't change either.

Figure 17.5 shows the temperature history for a block of ice as it's warmed to the melting point and then beyond, all with a fixed rate of energy input. Note that the temperature rises steadily while the water is in the solid, liquid, or gas phase, but remains constant during the solid-liquid and liquid-gas transitions.

The energy per unit mass required to change phase is called a **heat of transformation** L; for the solid-liquid change it's the **heat of fusion** L_f, and for liquid-gas it's the **heat of vaporization** L_v. Less familiar is the **heat of sublimation** for the transition from solid directly to gas. These quantities have units of J/kg, so the energy required to change the phase of a mass m is

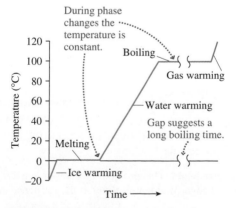

FIGURE 17.5 Temperature versus time for what's initially a block of ice at −20°C, supplied with energy at a constant rate. The process takes place at atmospheric pressure.

$$Q = Lm \quad \text{(heat of transformation)} \qquad (17.5)$$

TABLE 17.1 Heats of Transformation (at Atmospheric Pressure)

Substance	Melting Point (K)	L_f (kJ/kg)	Boiling Point (K)	L_v (kJ/kg)
Alcohol, ethyl	159	109	351	879
Copper	1357	205	2840	4726
Lead	601	24.7	2013	858
Mercury	234	11.3	630	296
Oxygen	54.8	13.8	90.2	213
Sulfur	388	38.5	718	287
Water	273	334	373	2257
Uranium	1406	82.8	4091	1875

To reverse the change requires removing the same energy. Table 17.1 lists heats of transformation for some common materials. These quantities are typically quite large; water's heat of fusion, for example, is 334 kJ/kg or 80 cal/g—meaning it takes as much energy to melt 1 gram of ice as to heat the resulting water to 80°C.

GOT IT? 17.2 You bring a pot of water to boil and then forget about it. Ten minutes later you come back to the kitchen to find the water still boiling. Is its temperature (a) less than, (b) greater than, or (c) equal to 100°C?

EXAMPLE 17.3 **The Heat of Fusion: Meltdown!**

A nuclear power plant's reactor vessel cracks, and all the cooling water drains out. Although nuclear fission stops, radioactive decay continues to heat the reactor's 2.5×10^5 kg uranium core at the rate of 120 MW. Once the melting point is reached, how much energy will it take to melt the core? How long will the melting take?

INTERPRET Since this problem is about melting, it must involve the heat of fusion. We identify the material in question as uranium.

DEVELOP Our plan is to find uranium's heat of fusion in Table 17.1 and then use Equation 17.5, $Q = Lm$, to calculate the energy required for melting. We're given the rate of energy generation by radioactive decay, and from that we'll be able to get the time.

EVALUATE Using uranium's L_f value from Table 17.1 in Equation 17.5, we have

$$Q = L_f m = (2.5 \times 10^5 \text{ kg})(82.8 \text{ kJ/kg}) = 20.7 \text{ GJ}$$

With a heating rate of 120 MW or 0.12 GJ/s, the time to melt the uranium is $(20.7 \text{ GJ})/(0.12 \text{ GJ/s}) = 173$ s.

ASSESS The time to meltdown is just under 3 minutes! Failsafe emergency cooling systems are essential to prevent nuclear meltdowns. ∎

Often we're interested in the total energy needed to bring a material to its transition point and then to make the phase transition. Then we need to combine specific-heat considerations of Chapter 16 with the heats of transformation introduced here.

EXAMPLE 17.4 **Heating and Phase Change: Enough Ice?**

When 200 g of ice at $-10°C$ are added to 1.0 kg of water at 15°C, is there enough ice to cool the water to 0°C? If so, how much ice is left in the mixture?

INTERPRET This problem involves both a temperature rise and a phase change. We identify water as the substance involved.

DEVELOP Equation 16.3, $\Delta Q = mc\,\Delta T$, determines the energy for the temperature rise, and Equation 17.5, $Q = Lm$, determines the phase-change energy. But we don't know whether all the ice melts. So our plan is to find the energy that it *would* take to heat the ice to 0°C and then melt all of it; if *more* than that much is available in cooling the water to 0°C, we'll know that we end up with all water at $T > 0°C$. But if there isn't sufficient energy, then we'll have a mixture with both ice and water at 0°C, and we can use the energy extracted in cooling the water to find out how much ice melts.

EVALUATE We begin by evaluating the energy Q_1 to heat the ice and then melt it all, adding the energies from Equations 16.3 and 17.5 and then getting the specific heat and heat of fusion from Tables 16.1 and 17.1, respectively:

$$Q_1 = m_{ice}c_{ice}\,\Delta T_{ice} + m_{ice}L_f$$
$$= (0.20 \text{ kg})(2.05 \text{ kJ/kg·K})(10 \text{ K}) + (0.20 \text{ kg})(334 \text{ kJ/kg})$$
$$= 4.1 \text{ kJ} + 66.8 \text{ kJ} = 70.9 \text{ kJ}$$

Cooling the water to 0°C would extract energy Q_2 given by Equation 16.3:

$$Q_2 = m_{water}c_{water}\,\Delta T_{water} = (1.0 \text{ kg})(4.184 \text{ kJ/kg·K})(15 \text{ K}) = 62.8 \text{ kJ}$$

This is far more than the 4.1 kJ needed to bring the ice to 0°C, but not quite the 70.9 kJ needed to leave it all melted. So there's enough ice to

cont'd.

cool the water to 0°C, with some left over. How much? Our calculation of Q_1 shows that 4.1 kJ go into raising the ice temperature. Of the 62.8 kJ extracted from the water, the remaining 58.7 kJ go to melting ice. From Equation 17.5, the amount of ice melted is then

$$m_{\text{melted}} = \frac{Q}{L_f} = \frac{58.7 \text{ kJ}}{334 \text{ kJ/kg}} = 0.176 \text{ kg} = 176 \text{ g}$$

So we're left with 24 g of ice in 1176 g of water, all at 0°C.

ASSESS Make sense? Our 62.8 kJ was nearly enough to bring all the ice to the liquid phase, so it makes sense that only a small fraction of the ice remains. ∎

Phase Diagrams

Why can't mountaineers enjoy piping hot coffee? Because water's boiling point drops with the decreasing pressure at high altitudes. In general, the temperatures at which phase changes occur depend on pressure. A **phase diagram** shows the different phases on a plot of pressure versus temperature. Figure 17.6 is a phase diagram for a typical substance. Most phase diagrams are similar, although water's is slightly unusual for reasons we'll discuss in the next section.

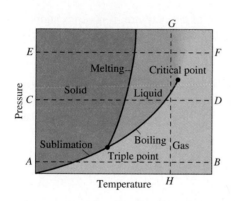

FIGURE 17.6 A phase diagram showing solid, liquid, and gas phases on a plot of pressure versus temperature.

The phase diagram divides pressure-temperature space into regions corresponding to solid, liquid, and gas phases. Lines separating these regions mark the phase transitions. Everyday experience suggests that heating takes a substance from solid, to liquid, to gas— as with water in Fig. 17.5. But Fig. 17.6 shows that this sequence doesn't always occur. At low pressure (line *AB* in Fig. 17.6) the substance goes directly from solid to gas. This is **sublimation**. We don't see this with water because normal atmospheric pressure is too high. For carbon dioxide, though, atmospheric pressure is low in the phase diagram, which is why "dry ice" turns directly into gaseous CO_2 without becoming liquid. At higher pressures (line *CD*) we get the familiar solid-liquid-gas sequence. Higher still (line *EF*), we're above the **critical point**, where the abrupt distinction between liquid and gas disappears. Instead, the substance starts out as a thick fluid whose properties change gradually from liquidlike to gaslike as it's heated.

We think of changing phase by applying heat, but Fig. 17.6 shows we can also change phase by changing pressure. Lowering pressure along line *GH*, for example, takes the substance from liquid to gas without any heat input. You may have seen a demonstration of water boiling vigorously at room temperature in a closed container pumped down to low pressure.

Don't let Fig. 17.6 fool you into thinking that phase transitions occur instantaneously. Those heats of transformation are large, and a substance moving, say, along line *CD* in response to heating will linger at each phase transition until all of it has changed phase; that's what the level portions of Fig. 17.5 showed.

The dividing curves in Fig. 17.6 show where two phases can coexist simultaneously, like ice floating in water at 0°C and atmospheric pressure. It's because phase changes occur along curves that terms like "melting point" and "boiling point" are meaningless unless pressure is specified. But there's one unique **triple point** where solid, liquid, and gas all coexist in equilibrium. Here temperature and pressure have unique, unambiguous values—which is why the 273.16-K triple point of water is used to define the kelvin scale.

17.3 Thermal Expansion

We've seen how heating causes changes in temperature and phase. But heating also results in pressure or volume changes. For a gas at constant pressure, for example, the ideal-gas law shows that volume increases in direct proportion to temperature. The volume and pressure relations for liquids and solids aren't so simple. Because their molecules are closely spaced, liquids and solids aren't very compressible, so thermal expansion is less pronounced.

We characterize the change in the volume with temperature using the **coefficient of volume expansion** β, defined as the fractional change in volume when a substance undergoes a small temperature change ΔT:

$$\beta = \frac{\Delta V/V}{\Delta T} \tag{17.6}$$

This equation assumes that β is independent of temperature; if it varies significantly, then we would need to define β in terms of the derivative dV/dT (Problem 69). Our definition of β also assumes constant pressure; we could entirely inhibit thermal expansion with appropriate pressure increases.

Often we want to know how one linear dimension of a solid changes with temperature. This is especially true with long structures, where the absolute change is greatest along the long dimension (Fig. 17.7). We then speak of the **coefficient of linear expansion** α, defined by

$$\alpha = \frac{\Delta L/L}{\Delta T} \tag{17.7}$$

The volume- and linear-expansion coefficients are related in a simple way: $\beta = 3\alpha$, as you can show in Problem 72. However, the linear-expansion coefficient α is really meaningful only with solids, because liquids and gases deform and don't expand proportionately in all directions. Table 17.2 lists the expansion coefficients for some common substances.

FIGURE 17.7 Thermal expansion distorted these tracks, causing a derailment. Expansion of long structures like this is best described using the coefficient of linear expansion.

TABLE 17.2 Expansion Coefficients*

Solids	α (K^{-1})	Liquids and Gases	β (K^{-1})
Aluminum	24×10^{-6}	Air	3.7×10^{-3}
Brass	19×10^{-6}	Alcohol, ethyl	75×10^{-5}
Copper	17×10^{-6}	Gasoline	95×10^{-5}
Glass (Pyrex)	3.2×10^{-6}	Mercury	18×10^{-5}
Ice	51×10^{-6}	Water, 1°C	-4.8×10^{-5}
Invar†	0.9×10^{-6}	Water, 20°C	20×10^{-5}
Steel	12×10^{-6}	Water, 50°C	50×10^{-5}

*At approximately room temperature unless noted.

†Invar, consisting of 64% iron and 36% nickel, is an alloy designed to minimize thermal expansion.

GOT IT? 17.3 The figure shows a donut-shaped object. If it's heated, will the hole get (a) larger or (b) smaller?

EXAMPLE 17.5 **Thermal Expansion: Spilled Gasoline**

A steel gas can holds 20 L at 10°C. It's filled to the brim with gas at 10°C. If the temperature now increases to 25°C, by how much does the can's volume increase? How much gas spills out?

INTERPRET This is a problem about thermal expansion. Since it involves volume, we identify the relevant quantity as the coefficient of volume expansion β.

DEVELOP Equation 17.6, $\beta = (\Delta V/V)/\Delta T$, determines the volume change. Our plan is to calculate the expanded volume of the tank and then of the gasoline. The difference will be the amount that spills out. Table 17.2 lists β for gasoline but α for steel; therefore we'll use the equation $\beta = 3\alpha$ for the steel.

EVALUATE First we use Equation 17.6 to evaluate the volume change ΔV of the can. Using $\beta = 3\alpha$, we have

$$\Delta V_{can} = \beta V \Delta T = (3)(12 \times 10^{-6}\,K^{-1})(20\,L)(15\,K) = 0.0108\,L$$

Similarly, for the gasoline,

$$\Delta V_{gas} = \beta V \Delta T = (95 \times 10^{-5}\,K^{-1})(20\,L)(15\,K) = 0.285\,L$$

We therefore lose 0.275 L.

ASSESS Make sense? The thermal-expansion coefficient for gasoline is so much greater than for steel that the can's expansion is negligible and the gas has nowhere to go. By the way, that spill wastes nearly 10 MJ of energy!

Thermal Expansion of Water

The entry for water at 1°C in Table 17.2 is remarkable, the negative expansion coefficient showing that water at this temperature actually *contracts* on heating. This unusual behavior occurs because ice has a relatively open crystal structure (Fig. 17.8) and therefore is less dense than liquid water. That's why ice floats. Immediately above the melting point, the intermolecular forces that bond H_2O molecules in ice still exert an influence, giving cold liquid water a lower density than at slightly higher temperatures. At 4°C water reaches its maximum density, and above this temperature the effect of molecular kinetic energy in keeping molecules apart wins out over intermolecular forces. From there on, water exhibits the more normal behavior of expansion with increasing temperature.

This unusual property of water near its melting point is reflected in its phase diagram, shown in Fig. 17.9. Note that the solid-liquid boundary extends leftward from the triple point, in contrast to the more typical behavior in Fig. 17.6. That means that ice at a fixed temperature will melt if the pressure is *increased*—an unusual property known as pressure melting.

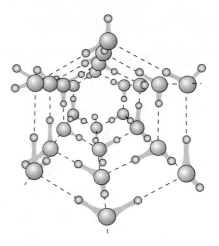

FIGURE 17.8 Water molecules in an ice crystal form an open structure, giving solid water a lower density than the liquid.

FIGURE 17.9 Phase diagram for water. Compare the solid-liquid boundary with that of Fig. 17.6.

APPLICATION **Aquatic Life and Lake Turnover**

The anomalous behavior of water has important consequences for life. If ice didn't float, then ponds, lakes, and even oceans would freeze solid from the bottom up, making aquatic life impossible. What actually happens, instead, is that a thin layer of ice forms on the surface,

insulating the water below and keeping it liquid; as a result, ice cover in temperate climates rarely exceeds a meter or so. Because water has the greatest density at 4°C, water at this temperature sinks to the bottom. At lake depths greater than a few meters, sunlight is inadequate to raise the temperature, which therefore remains year-round at 4°C.

Water's unusual density behavior also causes the twice-yearly turnover of lakes in temperate climates. In the summer, a lake's surface water is warm, but deep water remains at 4°C. In the winter, water just beneath the ice is at 0°C, while the bottom water is still at 4°C. Both situations are stable, with less dense and therefore more buoyant water at the surface. But in the spring, ice melts and the surface water warms. When that water reaches 4°C, there's no density variation and the lake water mixes freely. This is the spring overturning. A similar overturning occurs in the fall, as the surface water cools through 4°C. Turnover is important to aquatic life because it brings up nutrients that would otherwise be trapped in the deep water.

Big Picture

The big idea here is that matter responds to heating in a variety of ways in addition to changing temperature. Other responses include changes of phase and of volume and/or pressure. The ideal gas provides a particularly simple system for understanding volume and pressure changes. Analyzing ideal-gas behavior provides a link between the Newtonian mechanics of molecules and macroscopic thermodynamics, showing that temperature is a measure of the average molecular kinetic energy.

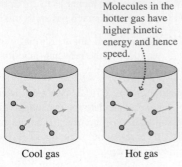

Molecules in the hotter gas have higher kinetic energy and hence speed.

Cool gas   Hot gas

Key Concepts and Equations

The **ideal-gas law** relates pressure, volume, temperature, and the number of molecules in a gas:

$$pV = NkT \qquad \text{(ideal-gas law)}$$

where **Boltzmann's constant** is $k = 1.38 \times 10^{-23}$ J/K.

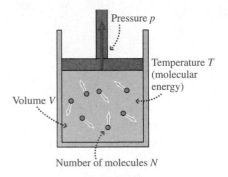

Pressure p

Temperature T (molecular energy)

Volume V

Number of molecules N

In terms of the number of moles n, the ideal-gas law is

$$pV = nN_A kT = nRT$$

where the **universal gas constant** $R = N_A k = 8.314$ J/K·mol.

Heats of transformation L describe the energy per unit mass needed to effect phase changes. The total energy required to change the phase of a mass m is given by

$$Q = Lm \qquad \text{(heat of transformation)}$$

Phase diagrams plot solid, liquid, and gas phases against temperature and pressure, and reveal the **triple point**, where all three phases can coexist, and the **critical point**, where the liquid-gas distinction disappears.

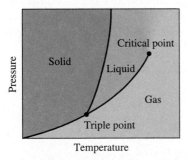

Pressure

Solid

Liquid

Critical point

Gas

Triple point

Temperature

The temperature of an ideal gas is a measure of the gas molecules' average kinetic energy:

$$\tfrac{1}{2} m \overline{v^2} = \tfrac{3}{2} kT \qquad \text{(temperature and molecular energy)}$$

Cases and Use

Thermal expansion results when pressure is held constant. It's described by the **coefficient of volume expansion** and its linear counterpart. The volume-expansion coefficient relates the fractional volume change $\Delta V/V$ to the temperature change ΔT:

$$\beta = \frac{\Delta V/V}{\Delta T} \qquad \text{(volume-expansion coefficient)}$$

while the **coefficient of linear expansion** relates the fractional length $\Delta L/L$ change to ΔT:

$$\alpha = \frac{\Delta L/L}{\Delta T} \qquad \text{(linear-expansion coefficient)}$$

ΔV

T, V $T + \Delta T$

For Thought and Discussion

1. If the volume of an ideal gas is increased, must the pressure drop proportionately? Explain.
2. According to the ideal-gas law, what should the volume of a gas be at absolute zero? Why is this result absurd?
3. Why are you supposed to check the pressure in a tire when the tire is cold?
4. The average *speed* of the molecules in a gas increases with increasing temperature. What about the average *velocity?*
5. Suppose you start running while holding a jar of air. Do you change the average speed of the air molecules? The average velocity? The temperature?
6. Do all molecules in a gas have the same speed? If not, how can the notion of temperature be meaningful?
7. Two different gases are at the same temperature, and both are at low enough densities that they behave like ideal gases. Do their molecules have the same thermal speeds? Explain.
8. Some people think that ice and snow must be at 0°C. Is this always true?
9. What is the temperature of water just under the ice layer of a frozen lake? At the bottom of a deep lake?
10. Some ice and water have been together in a glass for a long time. Is the water hotter than the ice?
11. Does it take more heat to melt a gram of ice at 0°C than to bring the resulting water to the boiling point? Once at the boiling point, does it take more heat to boil all the water than it did to bring it from the melting point to the boiling point?
12. Why does removing the plastic wrap from a package of frozen hamburger help it to thaw faster?
13. Why do we use the triple point of water for thermometer calibration? Why not just use the melting point or boiling point?
14. How is it possible to have boiling water at a temperature other than 100°C?
15. How does a pressure cooker work?
16. Suppose mercury and glass had the same coefficient of volume expansion. Could you build a mercury thermometer?
17. A bimetallic strip consists of thin pieces of brass and steel bonded together (Fig. 17.10). Such strips are often used in thermostats. What will happen when the strip is heated? *Hint:* Consult Table 17.2.

FIGURE 17.10 For Thought and Discussion 17

Exercises and Problems

Exercises

Section 17.1 Gases
18. Mars's atmospheric pressure is about 1% that of Earth, and its average temperature is around 215 K. Find the volume of 1 mol of the Martian atmosphere.
19. How many molecules are in an ideal-gas sample at 350 K that occupies 8.5 L when the pressure is 180 kPa?
20. What is the pressure of an ideal gas if 3.5 mol occupy 2.0 L at a temperature of −150°C?
21. An ideal gas occupies a volume V at 100°C. If the gas pressure is held constant, by what factor does the volume change (a) if the Celsius temperature is doubled and (b) if the kelvin temperature is doubled?
22. (a) If 2.0 mol of an ideal gas are at an initial temperature of 250 K and a pressure of 1.5 atm, what is the gas volume? (b) The pressure is now increased to 4.0 atm, and the gas volume drops to half its initial value. What is the new temperature?
23. A pressure of 1.0×10^{-10} Pa is readily achievable with laboratory vacuum apparatus. If the residual air in this "vacuum" is at 0°C, how many air molecules are in 1 L?
24. What is the thermal speed of hydrogen (H_2) molecules at 800 K?
25. In which gas are the molecules moving faster: hydrogen (H_2) at 75 K or sulfur dioxide (SO_2) at 350 K?

Section 17.2 Phase Changes
26. How much energy does it take to melt a 65-g ice cube?
27. It takes 200 J to melt an 8.0-g sample of one of the substances in Table 17.1. What is the substance?
28. If it takes 840 kJ to vaporize a sample of liquid oxygen, how large is the sample?
29. Carbon dioxide sublimes (changes from solid to gas) at 195 K. The heat of sublimation is 573 kJ/kg. How much heat must be extracted from 250 g of CO_2 gas at 195 K in order to solidify it?

30. Find the energy needed to convert 28 kg of liquid oxygen at its boiling point into gas.
31. What is the density, in moles per cubic meter, of air in a tire whose absolute pressure is 300 kPa at 34°C?

Section 17.3 Thermal Expansion
32. A copper wire is 20 m long on a winter day when the temperature is −12°C. By how much does its length increase on a 26°C summer day?
33. You have exactly 1 L of ethyl alcohol at room temperature (20°C). You put it in a refrigerator at 2°C. What's its new volume?
34. A Pyrex glass marble is 1.00000 cm in diameter at 20°C. What will its diameter be at 85°C?
35. At 0°C, the hole in a steel washer is 9.52 mm in diameter. To what temperature must it be heated in order to fit over a 9.55-mm-diameter bolt?
36. Suppose a single piece of welded steel railroad track stretched 5000 km across the continental United States. If the track were free to expand, by how much would its length change if the entire track went from a cold winter temperature of −25°C to a hot summer day at 40°C?

Problems
37. The solar corona is an extended atmosphere of hot $(2 \times 10^6 \text{ K})$ gas surrounding the cooler visible surface of the Sun. The gas pressure in the solar corona is about 0.03 Pa. What is the coronal density in particles per cubic meter? Compare with Earth's atmosphere.
38. A helium balloon occupies 8.0 L at 20°C and 1.0-atm pressure. The balloon rises to an altitude where the air pressure is 0.65 atm and the temperature is −10°C. What is its volume when it reaches equilibrium at the new altitude? *Note:* Neglect tension forces in the material of the balloon.

39. A compressed air cylinder stands 100 cm tall and has an internal diameter of 20.0 cm. At room temperature, the pressure is 180 atm. (a) How many moles of air are in the cylinder? (b) What volume would this air occupy at 1.0 atm and room temperature?

40. An aerosol can of whipped cream is pressurized at 440 kPa when it's refrigerated at 3°C. The can warns against temperatures in excess of 50°C. What is the maximum safe pressure for the can?

41. A 3000-mL flask is initially open while in a room containing air at 1.00 atm and 20°C. The flask is then closed and immersed in a bath of boiling water. When the air in the flask has reached thermodynamic equilibrium, the flask is opened and air is allowed to escape. The flask is then closed and cooled back to 20°C. (a) What is the maximum pressure reached in the flask? (b) How many moles escape when air is released from the flask? (c) What is the final pressure in the flask?

42. A stove burner supplies heat to a pan at the rate of 1500 W. How long will it take to boil away 1.1 kg of water, from the time the water is at its boiling point?

43. If a 1-megaton nuclear bomb were exploded deep in the Greenland ice cap, how much ice would it melt? Assume the ice is initially at about its freezing point, and consult Appendix C for the appropriate energy conversion.

44. How much ice can a 625-W microwave oven melt in 1.0 min if the ice is initially at 0°C?

45. At winter's end, Lake Superior's 82,000-km² surface is frozen to a depth of 1.3 m. The density of ice is 917 kg/m³. (a) How much energy does it take to melt the ice? (b) If the ice disappears in 3 weeks, what is the average power supplied to melt it?

46. A refrigerator extracts energy from its contents at the rate of 95 W. How long will it take to freeze 750 g of water already at 0°C?

47. An ice layer 50 cm thick covers a lake; soot from air pollution covers the ice and absorbs 75% of the incident sunlight, whose intensity averages 200 W/m². How long will it take to melt the ice, assuming an initial temperature of about 0°C? The density of ice is 917 kg/m³.

48. Repeat Example 17.4 if the initial mass of ice is 50 g.

49. How much energy does it take to melt 10 kg of ice initially at −10°C?

50. Water is brought to its boiling point and then allowed to boil away completely. If the energy needed to raise the water to the boiling point is one-tenth of that needed to boil it away, what was the initial temperature?

51. During a nuclear accident, 420 m³ of emergency cooling water at 20°C are injected into a reactor vessel where the reactor core is producing heat at the rate of 200 MW. If the water is allowed to boil at normal atmospheric pressure, how long will it take to boil the reactor dry?

52. What is the minimum amount of ice in Example 17.4 that will ensure a final temperature of 0°C?

53. A bowl contains 16 kg of punch (essentially water) at a warm 25°C. What is the minimum amount of ice at 0°C that will cool the punch to 0°C?

54. A 50-g ice cube at −10°C is placed in an equal mass of water. What must the initial water temperature be if the final mixture still contains equal amounts of ice and water?

55. A 40-kg block of aluminum is initially at 50°C. A jet of steam at 100°C hits the block and condenses, and the resulting water drops off before its temperature changes. How much steam must hit the block to raise its temperature to 100°C?

56. What power is needed to melt 20 kg of ice in 6.0 min?

57. You put 300 g of water into a 500-W microwave oven and accidentally set the time for 20 min instead of 2.0 min. If the water is initially at 20°C, how much is left at the end of 20 min?

58. If 4.5 × 10⁵ kg of emergency cooling water at 10°C are dumped into a malfunctioning nuclear reactor whose core is producing energy at the rate of 200 MW, and if no circulation or cooling of the water is provided, how long will it be before half the water has boiled away?

59. Describe the composition and temperature of the equilibrium mixture after 1.0 kg of ice at −40°C is added to 1.0 kg of water at 5.0°C.

60. A glass marble 1.000 cm in diameter is to be dropped through a hole in a steel plate. At room temperature the hole diameter is 0.997 cm. By how much must the steel temperature be raised so that the marble will fit through the hole?

61. A 2000-mL graduated cylinder is filled with liquid at 350 K. When the liquid is cooled to 300 K, the cylinder is full to only the 1925-mL mark. Use Table 17.2 to identify the liquid.

62. A steel ball bearing is encased in a Pyrex glass cube 1.0 cm on a side. At 330 K, the ball bearing fits tightly inside the cube. At what temperature will it have a clearance of 1.0 μm all around?

63. Gasoline comes from its underground tank at 10°C. On a summer day, how much gas can you put in your car's 60-L tank if the tank is not to overflow when the gas reaches the ambient temperature of 25°C?

64. A rod of length L_0 is clamped rigidly at both ends. Its temperature increases by an amount ΔT, and in the ensuing expansion it cracks to form two straight pieces, as shown in Fig. 17.11. Find an expression for the distance d shown in the figure, in terms of L_0, ΔT, and the coefficient of linear expansion α.

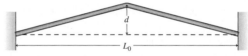

FIGURE 17.11 Problem 64

65. The thermal resistance of a refrigerator's walls is 0.12 K/W. During a power failure you put a 15-kg block of ice at 0°C in the refrigerator, bringing the interior temperature to 0°C. If room temperature is 20°C, how long will the ice last?

66. A solar-heated house stores energy in 5.0 tons of Glauber salt ($Na_2SO_4 \cdot 10\,H_2O$), a substance that melts at 90°F. The heat of fusion of Glauber salt is 104 Btu/lb, and the specific heats of the solid and liquid are, respectively, 0.46 Btu/lb·°F and 0.68 Btu/lb·°F. After a week of sunny weather, the storage medium is all liquid at 95°F. Then a cool, cloudy period sets in during which the house loses heat at an average rate of 20,000 Btu/h. (a) How long is it before the temperature of the storage medium drops below 60°F? (b) How much of this time is spent at 90°F?

67. Show that the coefficient of volume expansion of an ideal gas at constant pressure is the reciprocal of its kelvin temperature.

68. Water's coefficient of volume expansion in the temperature range from 0°C to about 20°C is given approximately by $\beta = a + bT + cT^2$, where T is in Celsius and $a = -6.43 \times 10^{-5}\,°C^{-1}$, $b = 1.70 \times 10^{-5}\,°C^{-2}$, and $c = -2.02 \times 10^{-7}\,°C^{-3}$. Show that water has its greatest density at approximately 4.0°C.

69. When the expansion coefficient varies with temperature, Equation 17.6 should be written $\beta = (1/V)(dV/dT)$. If a sample of water occupies 1.00000 L at 0°C, find its volume at 12°C. Use the information from Problem 68, and integrate the equation above.

70. Ignoring air resistance, find the height from which you must drop an ice cube at 0°C so it melts completely on impact. Assume no heat exchange with the environment.

71. The timekeeping of an old clock is regulated by a brass pendulum 20.0 cm long. If the clock is accurate at 20°C but is in a room at 18°C, how long will it be before the clock is in error by 1 minute? Will it be too fast or too slow?

72. Prove the equation $\beta = 3\alpha$ (Section 17.3) by considering a cube of side s and therefore volume $V = s^3$ that undergoes a small temperature change dT and corresponding length and volume changes ds and dV.

73. Venus's atmospheric pressure is 90 times that of Earth, and its average temperature is 730 K. Suppose you're on the team planning a mission to Venus to collect samples of the atmosphere and soil for analysis. The design engineers are planning a container for the atmospheric sample that can hold 1 L of gas. The atmospheric scientists need at least 1 mol of gas in order to do their experiments. Will the design work?

74. Your chemistry-major roommate is in the midst of an experiment. A 140-W heat source is being applied to 250 g of ethyl alcohol in order to evaporate it completely. Rather than sit around and wait for the whole process, your roommate, who has not had lunch, wants to know if there's at least 30 minutes of time before the alcohol entirely evaporates. In addition to a lecture on not leaving boiling alcohol unattended, how do you respond?

75. For your first case out of law school you're representing a microwave oven manufacturer. The plaintiff complains the oven was defective and caused a fire. The argument claims the 500-W oven can completely vaporize 500 g of ice at 0°C in 10 minutes. Is that correct?

76. Perhaps for a literature class you have read the book *Fahrenheit 451* or seen the film version. The setting (in England) is an Orwellian society where books are banned and all information is disseminated by a large TV screen in each home. Fire departments respond not to put out fires but to burn books, which combust at a temperature of 451°F—hence the title. Suppose the fire department decides to use the burning books to heat water for afternoon tea. This is done with a burner that transfers 80% of the heat from the burning book to 500 mL of water on top of the burner. The water is initially at 25°C and needs to be raised to 100°C. How many copies of a 500-page book would need to burn if the energy released in combustion is 5 J/page?

Answers to Chapter Questions

Answer to Chapter Opening Question

Water's solid phase is less dense than the liquid, which causes ice to float. Our world would be a very different place if ice were denser than water.

Answers to GOT IT? Questions

17.1 It increases by a factor of $\sqrt{2}$.

17.2 (c) Equal to 100°C.

17.3 (a) Larger, since all linear dimensions of the object expand equally.

18

Heat, Work, and the First Law of Thermodynamics

■ A jet aircraft engine converts the energy of burning fuel into mechanical energy. How does energy conservation apply in this process?

▶ **To Learn**

By the end of this chapter you should be able to

■ Explain how the first law of thermodynamics extends the conservation-of-energy principle to include thermal energy (18.1).

■ Describe quantitatively the effects of basic thermodynamic processes—isothermal, adiabatic, isobaric, and constant-volume—on an ideal gas (18.2).

■ Compute the specific heat of an ideal gas based on its molecular structure (18.3).

◀ **To Know**

■ This chapter joins the concept of work, introduced in Chapter 6 (6.1, 6.2), and the behavior of ideal gases as described in Chapter 17 (17.1).

■ You should be ready to extend the conservation-of-energy principle introduced in Chapter 7 (7.3).

In Chapter 7 we introduced the powerful idea of energy conservation, limited then to mechanical energy and conservative forces. Now we've learned that thermal processes involve energy—and that sets the stage for a broadening of the conservation-of-energy principle. Here we explore this broader principle and see how it describes energy interchanges in thermodynamic systems ranging from engines to Earth's atmosphere.

18.1 The First Law of Thermodynamics

Figure 18.1 shows two ways to raise the temperature in a beaker of water: by heating with a flame and by stirring vigorously with a spoon. Using the flame involves heat—energy in transit because of the temperature difference between flame and water. But there's no temperature difference between spoon and water; here the energy transfer occurs because the spoon does mechanical work on the water. We already know that doing work can increase the kinetic or potential energy of a macroscopic object; here we see it, instead, changing the **internal energy** associated ultimately with individual molecules. The point is that both processes—heating and mechanical work—result in exactly the same final state—

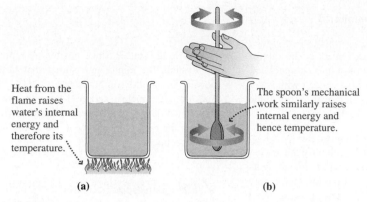

Heat from the flame raises water's internal energy and therefore its temperature.

The spoon's mechanical work similarly raises internal energy and hence temperature.

(a)   (b)

FIGURE 18.1 Two ways to raise temperature.

namely, water with a higher temperature and therefore greater internal energy. It's this common result that made possible Joule's quantitative identification of heat as a form of energy (Fig. 18.2).

Keep track of all the energy entering and leaving a system—both heat and work—and you'll find that the change in the system's internal energy depends only on the net energy transferred. In one sense this is hardly surprising; it just extends the idea of energy conservation to include heat. But in another way it's remarkable; it doesn't matter at all *how* the energy gets into the system—heat, work, or some combination of the two. This statement constitutes the **first law of thermodynamics**:

First law of thermodynamics The change in the internal energy of a system depends only on the net heat transferred to the system and the net work done by the system, independent of the particular processes involved.

Mathematically, the first law is

$$\Delta U = Q - W \qquad \text{(first law of thermodynamics)} \qquad (18.1)$$

where ΔU is the change in a system's internal energy, Q the heat transferred *to* the system, and W the work done *by* the system. A positive Q represents an energy gain, while a positive W represents an energy loss; that's the minus sign in Equation 18.1. This sign convention is used because the first law was developed in connection with engines, which take heat from a source and give out mechanical work.

The first law says that the change in a system's internal energy doesn't depend on how the energy gets transferred, but only on the net energy. Internal energy is therefore a **thermodynamic state variable**, meaning a quantity whose value doesn't depend on how a system got into its particular state. Temperature and pressure are also thermodynamic state variables; heat and work are not.

We're frequently concerned with *rates* of energy flow. Differentiating the first law with respect to time gives a statement about rates:

$$\frac{dU}{dt} = \frac{dQ}{dt} - \frac{dW}{dt} \qquad (18.2)$$

where dU/dt is the rate of change of a system's internal energy, dQ/dt the rate of heat transfer to the system, and dW/dt the rate at which the system does work.

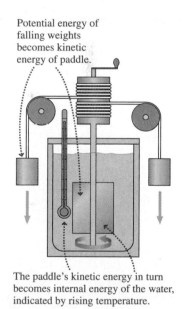

Potential energy of falling weights becomes kinetic energy of paddle.

The paddle's kinetic energy in turn becomes internal energy of the water, indicated by rising temperature.

FIGURE 18.2 Joule's apparatus for determining what he called "the mechanical equivalent of heat."

EXAMPLE 18.1 **The First Law of Thermodynamics: Thermal Pollution**

The reactor in a nuclear power plant supplies energy at the rate of 3.0 GW, boiling water to produce steam that turns a turbine-generator. The spent steam is then condensed through thermal contact with water taken from a river. If the power plant produces electrical energy at the rate of 1.0 GW, at what rate is heat transferred to the river?

INTERPRET This problem is about heat and mechanical energy, which are related by the first law of thermodynamics. We identify the system as the entire power plant, comprising the nuclear reactor, including its fuel, and the turbine-generator. We identify U as the internal energy stored in the fuel, W as the mechanical work that ends up as electrical energy, and Q as the heat transferred to the river.

DEVELOP Since we're dealing here with *rates*, Equation 18.2, $dU/dt = dQ/dt - dW/dt$, applies. The reactor extracts internal energy from its fuel, so the rate dU/dt is negative; the power plant delivers electrical energy to the outside world, so dW/dt is positive. Our plan is then to solve for dQ/dt, the rate of energy transfer to the river.

EVALUATE Solving, we have

$$\frac{dQ}{dt} = \frac{dU}{dt} + \frac{dW}{dt} = -3.0 \text{ GW} + 1.0 \text{ GW} = -2.0 \text{ GW}$$

ASSESS Make sense? Since positive Q represents heat transferred *to* the system, the minus sign shows that heat is transferred *from* the power plant to the river at the rate of 2 GW. The numbers here are typical for large nuclear and coal-burning power plants, and show that about two-thirds of the energy extracted from the fuel is wasted in heating the environment. We'll see in the next chapter just why this waste occurs.

✓**TIP** Identify the System

The first law of thermodynamics deals with energy flows into and out of a system. It's up to you to define the system, and how you do so affects the meanings of the terms in the first law. In this Example we included the nuclear reactor, with the internal energy of its fuel, as part of the system. If we had considered only the turbine-generator, then we would have had 3 GW of heat coming in from the reactor and no change in internal energy. But the result would be the same: 1 GW going out as electricity and 2 GW of heat dumped into the river.

18.2 Thermodynamic Processes

Although the first law applies to *any* system, it's easiest to understand when applied to an ideal gas. The ideal-gas law relates the temperature, pressure, and volume of a given gas sample: $pV = nRT$. The thermodynamic state is completely determined by any two of the quantities p, V, or T. We'll find it convenient to represent different states as points on a **pV diagram**—a graph whose vertical and horizontal axes represent pressure and volume, respectively.

Reversible and Irreversible Processes

Imagine a gas sample immersed in a large reservoir of water and allowed to come to equilibrium (Fig. 18.3). If we then raise the reservoir temperature very slowly, both water and gas temperatures will rise essentially in unison, and the gas will remain in equilibrium. Such a slow change is called a **quasi-static process**. Because a system undergoing a quasi-static process is always in thermodynamic equilibrium, its evolution from one state to another is described by a continuous sequence of points—a curve—in its pV diagram (Fig. 18.4).

We could reverse this heating process by slowly lowering the reservoir temperature; the gas would cool, reversing its path in the pV diagram. For that reason, a quasi-static process is also called a **reversible process**. A process like suddenly plunging a cool gas sample into hot water is, in contrast, **irreversible**. During an irreversible process the system is not in equilibrium, and thermodynamic variables like temperature and pressure don't have well-defined values. It therefore makes no sense to think of a path in the pV diagram. A process may be irreversible even though it returns a system to its original state. The distinction lies not in the end states but in the *process* that takes the system between states.

There are many ways to change the thermodynamic state of a system. Here we consider several important special cases involving an ideal gas. These illustrate the physical principles behind a myriad of technological devices and natural phenomena, from the operation of a gasoline engine to the propagation of a sound wave to the oscillations of a star.

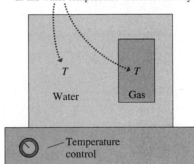

These temperatures stay the same as the water temperature increases slowly.

T

Water

T

Gas

Temperature control

FIGURE 18.3 A quasi-static, or reversible, process keeps water and gas always in equilibrium.

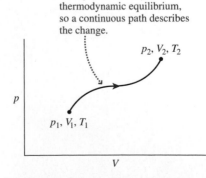

The system is always in thermodynamic equilibrium, so a continuous path describes the change.

p_2, V_2, T_2

p

p_1, V_1, T_1

V

FIGURE 18.4 The pV diagram of a system undergoing quasi-static change.

Our system consists of an ideal gas confined to a cylinder sealed with a movable piston (Fig. 18.5). The piston and cylinder walls are perfectly insulating—they block all heat transfer—and the bottom is a perfect conductor of heat. We can change the thermodynamic state of the gas mechanically by moving the piston, or thermally by transferring heat through the bottom. We'll consider only reversible processes, which we can describe by paths in the pV diagram for the gas.

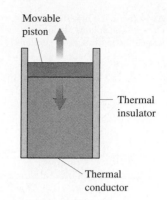

FIGURE 18.5 A gas-cylinder system with insulating walls and a conducting bottom.

Work and Volume Changes

We begin by developing a relation between volume change and work that holds for all processes. If A is the cross-sectional area of our piston and cylinder and p the gas pressure, then $F = pA$ is the force the gas exerts on the piston. If the piston moves a small distance Δx, the work done by the gas is $\Delta W = F\,\Delta x = pA\,\Delta x = p\,\Delta V$, where $\Delta V = A\,\Delta x$ is the change in gas volume (Fig. 18.6a). The work ΔW is positive when the volume increases, indicating that the gas does work on the piston, and negative when ΔV is negative, indicating that work is done on the gas. Pressure may vary with volume, in which case our result is strictly valid only in the limit of very small volume changes. To find the work associated with a larger change, we take the limit $\Delta V \rightarrow 0$ and integrate between the initial volume V_1 and final volume V_2:

$$ W = \int dW = \int_{V_1}^{V_2} p\,dV \qquad \text{(work done during volume change)} \qquad (18.3) $$

Figure 18.6b shows that the work done by the gas is the area under the pV curve.

We'll now explore several basic thermodynamic processes, in each case holding one thermodynamic variable constant.

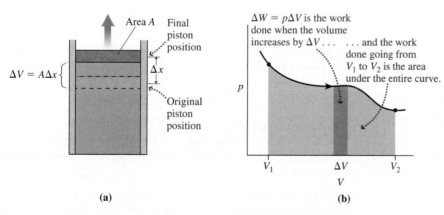

(a)   **(b)**

FIGURE 18.6 Work done as the piston rises is the area under the pV curve.

GOT IT? 18.1 Two identical gas-cylinder systems are taken from the same initial state to the same final state, but by different processes. Which are the same in both cases: (a) the work done on or by the gas; (b) the heat added or removed; or (c) the change in internal energy?

Isothermal Processes

An **isothermal process** occurs at constant temperature. Figure 18.7 shows one way to effect an isothermal process: We place our gas cylinder in thermal contact with a heat reservoir whose temperature is constant. We then move the piston to change the gas volume, slowly enough that the gas remains in equilibrium with the heat reservoir. The system

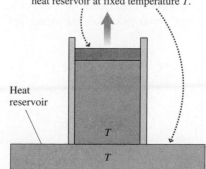

FIGURE 18.7 An isothermal process.

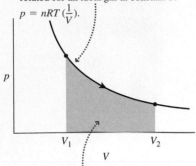

An isotherm is a hyperbola because pressure and volume are inversely related for an ideal gas at constant T:

$$p = nRT \left(\frac{1}{V}\right).$$

Work is the area under the pV curve:

$$W = \int_{V_1}^{V_2} p \, dV.$$

FIGURE 18.8 A pV diagram for an isothermal process.

moves from its initial state to its final state along a curve of constant temperature—an **isotherm**—in the pV diagram (Fig. 18.8). The work done in the process is given by Equation 18.3 and is equal to the area under the isotherm.

To find that work, we relate pressure and volume through the ideal-gas law: $p = (nRT)/V$. Then Equation 18.3 becomes

$$W = \int_{V_1}^{V_2} \frac{nRT}{V} \, dV$$

For an isothermal process, the temperature T is constant, giving

$$W = nRT \int_{V_1}^{V_2} \frac{dV}{V} = nRT \ln V \Big|_{V_1}^{V_2} = nRT \ln\left(\frac{V_2}{V_1}\right)$$

The internal energy of an ideal gas consists only of the kinetic energy of its molecules, which, in turn, depends only on temperature. Thus, there's no change in the internal energy of an ideal gas during an isothermal process. The first law of thermodynamics then gives $\Delta U = 0 = Q - W$, so

$$Q = W = nRT \ln\left(\frac{V_2}{V_1}\right) \qquad \text{(isothermal process)} \qquad (18.4)$$

Does this result $Q = W$ make sense? Recall that Q is the heat transferred to the gas and W is the work done by the gas. Therefore, our result says that when an ideal gas does work W on the external world with no change in temperature, it must absorb an equal amount of heat from outside. Similarly, if work is done on the gas, then the gas must transfer an equal amount of heat to the outside if its temperature is not to change.

EXAMPLE 18.2   An Isothermal Process: Bubbles!

A scuba diver is 25 m down, where the pressure is 3.5 atm or about 350 kPa. The air she exhales forms bubbles 8.0 mm in radius. How much work does each bubble do as it rises to the surface, assuming the bubbles remain at the uniform 300 K temperature of the water?

INTERPRET The constant 300 K temperature tells us we're dealing with an isothermal process.

DEVELOP Equation 18.4, $W = nRT \ln(V_2/V_1)$, determines the work. To use this equation, we need the quantity nRT and the volume ratio V_2/V_1. We know p and V (actually the radius, from which we can get V) at the 25-m depth, so we can use the ideal-gas law $pV = nRT$ to get nRT and also the bubble volume just before it reaches the surface. Then we'll have everything we need to apply Equation 18.4.

EVALUATE The ideal-gas law gives $nRT = pV = \frac{4}{3}\pi r^3 p$. The number of moles n doesn't change and R is a constant, so pV is itself constant in the isothermal process. That means $p_1 V_1 = p_2 V_2$, showing that the

volume expands by a factor of 3.5 as the pressure drops from 3.5 atm to 1 atm at the surface—so $V_2/V_1 = 3.5$. Then Equation 18.4 gives

$$W = nRT \ln\left(\frac{V_2}{V_1}\right) = \frac{4}{3}\pi r^3 p \ln 3.5$$

Using the 8-mm bubble radius and the 350-kPa pressure gives 0.94 J for the work. Note that we needed to use pressure in SI units here; to find the volume ratio, any units would do because V_2/V_1 followed from the pressure *ratio* p_1/p_2.

ASSESS Make sense? The work is positive because an expanding bubble pushes water outward and ultimately upward. It therefore raises the ocean's gravitational potential energy. When the bubble breaks, this excess potential energy becomes kinetic energy, appearing as small waves on the water surface. The bubble, in turn, gets its energy from heat that flows in to keep it at constant temperature. Energy is conserved! ∎

Constant-Volume Processes and Specific Heat

A **constant-volume process** (also called isometric, isochoric, or isovolumic) occurs in a rigid closed container whose volume can't change. We could tightly clamp the piston in Fig. 18.5 for a constant-volume process. Because the piston doesn't move, the gas does no work, and the first law becomes simply $\Delta U = Q$. To express this result in terms of a temperature change ΔT, we introduce the **molar specific heat at constant volume** C_V, defined by

$$Q = nC_V \Delta T \quad \text{(constant-volume process)} \qquad (18.5)$$

where n is the number of moles. This molar specific heat is like the specific heat defined in Chapter 16, except it's per mole rather than per unit mass. Using Equation 18.5 for Q in the statement $\Delta U = Q$ gives

$$\Delta U = nC_V \Delta T \quad \text{(any process)} \tag{18.6}$$

For an ideal gas, the internal energy is a function of temperature alone, so $\Delta U / \Delta T$ has the same value no matter what process the gas undergoes. Therefore Equation 18.6, relating the temperature change ΔT and internal-energy change ΔU, applies not only to a constant-volume process but to *any* ideal-gas process. Why, then, have we been so careful to label C_V the specific heat *at constant volume?* Although Equation 18.6, $\Delta U = nC_V \Delta T$, holds for any process, it's only when there's no work that the first law lets us write $Q = \Delta U$, and therefore only for a constant-volume process that Equation 18.5 holds.

Isobaric Processes and Specific Heat

Isobaric means constant pressure. Processes occurring in systems exposed to the atmosphere are essentially isobaric. In a reversible isobaric process, a system moves along an isobar, or curve of constant pressure, in its pV diagram (Fig. 18.9). The work done as the volume changes from V_1 to V_2 is the area under the isobar, or

$$W = p(V_2 - V_1) = p\,\Delta V \tag{18.7}$$

a result we could obtain formally by integrating Equation 18.3.

Solving the first law (Equation 18.1) for Q and using our expression for work give $Q = \Delta U + W = \Delta U + p\,\Delta V$. For an ideal gas, we've just found that the change in internal energy is given by $\Delta U = nC_V \Delta T$ for *any* process. Therefore $Q = nC_V \Delta T + p\,\Delta V$ for an ideal gas undergoing an isobaric process. We define the **molar specific heat at constant pressure** C_p as the heat required to raise 1 mol of gas by 1 K at constant pressure, or $Q = nC_p \Delta T$. Equating our two expressions for Q gives

$$nC_p \,\Delta T = nC_V \,\Delta T + p\,\Delta V \quad \text{(isobaric process)} \tag{18.8}$$

This is a useful form for calculating temperature changes in an isobaric process if we know both specific heats C_p and C_V. However, we really need only one of these specific heats because a simple relation holds between the two. The ideal-gas law, $pV = nRT$, allows us to write $p\,\Delta V = nR\,\Delta T$ for an isobaric process. Using this expression in Equation 18.8 gives $nC_p \,\Delta T = nC_V \,\Delta T + nR\,\Delta T$, so

$$C_p = C_V + R \quad \text{(molar specific heats)} \tag{18.9}$$

Does this make sense? Specific heat measures the heat needed to cause a given temperature change. In a constant-volume process, no work is done and all the heat goes into raising the internal energy and thus the temperature of an ideal gas. In a constant-pressure process, work *is* done and some of the added heat ends up as mechanical energy, leaving less available for raising the temperature. Therefore a constant-pressure process requires *more* heat for a given temperature change. Thus the specific heat at constant pressure is greater than at constant volume, as reflected in Equation 18.9.

Why didn't we distinguish specific heats at constant volume and constant pressure earlier? Because we were concerned mostly with solids and liquids, whose coefficients of expansion are far lower than those of gases. As a result, much less work is done by a solid or liquid than by a gas. Since work is what gives rise to the difference between C_V and C_p, the distinction is less significant for solids and liquids. As a practical matter, measured specific heats are usually at constant pressure.

Adiabatic Processes

In an **adiabatic process**, no heat flows between a system and its environment. The way to achieve this is to surround the system with perfect thermal insulation. Even without insulation, processes that occur quickly are often approximately adiabatic because they're over before significant heat transfer has had time to occur. In a gasoline engine, for example,

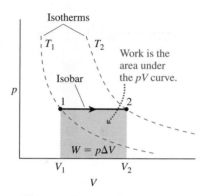

FIGURE 18.9 A pV diagram for an isobaric process; also shown are isotherms for the initial and final temperatures.

Molecules rebound with the same speed, and the gas's internal energy doesn't change.

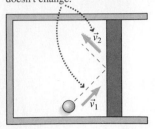

(a) Stationary piston

Rebounding molecules have lower speed as energy is transferred to the outward-moving piston. With the decrease in internal energy comes a drop in temperature.

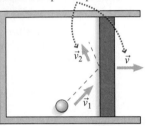

(b) Moving piston

FIGURE 18.10 Internal energy decreases during an adiabatic expansion, for the reason shown in (b).

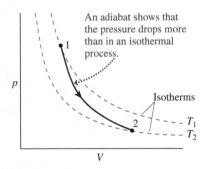

FIGURE 18.11 A pV curve for an adiabatic expansion (*dark curve*).

compression of the gasoline-air mixture and expansion of the combustion products are nearly adiabatic because they occur so rapidly that little heat flows through the cylinder walls.

Since the heat Q is zero in an adiabatic process, the first law becomes simply

$$\Delta U = -W \qquad \text{(adiabatic process)} \qquad (18.10)$$

This says that if a system does work on its surroundings, then—in the absence of heat transfer—its internal energy must decrease by the same amount. Microscopically, this energy loss occurs in a gas-cylinder system when molecules give up energy as they collide with the moving piston (Fig. 18.10).

As a gas expands adiabatically, its volume increases while its internal energy and temperature decrease. The ideal-gas law, $pV = nRT$, then requires that the pressure decrease as well—and by more than it would in an isothermal process where T remains constant. In a pV diagram, the path of an adiabatic process—called an **adiabat**—is therefore steeper than the isotherms (Fig. 18.11).

Tactics 18.1 details the math involved in finding the adiabatic path; the result is

$$pV^\gamma = \text{constant} \qquad \text{(adiabatic process)} \qquad (18.11a)$$

where $\gamma = C_p/C_V$ is the ratio of the specific heats. Because $C_p = C_V + R$, the ratio $\gamma = C_p/C_V$ is always greater than 1. As expected, an adiabatic process therefore results in a greater pressure change than would a comparable isothermal process, as reflected in the steeper adiabatic path in Fig. 18.11. Physically, the adiabatic path is steeper because the gas loses internal energy as it does work, so its temperature drops. Problem 66 shows how to rewrite Equation 18.11a in terms of temperature:

$$TV^{\gamma-1} = \text{constant} \qquad \text{(adiabatic process)} \qquad (18.11b)$$

GOT IT? 18.2 The ideal-gas law says $pV = nRT$, but Equation 18.11a says $pV^\gamma = \text{constant}$. Which is true: (a) the ideal-gas law, (b) Equation 18.11a, or (c) both? Explain.

TACTICS 18.1 Deriving the Adiabatic Equation

Equation 18.6 gives the infinitesimal change in internal energy for *any* process: $dU = nC_V \, dT$. The corresponding work is $dW = p \, dV$ so, with $Q = 0$ in an adiabatic process, the first law becomes $nC_V \, dT = -p \, dV$. We can eliminate dT by differentiating the ideal-gas law, now letting *both* p and V change: $nR \, dT = d(pV) = p \, dV + V \, dp$. Solving for dT, substituting in our first-law statement, and multiplying through by R lead to $C_V V \, dp + (C_V + R)p \, dV = 0$. But $C_V + R = C_p$; substituting this and dividing through by $C_V pV$ give

$$\frac{dp}{p} + \frac{C_p}{C_V}\frac{dV}{V} = 0$$

Defining $\gamma \equiv C_p/C_V$ and integrating give

$$\ln p + \gamma \ln V = \ln(\text{constant})$$

where we've chosen to call the constant of integration $\ln(\text{constant})$. Since $\gamma \ln V = \ln V^\gamma$, it follows by exponentiation that

$$pV^\gamma = \text{constant}$$

It's another exercise in calculus to integrate Equation 18.3 for the work done in an adiabatic process. You can do this yourself in Problem 62; the result is

$$W = \frac{p_1 V_1 - p_2 V_2}{\gamma - 1} \qquad (18.12)$$

EXAMPLE 18.3 **An Adiabatic Process: Diesel Power**

Fuel ignites in a diesel engine from the heat of compression as the piston moves toward the top of the cylinder; there's no spark plug as in a gasoline engine. Compression is fast enough that the process is essentially adiabatic. If the ignition temperature is 500°C, what compression ratio V_{max}/V_{min} is needed (Fig. 18.12)? Air's specific-heat ratio is $\gamma = 1.4$, and before compression the air is at 20°C.

INTERPRET We identify the thermodynamic process here as adiabatic compression.

DEVELOP The problem involves temperature and volume, so Equation 18.11b applies, giving $T_{min}V_{min}^{\gamma-1} = T_{max}V_{max}^{\gamma-1}$.

EVALUATE Solving for the compression ratio V_{max}/V_{min} gives

$$\frac{V_{max}}{V_{min}} = \left(\frac{T_{min}}{T_{max}}\right)^{1/(\gamma-1)} = \left(\frac{773 \text{ K}}{293 \text{ K}}\right)^{1/0.4} = 11$$

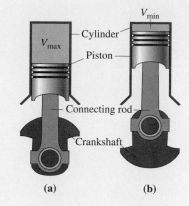

FIGURE 18.12 One cylinder of a diesel engine, shown with the piston (a) at the bottom of its stroke and (b) at the top. The compression ratio is V_{max}/V_{min}.

(a) **(b)**

ASSESS Practical diesel engines have higher ratios to ensure reliable ignition. Their high compression makes diesels heavier than their gasoline counterparts, but also more fuel efficient. Problem 67 explores the diesel engine further. ∎

APPLICATION **Smog Alert!**

The smog that blankets urban areas is an unfortunate manifestation of our prolific fossil-fueled energy consumption. Adiabatic processes in the atmosphere determine whether or not smog lingers over a city. Consider a volume of air that's heated, perhaps because it's over hot pavement that absorbs solar energy. The air becomes less dense, and its buoyancy makes it rise. As it ascends into regions of lower pressure, it expands, doing work against the surrounding atmosphere. Air is a poor heat conductor, so the process is essentially adiabatic. Therefore the gas cools as it does work.

Now, temperature in the atmosphere normally decreases with altitude. So here's the crucial question: Does the rising air cool faster or slower than the surrounding atmosphere? If it cools more slowly, then it continues to be warmer, and it continues to rise. Any pollution is carried high into the atmosphere where it's dispersed. But if the decrease in air temperature with altitude isn't great, or in an **inversion** where it's actually warmer aloft, the rising air will soon reach equilibrium with its surroundings and won't rise any higher. The effect is to trap air and its entrained pollutants near the surface, as shown in this photo of Los Angeles. Smog alert!

GOT IT? 18.3 Name the basic thermodynamic process involved when each of the following is done to a piston-cylinder system containing ideal gas, and tell also whether temperature, pressure, volume, and internal energy increase or decrease: (a) The piston is locked in place and a flame is applied to the bottom of the cylinder; (b) the cylinder is completely insulated and the piston is pushed downward; (c) the piston is exposed to atmospheric pressure and is free to move, while the cylinder is cooled by placing it on a block of ice.

Cyclic Processes

Many natural and technological systems undergo **cyclic processes**, in which the system returns periodically to the same thermodynamic state. Engineering examples include engines and refrigerators whose mechanical construction ensures cyclic behavior. Many natural oscillations, like those of a sound wave or a pulsating star, are essentially cyclic.

Cyclic processes often involve the four basic processes we've just explored, as summarized in Table 18.1. We've seen that the work done in any reversible process is just the area under the pV curve. A cyclic process returns to the same point in the pV diagram, so it involves both expansion and compression (Fig. 18.13). During expansion the gas does work on its surroundings; during compression work gets done on the gas. The net work is the difference between the two Figure 18.13c shows that this work is the area enclosed by the cyclic path in the pV diagram.

TABLE 18.1　Ideal-Gas Processes

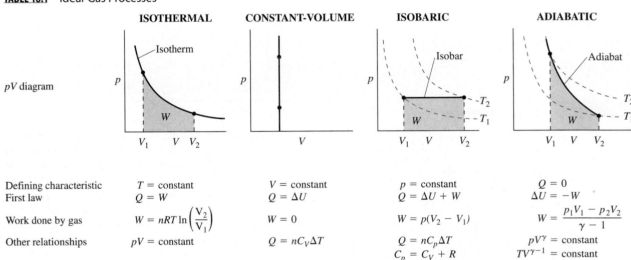

	ISOTHERMAL	**CONSTANT-VOLUME**	**ISOBARIC**	**ADIABATIC**
Defining characteristic	T = constant	V = constant	p = constant	$Q = 0$
First law	$Q = W$	$Q = \Delta U$	$Q = \Delta U + W$	$\Delta U = -W$
Work done by gas	$W = nRT \ln\left(\dfrac{V_2}{V_1}\right)$	$W = 0$	$W = p(V_2 - V_1)$	$W = \dfrac{p_1 V_1 - p_2 V_2}{\gamma - 1}$
Other relationships	pV = constant	$Q = nC_V \Delta T$	$Q = nC_p \Delta T$ $C_p = C_V + R$	pV^γ = constant $TV^{\gamma-1}$ = constant

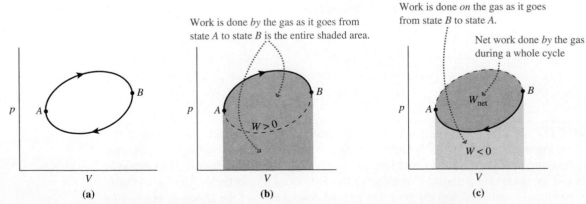

FIGURE 18.13　(a) A pV diagram for a cyclic process. (b), (c) Work done in one cycle is the area inside the closed path.

EXAMPLE 18.4 **A Cyclic Process: Finding the Work**

An ideal gas with $\gamma = 1.4$ occupies 4.0 L at 300 K and 100 kPa pressure. It's compressed adiabatically to one-fourth of its original volume, then cooled at constant volume back to 300 K, and finally allowed to expand isothermally to its original volume. How much work is done on the gas?

INTERPRET This problem involves a cyclic process, and we identify three separate thermodynamic processes that make up the cycle: adiabatic, constant-volume, and isothermal.

DEVELOP Here it helps to draw a pV diagram, shown in Fig. 18.14. Our plan is to use equations in Table 18.1 to determine the work for each of the basic processes and then combine them to get the net work. For the adiabatic process AB, Table 18.1 gives $W_{AB} = (p_A V_A - p_B V_B)/(\gamma - 1)$; for the constant-volume process BC, $W_{BC} = 0$; and for the isothermal process CA, the work is $W_{CA} = nRT \ln(V_A/V_C)$.

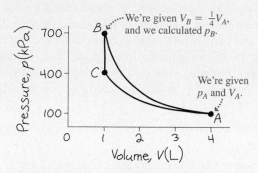

FIGURE 18.14 The cyclic process $ABCA$ of Example 18.4 includes adiabatic (AB), constant-volume (BC), and isothermal (CA) sections.

EVALUATE For the adiabatic process AB we're given all quantities except p_B. This we can get from the adiabatic equation $pV^\gamma = $ constant, or $p_B V_B^\gamma = p_A V_A^\gamma$. Solving gives $p_B = p_A (V_A/V_B)^\gamma = 696.4$ kPa, where

we used the given information $p_A = 100$ kPa, $\gamma = 1.4$, and a compression to one-fourth the original volume ($V_A/V_B = 4$). We now have enough information to find the work done over the adiabatic path:

$$W_{AB} = \frac{p_A V_A - p_B V_B}{\gamma - 1} = -741 \text{ J}$$

where, with pressures in kPa ($= 10^3$ Pa) and volumes in L ($= 10^{-3}$ m^3), the factors $10^{\pm 3}$ cancel and there's no need to convert. The work W_{AB} is *negative* because work is done *on* the gas when it's compressed.

In the expression $W_{CA} = nRT \ln(V_A/V_C)$ for the isothermal work, we can evaluate the quantity nRT at *any* point on the isothermal curve because T is constant. The ideal-gas law says that $nRT = pV$, and we know both p and V at point A. So $nRT = p_A V_A = 400$ J, where again we could multiply $p_A = 100$ kPa by $V_A = 4.0$ L to get an answer in SI units. The isothermal work is then

$$W_{CA} = nRT \ln\left(\frac{V_A}{V_C}\right) = (400 \text{ J})(\ln 4) = 555 \text{ J}$$

This is positive because the gas does work in expanding from C to A.

Combining our results for all three segments gives the net work:

$$W_{ABCA} = W_{AB} + W_{BC} + W_{CA} = -741 \text{ J} + 0 \text{ J} + 555 \text{ J} = -186 \text{ J}$$

ASSESS Make sense? The final answer is negative because we've done net work *on* the gas; that's always the case in going counterclockwise around a cyclic path in a pV diagram. Since the system returns to its original state, its internal energy undergoes no net change. That means all the work that's done on it must be transferred to its surroundings as heat. Since no heat flows during the adiabatic process AB, and since the gas *absorbs* heat during the isothermal expansion CA, the only time it transfers heat to its surroundings is during the constant-volume cooling process BC. ∎

18.3 Specific Heats of an Ideal Gas

We've found that the thermodynamic behavior of an ideal gas depends on the specific heats C_V and C_p. What are the values of those quantities?

Our ideal-gas model of Chapter 17 assumed the gas molecules were structureless point particles with only translational kinetic energy. The internal energy U of the gas is the sum of all those molecular kinetic energies. But the average kinetic energy is directly proportional to the temperature: $\frac{1}{2}m\overline{v^2} = \frac{3}{2}kT$. If we have n moles of gas, the internal energy is then $U = nN_A\left(\frac{1}{2}m\overline{v^2}\right) = \frac{3}{2}nN_A kT$, where N_A is Avogadro's number. But $N_A k = R$, the gas constant, so $U = \frac{3}{2}nRT$. Solving Equation 18.6 for the molar specific heat then gives

$$C_V = \frac{1}{n}\frac{\Delta U}{\Delta T} = \frac{3}{2}R \qquad (18.13)$$

For this gas of structureless particles, the adiabatic exponent γ is therefore

$$\gamma = \frac{C_p}{C_V} = \frac{C_V + R}{C_V} = \frac{\frac{5}{2}R}{\frac{3}{2}R} = \frac{5}{3} = 1.67$$

Some gases, notably the inert gases helium (He), neon (Ne), argon (Ar), and others in the last column of the periodic table, have adiabatic exponents and specific heats given by these equations. But others do not. At room temperature, for example, hydrogen (H_2), oxygen (O_2), and nitrogen (N_2) obey adiabatic laws with γ very nearly $\frac{7}{5}(= 1.4)$ and, correspondingly, specific heat $C_V = \frac{5}{2}R$. On the other hand, sulfur dioxide (SO_2) and nitrogen dioxide (NO_2) have specific-heat ratios close to 1.3 and therefore C_V of about $3.4R$.

What's going on here? A clue lies in the structure of individual gas molecules, reflected in their chemical formulas. The inert-gas molecules are **monatomic**, consisting of single atoms. To the extent that these atoms behave like structureless mass points, the only energy they can have is kinetic energy of translational motion. We can think of that kinetic energy as being a sum of *three* terms, each associated with motion in one of the three mutually perpendicular directions. We call each separate term in the energy of a system a **degree of freedom**, meaning a way that system can take on energy. So a monatomic molecule has three degrees of freedom.

In contrast, hydrogen, oxygen, and nitrogen molecules are **diatomic**, as shown in Fig. 18.15. Although a gas of such molecules should still obey the ideal-gas law $PV = nRT$, these molecules can have rotational as well as translational kinetic energy. Then the kinetic energy of a diatomic molecule consists of *five* terms, three for the three directions of translational motion and two for rotational motions about the two mutually perpendicular axes shown in Fig. 18.15. So a diatomic molecule has five degrees of freedom. We'll now see how this difference between *three* degrees of freedom for monatomic molecules and *five* for diatomic molecules accounts for the difference between their specific heats.

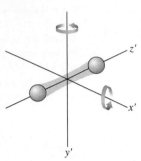

FIGURE 18.15 A diatomic molecule can have significant rotation about two perpendicular axes.

The Equipartition Theorem

We showed in Chapter 17 that the average kinetic energy associated with a gas molecule's motion in one direction is $\frac{1}{2}kT$. We then argued that all three directions are equally probable, making the molecular kinetic energy, on average, $\frac{3}{2}kT$. The argument from one direction to three is based on the assumption that random collisions will share energy equally among the possible motions. When a molecule can rotate as well as translate, energy should be shared also among possible rotational motions. The 19th-century Scottish physicist James Clerk Maxwell first proved this fact, which is known as the **equipartition theorem**:

> **Equipartition theorem** When a system is in thermodynamic equilibrium, the average energy per molecule is $\frac{1}{2}kT$ for each degree of freedom.

We've just seen that a diatomic molecule has five degrees of freedom: three translational and two rotational. The average energy of such a molecule is then $5\left(\frac{1}{2}kT\right) = \frac{5}{2}kT$, so the total internal energy in n moles of a diatomic gas is $U = nN_A\left(\frac{5}{2}kT\right) = \frac{5}{2}nRT$. Equation 18.6 then gives the molar specific heat at constant volume:

$$C_V = \frac{1}{n}\frac{\Delta U}{\Delta T} = \frac{5}{2}R \qquad \text{(diatomic molecule)}$$

Our result $C_p = C_V + R$ still holds, since it was derived from the first law of thermodynamics without regard to molecular structure, so $C_p = \frac{7}{2}R$ and $\gamma = C_p/C_V = \frac{7}{5} = 1.4$. These results describe the observed behavior of diatomic gases like hydrogen, oxygen, and nitrogen at room temperature.

A polyatomic molecule like NO_2 can rotate about any of three perpendicular axes (Fig. 18.16). It then has a total of six degrees of freedom, giving $U = 3nRT$ and corresponding specific heats $C_V = 3R$ and $C_p = C_V + R = 4R$. The adiabatic exponent is then $\gamma = \frac{4}{3} \approx 1.33$, reasonably close to the experimental value $\gamma = 1.29$ for NO_2.

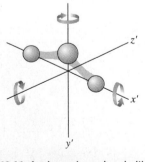

FIGURE 18.16 A triatomic molecule like NO_2 has three rotational degrees of freedom.

EXAMPLE 18.5 **Specific Heat: A Gas Mixture**

A gas mixture consists of 2.0 mol of oxygen (O_2) and 1.0 mol of argon (Ar). Find the volume specific heat of the mixture.

INTERPRET This problem is about specific heat and molecular structure. We identify the molecules involved as diatomic O_2 and monatomic Ar.

DEVELOP Equation 18.6, $\Delta U = nC_V \Delta T$, determines the volume specific heat, so we need to find how the internal energy U depends on temperature. Our plan is to use the equipartition theorem to get the energy per molecule for each gas, then find the total energy as a function of temperature, and from that the specific heat.

EVALUATE Being diatomic, O_2 has five degrees of freedom, so the equipartition theorem gives the average energy per molecule as $\frac{5}{2}kT$. Then the total energy in $n = 2$ moles of oxygen is $U_{O_2} = nN_A\left(\frac{5}{2}kT\right) = \frac{5}{2}nRT = 5.0RT$, where we used $N_A k = R$. Monotonic Ar has three degrees of freedom, so the internal energy in our 1 mole of argon is, similarly, $U_{Ar} = \frac{3}{2}nRT = 1.5RT$. The total internal energy is then $U = 6.5RT$, so Equation 18.6 gives

$$C_V = \frac{1}{n}\frac{\Delta U}{\Delta T} = \frac{6.5R}{3.0 \text{ mol}} = 2.2R$$

ASSESS Make sense? Our answer lies between the values $1.5R$ and $2.5R$ that we found for monatomic and diatomic gases, respectively. It's closer to $2.5R$ because there's more oxygen in the mixture. ∎

Quantum Effects

Relating molecular structure and gas behavior is a remarkable triumph for Newtonian physics. But hidden in our analysis is an assumption that Newtonian physics can't justify. Real atoms have size, so even monatomic molecules should rotate. Why not more degrees of freedom? The answer lies in quantum physics, which requires a certain minimum energy for a periodic motion such as rotation. At normal temperatures, the average thermal energy is too low to excite rotation of monatomic molecules, or of diatomic molecules about their long axis. So these molecules exhibit three and five degrees of freedom, respectively. That results in the volume specific heats $\frac{3}{2}R$ and $\frac{5}{2}R$ that we've seen. For diatomic molecules at higher temperatures, still another motion comes into play—the simple harmonic oscillation of the two atoms due to the springlike bond between them. That adds two more degrees of freedom, corresponding to the kinetic and potential energies of this oscillation, and the specific heat increases correspondingly. At very low temperatures, in contrast, there isn't enough thermal energy to excite any rotation in a diatomic gas, and it then exhibits the specific heat $C_V = \frac{3}{2}R$ that we normally associate with a monatomic gas. Figure 18.17 shows these effects for diatomic hydrogen (H_2).

Are you bothered by the strange restrictions quantum mechanics imposes on molecular rotation and vibration? You should be! Nothing in your experience suggests that a rotating object can't have any amount of energy you care to give it. But quantum mechanics deals with a realm much smaller than that of our daily experience. The quantization of energy is only one of many unusual things that occur in the quantum realm. We'll explore more quantum phenomena in Part 6.

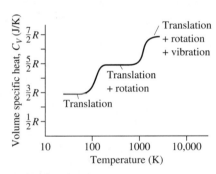

FIGURE 18.17 Volume specific heat of H_2 gas as a function of temperature. Below 20 K hydrogen is liquid, and above 3200 K it dissociates into individual atoms.

Big Picture

The big idea here is conservation of energy, now expanded to include heat. The expanded statement of energy conservation is the **first law of thermodynamics**, which relates the change in a system's internal energy to the heat flowing into the system and the work done by the system. The first law combines with the ideal-gas law to give a quantitative description of basic thermodynamic processes applied to ideal gases; these are described graphically using **pV diagrams**. The **equipartition theorem** states that in thermodynamic equilibrium, energy is shared equally among the possible energy modes of a system.

Key Concepts and Equations

Quantitatively, the first law of thermodynamics states

$$\Delta U = Q - W$$

Meaning of terms in the first law:

- ΔU is the change in a system's internal energy.
- Q is the heat transferred *to* the system.
  - Positive Q means a net heat input to the system.
  - Negative Q means heat leaves the system.
- W is the work done *by* the system.
  - Positive W means the system does work on its surroundings.
  - Negative W means work is done on the system.

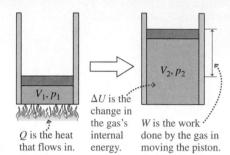

ΔU is the change in the gas's internal energy.

Q is the heat that flows in.

W is the work done by the gas in moving the piston.

In general, the work done by a system is related to the changes in pressure and volume:

$$W = \int_{V_1}^{V_2} p \, dV$$

Cases and Uses

Ideal-gas processes:

ISOTHERMAL	CONSTANT-VOLUME	ISOBARIC	ADIABATIC

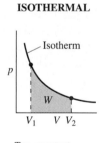 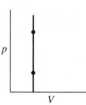 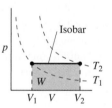 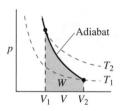

ISOTHERMAL	CONSTANT-VOLUME	ISOBARIC	ADIABATIC
$T = $ constant	$V = $ constant	$p = $ constant	$Q = 0$
$Q = W$	$Q = \Delta U$	$Q = \Delta U + W$	$\Delta U = -W$
$W = nRT \ln\left(\dfrac{V_2}{V_1}\right)$	$W = 0$	$W = p(V_2 - V_1)$	$W = \dfrac{p_1 V_1 - p_2 V_2}{\gamma - 1}$
$pV = $ constant	$Q = nC_V \Delta T$	$Q = nC_p \Delta T$	$pV^\gamma = $ constant
		$C_p = C_V + R$	$TV^{\gamma-1} = $ constant

The specific heats of an ideal gas follow from the **degrees of freedom** of each molecule:

Monatomic
3 degrees of freedom
$C_V = \frac{3}{2}R$

Diatomic
5 degrees of freedom
$C_V = \frac{5}{2}R$

For Thought and Discussion

1. The temperature of the water in a jar is raised by violently shaking the jar. Which of the terms Q and W in the first law of thermodynamics is involved in this case?
2. What is the difference between heat and internal energy?
3. Some water is tightly sealed in a perfectly insulated container. Is it possible to change the water temperature? Explain.
4. Are the initial and final equilibrium states of an irreversible process describable by points in a pV diagram? Explain.
5. Why can't an irreversible process be described by a path in a pV diagram?
6. Does the first law of thermodynamics apply to an irreversible process?
7. A quasi-static process begins and ends at the same temperature. Is the process necessarily isothermal?
8. Figure 18.18 shows two processes, A and B, that connect the same initial and final states, 1 and 2. For which process is more heat added to the system?

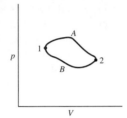

FIGURE 18.18 For Thought and Discussion 8

9. When you let air out of a tire, the air seems cool. Why? What kind of process is occurring?
10. Blow on the back of your hand with your mouth wide open. Your breath will feel hot. Now tighten your lips into a small opening and blow again. Now your breath feels cool. Why?
11. Water is boiled in an open pan. Of which of the four basic processes we considered is this an example?
12. Three identical gas-cylinder systems expand from the same initial state to final states that have the same volume. One system expands isothermally, one adiabatically, and one isobarically. Which does the most work? Which does the least work?
13. Why is the specific heat at constant pressure greater than at constant volume?
14. In what sense can a gas of diatomic molecules be considered an ideal gas, given that its molecules aren't point particles?

Exercises and Problems

Exercises

Section 18.1 The First Law of Thermodynamics

15. In a perfectly insulated container, 1.0 kg of water is stirred vigorously until its temperature rises by 7.0°C. How much work was done on the water?
16. In a closed but uninsulated container, 500 g of water are shaken violently until the temperature rises by 3.0°C. The mechanical work required in the process is 9.0 kJ. (a) How much heat is transferred during the shaking? (b) How much mechanical energy would have been required if the container had been perfectly insulated?
17. A 40-W heat source is applied to a gas sample for 25 s, during which time the gas expands and does 750 J of work on its surroundings. By how much does the internal energy of the gas change?
18. What is the rate of heat flow into a system whose internal energy is increasing at the rate of 45 W, given that the system is doing work at the rate of 165 W?
19. In a certain automobile engine, 17% of the total energy released in burning gasoline ends up as mechanical work. What is the engine's mechanical power output if its heat output is 68 kW?

Section 18.2 Thermodynamic Processes

20. An ideal gas expands from the state (p_1, V_1) to the state (p_2, V_2), where $p_2 = 2p_1$ and $V_2 = 2V_1$. The expansion proceeds along the straight diagonal path AB shown in Fig. 18.19. Find an expression for the work done by the gas during this process.
21. Repeat Problem 20 for a process that follows the path ACB in Fig. 18.19.

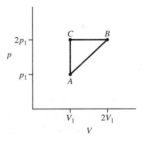

FIGURE 18.19 Exercises 20, 21 and Problem 70.

22. A balloon contains 0.30 mol of helium. It rises, while maintaining a constant 300 K temperature, to an altitude where its volume has expanded five times. How much work is done by the gas in the balloon during this isothermal expansion? Neglect tension forces in the balloon.
23. The balloon of Problem 22 starts at a pressure of 100 kPa and rises to an altitude where the pressure is 75 kPa, maintaining a constant 300 K temperature. (a) By what factor does its volume increase? (b) How much work does the gas in the balloon do?
24. How much work does it take to compress 2.5 mol of an ideal gas to half its original volume while maintaining a constant 300 K temperature?
25. By what factor must the volume of a gas with $\gamma = 1.4$ be changed in an adiabatic process if the kelvin temperature is to double?

Section 18.3 Specific Heats of an Ideal Gas

26. A gas mixture contains 2.5 mol of O_2 and 3.0 mol of Ar. What are the molar specific heats at constant volume and pressure for this mixture?

27. A mixture of monatomic and diatomic gases has specific-heat ratio $\gamma = 1.52$. What fraction of the molecules are monatomic?

28. What should be the approximate specific-heat ratio of a gas consisting of 50% NO_2 ($\gamma = 1.29$), 30% O_2 ($\gamma = 1.4$), and 20% Ar ($\gamma = 1.67$)?

29. By how much does the temperature of (a) an ideal monatomic gas and (b) an ideal diatomic gas (with molecular rotation but no vibration) change in an adiabatic process in which 2.5 kJ of work are done on each mole of gas?

Problems

30. Water flows over Niagara Falls (height 50 m) at the rate of about 10^6 kg/s. Suppose that all the water passes through a turbine connected to an electric generator producing 400 MW of electric power. If the water has negligible kinetic energy after leaving the turbine, by how much has its temperature increased between the top of the falls and the outlet of the turbine?

31. An ideal gas expands to 10 times its original volume, maintaining a constant 440 K temperature. If the gas does 3.3 kJ of work on its surroundings, (a) how much heat does it absorb, and (b) how many moles of gas are there?

32. A 0.25-mol sample of an ideal gas initially occupies 3.5 L. If it takes 61 J of work to compress the gas isothermally to 3.0 L, what is the temperature?

33. As the heart beats, blood pressure in an artery varies from a high of 125 mm of mercury to a low of 80 mm of mercury. These values are gauge pressures—that is, excesses over atmospheric pressure. An air bubble trapped in the artery has diameter 1.52 mm when the blood pressure is at its minimum. (a) What will be its diameter at maximum pressure? (b) How much work does the blood (and ultimately the heart) do in compressing this bubble, assuming the air remains at the same 37.0°C temperature as the blood?

34. It takes 600 J to compress a gas isothermally to half its original volume. How much work would it take to compress it by a factor of 10 starting from its original volume?

35. A gas undergoes an adiabatic compression during which its volume drops to half its original value. If the gas pressure increases by a factor of 2.55, what is its specific-heat ratio γ?

36. A gas with $\gamma = 1.4$ is at 100 kPa pressure and occupies 5.00 L. (a) How much work does it take to compress the gas adiabatically to 2.50 L? (b) What is its final pressure?

37. A gas sample undergoes the cyclic process $ABCA$ shown in Fig. 18.20, where AB lies on an isotherm. The pressure at point A is 60 kPa. Find (a) the pressure at B and (b) the net work done on the gas.

38. Repeat Problem 37 taking AB to be on an adiabat and using a specific-heat ratio of $\gamma = 1.4$.

39. A gasoline engine has a compression ratio of 8.5 (see Example 18.3 for the meaning of this term). If the fuel-air mixture enters the engine at 30°C, what will its temperature be at maximum compression? Assume the compression is adiabatic and the mixture has $\gamma = 1.4$.

40. By how much must the volume of a gas with $\gamma = 1.4$ be changed in an adiabatic process if the pressure is to double?

41. Volvo's B5340 engine, used in the V70 series cars, has a compression ratio of 10.2. If air at 320 K and atmospheric pressure fills an engine cylinder at its maximum volume, (a) what will be the air temperature at the point of maximum compression? (b) What will be the pressure at this point? Assume the compression is adiabatic, with $\gamma = 1.4$.

42. A gas expands isothermally from state A to state B, in the process absorbing 35 J of heat. It is then compressed isobarically to state C, where its volume equals that of state A. During the compression, 22 J of work are done on the gas. The gas is then heated at constant volume until it returns to state A. (a) Draw a pV diagram for this process. (b) How much work is done on or by the gas during the complete cycle? (c) How much heat is transferred to or from the gas as it goes from B to C to A?

43. A 2.0-mol sample of ideal gas with molar specific heat $C_V = \frac{5}{2}R$ is initially at 300 K and 100 kPa pressure. Determine the final temperature and the work done by the gas when 1.5 kJ of heat are added to the gas (a) isothermally, (b) at constant volume, and (c) isobarically.

44. Prove that the slope of an adiabat at a given point in a pV diagram is γ times the slope of the isotherm passing through the same point.

45. An ideal gas with $\gamma = 1.67$ starts at point A in Fig. 18.21, where its volume and pressure are 1.00 m³ and 250 kPa, respectively. It then undergoes an adiabatic expansion that triples its volume, ending at point B. It's then heated at constant volume to point C, and then compressed isothermally back to A. Find (a) the pressure at B, (b) the pressure at C, and (c) the net work done on the gas.

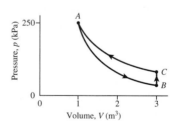

FIGURE 18.21 Problem 45

46. The gas of Example 18.4 starts at state A in Fig. 18.14 and is compressed adiabatically until its volume is 2.0 L. It's then cooled at constant pressure until it reaches 300 K, then allowed to expand isothermally back to state A. Find (a) the net work done on the gas and (b) the minimum volume reached.

47. The gas of Example 18.4 starts at state A in Fig. 18.14 and is heated at constant volume until its pressure has doubled. It's then compressed adiabatically until its volume is one-fourth its original value, then cooled at constant volume to 300 K, and finally allowed to expand isothermally to its original state. Find the net work done on the gas.

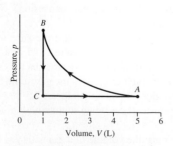

FIGURE 18.20 Problems 37 and 38

48. A 25-L sample of an ideal gas with $\gamma = 1.67$ is at 250 K and 50 kPa. The gas is compressed isothermally to one-third of its original volume, then heated at constant volume until its state lies on the adiabatic curve that passes through its original state, and then allowed to expand adiabatically to that original state. Find the net work involved. Is work done on or by the gas?

49. A 25-L sample of an ideal gas with $\gamma = 1.67$ is at 250 K and 50 kPa. The gas is compressed adiabatically until its pressure triples, then cooled at constant volume back to 250 K, and finally allowed to expand isothermally to its original state. (a) How much work is done on the gas? (b) What is the minimum volume reached? (c) Sketch this cyclic process in a pV diagram.

50. A bicycle pump consists of a cylinder 30 cm long when the pump handle is all the way out. The pump contains air $(\gamma = 1.4)$ at 20°C. If the pump outlet is blocked and the handle pushed until the internal length of the pump cylinder is 17 cm, by how much does the air temperature rise? Assume that no heat is lost.

51. A tightly sealed flask contains 5.0 L of air at 0°C and 100 kPa pressure. How much heat is required to raise the air temperature to 20°C? The molar specific heat of air at constant volume is $2.5R$.

52. External forces compress 21 mol of ideal monatomic gas; during the process the gas transfers 15 kJ of heat to its surroundings, yet its temperature rises by 160 K. How much work was done on the gas?

53. A gas with $\gamma = \frac{7}{5}$ is at 273 K when it's compressed isothermally to one-third of its original volume and then further compressed adiabatically to one-fifth of its original volume. What is its final temperature?

54. An ideal gas with $\gamma = 1.3$ is initially at 273 K and 100 kPa. The gas is compressed adiabatically to 240 kPa pressure. What is its final temperature?

55. The curved path in Fig. 18.22 lies on the 350 K isotherm for an ideal gas with $\gamma = 1.4$. (a) Calculate the net work done on the gas as it goes around the cyclic path $ABCA$. (b) How much heat flows into or out of the gas on the segment AB?

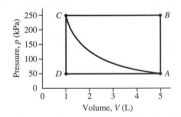

FIGURE 18.22 Problems 55 and 56

56. Repeat part (a) of Problem 55 for the path $ACDA$ in Fig. 18.22. (b) How much heat flows into or out of the gas on the segment CD?

57. A gas mixture contains monatomic argon and diatomic oxygen. An adiabatic expansion that doubles its volume results in the pressure dropping to one-third of its original value. What fraction of the molecules are argon?

58. How much of a triatomic gas with $C_V = 3R$ would you have to add to 10 mol of monatomic gas to get a mixture whose thermodynamic behavior was like that of a diatomic gas?

59. An 8.5-kg rock at 0°C is dropped into a well-insulated vat containing a mixture of ice and water at 0°C. When equilibrium is reached, there are 6.3 g less ice. From what height was the rock dropped?

60. A piston-cylinder arrangement containing 0.30 mol of nitrogen at high pressure is in thermal equilibrium with an ice-water bath containing 200 g of ice. The pressure of the ambient air is 1.0 atm. The gas is allowed to expand isothermally until it's in pressure balance with its surroundings. After the process is complete, the bath contains 210 g of ice. What was the original gas pressure?

61. A piston-cylinder arrangement contains 1.0 g of water at 50°C. The piston, which has negligible mass, is free to move and is exposed to atmospheric pressure (1.0 atm). The system is heated to 200°C. Assuming that liquid water is incompressible and that steam is an ideal gas with volume specific heat $4.3R$, determine (a) the work done by the system and (b) the heat added to it.

62. Show that the application of Equation 18.3 to an adiabatic process results in Equation 18.12.

63. A horizontal piston-cylinder system containing n mol of ideal gas is surrounded by air at temperature T_0 and pressure P_0. If the piston is displaced slightly from equilibrium, show that it executes simple harmonic motion with angular frequency $\omega = Ap_0 / \sqrt{MnRT_0}$, where A and M are the piston area and mass, respectively. Assume the gas temperature remains constant.

64. An ideal gas is taken clockwise around the circular path shown in Fig. 18.23. (a) How much work does the gas do? (b) If there are 1.3 mol of gas, what is the maximum temperature reached?

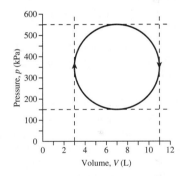

FIGURE 18.23 Problem 64

65. Warm winds called Chinooks (a Native-American term meaning "snow eaters") sometimes sweep across the plains just east of the Rocky Mountains. These winds carry air from high in the mountains down to the plains rapidly enough that the air has no time to exchange significant heat with its surroundings (Fig. 18.24). On the day of a Chinook wind, pressure and temperature high in the Colorado Rockies are 62.0 kPa and -11.0°C, respectively. (a) What will the air temperature be when the Chinook has carried it to the plain where pressure is 86.5 kPa? (b) How much work is done on what was initially a cubic meter of air as it descends to the plain?

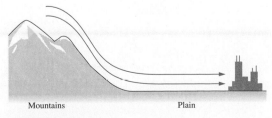

Mountains   Plain

FIGURE 18.24 Problem 65

66. Use the ideal-gas law to eliminate pressure in Equation 18.11a, and show that the result can be written as Equation 18.11b.

67. Figure 18.25 shows the thermodynamic cycle of a diesel engine. Note that this cycle differs from that of a gasoline engine in that combustion takes place isobarically. As in Example 18.3, the compression ratio r is the ratio of maximum to minimum volume: $r = V_1/V_2$. In addition, the so-called *cutoff ratio* is defined by $r_c = V_3/V_2$. Find an expression for the engine's efficiency, in terms of the ratios r and r_c and the specific-heat ratio γ. Although your expression suggests that the diesel engine might be less efficient than the gasoline engine (see Problem 54 of Chapter 19), the diesel's higher compression ratio more than compensates, giving it a higher efficiency (the ratio of the work delivered to the heat extracted during the combustion phase).

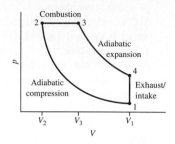

FIGURE 18.25 Problem 67

68. In a reversible process, a volume of air $V_0 = 17$ m³ at pressure $p_0 = 1$ atm is compressed such that the pressure and volume are related by $(p/p_0)^{-2} = V/V_0$. How much work is done by the gas in reaching a final pressure of 1.4 atm? (*Hint:* You may need to do an integral.)

69. A real gas is more accurately described using the Van der Waals equation: $[p + a(n/V)^2](V - nb) = nRT$, where a and b are constants. Find the expression, corresponding to Equation 18.4, for the work done by a Van der Waals gas undergoing an isothermal expansion from V_1 to V_2.

70. Repeat Exercise 20 for an expansion along the path $p = p_1[1 + (V - V_1)^2/V_1^2]$ in the pV diagram.

71. When air rises in the atmosphere, it expands adiabatically and its temperature falls with altitude. (This is called the adiabatic lapse rate.) Express dT in terms of dp for an adiabatic process, and use the hydrostatic equation (Equation 15.2) to express dp in terms of dy. Then, calculate this lapse rate dT/dy. Take the average molecular weight of air to be 29 g, $\gamma = 1.4$, and remember that the altitude y is the negative of the depth h in Equation 15.2.

72. The nuclear power plant at which you're the public affairs manager has a backup high-temperature gas-turbine system. It's a so-called combined-cycle system that maximizes the conversion of thermal energy into useful work. One such turbine produces electrical energy at the rate of 360 MW, while extracting energy from natural gas at the rate of 670 MW. The local town council has raised concern over waste thermal energy dumped into the environment. The council adopted a standard that the thermal waste power must not exceed 400 MW and all power generation must be at least 50% efficient. Does the turbine meet this standard, which was set after minimal technical research by the council?

73. Your class on alternative habitats has an assignment to design an underwater habitat. A small diving bell will be lowered to the habitat. A hatch at the bottom of the bell is open, so water can enter to compress the air and thus keep the air pressure inside equal to the pressure of the surrounding water. The bell is lowered slowly enough so that the inside air remains at the same temperature as the water. But the water temperature increases with depth in such a way that the air pressure and volume are related by $p = p_0\sqrt{V_0/V}$, where $V_0 = 17$ m³ and $p_0 = 1$ atm are the surface values. Suppose the diving bell's air volume cannot be less than 8.66 m³ and the pressure must not exceed 1.5 atm when submerged. Are these criteria met?

74. The phone rings. It's your roommate calling from the chemistry building. "I'm not sure I figured out the right equation for γ given the volume and temperature change of a gas in an adiabatic process. The final volume was 0.177 of the initial volume, and the temperature changed by a factor of 2. I get $\gamma = -0.6$. I don't think it's correct. Help. What is γ?"

75. A magician friend asks for your help in designing an illusion involving a "frozen" balloon. He will take the balloon containing 5.0 L of air at 0 °C out of a freezer and heat it with a small explosive charge. He needs to know how much heat energy will be required to raise the temperature to 20°C. Assume the pressure in the balloon remains at atmospheric pressure, and take $C_p = 3.5 R$. Neglecting the tension in the balloon's latex, what do you tell him?

Answers to Chapter Questions

Answer to Chapter Opening Question

Energy is conserved, provided thermal energy is included. The engine produces both mechanical energy and thermal energy of its exhaust gases; together, they sum to the energy released in combustion.

Answers to GOT IT? Questions

18.1 (c). Only the internal energy is the same, since it's a thermodynamic state variable unique to a point in the pV diagram.

18.2 (c). Superficially they seem to be contradictory, but that's only if you read the ideal-gas law as stating that pV remains constant. That's true in an isothermal process, but not in an adiabatic process to which Equation 18.11a applies. The ideal-gas law is true in both cases, and in an adiabatic process T changes in such a way that pV^γ, rather than pV, remains constant.

18.3 (a) Constant-volume; T and p increase, V doesn't change, U increases as heat flows into the gas. (b) Adiabatic; T and p increase, V decreases, U increases as work is done on the gas. (c) Isobaric; T decreases, p doesn't change, V decreases, U decreases as heat flows out of the gas.

19 The Second Law of Thermodynamics

■ Most of the energy extracted from the fuel in power plants is dumped to the environment as waste heat, here using a large cooling tower. Why is so much energy wasted?

The first law of thermodynamics relates heat and other forms of energy. Much of our world depends on this relationship. Cars extract energy from the heat of burning gasoline. Electricity that powers lights and motors and computers originates in heat released by burning fuels or fissioning uranium. Our own bodies run on energy that begins as heat in the Sun's core. But the first law doesn't tell the whole story. Heat and mechanical energy aren't the same, and the difference makes the conversion of heat to work a more subtle task than the first law would imply.

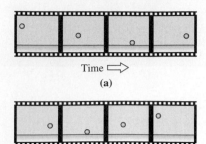

Time ⟹

(a)

(b)

FIGURE 19.1 A movie of a bouncing ball makes sense whether it's shown forward or backward.

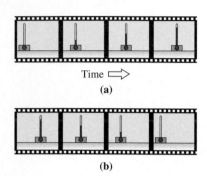

Time ⟹

(a)

(b)

FIGURE 19.2 (a) A block warming (note thermometer) as friction dissipates its kinetic energy. (b) The reverse sequence would never happen, even though it doesn't violate energy conservation.

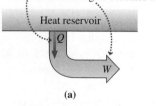

All the heat Q extracted from the reservoir of a perfect heat engine becomes work.

Heat reservoir

Q

W

(a)

Extract heat Q_h from the high-temperature reservoir of a real heat engine.

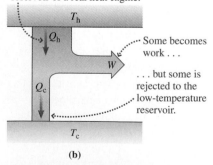

T_h

Q_h

Some becomes work . . .

W

. . . but some is rejected to the low-temperature reservoir.

Q_c

T_c

(b)

FIGURE 19.3 (a) Energy-flow diagram for a perfect heat engine. (b) A real engine delivers as work only a fraction of the energy extracted from the high-temperature reservoir.

19.1 Reversibility and Irreversibility

Figure 19.1 shows a movie of a bouncing ball. Run it backward and it still makes sense. Figure 19.2 shows a block sliding along a table, slowing to a stop because of friction—and warming in the process. Play this film backward and it makes no sense. You'll never see a block at rest suddenly start to move, cooling as it goes. Yet energy would be conserved if it did, so the first law of thermodynamics would be satisfied. Beat an egg, blending yolk and white. Reverse the beater, and you'll never see them separate again. Put cups of cold and hot water in contact; the hot water gets colder and the cold gets hotter. The opposite never occurs—although energy would still be conserved.

Why are these events **irreversible**? In each case we start with matter in an organized state. The molecules of the sliding block share a common motion. The yolk molecules are all in one place. The hot water has more energetic molecules. Of all the possible states, these *organized* ones are relatively rare. There are many more *disorganized* states—for example, all the possible arrangements of molecules in a scrambled egg. As a system evolves, chances are it will end up less organized, simply because there are far more such states available to it. It's very unlikely to assume spontaneously a more organized state.

A key word here is "spontaneous." We could restore organization—for example, by putting one cup of water in the refrigerator and the other in the microwave—but that requires a rather deliberate and energy-consuming process.

Irreversibility is a probabilistic notion. Events that *could* occur without violating the principles of Newtonian physics nevertheless *don't* occur because they're too improbable. As a practical consequence, harnessing the internal energy associated with random molecular motions is difficult because those motions won't spontaneously become organized. That makes much of the world's energy unavailable for doing useful work.

GOT IT? 19.1 Which of these processes is irreversible: (a) stirring sugar into coffee, (b) building a house, (c) demolishing a house with a wrecking ball, (d) demolishing a house by taking it apart piece by piece, (e) harnessing the energy of falling water to drive machinery, (f) harnessing the energy of falling water to heat a house?

19.2 The Second Law of Thermodynamics

Heat Engines

It's impossible to convert *all* the internal energy of a system to useful work. But devices called **heat engines** can extract *some* of that internal energy. Examples include gasoline and diesel engines, fossil-fueled and nuclear power plants, and jet aircraft engines.

Figure 19.3a is an energy-flow diagram for a "perfect" heat engine—one that extracts heat from a heat reservoir and converts it all to work. Such an engine would do exactly what we argued against in the preceding section: It would convert the random energy of thermal motion entirely to the ordered motion associated with mechanical work. In fact a perfect heat engine is impossible, for the same reason that we can't unscramble an egg or make a block accelerate spontaneously at the expense of its internal energy. This fact is one statement of the **second law of thermodynamics**:

> **Second law of thermodynamics (Kelvin-Planck statement)** It is impossible to construct a heat engine operating in a cycle that extracts heat from a reservoir and delivers an equal amount of work.

The phrase "in a cycle" means that a practical engine must go through a repeated sequence of steps, as in the back-and-forth motions of the pistons in a gasoline engine.

A simple heat engine consists of a gas-cylinder system and a heat reservoir, the latter kept hot, perhaps, by burning a fuel. With the gas initially at high pressure, we place the cylinder in contact with the heat reservoir. The gas expands and does work W on the piston. In this isothermal process, the gas extracts heat $Q = W$ from the reservoir. Eventually the gas reaches pressure equilibrium and stops expanding. The piston must then be returned to its original position if it's to do more work.

If we just push the piston back, we'll have to do as much work as we got during the expansion, and our engine won't produce any net work. Instead we can cool the gas to reduce its volume, through thermal contact with a cool reservoir. But then some energy leaves the system as heat rather than work, as shown conceptually in Fig. 19.3b. Our engine extracts heat from a source and delivers mechanical work, but over a full cycle work delivered is less than the heat extracted. The remaining energy is rejected to the lower-temperature reservoir, usually the environment. In practical terms, much of the energy released from fuels in car engines and power plants ends up as waste heat.

The second law of thermodynamics says we can't build a perfect heat engine. But how close can we come? We define the **efficiency** e of an engine as the ratio of what we want from the engine—namely, the work W—to what we have to supply (and usually pay for)—namely, the heat energy Q_h: $e = W/Q_h$. Since the process is cyclic, there's no net change in internal energy over one cycle. Then the first law of thermodynamics, $\Delta U = Q - W$, ensures that the work done is the difference between the heat Q_h extracted from the high-temperature reservoir and the heat Q_c rejected to the cool reservoir. So the efficiency is

$$e = \frac{W}{Q_h} = \frac{Q_h - Q_c}{Q_h} = 1 - \frac{Q_c}{Q_h} \tag{19.1}$$

Figure 19.4 shows a heat engine whose efficiency we can calculate. The engine consists of a cylinder containing an ideal gas, sealed by a movable piston. The piston is connected to a rod that turns a wheel. The engine gets its energy from a heat reservoir at a high temperature T_h, and it rejects heat to a cooler reservoir at temperature T_c. Figure 19.5 shows how the engine works in a cycle of four steps, starting with the piston in its leftmost position (state A in Fig. 19.5), where the gas volume is a minimum:

1. **Isothermal expansion:** The high-temperature reservoir is placed in thermal contact with the cylinder. The gas absorbs heat Q_h from the hot reservoir and expands isothermally along path AB. Since temperature remains constant, so does internal energy. The first law then shows that the engine does work $W = Q$ on the piston and wheel.
2. **Adiabatic expansion:** At B we remove the hot reservoir. The expansion becomes adiabatic and follows path BC. We design the engine so the gas has cooled to T_c when the piston reaches its rightmost position (state C), the point of maximum gas volume.
3. **Isothermal compression:** At C we bring the cool reservoir into contact with the cylinder. The wheel's inertia keeps it turning, so the piston does work on the gas, compressing it isothermally from state C to D. This work ends up as heat rejected to the cool reservoir.
4. **Adiabatic compression:** At D we remove the cool reservoir and the compression continues adiabatically until the gas temperature is once again at T_h and the engine is back at state A.

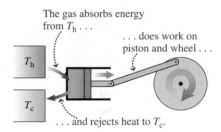

The gas absorbs energy from T_h . . .
. . . does work on piston and wheel . . .
. . . and rejects heat to T_c.

FIGURE 19.4 A simple heat engine.

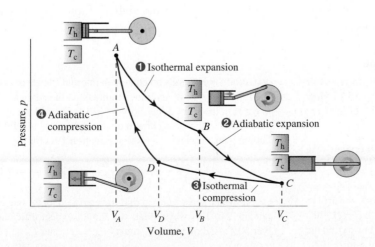

FIGURE 19.5 A pV diagram for the Carnot engine.

This cyclic process of two isothermal and two adiabatic steps is called a **Carnot cycle** and the engine a **Carnot engine**, after the French engineer Sadi Carnot (1796–1832). The particular configuration of the engine is not important, nor is the choice of an ideal gas as the engine's **working fluid**. What distinguishes the Carnot cycle from others is the sequence of thermodynamic processes and the fact that these processes are reversible. The Carnot engine is an example of a **reversible engine**—one in which thermodynamic equilibrium is maintained so that all steps could, in principle, be reversed.

What's the efficiency of a Carnot engine? To find out, we need the heats Q_h and Q_c absorbed and rejected during the isothermal parts of the cycle shown in Fig. 19.5. Equation 18.4 gives the heat Q_h absorbed during the isothermal expansion AB:

$$Q_h = nRT_h \ln\left(\frac{V_B}{V_A}\right)$$

and the heat Q_c rejected during the isothermal compression CD:

$$Q_c = -nRT_c \ln\left(\frac{V_D}{V_C}\right) = nRT_c \ln\left(\frac{V_C}{V_D}\right)$$

We put the minus sign here because our statement of the first law describes Q as the heat *absorbed*, while Equation 19.1 for the engine efficiency requires that Q_c be the heat *rejected*. To calculate engine efficiency according to Equation 19.1, we need the ratio Q_c/Q_h:

$$\frac{Q_c}{Q_h} = \frac{T_c \ln(V_C/V_D)}{T_h \ln(V_B/V_A)} \tag{19.2}$$

This expression can be simplified by applying Equation 18.11b to the adiabatic processes BC and DA in the Carnot cycle: $T_h V_B^{\gamma-1} = T_c V_C^{\gamma-1}$ and $T_h V_A^{\gamma-1} = T_c V_D^{\gamma-1}$. Dividing these two equations gives

$$\left(\frac{V_B}{V_A}\right)^{\gamma-1} = \left(\frac{V_C}{V_D}\right)^{\gamma-1} \quad \text{or} \quad \frac{V_B}{V_A} = \frac{V_C}{V_D}$$

so Equation 19.2 becomes simply $Q_c/Q_h = T_c/T_h$. Using this result in Equation 19.1 then gives the efficiency of the Carnot engine:

$$e_{\text{Carnot}} = 1 - \frac{T_c}{T_h} \quad \text{(Carnot engine efficiency)} \tag{19.3}$$

where the temperatures are measured on an absolute scale (Kelvin or Rankine). Equation 19.3 shows that the Carnot engine's efficiency depends only on the highest and lowest temperatures of its working fluid. In practice, the low temperature is usually that of the environment; then maximizing efficiency requires making the high temperature as high as possible. Real engines trade off efficiency with the ability of materials to withstand high temperature and pressure.

EXAMPLE 19.1 **Calculating Efficiency: A Carnot Engine**

A Carnot engine extracts 240 J from its high-temperature reservoir during each cycle, and rejects 100 J to the environment at 15°C. How much work does the engine do in one cycle? What's its efficiency? What's the temperature of the hot reservoir?

INTERPRET This problem is about a Carnot engine, which operates via the Carnot cycle.

DEVELOP Equation 19.3, $e_{\text{Carnot}} = 1 - T_c/T_h$, relates the two temperatures and the efficiency. Here $Q_h = 240$ J, $Q_c = 100$ J, and $T_c = 15$°C or 288 K. The first law of thermodynamics (Equation 18.1, $\Delta U = Q - W$) relates work and heat flows. So our plan is to use the first law to find the work, then find the efficiency, and then use Equation 19.3 to find T_h.

EVALUATE Since there's no change in internal energy over one cycle, the first law reads $Q = W$ and thus requires that the net heat, 240 J − 100 J, be equal to the work done. So $W = 140$ J. The efficiency is the ratio of work delivered to heat extracted, so $e = W/Q_h = 140$ J/240 J $= 58.3\%$. Knowing the efficiency, we solve Equation 19.3 for T_h:

$$T_h = \frac{T_c}{1 - e} = \frac{288 \text{ K}}{1 - 0.583} = 691 \text{ K} = 418\text{°C}$$

ASSESS Make sense? Given that the engine rejects somewhat less than half the 240 J as waste heat, we should expect efficiency somewhat over 50%. And of course T_h must be greater than T_c, as our calculation confirms. ∎

Engines, Refrigerators, and the Second Law

Why so much emphasis on the Carnot engine? Because understanding this device will help answer the broader question of how much work we can ever hope to extract from thermal energy. That, in turn, will help us understand practical limitations on humankind's attempts to harness ever more energy, and will ultimately bring a deeper understanding of the second law of thermodynamics.

Why is Carnot's engine special? Couldn't we build a better engine with greater efficiency? The answer is no. The special role of the Carnot cycle is embodied in **Carnot's theorem**:

> **Carnot's theorem** All Carnot engines operating between temperatures T_h and T_c have the same efficiency (given by Equation 19.3), and no other engine operating between the same two temperatures can have a greater efficiency.

To prove Carnot's theorem, we introduce the **refrigerator**. A refrigerator is the opposite of an engine: It extracts heat from a cool reservoir and rejects it to a hotter one, taking in work in the process (Fig. 19.6). A refrigerator forces heat to flow from cold to hot, but to do so it requires work. A household refrigerator cools its contents and warms the house (you can feel the heat coming out the back), but it uses electricity. That heat doesn't flow spontaneously from cold to hot constitutes another statement of the second law of thermodynamics:

> **Second law of thermodynamics (Clausius statement)** It is impossible to construct a refrigerator operating in a cycle whose sole effect is to transfer heat from a cooler object to a hotter one.

The Clausius statement rules out a perfect refrigerator (Fig. 19.7).

Suppose the Clausius statement were false. Then we could build the device of Fig. 19.8a, consisting of a reversible Carnot engine and a perfect refrigerator. In each cycle the engine would extract, say, 100 J from the hot reservoir, put out 60 J of useful work, and reject 40 J to the cool reservoir. The perfect refrigerator could transfer the 40 J back to the hot reservoir. The net effect would be to extract 60 J from the hot reservoir and convert it entirely to work (Fig. 19.8b)—and we would have a perfect heat engine, in violation of the Kelvin-Planck statement of the second law. A similar argument (Problem 38) shows that if a perfect heat engine is possible, then so is a perfect refrigerator. So the Clausius and Kelvin-Planck statements of the second law are equivalent, in that if one is false, then so is the other.

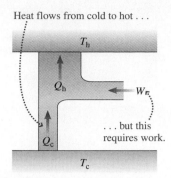

FIGURE 19.6 Energy-flow diagram for a real refrigerator.

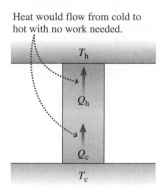

FIGURE 19.7 A perfect refrigerator is impossible.

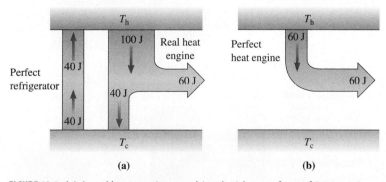

FIGURE 19.8 (a) A real heat engine combined with a perfect refrigerator is equivalent to (b) a perfect heat engine.

Because the Carnot engine is reversible, we could run it backward and reverse its path in Fig. 19.5. The engine would extract heat from the cool reservoir, take in work, and reject heat to the hot reservoir. It would be a refrigerator. Although real refrigerators aren't designed exactly like engines, the two are, in principle, interchangeable.

We're now ready to prove Carnot's assertion that Equation 19.3 gives the maximum engine efficiency. Consider again the Carnot engine in Fig. 19.8a. It extracts 100 J of heat and delivers 60 J of work, so it's 60% efficient. Suppose we had another engine operating

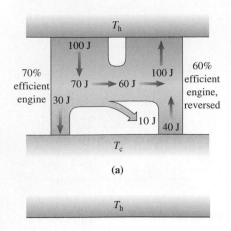

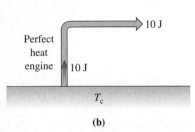

FIGURE 19.9 (a) A 60% efficient reversible engine run as a refrigerator, along with a hypothetical engine with 70% efficiency. (b) The combination is equivalent to a perfect heat engine.

between the same two reservoirs, but with 70% efficiency. Since the Carnot engine is reversible, we can run it as a refrigerator. If we put the two together, we get the device of Fig. 19.9a. Its net effect is to extract 10 J from the cool reservoir and deliver 10 J of work—so it's a perfect heat engine, in violation of the second law (Fig. 19.9b). It's therefore impossible to make an engine that's more efficient than a Carnot engine, and thus Equation 19.3 gives the maximum possible efficiency for *any* heat engine operating between the same two fixed temperatures. For that reason the Carnot efficiency of Equation 19.3 is also called the **thermodynamic efficiency**.

Irreversible engines, because they involve processes that dissipate organized motion, are necessarily *less* efficient. So are reversible engines, if their heat exchange doesn't take place solely at the highest and lowest temperatures. The ordinary gasoline engine is a case in point; even if it could be made perfectly reversible, its efficiency would be less than that of a comparable Carnot engine (see Problem 54).

19.3 Applications of the Second Law

The world abounds with thermal energy, but the second law of thermodynamics limits our ability to turn that energy to our own uses. Any device we construct that involves the interchange of heat and work is a heat engine or refrigerator, subject to the second law.

Limitations on Heat Engines

Most of our electricity is produced in large power plants that are heat engines powered by the fossil fuels coal, oil, or natural gas, or by nuclear fission. Figure 19.10 diagrams such a power plant. The working fluid is water, heated in a boiler and converted to steam at high pressure. The steam expands adiabatically to spin a fanlike turbine. The turbine turns a generator that converts mechanical work to electrical energy.

Steam leaving the turbine is still gaseous and is hotter than the water supplied to the boiler. Here's where the second law applies! Had the water returned from the turbine in its original state, we would have extracted as work all the energy acquired in the boiler, in violation of the second law. So we run the steam through a **condenser**, where it contacts pipes carrying cool water, typically from a river, lake, or ocean. The condensed steam, now cool water, is fed back into the boiler to repeat the cycle.

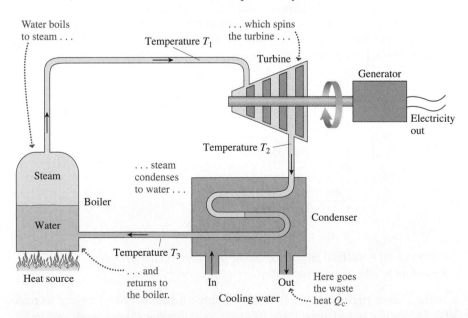

FIGURE 19.10 Schematic diagram of an electric power plant.

The maximum steam temperature in a power plant is limited by the materials used in its construction. For a conventional fossil-fueled plant, current technology permits high temperatures around 650 K. Potential damage to nuclear fuel rods limits the temperature in a nuclear plant to around 570 K. Cooling-water temperature averages about 40°C (310 K), so the maximum possible efficiencies for these power plants, given by Equa-tion 19.3, are

$$e_{fossil} = 1 - \frac{310 \text{ K}}{650 \text{ K}} = 52\% \quad \text{and} \quad e_{nuclear} = 1 - \frac{310 \text{ K}}{570 \text{ K}} = 46\%$$

Temperature differences between the exhausted steam and the cooling water, mechanical friction, and the need to divert energy for driving pumps and pollution-control devices all reduce efficiency further, to about 40% for fossil-fueled plants and 34% for nuclear plants. This means that, when we make electricity, roughly two-thirds of the fuel energy ends up as waste heat.

A typical large power plant produces 1 GW of electricity, so another 2 GW of waste heat goes into the cooling water. The resulting temperature rise can cause serious ecological problems. The huge cooling towers you see at power plants reduce such "thermal pollution" by transferring much of the waste heat to the atmosphere (see this chapter's opening photo). The need for power-plant cooling water nevertheless remains so great that a substantial fraction of all rainwater falling on the United States eventually finds its way through the condensers of power plants (see Problem 29).

EXAMPLE 19.2 **Improving Efficiency:**
A Combined-Cycle Power Plant

The gas turbine in a combined-cycle power plant (see the Application below) operates at 1450°C. Its waste heat at 500°C is the input for a conventional steam cycle, with its condenser fed by river water at 8°C. Find the thermodynamic efficiency of the combined cycle, and compare with the efficiencies of the individual components if they were operated independently.

INTERPRET This problem is about the thermodynamic efficiency of a combined-cycle power plant. As described in the Application, that means a plant using a high-temperature gas turbine whose waste heat becomes the energy input to a conventional steam turbine.

DEVELOP Figure 19.11 is a conceptual diagram of the combined-cycle plant, based on the Application. Equation 19.3, $e = 1 - T_c/T_h$, gives the thermodynamic efficiencies of each cycle and of the combination. We identify the 1450°C = 1723 K temperature as T_h in Equation 19.3 for the gas turbine. The intermediate temperature 500°C = 773 K serves as T_c for the gas turbine but as T_h for the steam cycle. Finally, the 8°C or 281-K cooling-water temperature is T_c for the steam cycle.

EVALUATE To treat the entire plant as a single heat engine in Equation 19.3, we use the highest and lowest temperatures:

$$e_{combined} = 1 - \frac{T_c}{T_h} = 1 - \frac{281 \text{ K}}{1723 \text{ K}} = 0.84 = 84\%$$

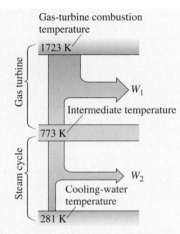

FIGURE 19.11 Conceptual diagram of a combined-cycle power plant.

Friction and other losses would reduce this figure substantially, but a combined-cycle plant operating at these temperatures could have a practical efficiency near 60%. The efficiencies of the individual components also follow from Equation 19.3:

$$e_{gas \, turbine} = 1 - \frac{773 \text{ K}}{1723 \text{ K}} = 55\% \quad \text{and} \quad e_{steam} = 1 - \frac{281 \text{ K}}{773 \text{ K}} = 64\%$$

ASSESS Make sense? Because of its extreme temperatures, the combined cycle gives an efficiency that's better than either of its parts! You can learn more about combined-cycle power plants in the Application below, and by working Problem 30. ∎

APPLICATION **Combined-Cycle Power Plants**

Improving power-plant efficiency helps reduce air pollution and greenhouse-gas emissions, not to mention the cost of electricity. Modern

combined-cycle power plants achieve efficiencies approaching 60% by combining a conventional steam system like that of Fig. 19.10 with a *gas turbine* similar to a jet aircraft engine. Gas turbines operate at high temperatures—between 1000 K and 2000 K—but they aren't very efficient because their exhaust temperature (T_c in Equation 19.3) is also high. In a combined-cycle plant, the hot exhaust from a gas turbine drives a conventional steam cycle whose condenser operates at near the ambient temperature. The overall effect is the same as a single heat engine operating between the gas turbine's high combustion temperature and the low temperature of the environment (see Problem 30). The second law still limits the efficiency, but with high T_h and low T_c, the efficiency can be much greater than in a conventional plant. Combined-cycle plants burning natural gas are becoming more widespread, but their expense relative to coal plants makes them less popular in two of the world's most coal-rich and polluting countries: the United States and China. The photo shows a gas-fired combined-cycle plant.

Gasoline and diesel engines provide another pervasive example of heat engines. A typical automobile engine has a theoretical maximum efficiency around 50%, but irreversible thermodynamic processes make the actual efficiency much lower. Mechanical friction dissipates additional energy, with the end result that less than 20% of the fuel energy reaches the driving wheels. Problems 54 and 55 explore the thermodynamics of the gasoline engine.

We wouldn't be so concerned with efficiency if we didn't have to pay for fuel. Engines with "free" fuel include solar-thermal power plants that concentrate sunlight to boil a fluid that drives a turbine, and ocean thermal-energy conversion (OTEC) schemes that extract useful work from the modest temperature difference between tropical surface waters and the deep ocean. Although pilot projects of both types have been constructed, neither has yet proven technologically or economically practical.

Refrigerators and Heat Pumps

Figure 19.6 showed that a refrigerator works like an engine in reverse: It takes in mechanical work and transfers heat from cooler to hotter. The heat you feel at the back of your kitchen refrigerator includes both energy removed from the refrigerator's contents and energy supplied as electricity to run the appliance. An efficient refrigerator should maximize what we want—namely, cooling, as measured by the heat Q_c removed—compared with what we have to supply—namely, mechanical work W or its equivalent in electrical energy. The **coefficient of performance** (COP) quantifies this ratio:

$$\text{COP} = \frac{Q_c}{W} = \frac{Q_c}{Q_h - Q_c} = \frac{T_c}{T_h - T_c} \quad \left(\begin{array}{l} \text{refrigerator;} \\ \text{for heat pump c} \rightarrow \text{h in numerator} \end{array} \right) \quad (19.4)$$

The second equality here follows from the first law of thermodynamics, and the third expresses the equality $Q_c/Q_h = T_c/T_h$ that we found in deriving Equation 19.3. The first equality in Equation 19.4 is a general definition of the COP; the final equality gives a theoretical maximum that real refrigerators can approach but not exceed. When the temperatures T_h and T_c are close, Equation 19.4 gives a high COP—meaning that the refrigerator takes relatively little work to do its job. But as the difference increases, the COP drops and we have to supply more work.

EXAMPLE 19.3 **The COP: A Home Freezer**

A typical home freezer operates between a low of 0°F (-18°C or 255 K) and a high of 86°F (30°C or 303 K). What's its maximum possible COP? With this COP, how much electrical energy would it take to freeze 500 g of water initially at 0°C?

INTERPRET This problem is about a refrigerator—in this case a freezer. We identify T_h and T_c with the values 303 K and 255 K, respectively.

DEVELOP Equation 19.4, COP $= T_c/(T_h - T_c)$, will determine the COP. Then we'll use Equation 17.5, $Q = Lm$, to find the heat Q_c that the freezer must extract to freeze the water. From there we'll be able to use the first equality in Equation 19.4 to find the work—equivalently, the electrical energy—required.

EVALUATE Equation 19.4 gives

$$\text{COP} = \frac{T_c}{T_h - T_c} = \frac{255 \text{ K}}{303 \text{ K} - 255 \text{ K}} = 5.31$$

From Equation 17.5, we find the heat that needs to be removed in freezing 500 g of ice: $Q_c = Lm = (0.50 \text{ kg})(334 \text{ kJ/kg}) = 167 \text{ kJ}$, where we found the heat of fusion in Table 17.1. The COP is the ratio of the heat removed to the work or electrical energy required, so we have $W = Q_c/\text{COP} = 167 \text{ kJ}/5.31 = 31 \text{ kJ}$.

ASSESS Make sense? A COP of 5.3 means that each unit of work can transfer 5.3 units of heat from inside the freezer—so the electrical-energy requirement here is modest. A practical freezer operating between these temperatures would have a lower COP and require correspondingly more electrical energy. ∎

Refrigerators aren't confined to kitchens. An air conditioner is a refrigerator that cools a room or building. A **heat pump** is a refrigerator that cools in summer and heats in winter (Fig. 19.12). In warmer climates, heat pumps exchange energy between a building and the outside air; in cooler climates, they exchange energy through pipes buried in the ground where the temperature is typically around 10°C year-round. Heat pumps require electricity, but like any refrigerator they transfer more heat energy than they use in electricity—although that gain is somewhat offset by the inefficiency of the power plant producing the electricity. Because the desired output is heat at the higher temperature, the COP of a heat pump is often defined as Q_h/W rather than Q_c/W as for a refrigerator; that means Q_c and T_c become Q_h and T_h, respectively, in the numerator of Equation 19.4 for a heat pump. Several end-of-chapter problems explore the physics and economics of heat pumps.

In summer the heat pump cools the house by extracting energy and rejecting it to the outdoor environment.

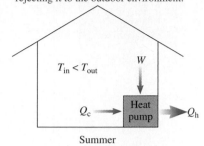

Summer

GOT IT? 19.2 A clever engineer decides to increase the efficiency of a Carnot engine by cooling the low-temperature reservoir using a refrigerator with the maximum possible COP. Will the overall efficiency of this system (a) exceed, (b) be less than, or (c) equal that of the original engine alone?

In winter the pump extracts energy from outside and transfers it to the inside.

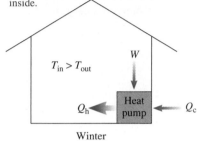

Winter

FIGURE 19.12 A heat pump.

19.4 Entropy and Energy Quality

If offered a joule of energy, would you rather have it delivered in the form of mechanical work, heat at 1000 K, or heat at 300 K? Your answer might depend on what you want to do. To lift or accelerate a mass, you'd be smart to take your energy as work. But if you want to keep warm, heat at 300 K would be perfectly acceptable.

If, on the other hand, you're not sure what you want to do with the energy, then which should you take? The second law of thermodynamics makes the answer clear: You should take the work. Why? Because you could use it directly as mechanical energy, or you could, through friction or other irreversible processes, use it to raise the temperature of something.

If you chose 300 K heat for your joule of energy, then you could supply a full joule only to objects cooler than 300 K. You couldn't do mechanical work unless you ran a heat engine. With its maximum temperature only a little above ambient, your engine would be very inefficient, and you could extract only a small fraction of a joule of mechanical energy. You would be better off with 1000-K heat since you could transfer it to anything cooler than 1000 K, or you could run a heat engine to produce up to 0.70 joule of mechanical energy.

Taking your energy in the form of work gives you the most options. Anything you can do with a joule of energy, you can do with the work. Heat is less versatile, with 300-K heat

APPLICATION **Energy Quality, End Use, and Cogeneration**

Smart use of energy means matching the quality of energy sources to our energy needs. It makes little sense to produce high-quality electrical energy and then use it to heat water for a shower. You'd be much better off burning fuel in a water heater where nearly all its energy gets transferred to the water than burning it in a 40% efficient power plant to make electricity, in the process discharging 60% of the fuel energy to the environment as waste heat. Yet electricity is such a convenient energy source that we tend to use it for everything, including low-grade heat. Is there a way to avoid this thermodynamic folly? There is, and it involves using the waste heat produced in generating electricity to supply low-quality energy needs—a process known as **cogeneration**. In parts of Europe, whole cities are heated with the waste heat from electric power plants. In the United States, industrial and institutional energy users are beginning to combine their heating plants with smaller electric generators to extract the most possible high-quality energy before using the lower-quality remainder for industrial-process heat or space heating. The photo shows a 600-kW cogenerating unit in the heating plant at Middlebury College.

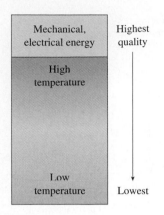

FIGURE 19.13 Energy quality measures the versatility of different energy forms.

the least useful of the three. We're not talking here about the quantity of energy—we have exactly 1 joule in each case—but about **energy quality**, indicated by the ability to do a variety of useful tasks (Fig. 19.13). We can readily convert an entire amount of energy from higher to lower quality, but the second law of thermodynamics prevents us from going in the opposite direction with 100% efficiency.

Entropy

Can we quantify the notion of energy quality? Imagine mixing glasses of hot and cold water (Fig. 19.14). Soon they come to equilibrium at a lukewarm temperature, with no overall energy loss. But something has been lost—namely, the ability to do useful work. In the initial state we could have run a heat engine between the hot and cold water. In the final state there's no temperature difference and therefore no way to run a heat engine. The quantity of energy hasn't changed, but its quality has decreased. Can we find some property whose value reflects this change?

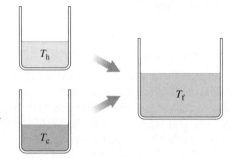

FIGURE 19.14 Mixing hot and cold water gives lukewarm water at final temperature T_f. There's no energy loss, but the system loses the ability to do work.

Consider an ideal gas undergoing a Carnot cycle of two isothermal and two adiabatic processes (recall Fig. 19.5). In deriving Equation 19.3 for the efficiency of this cycle, we found that $Q_c/Q_h = T_c/T_h$, where Q_c was the heat *rejected* from the system to the low-temperature reservoir at T_c, and Q_h the heat *added* from the reservoir at T_h.

We now change the definition of Q_c so it also means heat *added* to the system. This redefinition just changes the sign of Q_c, so now our equality can be written

$$\frac{Q_c}{T_c} + \frac{Q_h}{T_h} = 0$$

We can generalize this result to any reversible cycle by approximating the cycle as a sequence of adiabatic and isothermal steps (Fig. 19.15). Figure 19.15 shows how summing over the individual isothermal segments then gives $\Sigma Q/T = 0$. We can approximate the closed cycle ever closer by more and more adiabatic and isothermal segments. In the limit the approximation becomes exact and the sum becomes an integral:

$$\oint \frac{dQ}{T} = 0 \tag{19.5}$$

where the circle indicates that we integrate over a *closed* path.

Equation 19.5 describes a quantity that doesn't change when we take a system around a cyclic path. This quantity is called **entropy**, S. If we take a system around a path that isn't closed, then its entropy will, in general, change. That change is given by integrating over the path in question:

$$\Delta S = \int_1^2 \frac{dQ}{T} \qquad \text{(change in entropy)} \tag{19.6}$$

$\frac{Q_c}{T_c} + \frac{Q_h}{T_h} = 0$ for the highlighted cycle or any other cycle, so $\Sigma Q/T$ must be zero around the path.

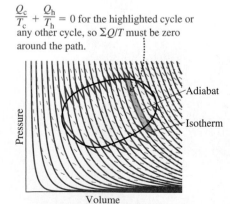

FIGURE 19.15 An arbitrary cycle approximated by isothermal (dashed curves) and adiabatic (solid curves) steps. Heat transfer occurs only during the isothermal steps.

where ΔS is the change in entropy between the two thermodynamic states 1 and 2. Note that entropy has the units J/K.

Suppose we take a system part way around a closed path, from state 1 to state 2 in Fig. 19.16; Equation 19.6 gives the corresponding entropy change ΔS_{12}. If we go back to state 1 by *any* path, the resulting entropy change ΔS_{21} must be $-\Delta S_{12}$, so that there's no entropy change around the closed path. Thus the entropy change of Equation 19.6 is independent of path; it depends only on the initial and final states. Like pressure and temperature, entropy is a thermodynamic state variable—a quantity that characterizes a given state independently of how the system got into that state.

Now, Equation 19.6 applies only to reversible processes, since irreversible processes take a system out of thermodynamic equilibrium and therefore aren't described by paths in the pV diagram. But because entropy depends only on the initial and final states, we can calculate the entropy change in an *irreversible* process by using Equation 19.6 for a *reversible* process that goes between the same two states. Doing so for a couple of examples will help you understand the meaning of entropy.

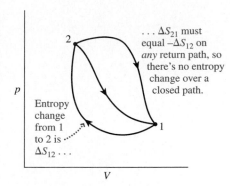

FIGURE 19.16 Entropy change is path-independent.

Irreversible Heat Transfer

Mixing hot and cold water in Fig. 19.14 is an irreversible process, as we described in Section 19.1. But we could produce the same change reversibly, by putting each glass in contact with a separate heat reservoir, then cooling or heating until each is at the final temperature T_f, and then mixing them. The entropy change for the cold water would then be $\Delta S_c = \int_{T_c}^{T_f} dQ/T$. Rather than evaluate this integral, note that there must be some intermediate temperature T_1 that lies between T_c and T_f and for which $\Delta S_c = Q_c/T_1$, where Q_c is the heat absorbed as the cold water warms to T_f. We've marked T_1 on Fig. 19.17. Similarly, the entropy change for the hot water is $\Delta S_h = Q_h/T_2$, where T_2 lies between T_f and T_h. Then the entropy change for the whole system is $\Delta S = Q_c/T_1 + Q_h/T_2$. But the heat lost from the hot water in the actual irreversible process is equal to the heat gained by the cold water, so $Q_h = -Q_c$, and the entropy change is

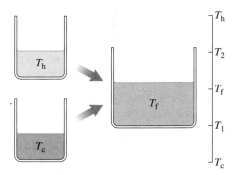

$$\Delta S = \frac{Q_c}{T_1} - \frac{Q_c}{T_2} = Q_c\left(\frac{1}{T_1} - \frac{1}{T_2}\right)$$

Now $T_1 < T_2$ (see Fig. 19.17), and that makes ΔS a *positive* quantity. Therefore entropy has *increased* during the irreversible mixing process.

FIGURE 19.17 Repeat of Fig. 19.14, showing intermediate temperatures T_1 and T_2 used in calculating entropy change.

Adiabatic Free Expansion

Figure 19.18 shows another irreversible process that results in a loss of energy quality. In Fig. 19.18a, a partition confines an ideal gas to one side of an insulated box; on the other side is vacuum. Since the box is insulated, no heat can flow into or out of the gas, so the situation is adiabatic. Remove the partition, and the gas expands freely to fill the box (Fig. 19.18b). The process is therefore an **adiabatic free expansion**. In the vacuum there's no pressure to oppose the gas, so it does no work and therefore its internal energy doesn't change. Figure 19.18c shows how we could instead have used the expanding gas to turn a paddle wheel, extracting useful work. We can't do that with the uniform-pressure gas of Fig. 19.18b, so the free expansion results in the system's losing its ability to do work.

Now let's find the entropy change for the irreversible process of Fig. 19.18a → b. Since the internal energy of the gas doesn't change, neither does the temperature. But we've seen that we can calculate the entropy change by considering a *reversible* process that takes the gas between the same two states. An isothermal expansion is such a process, for which Equation 18.4 gives the heat that would be added: $Q = nRT \ln(V_2/V_1)$. Since the temperature is constant, the entropy change of Equation 19.6 becomes

$$\Delta S = \int \frac{dQ}{T} = \frac{1}{T} \int dQ = \frac{Q}{T} = nR \ln\left(\frac{V_2}{V_1}\right)$$

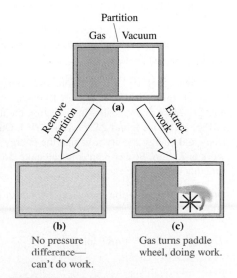

FIGURE 19.18 Two ways for a gas to expand into a vacuum.

The final volume V_2 is larger than V_1, so entropy has *increased*. Although we computed this result for the reversible process, we emphasize that it holds for *any* process that takes the system between the same initial and final states—including our irreversible adiabatic free expansion.

Entropy and the Availability of Work

Entropy increases during both the irreversible processes we just considered. Energy quality deteriorates in both processes, in that the systems lose ability to do work. Suppose we had let the gas in Fig. 19.18 undergo a reversible isothermal expansion instead of the adiabatic free expansion. Then it would have done work equal to the heat gained, which we just calculated:

$$W = Q = nRT \ln\left(\frac{V_2}{V_1}\right).$$

After the irreversible free expansion, the gas can no longer do this work, even though its energy is unchanged. Comparing with the entropy change we calculated above, we see that the energy that becomes unavailable to do work is $E_{unavailable} = T \Delta S$. This is an example of a more general relation between entropy and the quality of energy:

During an irreversible process in which the entropy of a system increases by ΔS, energy $E = T_{min} \Delta S$ becomes unavailable to do work, where T_{min} is the coolest temperature available to the system.

This statement shows that entropy provides our measure of energy quality. Given two systems with identical energy content, the one with the lower entropy contains the higher-quality energy. An entropy increase corresponds to a degradation in energy quality, as energy becomes unavailable to do work.

EXAMPLE 19.4 **Increasing Entropy: The Loss of Energy Quality**

A 2.0-L cylinder contains 5.0 mol of compressed gas at 300 K. If the cylinder is discharged into a 150-L vacuum chamber and its temperature remains 300 K, how much energy has become unavailable to do work?

INTERPRET This problem asks about the loss of energy quality during an irreversible and therefore entropy-increasing process, namely an adiabatic free expansion.

DEVELOP Figure 19.19 is a before-and-after sketch of the situation, which is similar to that of Fig. 19.18 except that here the gas volume changes more dramatically. In analyzing the adiabatic free expansion of Fig. 19.18, we found $\Delta S = nR \ln(V_2/V_1)$. Our statement relating entropy and energy quality says that the energy made unavailable to do work is $T_{min} \Delta S$. So our plan is to calculate ΔS and multiply by T_{min} to find that unavailable energy.

EVALUATE Because the temperature doesn't change, T_{min} is the 300-K temperature we're given, and we have

$$E_{unavailable} = T \Delta S = nRT \ln\left(\frac{V_2}{V_1}\right)$$

$$= (5.0 \text{ mol})(8.314 \text{ J/K} \cdot \text{mol})(300 \text{ K}) \ln\left(\frac{152 \text{ L}}{2.0 \text{ L}}\right) = 54 \text{ kJ}$$

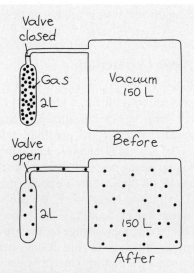

FIGURE 19.19 Our sketch for Example 19.4.

ASSESS Make sense? Yes: This is the work we could have gained from a reversible isothermal expansion. By letting the gas undergo an irreversible process, we gave up the possibility of extracting this work. ∎

Entropy and the Second Law of Thermodynamics

We started this chapter arguing that natural processes are generally irreversible, going from ordered states to disordered states. It's this loss of order that makes energy unavailable to do work. Entropy is a measure of this disorder. Given the tendency of systems to evolve toward disordered states, we can make a statement about entropy that is, in fact, a general statement of the second law of thermodynamics:

Second law of thermodynamics The entropy of a closed system can never decrease.

At best, the entropy of a closed system remains constant—and this happens only in an ideal, reversible process. If anything irreversible occurs—a slight amount of friction or a deviation from exact thermodynamic equilibrium—then entropy increases. There's no going back. As entropy increases, energy becomes unavailable to do work, and nothing within the closed system can restore that energy to its original quality. This statement of the second law in terms of entropy is equivalent to our previous statements about the impossibility of perfect heat engines and refrigerators, for the operation of either device would require a decrease in entropy.

What about a system that isn't closed? Can't we decrease its entropy? Yes—but only by supplying high-quality energy from outside. Running a refrigerator decreases the entropy of its contents, but this requires electrical energy to make heat flow in the direction it doesn't normally go. That high-quality electrical energy deteriorates into additional heat that's rejected to the refrigerator's environment. If we consider the entire system, not just the refrigerator's contents, this means that overall entropy has increased.

Any system whose entropy seems to decrease—that gets more rather than less organized—cannot be a closed system. If we enlarge a system's boundaries to encompass the entire universe, then we definitely have a closed system. Therefore the ultimate statement of the second law becomes

Second law of thermodynamics The entropy of the universe can never decrease.

As examples of this broad statement, consider the growth of a living thing from the random mix of molecules in its environment, or the construction of a skyscraper from materials that were originally dispersed about Earth, or the appearance of ordered symbols on a printed page from a bottle of ink. All these are processes in which matter goes from near chaos to a highly organized state—akin to separating yolk and white from a scrambled egg. They are certainly processes in which entropy decreases. But Earth isn't a closed system. It gets high-quality energy from the Sun, energy that is ultimately responsible for life and all its actions. If we consider the Earth-Sun system, then the entropy decrease associated with life and civilization is more than balanced by the entropy increase associated with the degradation of high-quality solar energy. We living things represent a remarkable phenomenon—the organization of matter in a universe governed by a tendency toward disorder. But we can't escape the second law of thermodynamics. Our highly organized selves and society, and the entropy decreases they represent, come into being only at the expense of greater entropy increases elsewhere.

GOT IT? 19.3 In each of the following processes, does the entropy of the named system alone increase, decrease, or stay the same? (a) A balloon deflates; (b) cells differentiate in a growing embryo, forming different physiological structures; (c) an animal dies, and its remains gradually decay; (d) an earthquake demolishes a building; (e) a plant utilizes sunlight, carbon dioxide, and water to manufacture sugar; (f) a power plant burns coal and produces electrical energy; (g) a car's friction-based brakes stop the car.

Big Picture

The big idea behind this chapter is the **second law of thermodynamics**—ultimately, the statement that systems tend naturally toward disorder, or states of higher **entropy**. The second law is manifest in our practical world in forbidding the construction of perfect heat engines and perfect refrigerators—and therefore preventing us from extracting as useful work all the energy that's contained in random thermal motions. Ultimately, the second law says that the entropy of any closed system, including the entire universe, cannot decrease.

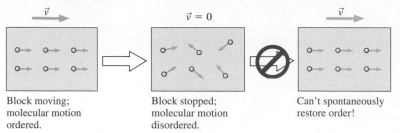

Block moving; molecular motion ordered.

Block stopped; molecular motion disordered.

Can't spontaneously restore order!

Key Concepts and Equations

Entropy is a quantitative measure of energy quality and of disorder; the higher the entropy, the lower the energy quality and the greater the disorder. The highest-quality energy is mechanical or electrical energy, followed by the internal energy of systems at high temperature, and finally low-temperature internal energy. Whenever entropy increases, energy becomes unavailable to do work.

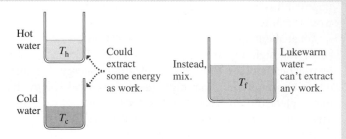

Hot water T_h — Could extract some energy as work. Instead, mix. Lukewarm water – can't extract any work. T_f

Cold water T_c

- $\Delta S = \int_1^2 \dfrac{dQ}{T}$ gives the entropy change as a system goes from state 1 to state 2.

- $E_{\text{unavailable}} = T_{\min} \Delta S$ is the energy that becomes unavailable as a result of entropy increase ΔS.

Cases and Uses

The second law sets the maximum possible efficiency of any heat engine as that of the **Carnot engine**, an engine that combines adiabatic and isothermal processes.

$$e = \frac{W}{Q_h} \le e_{\max} = 1 - \frac{T_c}{T_h}$$

This defines an engine's efficiency.   This is the maximum possible efficiency.

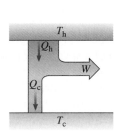

T_h

Q_h

W

Q_c

T_c

Energy-flow diagram for an engine

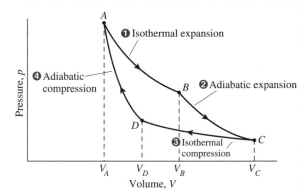

❶ Isothermal expansion

❹ Adiabatic compression

❷ Adiabatic expansion

❸ Isothermal compression

pV diagram for Carnot engine

Similarly, the second law limits the **coefficient of performance** of a refrigerator to

$$\text{COP} = \frac{Q_c}{W} \le \text{COP}_{\max} = \frac{T_c}{T_h - T_c}$$

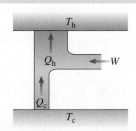

T_h

Q_h ← W

Q_c

T_c

For Thought and Discussion

1. Could you cool the kitchen by leaving the refrigerator door open? Explain.
2. Could you heat the kitchen by leaving the oven open? Explain.
3. Should a car get better mileage in the summer or the winter? Explain.
4. Is there a limit to the maximum temperature that can be achieved by focusing sunlight with a lens? If so, what is it?
5. Name some irreversible processes that occur in a real engine.
6. A power company claims that electric heat is 100% efficient. Discuss this claim.
7. A hydroelectric power plant, using the energy of falling water, can operate with an efficiency arbitrarily close to 100%. Why?
8. To maximize the COP of a refrigerator, should you strive for a large or a small temperature difference? Explain.
9. The manufacturer of a heat pump claims that the device will heat your home using only energy already available in the ground. Is this true?
10. Does sunlight represent high- or low-quality energy? Explain. *Note:* See also For Thought and Discussion 4.
11. The heat Q added during adiabatic free expansion is zero. Why can't we then argue from Equation 19.6 that the entropy change is zero?
12. Energy is conserved, so why can't we recycle it as we do materials?
13. Why doesn't the evolution of human civilization violate the second law of thermodynamics?

Exercises and Problems

Exercises

Sections 19.2 and 19.3 The Second Law of Thermodynamics and Its Applications

14. What are the efficiencies of reversible heat engines operating between (a) the normal freezing and boiling points of water, (b) the 25°C temperature at the surface of a tropical ocean and deep water at 4°C, and (c) a 1000°C candle flame and room temperature?
15. A cosmic heat engine might operate between the Sun's 5600 K surface and the 2.7 K temperature of intergalactic space. What would be its efficiency?
16. A reversible Carnot engine operating between helium's melting point and its 4.25K boiling point has an efficiency of 77.7%. What is the melting point?
17. A Carnot engine absorbs 900 J of heat each cycle and provides 350 J of work. (a) What is its efficiency? (b) How much heat is rejected each cycle? (c) If the engine rejects heat at 10°C, what is its maximum temperature?
18. What is the COP of a reversible refrigerator operating between 0°C and 30°C?
19. How much work does a refrigerator with a COP of 4.2 require to freeze 670 g of water already at its freezing point?

Section 19.4 Entropy and Energy Quality

20. Calculate the entropy change associated with melting 1.0 kg of ice at 0°C.
21. You heat 250 g of water from 10°C to 95°C. By how much does the entropy of the water increase?
22. A 2.0-kg sample of water is heated to 35°C. If the entropy change is 740 J/K, what was the initial temperature?
23. Melting a block of lead already at its melting point results in an entropy increase of 900 J/K. What is the mass of the lead? *Hint:* Consult Table 17.1.
24. How much energy becomes unavailable for work in an isothermal process at 440 K, if the entropy increase is 25 J/K?

Problems

25. A Carnot engine extracts 890 J from a 550 K reservoir during each cycle and rejects 470 J to a cooler reservoir. (a) How much work does it do during each cycle? (b) What is its efficiency? (c) What is the temperature of the cool reservoir? (d) If the engine undergoes 22 cycles per second, what is its mechanical power output?

26. The maximum temperature in a nuclear power plant is 570 K. The plant rejects heat to a river whose temperature is 0°C in the winter and 25°C in the summer. What are the maximum possible efficiencies for the plant in these seasons?
27. A power plant's electrical output is 750 MW. Cooling water at 15°C flows through the plant at 2.8×10^4 kg/s, and its temperature rises by 8.5°C. Assuming that the plant's only energy loss is to the cooling water and that the cooling water is effectively the low-temperature reservoir, find (a) the rate of energy extraction from the fuel, (b) the plant's efficiency, and (c) its highest temperature.
28. A power plant extracts energy from steam at 250°C and delivers 800 MW of electric power. It discharges waste heat to a river at 30°C. The overall efficiency of the plant is 28%. (a) How does this efficiency compare with the maximum possible at these temperatures? (b) What is the rate of waste-heat discharge to the river? (c) How many houses, each requiring 18 kW of heating power, could be heated with the waste heat from this plant?
29. The electric power output of all the thermal electric power plants in the United States is about 2×10^{11} W, and these plants operate at an average efficiency around 33%. What is the rate at which all these plants use cooling water, assuming an average 5°C rise in cooling-water temperature? Compare with the 1.8×10^7 kg/s average flow at the mouth of the Mississippi River.
30. Consider a Carnot engine operating between temperatures T_h and T_i, where T_i is intermediate between T_h and the ambient temperature T_c (Fig. 19.20). It should be possible to operate a second engine between T_i and T_c. Show that the maximum overall efficiency of such a two-stage engine is the same as that of a single engine operating between T_h and T_c. *Note:* This problem shows why combined-cycle power plants achieve their high efficiencies.

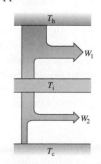

FIGURE 19.20 Problem 30

31. An industrial freezer operates between 0°C and 32°C, consuming electrical energy at the rate of 12 kW. Assuming the freezer is perfectly reversible, (a) what is its COP? (b) How much water at 0°C can it freeze in 1 hour?

32. Use appropriate energy-flow diagrams to analyze the situation in Got It? 19.2; that is, show that using a refrigerator to cool the low-temperature reservoir can't increase the overall efficiency of a Carnot engine when the work input to the refrigerator is included.

33. Cooling water circulates through a reversible Carnot engine at 3.2 kg/s. The water enters at 23°C and leaves at 28°C; the average temperature is essentially that of the engine's cool reservoir. If the engine's mechanical power output is 150 kW, what are (a) its efficiency and (b) its highest temperature?

34. It costs $180 to heat a house with electricity in a typical winter month. (Electric heat simply converts all the incoming electrical energy to heat.) What would the monthly heating bill be following conversion to an electrically powered heat-pump system with COP = 3.1?

35. A 4.0-L sample of water at 9.0°C is put into a refrigerator. The refrigerator's 130-W motor then runs for 4.0 min to cool the water to the refrigerator's low temperature of 1.0°C. (a) What is the COP of the refrigerator? (b) How does this compare with the maximum possible COP if the refrigerator exhausts heat at 25°C?

36. A refrigerator maintains an interior temperature of 4°C while the temperature near its heat exhaust is 30°C. The refrigerator's insulation is imperfect, and heat leaks into the refrigerator at the rate of 340 W. Assuming the refrigerator is reversible, at what rate must it consume electrical energy to maintain a constant 4°C interior?

37. A store is heated by an oil furnace that supplies 30 kWh of heat from each gallon. The store's owners are considering switching to a heat-pump system. Oil costs $1.75/gallon, and electricity costs 16.5¢/kWh. What is the minimum heat-pump COP that will result in a savings in heating costs?

38. Use appropriate energy-flow diagrams to show that the existence of a perfect heat engine would permit the construction of a perfect refrigerator, thus violating the Clausius statement of the second law.

39. A heat pump extracts energy from the ground at 10°C and transfers it to water at 70°C to heat a building. Find (a) its COP and (b) its electric power consumption if it supplies heat at the rate of 20 kW. (c) Compare the hourly operating cost of the pump with an oil furnace if electricity costs 15.5¢/kWh and oil costs $1.95/gallon and releases about 30 kWh/gal when burned.

40. A reversible engine contains 0.20 mol of ideal monatomic gas, initially at 600 K and confined to 2.0 L. The gas undergoes the following cycle:
    - Isothermal expansion to 4.0 L.
    - Isovolumic cooling to 300 K.
    - Isothermal compression to 2.0 L.
    - Isovolumic heating to 600 K.
    (a) Calculate the net heat added during the cycle and the net work done. (b) Determine the engine's efficiency, defined as the ratio of the work done to only the heat *absorbed* during the cycle.

41. (a) Determine the efficiency for the cycle shown in Fig. 19.21, using the definition given in Problem 40. (b) Compare with the efficiency of a Carnot engine operating between the same temperature extremes. Why are the two efficiencies different?

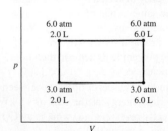

42. A 0.20-mol sample of an ideal gas goes through the Carnot cycle of Fig. 19.22. Calculate (a) the heat Q_h absorbed, (b) the heat Q_c rejected, and (c) the work done. (d) Use these quantities to determine the efficiency. (e) Find the maximum and minimum temperatures, and show explicitly that the efficiency as defined in Equation 19.1 is equal to the Carnot efficiency of Equation 19.3.

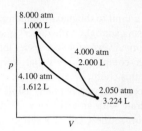

FIGURE 19.22 Problem 42

43. A shallow pond contains 94,000 kg of water. In winter it's entirely frozen. By how much does the entropy of the pond increase when the ice, already at 0°C, melts and then heats to its summer temperature of 15°C?

44. The temperature of n moles of ideal gas is changed from T_1 to T_2 while the gas volume is held constant. Show that the corresponding entropy change is $\Delta S = nC_V \ln(T_2/T_1)$.

45. The temperature of n moles of ideal gas is changed from T_1 to T_2 while the gas pressure is held constant. Show that the corresponding entropy change is $\Delta S = nC_p \ln(T_2/T_1)$.

46. A 5.0-mol sample of an ideal diatomic gas $\left(C_V = \frac{5}{2}R\right)$ is initially at 1.0 atm pressure and 300 K. What is the entropy change if the gas is heated to 500 K (a) at constant volume, (b) at constant pressure, and (c) adiabatically?

47. The interior of a house is maintained at 20°C while the outdoor temperature is −10°C. The house loses heat at the rate of 30 kW. At what rate does the entropy of the universe increase because of this irreversible heat flow?

48. A 250-g sample of water at 80°C is mixed with 250 g of water at 10°C. Find the entropy changes for (a) the hot water, (b) the cool water, and (c) the system.

49. In an adiabatic free expansion, 8.7 mol of ideal gas at 450 K expand 10-fold in volume. How much energy becomes unavailable to do work?

50. Ideal gas occupying 1.0 cm³ is placed in a 1.0-m³ vacuum chamber, where it expands adiabatically. If 6.5 J of energy become unavailable to do work, what was the initial gas pressure?

51. Find the entropy change when a 2.4-kg aluminum pan at 155°C is plunged into 3.5 kg of water at 15°C.

52. You're lying in a bathtub of water at 42°C. Suppose, in violation of the second law of thermodynamics, that the water spontaneously cooled to room temperature (20°C) and that the energy so released was transformed into your gravitational potential energy. Estimate the height to which you would rise above the bathtub.

53. An engine with mechanical power output 8.5 kW extracts heat from a source at 420 K and rejects it to a 1000-kg block of ice at its melting point. (a) What is its efficiency? (b) How long can it maintain this efficiency if the ice is not replenished?

54. Gasoline engines operate approximately on the **Otto cycle**, consisting of two adiabatic and two constant-volume segments. The Otto cycle for a particular engine is shown in Fig. 19.23. (a) If the gas in the engine has specific-heat ratio γ, find the engine's efficiency, assuming all processes are reversible. (b) Find the maximum temperature in terms of the minimum temperature T_{min}.

(c) How does the efficiency compare with that of a Carnot engine operating between the same two temperature extremes? *Note:* Figure 19.23 neglects the intake of fuel-air and the exhaust of combustion products, which together involve essentially no net work.

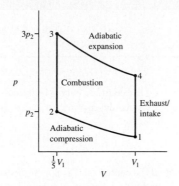

FIGURE 19.23 Problems 54 and 55

55. The compression ratio r of an engine is the ratio of maximum to minimum gas volume. For the engine of Problem 54, Fig. 19.23 shows that the compression ratio is 5. (a) Find an expression for the engine efficiency as a function of the compression ratio. Assume that pressure continues to triple during the combustion phase, as shown in Fig. 19.23. (b) Make a graph of efficiency versus r, and on the same plot show also the efficiency of a Carnot engine operating between the same temperature extremes.

56. A 5.0-mol sample of ideal monatomic gas undergoes the cycle shown in Fig. 19.24, in which the process BC is isothermal. Calculate the entropy change associated with each of the three steps, and show explicitly that there is zero net entropy change over the full cycle.

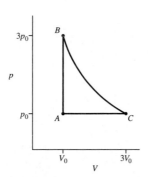

FIGURE 19.24 Problem 56

57. The McNeil Generating Station in Burlington, Vermont, is one of the world's largest wood-fired electric power plants. McNeil produces steam at 950°F to drive its turbines, and condensed steam returns to the boiler as water at 90°F. (Note the temperatures in Fahrenheit, used in U.S. engineering situations.) Find McNeil's maximum thermodynamic efficiency, and compare with its actual efficiency of 25%. *Note:* Some of the difference comes from having to evaporate moisture out of the wood-chip fuel.

58. A 500-g copper block at 80°C is dropped into 1.0 kg of water at 10°C. Find (a) the final temperature and (b) the entropy change of the system.

59. A Carnot engine extracts heat from a block of mass m and specific heat c that is initially at temperature T_{h0} but that has no heat source to maintain that temperature. The engine rejects heat to a reservoir at a constant temperature T_c. The engine is operated so its mechanical power output is proportional to the temperature difference $T_h - T_c$:

$$P = P_0 \frac{T_h - T_c}{T_{h0} - T_c}$$

where T_h is the instantaneous temperature of the hot block and P_0 is the initial power output. (a) Find an expression for T_h as a function of time, and (b) determine how long it takes for the engine's power output to reach zero.

60. Find an expression for the entropy change of the whole system in Fig. 19.14, in terms of T_c and T_h, by evaluating the integrals in the paragraph under the heading on Irreversible Heat Transfer on page 319. Assume both glasses contain the same mass, m, of water, whose specific heat c is constant, and neglect the heat capacity of the glasses. Prove that your expression is positive.

61. Problem 77 of Chapter 16 provided an approximate expression for the specific heat of copper at low absolute temperatures; $c = 31(T/343 \text{ K})^3$ J/g·K. Use this to find the entropy change in a 40-g sample of copper which is cooled from 25 K to 10 K. Why is the change negative?

62. The molar specific heat at constant pressure for a certain gas is given by $c_p = a + bT + cT^2$, where $a = 33.6$ J/mol·K, $b = 2.93 \times 10^{-3}$ J/mol·K^2, and $c = 2.13 \times 10^{-5}$ J/mol·K^3. What is the change in entropy of 2 moles of this gas when heated from 20°C to 200°C?

63. You have 50 kg of steam at 100°C, but no heat source to maintain it in that condition. You also have a heat reservoir at 0°C. Suppose you operate a reversible heat engine with this system, so the steam gradually condenses and cools until it reaches 0°C. (a) Calculate the total entropy change of the steam and subsequent water. (b) Calculate the total entropy change of the reservoir. (c) Find the total amount of work that the engine can do. *Hint:* Consider the entropy change of the entire system. This change occurs as the result of a reversible process that supplies useful work. But if it had occurred irreversibly, how much energy would have become unavailable to do work?

64. A city council member contacts you, the public-affairs manager for a nuclear power plant, concerning the environmental effects of discharging hot water into the river that flows by the plant. The flow rate of the river is 3.4×10^4 kg/s. The thermal-energy input rate to the river is 100 MW. The council member says a 5°C increase in temperature is acceptable. Does your plant meet this standard?

65. A friend has a scheme to run an engine that will deposit its heat into a block of ice. The engine has a power output of 8.5 kW and extracts heat from a source at 420 K and rejects heat to a 1000-kg block of ice at its melting point. Your friend claims his gadget will run for an 8-hour workday. Is he right? *Hint:* Calculate the efficiency and then how long the ice will last without replenishment.

Answers to Chapter Questions

Answer to Chapter Opening Question
The second law of thermodynamics prevents us from converting thermal energy to mechanical energy with 100% efficiency, and practical limits on temperature make it hard to achieve 50% efficiency in conventional power plants.

Answers to GOT IT? Questions
19.1 (a), (c), and (f).
19.2 (c); see Problem 32 for a proof.
19.3 (a) increase; (b) decrease; (c) increase; (d) increase; (e) decrease; (f) increase; (g) increase.

Thermodynamics

Thermodynamics is the study of heat, temperature, and related phenomena—and their relation to the all-important concept of energy. Thermodynamics provides a macroscopic description in terms of parameters like temperature and pressure.

This contrasts with **statistical mechanics**, which provides a microscopic description in terms of the properties and behavior of molecules.

Thermodynamic equilibrium occurs when two systems are brought into thermal contact and no further changes occur in any macroscopic properties. The **zeroth law of thermodynamics** says that two systems each in thermodynamic equilibrium with a third are also in thermodynamic equilibrium with each other. This law allows us to establish temperature scales and construct thermometers.

Systems *A* and *C* are each in thermodynamic equilibrium with *B*.

If *A* and *C* are placed in thermal contact, their macroscopic properties don't change—showing that they're already in equilibrium.

(a)   **(b)**

Heat is energy that's flowing because of a temperature difference. Important heat-transfer mechanisms include **conduction, convection,** and **radiation**. A system is in **thermal-energy balance** at a fixed temperature when its energy input balances heat transfer to its surroundings.

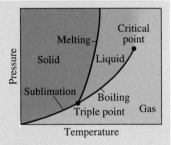

Incident sunlight

Outgoing infrared

Earth's energy balance

Ideal gases exhibit a simple relation among temperature, pressure, and volume:

$$pV = NkT = nRT$$

This is the **ideal gas law**, with $k = 1.38 \times 10^{-23}$ J/K and $R = 8.314$ J/K·mol

Real substances undergo **phase changes** among liquid, solid, and gaseous phases. Substantial **heats of transformation** describe the energies involved in phase changes.

Pressure

Melting — Critical point

Solid   Liquid

Sublimation   Boiling

Triple point   Gas

Temperature

The **first law of thermodynamics** relates the change ΔU in a system's internal energy to the heat Q added *to* the system and the work W done *by* the system:

$$\Delta U = Q - W$$

For an ideal gas, **reversible thermodynamic processes** are described by curves in the pressure–volume diagram. Common processes include **isothermal** (constant temperature), **constant volume, constant pressure,** and **adiabatic** (no heat flow).

Entropy is a measure of disorder. The **second law of thermodynamics** states that the entropy of a closed system can never decrease. Applied to the heat engines that provide most of humankind's electrical and transportation energy, the second law shows that it's impossible to extract as useful work all the random internal energy of hot objects.

Maximum efficiency (Carnot):

$$e = \frac{W}{Q_h} = 1 - \frac{Q_c}{Q_h} = 1 - \frac{T_c}{T_h}$$

Part Three challenge problem

The ideal Carnot engine shown in the figure operates between a heat reservoir and a block of ice with mass M. An external energy source maintains the reservoir at a constant temperature T_h. At time $t = 0$, the ice is at its melting point T_0, but it's insulated from everything except the engine, so it's free to change state and temperature. The engine is operated in such a way that it extracts heat from the reservoir at a constant rate P_h. (a) Find an expression for the time t_1 at which the ice is all melted, in terms of the quantities given and any other appropriate thermodynamic parameters. (b) Find an expression for the mechanical power output of the engine as a function of time for times $t > t_1$. (c) Your expression in part (b) holds only up to some maximum time t_2. Why? Find an expression for t_2.

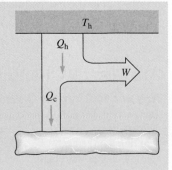

T_h

Q_h

W

Q_c

Mathematics

A-1 Algebra and Trigonometry

Quadratic Formula

If $ax^2 + bx + c = 0$, then $x = \dfrac{-b \pm \sqrt{b^2 - 4ac}}{2a}$.

Circumference, Area, Volume

Where $\pi \simeq 3.14159 \ldots$

circumference of circle	$2\pi r$
area of circle	πr^2
surface area of sphere	$4\pi r^2$
volume of sphere	$\frac{4}{3}\pi r^3$
area of triangle	$\frac{1}{2}bh$
volume of cylinder	$\pi r^2 l$

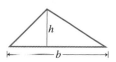

Trigonometry

definition of angle (in radians): $\theta = \dfrac{s}{r}$

2π radians in complete circle

1 radian $\simeq 57.3°$

Trigonometric Functions

$\sin\theta = \dfrac{y}{r}$

$\cos\theta = \dfrac{x}{r}$

$\tan\theta = \dfrac{\sin\theta}{\cos\theta} = \dfrac{y}{x}$

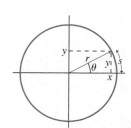

Values at Selected Angles

$\theta \rightarrow$	0	$\dfrac{\pi}{6}$ (30°)	$\dfrac{\pi}{4}$ (45°)	$\dfrac{\pi}{3}$ (60°)	$\dfrac{\pi}{2}$ (90°)
$\sin\theta$	0	$\dfrac{1}{2}$	$\dfrac{\sqrt{2}}{2}$	$\dfrac{\sqrt{3}}{2}$	1
$\cos\theta$	1	$\dfrac{\sqrt{3}}{2}$	$\dfrac{\sqrt{2}}{2}$	$\dfrac{1}{2}$	0
$\tan\theta$	0	$\dfrac{\sqrt{3}}{3}$	1	$\sqrt{3}$	∞

Graphs of Trigonometric Functions

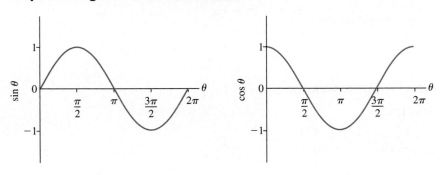

Trigonometric Identities

$$\sin(-\theta) = -\sin\theta$$

$$\cos(-\theta) = \cos\theta$$

$$\sin\left(\theta \pm \frac{\pi}{2}\right) = \pm\cos\theta$$

$$\cos\left(\theta \pm \frac{\pi}{2}\right) = \mp\sin\theta$$

$$\sin^2\theta + \cos^2\theta = 1$$

$$\sin 2\theta = 2\sin\theta\cos\theta$$

$$\cos 2\theta = \cos^2\theta - \sin^2\theta = 1 - 2\sin^2\theta = 2\cos^2\theta - 1$$

$$\sin(\alpha \pm \beta) = \sin\alpha\cos\beta \pm \cos\alpha\sin\beta$$

$$\cos(\alpha \pm \beta) = \cos\alpha\cos\beta \mp \sin\alpha\sin\beta$$

$$\sin\alpha \pm \sin\beta = 2\sin\left[\tfrac{1}{2}(\alpha \pm \beta)\right]\cos\left[\tfrac{1}{2}(\alpha \mp \beta)\right]$$

$$\cos\alpha + \cos\beta = 2\cos\left[\tfrac{1}{2}(\alpha + \beta)\right]\cos\left[\tfrac{1}{2}(\alpha - \beta)\right]$$

$$\cos\alpha - \cos\beta = -2\sin\left[\tfrac{1}{2}(\alpha + \beta)\right]\sin\left[\tfrac{1}{2}(\alpha - \beta)\right]$$

Laws of Cosines and Sines

Where A, B, C are the sides of an arbitrary triangle and α, β, γ the angles opposite those sides:

Law of cosines

$$C^2 = A^2 + B^2 - 2AB\cos\gamma$$

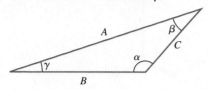

Law of sines

$$\frac{\sin\alpha}{A} = \frac{\sin\beta}{B} = \frac{\sin\gamma}{C}$$

Exponentials and Logarithms

$$e^{\ln x} = x, \quad \ln e^x = x \quad e = 2.71828\ldots$$

$$a^x = e^{x\ln a} \qquad \ln(xy) = \ln x + \ln y$$

$$a^x a^y = a^{x+y} \qquad \ln\left(\frac{x}{y}\right) = \ln x - \ln y$$

$$(a^x)^y = a^{xy} \qquad \ln\left(\frac{1}{x}\right) - -\ln x$$

$$\log x \equiv \log_{10} x = \ln(10)\ln x \simeq 2.3\ln x$$

Approximations

For $|x| \ll 1$, the following expressions provide good approximations to common functions:

$$e^x \simeq 1 + x$$

$$\sin x \simeq x$$

$$\cos x \simeq 1 - \tfrac{1}{2}x^2$$

$$\ln(1 + x) \simeq x$$

$$(1 + x)^p \simeq 1 + px \quad \text{(binomial approximation)}$$

Expressions that don't have the forms shown may often be put in the appropriate form. For example:

$$\frac{1}{\sqrt{a^2 + y^2}} = \frac{1}{a\sqrt{1 + \dfrac{y^2}{a^2}}} = \frac{1}{a}\left(1 + \frac{y^2}{a^2}\right)^{-1/2} \simeq \frac{1}{a}\left(1 - \frac{y^2}{2a^2}\right) \quad \text{for } y^2/a^2 \ll 1, \text{ or } y^2 \ll a^2.$$

Vector Algebra

Vector Products

$$\vec{A} \cdot \vec{B} = AB\cos\theta$$

$$|\vec{A} \times \vec{B}| = AB\sin\theta, \text{ with direction of } \vec{A} \times \vec{B} \text{ given by the right-hand rule:}$$

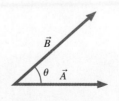
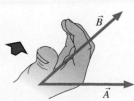

Unit Vector Notation

An arbitrary vector $\vec{A}$ may be written in terms of its components A_x, A_y, A_z and the unit vectors $\hat{\imath}$, $\hat{\jmath}$, $\hat{k}$ that have length 1 and lie along the x, y, z axes:

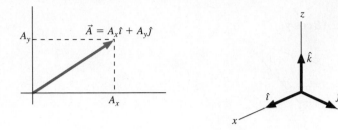

In unit vector notation, vector products become

$$\vec{A} \cdot \vec{B} = A_x B_x + A_y B_y + A_z B_z$$

$$\vec{A} \times \vec{B} = (A_y B_z - A_z B_y)\hat{\imath} + (A_z B_x - A_x B_z)\hat{\jmath} + (A_x B_y - A_y B_x)\hat{k}$$

Vector Identities

$$\vec{A} \cdot \vec{B} = \vec{B} \cdot \vec{A}$$

$$\vec{A} \times \vec{B} = -\vec{B} \times \vec{A}$$

$$\vec{A} \cdot (\vec{B} \times \vec{C}) = \vec{B} \cdot (\vec{C} \times \vec{A}) = \vec{C} \cdot (\vec{A} \times \vec{B})$$

$$\vec{A} \times (\vec{B} \times \vec{C}) = (\vec{A} \cdot \vec{C})\vec{B} - (\vec{A} \cdot \vec{B})\vec{C}$$

A-2 Calculus

Derivatives

Definition of the Derivative

If y is a function of x, then the **derivative of y with respect to x** is the ratio of the change Δy in y to the corresponding change Δx in x, in the limit of arbitrarily small Δx:

$$\frac{dy}{dx} = \lim_{\Delta x \to 0} \frac{\Delta y}{\Delta x}$$

Algebraically, the derivative is the rate of change of y with respect to x; geometrically, it is the slope of the y versus x graph—that is, of the tangent line to the graph at a given point:

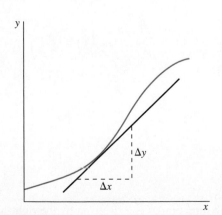

Derivatives of Common Functions

$$\frac{da}{dx} = 0 \quad (a \text{ is a constant})$$

$$\frac{d}{dx}\tan x = \frac{1}{\cos^2 x}$$

$$\frac{dx^n}{dx} = nx^{n-1} \quad (n \text{ need not be an integer})$$

$$\frac{de^x}{dx} = e^x$$

$$\frac{d}{dx}\sin x = \cos x$$

$$\frac{d}{dx}\ln x = \frac{1}{x}$$

$$\frac{d}{dx}\cos x = -\sin x$$

Derivatives of Sums, Products, and Functions of Functions

1. Derivative of a constant times a function

$$\frac{d}{dx}[af(x)] = a\frac{df}{dx} \quad (a \text{ is a constant})$$

2. Derivative of a sum

$$\frac{d}{dx}[f(x) + g(x)] = \frac{df}{dx} + \frac{dg}{dx}$$

3. Derivative of a product

$$\frac{d}{dx}[f(x)g(x)] = g\frac{df}{dx} + f\frac{dg}{dx}$$

Examples

$$\frac{d}{dx}(x^2\cos x) = \cos x\frac{dx^2}{dx} + x^2\frac{d}{dx}\cos x = 2x\cos x - x^2\sin x$$

$$\frac{d}{dx}(x\ln x) = \ln x\frac{dx}{dx} + x\frac{d}{dx}\ln x = (\ln x)(1) + x\left(\frac{1}{x}\right) = \ln x + 1$$

4. Derivative of a quotient

$$\frac{d}{dx}\left[\frac{f(x)}{g(x)}\right] = \frac{1}{g^2}\left(g\frac{df}{dx} - f\frac{dg}{dx}\right)$$

Example

$$\frac{d}{dx}\left(\frac{\sin x}{x^2}\right) = \frac{1}{x^4}\left(x^2\frac{d}{dx}\sin x - \sin x\frac{dx^2}{dx}\right) = \frac{\cos x}{x^2} - \frac{2\sin x}{x^3}$$

5. Chain rule for derivatives

If f is a function of u and u is a function of x, then

$$\frac{df}{dx} = \frac{df}{du}\frac{du}{dx}$$

Examples

a. Evaluate $\dfrac{d}{dx}\sin(x^2)$. Here $u = x^2$ and $f(u) = \sin u$, so

$$\frac{d}{dx}\sin(x^2) = \frac{d}{du}\sin u\frac{du}{dx} = (\cos u)\frac{dx^2}{dx} = 2x\cos(x^2)$$

b. $\dfrac{d}{dt}\sin\omega t = \dfrac{d}{d\,\omega t}\sin\omega t\dfrac{d}{dt}\omega t = \omega\cos\omega t \quad (\omega \text{ is a constant})$

c. Evaluate $\dfrac{d}{dx}\sin^2 5x$. Here $u = \sin 5x$ and $f(u) = u^2$, so

$$\frac{d}{dx}\sin^2 5x = \frac{d}{du}u^2\frac{du}{dx} = 2u\frac{du}{dx} = 2\sin 5x\frac{d}{dx}\sin 5x$$

$$= (2)(\sin 5x)(5)(\cos 5x) = 10\sin 5x\cos 5x = 5\sin 2x$$

Second Derivative

The second derivative of y with respect to x is defined as the derivative of the derivative:

$$\frac{d^2y}{dx^2} = \frac{d}{dx}\left(\frac{dy}{dx}\right)$$

Example

If $y = ax^3$, then $dy/dx = 3ax^2$, so

$$\frac{d^2y}{dx^2} = \frac{d}{dx}3ax^2 = 6ax$$

Partial Derivatives

When a function depends on more than one variable, then the partial derivatives of that function are the derivatives with respect to each variable, taken with all other variables held constant. If f is a function of x and y, then the partial derivatives are written

$$\frac{\partial f}{\partial x} \quad \text{and} \quad \frac{\partial f}{\partial y}$$

Example

If $f(x, y) = x^3\sin y$, then

$$\frac{\partial f}{\partial x} = 3x^2\sin y \quad \text{and} \quad \frac{\partial f}{\partial y} = x^3\cos y$$

Integrals

Indefinite Integrals

Integration is the inverse of differentiation. The **indefinite integral**, $\int f(x)\,dx$, is defined as a function whose derivative is $f(x)$:

$$\frac{d}{dx}\left[\int f(x)\,dx\right] = f(x)$$

If $A(x)$ is an indefinite integral of $f(x)$, then because the derivative of a constant is zero, the function $A(x) + C$ is also an indefinite integral of $f(x)$, where C is any constant. Inverting the derivatives of common functions listed in the preceding section gives the integrals that follow (a more extensive table appears at the end of this appendix).

$$\int a\,dx = ax + C \qquad\qquad \int \cos x\,dx = \sin x + C$$

$$\int x^n\,dx = \frac{x^{n+1}}{n+1} + C, \quad n \neq -1 \qquad \int e^x\,dx = e^x + C$$

$$\int \sin x\,dx = -\cos x + C \qquad\qquad \int x^{-1}\,dx = \ln x + C$$

Definite Integrals

In physics we're most often interested in the **definite integral**, defined as the sum of a large number of very small quantities, in the limit as the number of quantities grows arbitrarily large and the size of each arbitrarily small:

$$\int_{x_1}^{x_2} f(x)\, dx \equiv \lim_{\substack{\Delta x \to 0 \\ N \to \infty}} \sum_{i=1}^{N} f(x_i)\, \Delta x$$

where the terms in the sum are evaluated at values x_i between the limits of integration x_1 and x_2; in the limit $\Delta x \to 0$, the sum is over all values of x in the interval.

The key to evaluating the definite integral is provided by the **fundamental theorem of calculus**. The theorem states that, if $A(x)$ is an *indefinite* integral of $f(x)$, then the *definite integral* is given by

$$\int_{x_1}^{x_2} f(x)\, dx = A(x_2) - A(x_1) \equiv A(x) \Big|_{x_1}^{x_2}$$

Geometrically, the definite integral is the area under the graph of $f(x)$ between the limits x_1 and x_2.

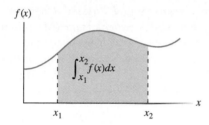

Evaluating Integrals

The first step in evaluating an integral is to express all varying quantities within the integral in terms of a single variable; Chapter 9 outlines a general strategy for setting up an integral. Once you've set up an integral, you can evaluate it yourself or look it up in tables. Two common techniques can help you evaluate integrals or convert them to forms listed in tables:

1. **Change of variables**

   An unfamiliar integral can often be put into familiar form by defining a new variable. For example, it is not obvious how to integrate the expression

   $$\int \frac{x\, dx}{\sqrt{a^2 + x^2}}$$

   where a is a constant. But let $z = a^2 + x^2$. Then

   $$\frac{dz}{dx} = \frac{da^2}{dx} + \frac{dx^2}{dx} = 0 + 2x = 2x$$

   so $dz = 2x\, dx$. Then the quantity $x\, dx$ in our unfamiliar integral is just $\frac{1}{2} dz$, while the quantity $\sqrt{a^2 + x^2}$ is just $z^{1/2}$. So the integral becomes

   $$\int \frac{1}{2} z^{-1/2}\, dz = \frac{\frac{1}{2} z^{1/2}}{1/2} = \sqrt{z}$$

   where we have used the standard form for the integral of a power of the independent variable. Substituting back $z = a^2 + x^2$ gives

   $$\int \frac{x\, dx}{\sqrt{a^2 + x^2}} = \sqrt{a^2 + x^2}$$

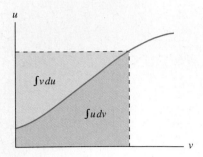

2. Integration by parts

The quantity $\int u\,dv$ is the area under the curve of u as a function of v between specified limits. In the figure, that area can also be expressed as the area of the rectangle shown minus the area under the curve of v as a function of u. Mathematically, this relation among areas may be expressed as a relation among integrals:

$$\int u\,dv = uv - \int v\,du \qquad (\text{integration by parts})$$

This expression may often be used to transform complicated integrals into simpler ones.

Example

Evaluate $\int x \cos x\,dx$. Here let $u = x$, so $du = dx$. Then $dv = \cos x\,dx$, so we have $v = \int dv = \int \cos x\,dx = \sin x$. Integrating by parts then gives

$$\int x \cos x\,dx = (x)(\sin x) - \int \sin x\,dx = x \sin x + \cos x$$

where the $+$ sign arises because $\int \sin x\,dx = -\cos x$.

Table of Integrals

More extensive tables are available in many mathematical and scientific handbooks; see, for example, **Handbook of Chemistry and Physics** (Chemical Rubber Co.) or Dwight, **Tables of Integrals and Other Mathematical Data** (Macmillan). Some math software, including *Mathematica* and *Maple*, can also evaluate integrals symbolically. Wolfram Research provides *Mathematica*-based integration at *http://integrals.wolfram.com*.

In the expressions below, a and b are constants. An arbitrary constant of integration may be added to the right-hand side.

$$\int e^{ax}\,dx = \frac{e^{ax}}{a}$$

$$\int \sin ax\,dx = -\frac{\cos ax}{a}$$

$$\int \cos ax\,dx = \frac{\sin ax}{a}$$

$$\int \tan ax\,dx = -\frac{1}{a}\ln(\cos ax)$$

$$\int \sin^2 ax\,dx = \frac{x}{2} - \frac{\sin 2ax}{4a}$$

$$\int \cos^2 ax\,dx = \frac{x}{2} + \frac{\sin 2ax}{4a}$$

$$\int x \sin ax\,dx = \frac{1}{a^2}\sin ax - \frac{1}{a}x \cos ax$$

$$\int x \cos ax\,dx = \frac{1}{a^2}\cos ax + \frac{1}{a}x \sin ax$$

$$\int \frac{dx}{\sqrt{a^2 - x^2}} = \sin^{-1}\left(\frac{x}{a}\right)$$

$$\int \frac{dx}{\sqrt{x^2 \pm a^2}} = \ln\left(x + \sqrt{x^2 \pm a^2}\right)$$

$$\int \frac{dx}{x^2 + a^2} = \frac{1}{a}\tan^{-1}\left(\frac{x}{a}\right)$$

$$\int \frac{x\,dx}{\sqrt{a^2 - x^2}} = -\sqrt{a^2 - x^2}$$

$$\int \frac{x\,dx}{\sqrt{x^2 \pm a^2}} = \sqrt{x^2 \pm a^2}$$

$$\int \frac{dx}{(x^2 \pm a^2)^{3/2}} = \frac{\pm x}{a^2\sqrt{x^2 \pm a^2}}$$

$$\int xe^{ax}\,dx = \frac{e^{ax}}{a^2}(ax - 1)$$

$$\int x^2 e^{ax}\,dx = \frac{x^2 e^{ax}}{a} - \frac{2}{a}\left[\frac{e^{ax}}{a^2}(ax - 1)\right]$$

$$\int \frac{dx}{a + bx} = \frac{1}{b}\ln(a + bx)$$

$$\int \frac{dx}{(a + bx)^2} = -\frac{1}{b(a + bx)}$$

$$\int \ln ax\,dx = x \ln ax - x$$

The International System of Units (SI)

This material is from the U.S. edition of the English translation of the seventh edition of "Le Système International d'Unités (SI)," the definitive publication in the French language issued in 1991 by the International Bureau of Weights and Measures (BIPM). The year the definition was adopted is given in parentheses.

length (meter): The meter is the length of the path traveled by light in vacuum during a time interval of 1/299 792 458 of a second. (1983)

mass (kilogram): The kilogram is equal to the mass of the international prototype of the kilogram. (1889)

time (second): The second is the duration of 9 192 631 770 periods of the radiation corresponding to the transition between the two hyperfine levels of the ground state of the cesium-133 atom. (1967)

electric current (ampere): The ampere is that constant current which, if maintained in two straight parallel conductors of infinite length, of negligible circular cross section, and placed 1 meter apart in vacuum, would produce between these conductors a force equal to 2×10^{-7} newton per meter of length. (1948)

temperature (kelvin): The kelvin, unit of thermodynamic temperature, is the fraction 1/273.16 of the thermodynamic temperature of the triple point of water. (1967)

amount of substance (mole): The mole is the amount of substance of a system that contains as many elementary entities as there are atoms in 0.012 kilogram of carbon-12. (1971)

luminous intensity (candela): The candela is the luminous intensity, in a given direction, of a source that emits monochromatic radiation of frequency 540×10^{12} hertz and that has a radiant intensity in that direction of $(1/683)$ watt per steradian. (1979)

SI Base and Supplementary Units

Quantity	SI Unit Name	SI Unit Symbol
Base Unit		
Length	meter	m
Mass	kilogram	kg
Time	second	s
Electric current	ampere	A
Thermodynamic temperature	kelvin	K
Amount of substance	mole	mol
Luminous intensity	candela	cd
Supplementary Units		
Plane angle	radian	rad
Solid angle	steradian	sr

SI Prefixes

Factor	Prefix	Symbol
10^{24}	yotta	Y
10^{21}	zetta	Z
10^{18}	exa	E
10^{15}	peta	P
10^{12}	tera	T
10^{9}	giga	G
10^{6}	mega	M
10^{3}	kilo	k
10^{2}	hecto	h
10^{1}	deka	da
10^{0}	—	—
10^{-1}	deci	d
10^{-2}	centi	c
10^{-3}	milli	m
10^{-6}	micro	μ
10^{-9}	nano	n
10^{-12}	pico	p
10^{-15}	femto	f
10^{-18}	atto	a
10^{-21}	zepto	z
10^{-24}	yocto	y

Some SI Derived Units with Special Names

Quantity	Name	Symbol	SI Unit Expression in Terms of Other Units	SI Unit Expression in Terms of SI Base Units
Frequency	hertz	Hz		s^{-1}
Force	newton	N		$m \cdot kg \cdot s^{-2}$
Pressure, stress	pascal	Pa	N/m^2	$m^{-1} \cdot kg \cdot s^{-2}$
Energy, work, heat	joule	J	$N \cdot m$	$m^2 \cdot kg \cdot s^{-2}$
Power	watt	W	J/s	$m^2 \cdot kg \cdot s^{-3}$
Electric charge	coulomb	C		$s \cdot A$
Electric potential, potential difference, electromotive force	volt	V	J/C	$m^2 \cdot kg \cdot s^{-3} \cdot A^{-1}$
Capacitance	farad	F	C/V	$m^{-2} \cdot kg^{-1} \cdot s^4 \cdot A^2$
Electric resistance	ohm	Ω	V/A	$m^2 \cdot kg \cdot s^{-3} \cdot A^{-2}$
Magnetic flux	weber	Wb	$V \cdot s$	$m^2 \cdot kg \cdot s^{-2} \cdot A^{-1}$
Magnetic field	tesla	T	Wb/m^2	$kg \cdot s^{-2} \cdot A^{-1}$
Inductance	henry	H	Wb/A	$m^2 \cdot kg \cdot s^{-2} \cdot A^{-2}$
Radioactivity	becquerel	Bq	1 decay/s	s^{-1}
Absorbed radiation dose	gray	Gy	J/kg, 100 rad	$m^2 \cdot s^{-2}$
Radiation dose equivalent	sievert	Sv	J/kg, 100 rem	$m^2 \cdot s^{-2}$

The listings below give the SI equivalents of non-SI units. To convert from the units shown to SI, multiply by the factor given; to convert the other way, divide. For conversions within the SI system, see the table of SI prefixes in Appendix B, Chapter 1, or the inside front cover. Conversions that are not exact by definition are given to, at most, four significant figures.

Length

1 inch (in) = 0.0254 m

1 foot (ft) = 0.3048 m

1 yard (yd) = 0.9144 m

1 mile (mi) = 1609.34 m

1 nautical mile = 1852 m

1 angstrom ($\mathring{A}$) = 10^{-10} m

1 light year (ly) = 9.46×10^{15} m

1 astronomical unit (AU) = 1.5×10^{11} m

1 parsec = 3.09×10^{16} m

1 fermi = 10^{-15} m = 1 fm

Mass

1 slug = 14.59 kg

1 metric ton (tonne; t) = 1000 kg

1 unified mass unit (u) = 1.661×10^{-27} kg

Force units in the English system are sometimes used (incorrectly) for mass. The units given below are actually equal to the number of kilograms multiplied by g, the acceleration of gravity.

1 pound (lb) = weight of 0.454 kg

1 ton = 2000 lb = weight of 908 kg

1 ounce (oz) = weight of 0.02835 kg

Time

1 minute (min) = 60 s

1 hour (h) = 60 min = 3600 s

1 day (d) = 24 h = 86 400 s

1 year (y) = 365.2422 d* = 3.156×10^7 s

Area

1 hectare (ha) = 10^4 m^2

1 square inch (in^2) = 6.452×10^{-4} m^2

1 square foot (ft^2) = 9.290×10^{-2} m^2

1 acre = 4047 m^2

1 barn = 10^{-28} m^2

1 shed = 10^{-30} m^2

Volume

1 liter (L) = 1000 cm^3 = 10^{-3} m^3

1 cubic foot (ft^3) = 2.832×10^{-2} m^3

1 cubic inch (in^3) = 1.639×10^{-5} m^3

1 fluid ounce = 1/128 gal = 2.957×10^{-5} m^3

1 barrel = 42 gal = 0.1590 m^3

1 gallon (U.S.; gal) = 3.785×10^{-3} m^3

1 gallon (British) = 4.546×10^{-3} m^3

Angle, Phase

1 degree (°) = $\pi/180$ rad = 1.745×10^{-2} rad

1 revolution (rev) = 360° = 2π rad

1 cycle = 360° = 2π rad

* The length of the year changes very slowly with changes in Earth's orbital period.

Speed, Velocity

$1 \text{ km/h} = (1/3.6) \text{ m/s} = 0.2778 \text{ m/s}$ $1 \text{ ft/s} = 0.3048 \text{ m/s}$

$1 \text{ mi/h (mph)} = 0.4470 \text{ m/s}$ $1 \text{ ly/y} = 3.00 \times 10^8 \text{ m/s}$

Angular Speed, Angular Velocity, Frequency, and Angular Frequency

$1 \text{ rev/s} = 2\pi \text{ rad/s} = 6.283 \text{ rad/s } (\text{s}^{-1})$ $1 \text{ rev/min (rpm)} = 0.1047 \text{ rad/s } (\text{s}^{-1})$

$1 \text{ Hz} = 1 \text{ cycle/s} = 2\pi \text{ s}^{-1}$

Force

$1 \text{ dyne} = 10^{-5} \text{ N}$ $1 \text{ pound (lb)} = 4.448 \text{ N}$

Pressure

$1 \text{ dyne/cm}^2 = 0.10 \text{ Pa}$ $1 \text{ lb/in}^2 \text{ (psi)} = 6.895 \times 10^3 \text{ Pa}$

$1 \text{ atmosphere (atm)} = 1.013 \times 10^5 \text{ Pa}$ $1 \text{ in } H_2O \text{ (60°F)} = 248.8 \text{ Pa}$

$1 \text{ torr} = 1 \text{ mm Hg at 0°C} = 133.3 \text{ Pa}$ $1 \text{ in Hg (60°F)} = 3.377 \times 10^3 \text{ Pa}$

$1 \text{ bar} = 10^5 \text{ Pa} = 0.987 \text{ atm}$

Energy, Work, Heat

$1 \text{ erg} = 10^{-7} \text{ J}$ $1 \text{ Btu}^* = 1.054 \times 10^3 \text{ J}$

$1 \text{ calorie}^* \text{ (cal)} = 4.184 \text{ J}$ $1 \text{ kWh} = 3.6 \times 10^6 \text{ J}$

$1 \text{ electronvolt (eV)} = 1.602 \times 10^{-19} \text{ J}$ $1 \text{ megaton (explosive yield; Mt)}$

$1 \text{ foot-pound (ft} \cdot \text{lb)} = 1.356 \text{ J}$ $= 4.18 \times 10^{15} \text{ J}$

Power

$1 \text{ erg/s} = 10^{-7} \text{ W}$ $1 \text{ Btu/h (Btuh)} = 0.293 \text{ W}$

$1 \text{ horsepower (hp)} = 746 \text{ W}$ $1 \text{ ft} \cdot \text{lb/s} = 1.356 \text{ W}$

Magnetic Field

$1 \text{ gauss (G)} = 10^{-4} \text{ T}$ $1 \text{ gamma } (\gamma) = 10^{-9} \text{ T}$

Radiation

$1 \text{ curie (ci)} = 3.7 \times 10^{10} \text{ Bq}$ $1 \text{ rad} = 10^{-2} \text{ Gy}$

     $1 \text{ rem} = 10^{-2} \text{ Sv}$

Energy Content of Fuels

Energy Source	Energy Content
Coal	$29 \text{ MJ/kg} = 7300 \text{ kWh/ton} = 25 \times 10^6 \text{ Btu/ton}$
Oil	$43 \text{ MJ/kg} = 39 \text{ kWh/gal} = 1.3 \times 10^5 \text{ Btu/gal}$
Gasoline	$44 \text{ MJ/kg} = 36 \text{ kWh/gal} = 1.2 \times 10^5 \text{ Btu/gal}$
Natural gas	$55 \text{ MJ/kg} = 30 \text{ kWh/100 ft}^3 = 1000 \text{ Btu/ft}^3$
Uranium (fission)	
Normal abundance	$5.8 \times 10^{11} \text{ J/kg} = 1.6 \times 10^5 \text{ kWh/kg}$
Pure U-235	$8.2 \times 10^{13} \text{ J/kg} = 2.3 \times 10^7 \text{ kWh/kg}$
Hydrogen (fusion)	
Normal abundance	$7 \times 10^{11} \text{ J/kg} = 3.0 \times 10^4 \text{ kWh/kg}$
Pure deuterium	$3.3 \times 10^{14} \text{ J/kg} = 9.2 \times 10^7 \text{ kWh/kg}$
Water	$1.2 \times 10^{10} \text{ J/kg} = 1.3 \times 10^4 \text{ kWh/gal} = 340 \text{ gal gasoline/gal}$
H_2O	
100% conversion, matter to energy	$9.0 \times 10^{16} \text{ J/kg} = 931 \text{ MeV/u} = 2.5 \times 10^{10} \text{ kWh/kg}$

* Values based on the thermochemical calorie; other definitions vary slightly.

The Elements

The atomic weights of stable elements reflect the abundances of different isotopes; values given here apply to elements as they exist naturally on Earth. For stable elements, parentheses express uncertainties in the last decimal place given. For elements with no stable isotopes (indicated in **boldface**), at most three isotopes are given; for elements 99 and beyond, only the longest-lived isotope is given. (Exceptions are the unstable elements thorium, protactinium, and uranium, for which atomic weights reflect natural abundances of long-lived isotopes.) See also the periodic table inside the back cover.

Atomic Number	Names	Symbol	Atomic Weight
1	Hydrogen	H	1.00794 (7)
2	Helium	He	4.002602 (2)
3	Lithium	Li	6.941 (2)
4	Beryllium	Be	9.012182 (3)
5	Boron	B	10.811 (5)
6	Carbon	C	12.011 (1)
7	Nitrogen	N	14.00674 (7)
8	Oxygen	O	15.9994 (3)
9	Fluorine	F	18.9984032 (9)
10	Neon	Ne	20.1797 (6)
11	Sodium (Natrium)	Na	22.989768 (6)
12	Magnesium	Mg	24.3050 (6)
13	Aluminum	Al	26.981539 (5)
14	Silicon	Si	28.0855 (3)
15	Phosphorus	P	30.973762 (4)
16	Sulfur	S	32.066 (6)
17	Chlorine	Cl	35.4527 (9)
18	Argon	Ar	39.948 (1)
19	Potassium (Kalium)	K	39.0983 (1)
20	Calcium	Ca	40.078 (4)
21	Scandium	Sc	44.955910 (9)
22	Titanium	Ti	47.88 (3)
23	Vanadium	V	50.9415 (1)
24	Chromium	Cr	51.9961 (6)
25	Manganese	Mn	54.93805 (1)
26	Iron	Fe	55.847 (3)
27	Cobalt	Co	58.93320 (1)
28	Nickel	Ni	58.69 (1)
29	Copper	Cu	63.546 (3)
30	Zinc	Zn	65.39 (2)
31	Gallium	Ga	69.723 (1)
32	Germanium	Ge	72.61 (2)
33	Arsenic	As	74.92159 (2)

cont'd.

Atomic Number	Names	Symbol	Atomic Weight
34	Selenium	Se	78.96 (3)
35	Bromine	Br	79.904 (1)
36	Krypton	Kr	83.80 (1)
37	Rubidium	Rb	85.4678 (3)
38	Strontium	Sr	87.62 (1)
39	Yttrium	Y	88.90585 (2)
40	Zirconium	Zr	91.224 (2)
41	Niobium	Nb	92.90638 (2)
42	Molybdenum	Mo	95.94 (1)
43	**Technetium**	**Tc**	**97, 98, 99**
44	Ruthenium	Ru	101.07 (2)
45	Rhodium	Rh	102.90550 (3)
46	Palladium	Pd	106.42 (1)
47	Silver	Ag	107.8682 (2)
48	Cadmium	Cd	112.411 (8)
49	Indium	In	114.82 (1)
50	Tin	Sn	118.710 (7)
51	Antimony (Stibium)	Sb	121.75 (3)
52	Tellurium	Te	127.60 (3)
53	Iodine	I	126.90447 (3)
54	Xenon	Xe	131.29 (2)
55	Cesium	Cs	132.90543 (5)
56	Barium	Ba	137.327 (7)
57	Lanthanum	La	138.9055 (2)
58	Cerium	Ce	140.115 (4)
59	Praseodymium	Pr	140.90765 (3)
60	Neodymium	Nd	144.24 (3)
61	**Promethium**	**Pm**	**145, 147**
62	Samarium	Sm	150.36 (3)
63	Europium	Eu	151.965 (9)
64	Gadolinium	Gd	157.25 (3)
65	Terbium	Tb	158.92534 (3)
66	Dysprosium	Dy	162.50 (3)
67	Holmium	Ho	164.93032 (3)
68	Erbium	Er	167.26 (3)
69	Thulium	Tm	168.93421 (3)
70	Ytterbium	Yb	173.04 (3)
71	Lutetium	Lu	174.967 (1)
72	Hafnium	Hf	178.49 (2)
73	Tantalum	Ta	180.9479 (1)
74	Tungsten (Wolfram)	W	183.85 (3)
75	Rhenium	Re	186.207 (1)
76	Osmium	Os	190.2 (1)
77	Iridium	Ir	192.22 (3)
78	Platinum	Pt	195.08 (3)
79	Gold	Au	196.96654 (3)
80	Mercury	Hg	200.59 (3)
81	Thallium	Tl	204.3833 (2)
82	Lead	Pb	207.2 (1)
83	Bismuth	Bi	208.98037 (3)

cont'd.

Atomic Number	Names	Symbol	Atomic Weight
84	Polonium	Po	209, 210
85	Astatine	At	210, 211
86	Radon	Rn	211, 220, 222
87	Francium	Fr	223
88	Radium	Ra	223, 224, 226
89	Actinium	Ac	227
90	Thorium	Th	232.0381 (1)
91	Protactinium	Pa	231.03588 (2)
92	Uranium	U	238.0289 (1)
93	Neptunium	Np	237, 239
94	Plutonium	Pu	239, 242, 244
95	Americium	Am	241, 243
96	Curium	Cm	245, 247, 248
97	Berkelium	Bk	247, 249
98	Californium	Cf	249, 250, 251
99	Einsteinium	Es	252
100	Fermium	Fm	257
101	Mendelevium	Md	258
102	Nobelium	No	259
103	Lawrencium	Lr	262
104	Rutherfordium	Rf	263
105	Dubnium	Db	268
106	Seaborgium	Sg	266
107	Bohrium	Bh	272
108	Hassium	Hs	277
109	Meitnerium	Mt	276
110	Darmstadtium	Ds	281
111	Roentgenium	Rg	280
112	—	—	285
113	—	—	284
114	—	—	289
115	—	—	288
116	—	—	292

Astrophysical Data

Sun, Planets, Principal Satellites

Body	Mass (10^{24} kg)	Mean Radius (10^6 m Except as Noted)	Surface Gravity (m/s²)	Escape Speed (km/s)	Sidereal Rotation Period[*] (days)	Mean Distance from Central Body[†] (10^6 km)	Orbital Period	Orbital Speed (km/s)
Sun	1.99×10^6	696	274	618	36 at poles 27 at equator	2.6×10^{11}	200 My	250
Mercury	0.330	2.44	3.70	4.25	58.6	57.6	88.0 d	48
Venus	4.87	6.05	8.87	10.4	−243	108	225 d	35
Earth	5.97	6.37	9.81	11.2	0.997	150	365.3 d	30
Moon	0.0735	1.74	1.62	2.38	27.3	0.385	27.3 d	1.0
Mars	0.642	3.38	3.74	5.03	1.03	228	1.88 y	24.1
Phobos	9.6×10^{-9}	9–13 km	0.001	0.008	0.32	9.4×10^{-3}	0.32 d	2.1
Deimos	2×10^{-9}	5–8 km	0.001	0.005	1.3	23×10^{-3}	1.3 d	1.3
Jupiter	1.90×10^3	69.1	26.5	60.6	0.414	778	11.9 y	13.0
Io	0.0888	1.82	1.8	2.6	1.77	0.422	1.77 d	17
Europa	0.479	1.57	1.3	2.0	3.55	0.671	3.55 d	14
Ganymede	0.148	2.63	1.4	2.7	7.15	1.07	7.15 d	11
Callisto	0.107	2.40	1.2	2.4	16.7	1.88	16.7 d	8.2
and 13 smaller satellites								
Saturn	569	56.8	11.8	36.6	0.438	1.43×10^3	29.5 y	9.65
Tethys	0.0007	0.53	0.2	0.4	1.89	0.294	1.89 d	11.3
Dione	0.00015	0.56	0.3	0.6	2.74	0.377	2.74 d	10.0
Rhea	0.0025	0.77	0.3	0.5	4.52	0.527	4.52 d	8.5
Titan	0.135	2.58	1.4	2.6	15.9	1.22	15.9 d	5.6
and 12 smaller satellites								
Uranus	86.6	25.0	9.23	21.5	−0.65	2.87×10^3	84.1 y	6.79
Ariel	0.0013	0.58	0.3	0.4	2.52	0.19	2.52 d	5.5
Umbriel	0.0013	0.59	0.3	0.4	4.14	0.27	4.14 d	4.7
Titania	0.0018	0.81	0.2	0.5	8.70	0.44	8.70 d	3.7
Oberon	0.0017	0.78	0.2	0.5	13.5	0.58	13.5 d	3.1
and 11 smaller satellites								
Neptune	103	24.0	11.9	23.9	0.768	4.50×10^3	165 y	5.43
Triton	0.134	1.9	2.5	3.1	5.88	0.354	5.88 d	4.4
and 7 smaller satellites								
Pluto	0.015	1.2	0.4	1.2	−6.39	5.91×10^3	249 y	4.7
Charon	0.001	0.6			−6.39	0.02	6.39 d	0.2
and 2 smaller satellites								

[*]Negative rotation period indicates retrograde motion, in opposite sense from orbital motion. Periods are sidereal, meaning the time for the body to return to the same orientation relative to the distant stars rather than the Sun.

[†]Central body is galactic center for Sun, Sun for planets, and planet for satellites.

Answers to Odd-Numbered Exercises and Problems

Chapter 1
11. 100,000 bigger
13. 0.108 782 775 7 ns
15. 10^8
17. 0.62 rad
19. 28 g
21. 10^6
23. 8.59 m²/L
25. 3.6 km/h
27. 57.3°
29. 7.4×10^6 m/s²
31. 4×10^6
33. 41 m
35. ≈5.20
37. 2×10^6 cows
39. About 0.1%
41. 9×10^{26}
43. 0.025 cm
45. 7500
47. (a) ≈±5%   (b) ±1%   (c) ±0.5%
49. 42.195 km
51. 30 tuners in Chicago
53. $8.12/lb., $10.06/bag including shipping
55. About ¼ kg

Chapter 2
13. 10.16 m/s
15. (a) 24 km north   (b) 9.6 km/h north
    (c) .16 km/h south (d) 0   (e) 0
17. 7.84 m/s
19. 1 m/s = 2.24 mi/h
21. (a) 2.0 m/s   (b) 0   (c) −5.0 m/s
    (d) 1.2 m/s   (e) 0.17 m/s
23. (a) $b - 2ct$ (b) 8.4 s after launch
25. 0.354 m/s²
27. While falling, −9.82 m/s². While
    stopping, +84.0 m/s².
29. 17.4 m/s²
31. $v = \dfrac{dx}{dt} = \dfrac{d}{dt}\left(x_0 + v_0 t + \dfrac{1}{2}at^2\right) = v_0 + at$
33. (a) 46 m/s²   (b) 61 s
35. −26.9 ft/s²
37. 12 km/s
39. 94.9 m
41. (a) 123 m   (b) 39.2 m/s, 44.1 m
    (c) 9.8 m/s, 118 m   (d) 19.6 m/s, 103 m
43. 10.5 m/s
45. 48 mi/h
47. (a) 84 km/h   (b) 55 h
49. 2.6 h later; 1800 km from New York
51. $\bar{v} = bt^3$
53. (a) 28.1 m/s   (b) 22.5 m/s²
    (c) 9.38 m/s   (d) 11.25 m/s²

55. 55%
57. (a) 0.014 s   (b) 0.51 m
59. 886 m
61. (a) 24.8 m/s   (b) 179 m
63. 4.6×10^{-3} m/s²
65. 10.59 m/s
67. 273 m
69. (a) Diver A: −7.88 m/s; Diver B:
    −7.67 m/s   (b) Diver A hits first about
    0.162 s before Diver B.
71. 3.85 s; 6.24 m/s
73. 5.0 s; 17 km/h.
75. (a) $\bar{v} = L/t = \dfrac{1}{2}(v_1 + v_2)$

    (b) $\bar{v} = \dfrac{L}{t} = \dfrac{2v_1 v_2}{v_1 + v_2}$
77. 70.7%
79. −0.309 m/s²
81. (a) $v(t) = v_0 + a_0 t + \dfrac{1}{2}bt^2$

    (b) $x(t) = x_0 + v_0 t + \dfrac{1}{2}a_0 t^2 + \dfrac{1}{6}bt^3$
83. (a) 9.1 m/s   (b) 9.1 m/s   (c) 0.18 s
85. 15 drops/s
87. Walking ≈933 hours, or 39 days;
    bicycling ≈180 hours, or 7.5 days;
    car ≈51 hours, or 2.1 days

Chapter 3
17. 266m, 34° N of W
19. 702 km, 21.3° W of N
21. $C = 5.0$ m; 127° clockwise from $\vec{A}$
23. 36.4 m; 20.9°
25. $\vec{a} = 3.0\hat{i} + 15\hat{j}$ m/s²
27. (a) 5.4 mi at 32° E of N
    (b) 15 mi/h at 32° E of N
29. Southwest
31. 0°
33. $(259$ km/h$)\hat{i} + (65.9$ km/h$)\hat{j}$
35. 196 km/h
37. 6.0 m/s² at 53° below x axis
39. 0.728 s for both
41. 1.54 m
43. 3.44×10^6 m/s
45. 60.7 mi/h
47. 12 h
49. The two possible solutions are $\theta_B = 117°$,
    or 243°.
51. 0.869 μm/s
53. 5.70 m/s²
55. (a) 0.32 cm/s   (b) 3.36×10^{-2} cm/s²
    (c) 90°

57. They are equal.
59. 0.5 m/s²
61. 5.7 m/s
63. 8.32 m/s; 61°
65. 30° = 5.10 s; 60° = 8.84 s
69. 70.4 km/h; the motorcyclist was probably
    speeding.
71. 31.2° or 65.7°
73. 89 m/s
75. 90 km
77. 7.2 m/s at 77° to horizontal
79. 38° below the x axis
81. $\sqrt{2gh + gR^2/2h}$
83. The rocket will produce a change in
    velocity corresponding to a 24.2° change
    in angle. However, the displacement is
    only 1008 km and is not enough to avoid
    collision, unless the asteroid "coasts" for
    additional time after the rocket ceases
    firing.
85. Including air resistance, the optimum
    angle is less than 45 degrees. The density
    of air is less and thus the air resistance is
    less, when the atmospheric pressure is
    lower (higher altitude, as in Denver) or
    the temperature is higher.

Chapter 4
13. 3.75 MN
15. 1.53×10^3 kg
17. Quadruples
19. 30.7 kN
21. 7.72 cm
23. Venus
25. (a) 3.33 N   (b) 12.0 oz
27. 620 N
29. −1.76 kN
31. (a) 4.41 MN   (b) 4.91 MN
33. Apparent weight is 55% of actual
    weight
35. 1.3×10^{-23} m
37. 15.9 cm
39. 5.77N at 72.2° to the x axis
41. (a) 725 N   (b) 725 N   (c) 725 N
    (d) 851 N   (e) 599 N
43. 0.531 s
45. 6 N
47. (a) 5.26 kN   (b) 1.08 kN   (c) 494 N
    (d) 589 N
49. 830 g
51. 4.29 cm

53. (a) 42.1 kN   (b) 55.9 kN   (c) 55.9 kN
   (d) 42.1 kN   (e) 28.4 kN
55. $n \le 9$
59. 900 N
61. 14.9 kN
63. 0.30 MN
65. (a) $a = 1.0$ m/s^2 upward
   (b) 756 N
67. $F_{\text{net}} = \dfrac{ma}{\left(1 - u^2/c^2\right)^{3/2}}$
69. 20.5 g

Chapter 5
13. $(4.02\ \text{N})\hat{\imath} + (1.66\ \text{N})\hat{\jmath}$
15. 9.40°
17. 880 N
19. Right-hand mass 2.5 times left-hand mass
21. (a) 3.86 m/s^2 (b) 533 N
25. 132 m
27. 493 km/h
29. 0.18
31. 0.12
33. 43 cm
35. 98 N in horizontal string, 139 N in the
   other string
37. (a) $T = m_2 g$ (b) $\tau = 2\pi\sqrt{\dfrac{m_1 R}{m_2 g}}$
39. (a) Horizontal: 357 N; vertical: 441 N
   (b) 39°
41. 2.89 m/s
43. 342 N
45. 11.4 m/s^2
47. 2.68 cm
49. 107 m
51. 59.3 mi/h
55. 0.46
59. (a) lower at equator, where a net
   downward force is needed to keep you
   accelerating with Earth's rotation
   (b) 0.34%
63. (a) 0.6 mg (b) 37°
65. $v_y = -(mg/b)\left(1 - e^{-bt/m}\right)$
67. $T_0 = T_s \cos\theta_s,\ mg = 2T_s \sin\theta_s,$
   $d^2y/dx^2 = (m/LT_0)\left(1 + (dy/dx)^2\right)^{1/2}$
69. 11.5 m/s, considerably slower than
   a bullet
71. 7.64 km

Chapter 6
13. 494 J
15. 9.6 MJ
17. 1.1 mgh
19. (a) 86.9 u^2 (b) 20.3 u^2
21. 1.9 m
23. (a) 1.0 J   (b) 3.0 J
25. 29.9 cm
27. 3.7×10^{-13} J
29. 2.25 kJ
31. (a) 23.5 J   (b) 17.7 m/s
33. 97 W
35. (a) 60 kW   (b) 1 kW   (c) 42 W
37. 9.4 MJ
39. 430 W

41. 22 s
43. (a) 400 J   (b) 31 kg
45. 4.4 MN
47. (a) all three are equal to 1
   (b) all three are equal to 0
   (c) $\vec{A} \cdot \vec{B} = A_x B_x + A_y B_y + A_z B_z$
49. 65.8°, 49.6°, 115°
51. (a) 360 J   (b) 350 J   (c) 357.5 J
   (d) 359.375 J
53. (a) 33 J   (b) 60 J   (c) 78 J
57. 22.9 m
59. 220 J
61. 0.634 kg/s, 10.1 gal/min
63. 2.1 MJ
65. 7.7°
67. (a) 450 W   (b) 8.0 kJ
69. 42 kJ
   $\dfrac{1}{3}(18\text{W/s}^2)\left[(20\text{s})^3 - (10\text{s})^3\right] = 42$ kJ
75. (a) 33 J   (b) 167 J
77. 135 J
79. 8.7 MJ, not efficient
81. The floor exerts an average force on the
   leg of 2.82 kN, or 36 times its weight.

Chapter 7
11. (a) $-2\ \mu mgL$,   (b) $-\sqrt{2}\ \mu mgL$
13. (a) 60.4 kJ   (b) 109 kJ   (c) $U_1 = 0$
15. -2.24 MJ
17. 842 m
19. 55 cm
21. 50 m/s (180 km/h)
23. 84 cm
25. (a) 4.85 m/s   (b) 7.00 m/s   (c) 11 m
27. (a) -2 N   (b) 0   (c) 8 N   (d) 1 N
   (e) -4 N   (f) 0
29. (a) 4.2×10^{13} J   (b) 11 h
31. $U = mgy$
33. (a) 1.07 J   (b) 1.12 J
35. 778 J, 4.9%
37. (a) $U(x_2) - U(x_1) = A\left(\dfrac{1}{x_2} - \dfrac{1}{x_1}\right)$
39. 95.3 m
41. 4.73 m/s
43. (a) 16 m/s   (b) 29%
45. 12.9 km/h
47. 1.59 m
49. 4.22 m and 0.488 m
51. 2.82 Å
53. 20.1 m/s, 30.4 m/s
55. 1.39 m
57. 62.5 cm from left end of frictional zone
59. 2.87 m
61. 14 m
63. $v = \sqrt{2K/m} = \sqrt{2(U_0 - U/m)}$
   $= 2(a/3m)^{1/2} x^{3/4}$
65. (a) Yes, 1.60 J   (b) No, 80.9 cm
67. 2.14×10^{-30} J $\cdot$ m

69. $P = \dfrac{W}{t} = \dfrac{\dfrac{1}{2}mv^2 + mgd}{2\dfrac{d}{\sqrt{2g(h-d)}}}$
   $= \dfrac{\dfrac{1}{2}m(v^2 + 2gd)\sqrt{2g(h-d)}}{2d}$
   $= \dfrac{m(v^2 + 2gd)\sqrt{2g(h-d)}}{4d}$

Chapter 8
13. 2.73×10^{-3} m/s^2; acceleration due to
   Earth's gravity at 3.85×10^8 m
15. (a) 3.70 m/s^2 (b) 1.35 m/s^2
17. 46 nN
19. ≈ 1690 km
21. 1.87 y
23. 109 min
25. 1.77×10^{32} J
27. 1670 km
29. 58 MJ
31. 7.7 Mm
33. 8.85×10^5 m
35. $0.414 R_p$
39. (a) 4.56×10^6 m/s, or about 1.5% the
   speed of light   (b) 8.77 s
41. $(T_A/T_B) = (r_A/r_B)^{3/2} = 2.83$
43. 616 km/s
45. $R_E/99 = 64$ km; underestimate
47. $\sqrt{2}$
49. (a) 11.2 km/s   (b) 9.74 km/s
   (c) no
51. $v = \sqrt{2GM\left(\dfrac{3}{R} + \dfrac{1}{r}\right)}$
53. 15 km/s, 23 km/s
55. 7.96 km/s
57. (a) 0.601 AU   (b) 5.3×10^{11} J
   (c) 38.4 km/s
63. The ratio is about 0.46 for the tide-
   producing force, but about 180 for the
   gravitational force.
65. $R < 2.9$ km
67. The distance from the center of the
   Earth 3.47×10^8 m. The distance
   from the surface of the Earth is
   3.40×10^8 m

Chapter 9
13. 2m
15. $\sqrt{3}\,L/6$ along any altitude
17. 2.1 m from the stern
19. $-(0.6$ m/s$)\hat{\imath} - (0.8$ m/s$)\hat{\jmath}$
21. 8.4 km/h
23. 1.23 J
25. $K_{\text{int}} = 3K_0,\ K_{\text{cm}} = K_0$
29. 3875 kg
31. 46 m/s
33. $v_{1f} = -11$ Mm/s; $v_{2f} = 6.9$ Mm/s
35. 0.115a above the vertex of the missing
   trangle
37. 72.2 cm
39. A distance $1/4h$ from the base along the
   axis of symmetry

41. (a) 99 cm   (b) 3.9 m/s

43. (a) $a_{x0} = \dfrac{v_0}{M}\left(\dfrac{dm}{dt}\right)$ (b) $v_c = v_0$

45. $-17.4°$

47. $-(46.7 \text{ m/s})\hat{\imath} + (67.7 \text{ m/s})\hat{\jmath}$

49. If the two masses are equal, 60°.

51. 3.0 mN

53. 35.2 m/s

55. 91.5 cm/s

59. 120°

61. $m_1 = \left(3 \pm \sqrt{8}\right)m_2 = 5.83 m_2$
    or $(5.83)^{-1}m_2$

63. Start with Equation 9.14 and solve for
    v_{1f}: $v_{1f} = v_{2f} - v_{1i} + v_{2i}$. Plug this into
    Equation 9.12a: $m_1 v_{1i} + m_2 v_{2i} = m_1(v_{2f} - v_{1i} + v_{2i}) + m_2 v_{2f}$. Solve for v_{2f}.

65. 18.6%

67. $v = 658$ m/s, $\theta_x = 11.5°$

71. 7.95 s

73. $\dfrac{K_{2f}}{K_{1i}} = \dfrac{4(m_2/m_1)}{(1 + m_2/m_1)^2}$

75. Along the bisector, a distance $\dfrac{4R}{3\theta}\sin\left(\dfrac{\theta}{2}\right)$
    from the center

77. $v_{\text{PrimitiveEarth}} = 32$ km/s, $v_{\text{Earth}} = 30$ km/s, so
    about an 8% change

79. $v_c = 38.7$ mi/hr, $v_t = 52.3$ mi/hr. Both are
    speeding.

81. 1.83×10^{13} J

Chapter 10

15. (a) 463 m/s   (b) $(463 \text{ m/s})\cos\theta$

17. 45.8 m/s

19. (a) 12.1 min   (b) 2.17×10^4 rev

21. 160 N·m

23. (a) 146 N   (b) 155 N

25. 0.072 N·m

27. 6.14×10^2 kg·m²

29. (a) 36.9 kg   (b) If not all the mass
    of the wheel is concentrated at the rim,
    the total mass is greater than this
    minimum.

31. (a) 9.69×10^{37} kg·m²
    (b) 2.59×10^{19} N·m

33. (a) 1.17×10^{-3} kg·m²   (b) 1.24 N·m

35. (a) 446 J   (b) 139 W

37. $\approx 0.089\%$

39. (a) 160 MJ   (b) 16 MW

41. 1/3

43. (a) 65.7 rpm   (b) 3.65 s

45. (a) 21.7 s⁻¹, 207 rpm
    (b) 34.7 s⁻¹, 331 rpm

47. -1.30 s⁻²

49. 0.147 N·m

53. (a) 15.6 kg·m²   (b) 0.303 s

55. (a) 7.21 h   (b) $\cong 1950$ rev

57. 0.36

59. 2.09 s⁻¹

61. 7.03 m/s

63. 17.1%

65. $0.498\, MR^2$

67. 32.9 m

69. (a) $M = \frac{2}{3}\rho_0 \pi R^2 w$;   (b) $I = \frac{3}{5}MR^2$

71. The object is in the x-y plane.
    Add these,

$$I_y = \int x^2\, dm \qquad I_x = \int y^2\, dm \qquad x^2 + y^2 = z^2$$

But $I_x + I_y = \int (x^2 + y^2)\, dm$

$$I_x + I_y = \int z^2\, dm = I_z$$

73. $\frac{3}{10}MR^2$

75. 184 N

77. $F > 10.6 \times 10^3$ N

Chapter 11

13. 3.55 rad/s² in the southeast direction

15. 18.8° west of north

17. (a) $-\hat{k}$ (b) $\hat{k}$ (c) $-\hat{\jmath} + \hat{\imath}$

19. (a) $8.1\hat{k}$ N·m   (b) $14.7\hat{k}$ N·m

21. 4.14 J·s

23. 2.31 J·s

25. 17.4 rpm

27. 2.5 d

29. 9.2 rpm

31. $-120°$ along the x axis

35. 36.9 J·s

37. 2.70×10^5 J·s, direction out of the
    plane of Fig. 11.13

39. 0.206 kg·m²

41. 0.630, or 63% of a full circle relative to
    the ground

43. (a) 0.537 s⁻¹   (b) 6.44 m/s   (c) 207 N

45. 3.14 rpm

47. 21.9 g

49. Sun's rotation 2.8%; Jupiter's orbital
    motion 60%.

51. (a) 142 rpm   (b) 26.5% lost

53. (a) $\omega = \dfrac{2\omega_0}{7}$ (b) $t = \dfrac{2\omega_0 R}{7\mu g}$

55. (a) $\tau = \dfrac{MgL\cos\theta}{2}$ (b) 48.2°

59. No, the acceleration is 0.6 rad/s²

61. 0.93 kg m²/s

Chapter 12

17. (a) $0 = \Sigma F_x = -F_1 + F_2 \sin\phi + F_3$
    $0 = \Sigma F_y = -F_2 \cos\phi + F_4$
    $0 = (\Sigma\tau_z)_O$
    $= -L_2 F_2 - L_1 F_3 \sin\phi + L_1 F_4 \cos\phi$
    (b) $0 = (\Sigma\tau_z)_P = -L_1 F_1 \sin\phi + (L_1 - L_2)F_2$

19. (a) $\tau_A = \dfrac{1}{2}mgL$
    (b) $\tau_B = 0$
    (c) $\tau_C = \dfrac{1}{4}mgL$

21. 0.20 m from the left end

23. 0.718 m on the opposite side of C from
    the worker

25. (a) $x = 47$ m   (b) unstable

27. $n_L = -1.17 \times 10^4$ N, which is downward
    and must be exerted by the bolt

29. 0.866

31. (a) 1298 N   (b) 1941 N

33. 50.2 kN

35. 6.05 kN

37. 1.17 m

39. 87 kg

41. 63.4°, unstable

43. The block slides before tipping.

45. 74.3 kg

47. $0.366\, mgs$

49. $m \leq m_L\dfrac{2\mu_s - \tan\theta}{2(\tan\theta - \mu_s)}$

51. $F_{app} = Mg\dfrac{\sin\theta}{1 + \cos\theta} = Mg\tan\left(\dfrac{\theta}{2}\right)$

53. Block will slide before it tips over if
    $\mu_s < \tan\alpha = 0.5$

57. $\mu = \dfrac{\sin 2\theta}{3 + \cos 2\theta}$

59. 836 N

61. 28°

63. (a) $0.438\dfrac{GmM}{R_E{}^2}$, 192° from North
    (b) $0.0356\dfrac{GmM}{R_E}$

65. 4170 N, tension

Chapter 13

17. 2.27 ms

19. (a) $x(t) = A\cos(\omega t + \phi)$
    $= (10 \text{ cm})\cos\left[(10\,\pi\text{s}^{-1})t\right]$
    (b) $x(t) = (2.5 \text{ cm})\sin\left[(5 \text{ s}^{-1})t\right]$

21. (a) 0.842 Hz   (b) 1.19 s
    (c) 1/32 m/s   (d) 1.40 N

23. 0.589 Hz, 1.70 s

25. (a) ≈ 4.29 s⁻¹   (b) 0.918 N/m
    (c) 0.817 m

27. (a) 2.21 s⁻¹   (b) 2.84 s   (c) 0.632 m

29. 1.21 s

31. 1.64 s

33. Seven oscillations in the x direction are
    completed for four in the y direction.

35. 20.0 cm

37. 0.248 s

39. 64.8 km/h

41. 0.694 s

45. (a) $T = 2\pi\sqrt{\dfrac{L}{|\vec{g} - \vec{a}_0|}}$

    (b) $T = 2\pi\sqrt{\dfrac{L}{g + \frac{1}{2}g}} = 2\pi\sqrt{\dfrac{2L}{3g}}$

    (c) $T = 2\pi\sqrt{\dfrac{L}{g - \frac{1}{2}g}} = 2\pi\sqrt{\dfrac{2L}{g}}$

    (d) $T = \infty$

47. 0.342 s

49. $R = \sqrt{\dfrac{2\kappa}{k}}$

51. 5.04 g

53. 0.147%

61. $T = 2\pi\sqrt{\dfrac{m}{k}} = 2\pi\sqrt{\dfrac{m}{2mga}} = \dfrac{2\pi}{\sqrt{2ga}}$

65. 1.53 s

67. (a) $E_2 = \dfrac{1}{4}E_1$; (b) $a_{2max} = \dfrac{1}{4}a_{1max}$

69. 27.16°

71. $\dfrac{d^2x}{dt^2} = -\dfrac{10}{7}gax$, therefore $T = 2\pi\sqrt{\dfrac{7}{10ga}}$

73. 0.540 Hz, −0.68 m/s, −0.109 rad

79. 2.1 m/s²

Chapter 14

17. (a) 0.192s (b) 6.54 cm
19. (a) 300 m (b) 1.58 m (c) 3 cm
    (d) 7.5 μm (e) 500 nm (f) 3 Å (See Appendix C on units.)
21. (a) $3.02 \times 10^7\ \text{s}^{-1}$ (b) $2.03 \times 10^4\ \text{m}^{-1}$
    (c) 1.49×10^3 m/s
23. (a) 1.3 cm (b) 9.11 cm (c) 0.203 s^{-1}
    (d) 44.9 cm/s (e) In the negative x direction
25. 250 m/s
27. (a) 7.6 N
29. 9.9 W
31. 343 m/s
33. 421 m/s
35. 942 Hz; in normal air, the frequency would be 686 Hz
37. 5.4 m
39. (a) 280 Hz (b) 70 Hz (c) 210 Hz
41. 567 Hz
43. 93.3 Hz
45. The galaxy is receding with speed $\approx 2.38 \times 10^4$ km/s
47. 30.4 m/s
49. $4\pi^2 A^2 F/\lambda$
51. ≈ 100 W
55. $v = \sqrt{kL(L - L_0)/m}$
57. 10 m
59. $L_0 = \dfrac{5}{7}L_1$
61. 190 m/s
65. 6.3 m
69. 25 m/s
71. 7.25 km
73. 41 m/s
75. 15.5 kHz
77. 60.033 Hz
79. 31.7 dB

Chapter 15

17. 10^{-14}
19. 1 torr = 133 Pa; 1 inch of mercury = 3.39 kPa
21. 101.3 kN
23. 21 N
25. 4.96 km
27. 890 Pa gauge
29. 93 cm higher in the eye
31. 2.53×10^3 kg/m³, too small for the specific gravity of the diamond
33. (a) no (b) yes
37. (a) 1.8×10^4 m³/s (b) 1.5 m/s

39. 1.75 m/s
41. 831 cm²
43. (a) 624 Pa gauge (b) 1.25 kPa
45. 3.6 mm
47. 8.11×10^3 kg
49. 59 g
51. 27 m
53. (a) 49 kg (b) 2500 kg
55. 14.4 kPa gauge
57. 14.3 m
59. (a) 1.5 m/s (b) 0.473 L/s
61. 70%
63. (a) $(1.63\%)p_{atm}$ (b) 16.8 cm
65. 14.6 kg
67. (a) 1.18 MW (b) 3.98 MW (c) 981
69. $t = \dfrac{A_0}{A_1}\sqrt{\dfrac{2h}{g}}$
71. (a) $\rho(h) = \dfrac{\rho_0}{h_0 g}e^{-h/h_0}$ (b) 5.7 km
75. 2.79 TN · m
77. yes, diluted 1:1 with water
79. $p = p_0 - \rho gh$ use $\rho g = 62.2$ lb/ft³, $h = 33$ ft, $p_0 = 18$ lb/in² and convert ft to in, $p = 3.8$ lb/in²

Chapter 16

15. 20° C
17. −40
19. 102.4°F
21. 32 kJ
23. 96.9 W
25. (a) 168 J/K (b) 480 J/kg·K
27. 0.293 W
29. 55 kW
31. −4.17 kW
33. $R_{air} = 5.56$ per in; $R_{conc} = 0.143$ per in; $R_{fg} = 3.45$ per in; $R_{glass} = 0.167$ to 0.2 per in; $R_{sf} = 5$ per in; $R_{pine} = 1.28$ per in
35. 2240 W
37. 22 mm²
39. (a) 138 kPa (b) 33.4 kPa (c) 233 kPa
41. −10.0°C
43. 0.36 kg
45. (a) 23 kJ (b) 337 kJ (c) 65 kJ
47. 2.51 min
49. (a) 560 g (b) 0.27 K/s
51. 1.8 kg
53. 9.22 K
55. 197 g
57. −14.2 W
59. 6.47 gal
61. 23°C
63. (a) 1.15×10^3 K (b) 700 K
65. 24°C
67. 1151 K
69. (a) $204 (b) $23.76
71. 40.7 K
73. 7.5 K
75. 1.5 h
77. 2.9 J

Chapter 17

19. 3.17×10^{23}
21. (a) $1.27V_2$ (b) $2V_2$

23. 2.7×10^7
25. H_2
27. Lead
29. 143 kJ
31. 120 mol/m³
33. 0.985 L
35. 263°C.
37. 2.69×10^{25} m^{-3}, about 2.5×10^{10} times greater
39. (a) 235 mol (b) 5.65 m³
41. (a) 1.27 atm (b) 0.0268 mol
    (c) 0.786 atm
43. 1.25×10^{10} kg
45. (a) 3.26×10^{19} J to melt at 0°C
    (b) 1.80×10^{13} W
47. 11.8 d
49. 3.55 MJ
51. 1.51 h
53. 5.0 kg
55. 0.798 kg
57. 79 g
59. 1.2 kg ice, 0.80 kg water, all at 0°C
61. ethyl alcohol
63. 59.2 L
65. 8.35 h
69. 1.00034 L
71. 36.5 d, fast
73. Yes. On Venus, one mole would occupy 0.67 L, so a 1 L container is big enough.
75. No; it would take 50 minutes.

Chapter 18

15. 29 kJ
17. Increase by 250 J
19. 13.9 kW
21. $2\rho_1 V_1$
23. (a) 1.33 (b) 215 J
25. $V = 0.18V_0$
27. 57.7%
29. (a) 200 K (b) 120 K
31. (a) 3.3 kJ (b) 0.392 mol
33. (a) 1.49 mm (b) 10.7 μJ
35. 1.35
37. (a) 300 kPa (b) −243 J
39. 440°C
41. (a) 810 K (b) 25.8 atm
43. (a) 300 K, 1.5 kJ (b) 336 K 0 J
    (c) 326 K, 429 J
45. (a) 39.9 kPa (b) 83.3 kPa (c) 80.2 kJ
47. −928 J
49. (a) 211 J (b) 12.9 L
51. 91.6 J
53. 335 K
55. (a) 598 J (b) 2500 J flows in
57. 79%
59. 25 m
61. (a) 218 J (b) 2.71 kJ
65. (a) 15.2°C (b) 15.5 kJ

67. $e = 1 - \dfrac{r^{1-\gamma}(r_c^{\gamma} - 1)}{\gamma(r_c - 1)}$

69. $W = nRT\ \ln\left[\dfrac{V_2 - nb}{V_1 - nb}\right] + an^2\left[\dfrac{1}{V_2} - \dfrac{1}{V_1}\right]$

71. -9.8 K/km

73. 1.4 atm, $V = 8.67$ L

75. 130 J

Chapter 19

15. 99.95%

17. (a) 39%   (b) 550 J   (c) 190°C

19. 53.3 kJ

21. 275 J/K

23. 21.9 kg

25. (a) 420 J   (b) 47.2%   (c) 290 K
    (d) 9.24 kW

27. (a) 1.75GW   (b) 43%   (c) 505 K

29. 2×10^7 kg/s, slightly more than the
    Mississippi's flow

31. (a) 8.53   (b) 1.10×10^3 kg

33. (a) 69%   (b) 967 K

35. (a) 4.3   (b) maximum COP = 11

37. 2.83

39. (a) 4.72   (b) 4.24 kW   (c) heat pump
    54¢/h, oil $1.30/h,

41. (a) 17.4%   (b) 83.3%.

43. 1.35×10^8 J/K

47. 11.7 J/k · s

49. 74.9 kJ

51. 163 J/K

53. (a) 35%   (b) 5.88 h

55. (a) $e_{otto} = 1 - r^{1-\gamma}$, $e_{carnot} = 1 - (1/3)r^{1-\gamma}$

57. 61%

59. (a) $Th = T_{h0}e^{-\rho_0 t/}[mc(T_{h0} - T_c)]$

    (b) $b = 0$ at $t = \dfrac{mc(T_{h0} - T_c)}{P_0} \ln\left(\dfrac{T_{h0}}{T_c}\right)$

61. -0.150 J/k; cooling produces a more
    ordered state

63. (a) -368 kJ/K   (b) 490 kJ/K
    (c) -65.3 kJ/K

65. No, 5.9 h

Chapter 20

15. 1.56×10^{20}

17. 2.3×10^{39}

19. 5.08 m

21. (a) $\hat{j}$ (b) $-\hat{i}$
    (c) $(\hat{i} + 3\hat{j})/\sqrt{10} \simeq 0.32\hat{i} + 0.95\hat{j}$

23. 3.81×10^9 N/C

25. (a) 2.2 MN/C   (b) 77 N

27. 5.15×10^{11} N/C

29. 39 pm

31. -6.50 μC/m

33. 3.3×10^{-12} kg

35. (a) 1.35 cm   (b) reverse direction,
    accelerates and exits field region at
    3.8×10^5 m/s

37. 1.04×10^{-11}

39. $-0.18\hat{i} + 0.64\hat{j}$ nN

41. $x = 5.45a$

43. $1.6\hat{i} - 0.33\hat{j}$ N

45. (a) Magnitude $4kqQ/a^2$; (b) directed
    toward the negative charge for positive Q,
    and vice versa

47. 4.14 cm

49. (a) $\vec{E} = \dfrac{2kqy}{(a^2 + y^2)^{3/2}}\hat{j}$ (b) $y = \pm a/\sqrt{2}$

53. (a) $\vec{E} = 4kq\hat{j}[4y(a^2 + 4y^2)^{-3/2}$
    $+ (2y - \sqrt{3a})^{-2}]$ for $y > \sqrt{3a}/2$

55. 6.94×10^{-2} m³

57. 2.8 Mm/s

59. -14 μC/m

61. 1.29×10^{-30} C·m

63. (a) $\vec{\tau} = -(2kQqa/x^2)\hat{k}$

    (b, c) $\vec{F} = (2kQqa/x^3)\hat{j}$, i.e., parallel to
    the dipole moment

65. 7.0 cm, 0.54 μC

67. (a) 2.5 μC/m   (b) 3.0×10^5 N/C
    (c) 1.8 N/C

73. $y = \dfrac{a}{\sqrt{2}}$

75. $-\hat{i}k\lambda_0(1.5 - 2\ln 2)/L$

77. 900 MN force, not negligible

79. 19.6 kN/C, upward

Chapter 21

21. $Q_C = -Q_B = 2Q_A$

23. 650 kN/C

25. ± 1.48 kN·m²/C

27. (a) $-q/\epsilon_0$ (b) $-2q/\epsilon_0$ (c) 0   (d) 0

29. 4.9×10^4 N·m²/C

31. (a) 1.21 MN/C   (b) 2.02 MN/C
    (c) 504 kN/C

33. Line symmetry

35. 49 kN/C

37. (a) 5.14×10^6 N/C   (b) 34.0 N/C

39. (a) 2.0 MN/C   (b) 7.2 kN/C

41. (a) $\rho = 0$ (b) $\sigma = 4.0$ mC/m²
    (c) Other charges would destroy the
    symmetry, making σ nonuniform.

43. 1.8 MN/C

45. $\pm E_0 \displaystyle\int_0^a y\, dy = \pm\dfrac{1}{2}E_0 a^2$

47. (a) 0   (b) 3.53 kN/C   (c) 1.42 μC

49. (a) $8kQ/R^2$, inward   (b) $kQ/4R^2$, inward
    (c) a would not change, b would
    become 0

51. (a) 3.6 MN/C   (b) 3.8 MN/C   (c) a
    would not change, b would nearly double

53. (a) 20.2 kN/C   (b) 1.68 kN/C

55. 6.3 μC/m³

57. (a) $E = \rho x/\epsilon_0$ (b) $E = \rho d/2\epsilon_0$

59. 18 N/C

61. (b) $-Q$

65. (a) $\pi\rho^0 a^3$ (b) $E(r) = \rho^0 r^2/4\epsilon_0 a$

67. $5\rho^0/3R^2$

69. $\vec{E} = \dfrac{\rho_0 R}{\epsilon_0}(e - 2)v/r$

71. $\vec{F} = m\vec{g} = -GmM_E\vec{r}/R_E^3$

Chapter 22

17. 1.92×10^{-17} J

21. 27.9 J

23. $-E_0 y$

25. 53 nC

27. (a) 442 kV   (b) 9.2 Mm/s

31. (a) 4 V   (b) $E_x = 1$ V/m, $E_y = -12$ V/m,
    $E_z = 3$ V/m

33. 3 kV

35. 5.6 kV/m

37. 80 mV

39. 9.37×10^7 m/s

43. 6.1 μC

45. $v = \sqrt{2keQ/mR}$

47. $V(R) - V(0) = -kQ/2R$

49. 21.0 V

51. $3\sqrt{3}kq/a$

53. $kq\{[(x-a)^2 + y^2]^{-1/2}$
    $+ [(x+a)^2 + y^2]^{-1/2}\}$

55. (a) 905 V   (b) 452 V

57. 16.2 cm, 809 nC

59. (a) $V(x, y) = -E_0(x + y)$
    $= -150(x + y)$ V/m
    (b) 150 V

63. $E = V_0/R$, radially otoward

65. (a) 43 kV   (b) 1.7 MV/m   (c) 540 V
    (d) 0

67. (a) -2.17 kV   (b) no change

69. (a) $x = -3$ m, 0 m, 1 m
    (b) $\vec{E} = (3x^2 + 4x - 3)\hat{i}$
    (c) $x = -1.87$ m, 0.535 m

73. 4.07×10^7 m/s, 6.96×10^5 m/s

75. $\dfrac{kq}{2a}\ln\left(\dfrac{\sqrt{2}+1}{\sqrt{2}-1}\right) \simeq \dfrac{0.881kq}{a}$

77. $V(x) = -\dfrac{k\lambda_0}{L^2}x\left[x\ln\left\{\dfrac{x-(L/2)}{x+(L/2)}\right\} + L\right]$

79. Yes

Chapter 23

15. $U = \dfrac{kq^2(2\sqrt{2}+1)}{2a\sqrt{2}}$

17. $v = \sqrt{\dfrac{2kq^2}{ma}}$

19. (a) 2.0 MV/m   (b) 9.9 kV   (c) 5.5 mJ

21. (a) 0.74 μC   (b) 40 kV

23. ± 14 C

27. 1.91 m²

29. 70 nF

31. Equal

33. (a) 6.0 μF,   (b) 0.55 μF,   (c) 0.83 μF,
    1.3 μF, 1.5 μF, 2.2 μF, 2.8 μF, 3.7 μF

35. 1 km³

37. 9×10^{30} J/m³

39. $\dfrac{1}{2}kQ^2\left(\dfrac{1}{a} - \dfrac{1}{b}\right)$

41. $C = \dfrac{2\pi\epsilon_0 L}{\ln(b/a)}$

45. 1 μF stores 15 times as much energy

47. (a) 30 μF   (b) 0.1 μF   (c) 0.01 μF

49. (a) 5.0 kW   (b) 250 μF   (c) 0.50 W

51. (a) Increases by factor of 2.5   (b) drops
    to 40% of its original value

53. (a) 2 series pairs in parallel or two
    parallel pairs in series   (b) 4 in series

55. 102 μJ

57. 75 V

59. (a) 3.5 mm   (b) 87 kV
61. $\pm 1\%$
63. $kQ^2/10R$
65. 2.22 GeV
67. 7.75 J
71. 1.41 fm
73. (b) $U = C_0 V_0^2(\kappa + 1)/4$

   (c) $F_x = \dfrac{1}{2}\dfrac{C_0 V_0^2(\kappa - 1)}{L}$

75. $U/L = \pi \rho^2 R^4/8\epsilon_0$

77. (a) $U = \dfrac{Q^2 x}{2A\epsilon_0}$ (b) $F = \dfrac{Q^2}{2A\epsilon_0}$ The force is the charge on one plate times the electric field due to the other plate. It is not the charge on one plate times the field due to both plates!

79. $u = \dfrac{\lambda^2}{8\pi^2\epsilon_0 r^2}$, $U = \dfrac{Q^2 \ln(b/a)}{4\pi\epsilon_0 L}$

81. 266 pF, within specs

Chapter 24
15. 2.9×10^5 C
17. 7.65 MA/m^2
19. 0.17 V/m
21. 2.2×10^{-6} $\Omega \cdot$m
23. 25 Ω
25. 2.34 mA
27. 25 mA
29. 1.38 kW
31. 160 μA
33. 240 Ω
35. 48 kΩ
37. (a) 0.48 pA   (b) 1.5×10^4
39. 0.31 mA
41. 0.723 mm/s, 3.26 mm/s, 3.62×10^7 m/s
43. (a) 4.29×10^6 A/m^2 (b) 3.37 A
45. $v_{Fe}/v_{Ag} = 6.1$
47. $d_{Al} = 1.26 d_{Cu}$
49. (a) 2.07 cm   (b) 2.60 cm   (c) aluminum
51. 34 mΩ
53. Resistor with more power has $\sqrt{2}$ time greater diameter
55. 0.625 A
57. 2.9 A
59. (a) 1.11 kW   (b) 13.0 Ω
61. 0.171 mm/s
65. 5 Ω
67. 5.59×10^{-5} $\Omega \cdot$m
69. 50 min

Chapter 25
17. 9 V
19. 216 kJ
21. 229 kΩ
23. 0.02 Ω
25. (a) $3 + 2/3 = 11/3$, $2 + 3/4 = 11/4$, $1 + 6/5 = 11/5$ Ω
   (b) $3 \times 3/(3+3) = 3/2$, $2 \times 4(2+4) = 4/3$, $1 \times 5/(1+5) = 5/6$ Ω
   (c) $1 + 2 + 3 = 6$ Ω
   (d) $(1^{-1} + 2^{-1} + 3^{-1})^{-1} = 6/11$ Ω
27. 2.07 A, 0.214 A, 1.86 A

29. 0
31. 0.662% lower
37. $\mathcal{E}R_2/(R_1 + R_2)$
39. 1.5 A
41. 30 A
43. 14.4 W
45. (a) 162 Ω (b) 125 mW
47. (a) $\dfrac{R_1\mathcal{E}}{R + 2R_1}$ (b) $\dfrac{1}{2}\mathcal{E}$
49. 2.45 W
51. $\dfrac{7}{5}R$
53. -4.77 mA, downward
55. 48.4 V, 57.3 V, 59.9 V
57. (a) 0.992 A   (b) 0.83%
59. (a) 0.35 s   (b) 0.17 s
61. 3.4 μF
63. (a) $I_1 = 25$ mA, $I_2 = 0$, $V_C = 0$
   (b) $I_1 = I_2 = 10$ mA, $V_C = 60$ V
   (c) $I_1 = 0$, $I_2 = 10$ mA, $V_C = 60$ V
   (d) $I_1 = I_2 = 0$, $V_C = 0$
65. (a) 4.51 V   (b) 35.2 Ω
67. 83 μs
71. 1.84 V, 20.8 Ω
75. 8.5 nA, downward
77. (a) as capacitance in parallel with a resistance   (b) 595 s
79. $I_2(t) = \dfrac{\mathcal{E}}{4R}\left[2 - e^{-\frac{2}{RC}(1 + \sqrt{2})t} - e^{-\frac{2}{RC}(1 - \sqrt{2})t}\right]$
81. 3.3 kΩ in parallel with 4.7 kΩ, this combination in series with 1.5 kΩ

Chapter 26
17. (a) 1.6 mT   (b) 2.3 mT
19. (a) 2.0×10^{-14} N   (b) 1.0×10^{-14} N   (c) 0
21. 400 km/s
23. 358 ns
25. $r_{proton} = 43 r_{electron}$
27. 1.3 μs
29. (a) 48.9 mT   (b) 0.734 N/m
31. 0.12 T
33. (a) 40.8 μT   (b) 5.09 nT
35. 4.02 A
37. 0.62 G
39. 482 mT
41. 24 A
45. (a) $-1.1\hat{i} + 1.5\hat{j} + 1.7\hat{k}$ mN
47. 40.1° or 140°
51. (a) 70.6 μm   (b) 444 μm
53. (a) 5.88 A   (b) upward from A to B
55. 377 mT
57. 1.97×10^{-25} J $= 1.23$ μe V
59. 732
61. 23° west of magnetic north
63. 18.2 mN
65. (a) $\mu_0 J_0 r^2/3R$ (b) $\mu_0 J_0 R^2/3r$
67. (a) $-\mu_0 J_s\hat{j}$ (b) 0
69. (a) 2.27×10^3 m^{-1} (b) 3.31 kW
73. (a) 8.0 μT   (b) 4.0 μT   (c) 0
75. (a) $B\pi R^2\mu_0 I/2w$ (b) $B \approx \mu_0 I/2\pi r$

77. (a) $\dfrac{1}{3J_0}$ (b) $B = \mu_0 J_0 R^2/6r$
   (c) $B = \dfrac{1}{2}\mu_0 J_0 r(1 - 2r/3R)$
79. (a) $\mu = IL^2/4\pi N$ (b) $N = 1$
81. 3.8 mm
83. (a) $b = a$ (b) $B_0 = 8 \mu_0 I/5^{3/2}a$
85. 1.25×10^{-5} N
87. $\mu_0 I \sqrt{2}/\pi a$
89. $F = 3 \mu_0\mu I/8\sqrt{2}a^2$, attractive
91. 3.8×10^9 A

Chapter 27
17. (a) 6.28×10^{-4} Wb   (b) 6.91×10^{-3} Wb
   (c) 2.51mA   (d) counterclockwise
19. 199
21. 0.4 H
23. 1350
25. 130 Ω
27. 100 mA
29. 0.510 μJ
31. 9.9×10^8 J/m^3
33. 1.08 T/ms
35. $B(t) = (bR/A)t^3/3$
37. $B^2\ell^3 v/R$
39. $16 B_0 x_0^3/3$
41. (a) 3 s   (b) clockwise
43. (a) -60 mA   (b) -32.4 mA
45. 20.6 V
47. (a) CCW   (b) $(Blv)^2/R$
49. (c) $v_{terminal} = \mathcal{E}/Bl$
51. $-(8.40 \times 10^{-18}$ N$)\hat{j} + (6.45 \times 10^{-16}$ N$)\hat{i}$
55. 11.0 s
57. (a) 0.11 H   (b) 20 mA
59. (a) 40 A/s   (b) 9.23 A/s
61. (a) 2.18 A   (b) 1.68 A
63. 1.77 ms
65. (a) 5.7 MJ   (b) 31 mΩ (c) 39.3s
67. Smaller by factor $1/4n^2 R^2$
69. (a) $P = I_0^2 R e^{-2Rt/L}$
71. (b) $v_t = mgR/w^2 B^2$
   (c) counterclockwise
75. $L = \mu_0 N^2(A - (A^2 - R^2)^{1/2})$
77. $I_2 = (\mathcal{E}_0/(R_1 + R_2)) \exp(-R_p t/L)$, where $R_p = R_1 R_2/(R_1 + R_2)$
79. No, it stores 80 J
81. $\mathcal{E} = \dfrac{1}{2}B_x l_\omega^2 \approx 19$ mV; not practical

Chapter 28
15. 8.91 V
17. (a) 350 mA   (b) 1.50 kHz
19. 45 mA
21. (a) 804 Ω (b) 48.2 Ω (c) 2.41 Ω
23. 1.47 μF
25. 8.23 kHz
27. 4.9 pF to 42 pF
29. 67 nH
31. 3.5 kΩ
33. 5.0 mA, 10 mA, 20 mA
35. 0.39 A
37. (a) 522   (b) 14 A
39. (a) 79.6 Hz   (b) 2 H   (c) $X_L = 4X_C$

41. (a) 10 mA   (b) 14 V   (c) 7.7 mA
43. 0.32 A
45. 0.199 μH
47. (a) $1/\sqrt{2}$ (b) 1/2   (c) $-1/\sqrt{2}$ (d) 1/2
49. 50.4
51. 6.20 Ω
53. (a) above   (b) current lags by approximately 49
55. (a) 0.333   (b) 4 W
57. (a) 5.5%   (b) 9.1%   (c) large
59. 3.73 mF
61. 2.73 V
63. 1.63 kHz
65. 36.2 nF, 11.5 nF
69. $R = 400\ \Omega$, $L = 67$ mH, $C = 0.094\ \mu$F
73. $V_p/\sqrt{3}$
75. 101 V, -190 V, 134 V
79. Yes, $I_{peak} = 1.45$ A

Chapter 29

13. 1.3 nA
15. $-\hat{k}$
19. 7.5 km
21. 2.57 s
23. 5000 km
25. x direction
27. 11.7%
29. 12 GW/m^2
31. 2.7×10^{-10} W/m^2
33. 3.18 $\mu W/m^2$
35. (a) 7.20×10^{11} V/m·s   (b) increasing
37. 15 kV/m
39. 1.1 pT
41. 90.5%
43. 18.8%
47. 1.1%
49. 4×10^{10}
51. $E_p = 40$ V/m, $B_p = 0.13\ \mu$T
53. (a) 2.5 J   (b) 8.33×10^{-9} kg·m/s
55. 6.0 mPa
57. 110 km^2
59. 1.05×10^{16} W
61. 211 mPa
63. Long, with cylindrical symmetry
67. $n = 6$
69. 2.75 m
71. about 10^{16} y

Chapter 30

13. (a) three   (b) along the reverse of the entry path
15. 240° CCW (or 120° CW)
17. 126 nm
19. 1.57
21. 67.5°
23. (a) 49.8°   (b) 42.2°   (c) 22.4°
25. (a) 61.3°   (b) 80.9°   (c) there is none
27. 1.07°
29. (a) 2   (b) 210°
33. (a) 40 cm   (b) 32.5 cm   (c) 25.0
35. 9.48°
37. 1.30 m
39. 42.2°
41. $\sqrt{2}$

43. 2.62×10^8 m/s
47. 36.5° to 38.1°
49. $v = c \sin \theta_c$
51. 1.96×10^8 m/s
53. 2.72 m
57. $n = 1.17$
61. 50.9°
67. 375 nm, yes
69. $\theta_c = 37°$, no

Chapter 31

17. Mirror surface 55° to horizontal
19. (a) Image height is $\frac{1}{4}$ object height
    (b) inverted
21. (a) At the center of curvature of the mirror (twice its focal length)   (b) real
23. $M = -2$ (i.e., a real, inverted image)
25. 21.1 cm
27. 27 cm
29. 40 cm
31. -0.857 mm
33. $(25\text{ cm})^{-1} + (-55\text{ cm})^{-1} = 2.18$ diopters
35. -1.25 diopters, the minus sign signifying a diverging lens
37. -200
39. (a) 24.3 cm behind the mirror   (b) 29 mm
    (c) virtual
41. 18 cm in front of the mirror
43. 7.42 cm
45. (a) 2.0 m   (b) 3.0 m in front of the mirror
47. 12 cm
49. (a) Real, inverted, 7.7 cm high
    (b) virtual, upright, 7.7 cm high
51. 29 cm or 41 cm from the candle
53. 11.1 cm
55. Real image, 109 cm on other side of lens
57. $L' = -67.9$ cm
59. 2
61. $n = 2$
63. 1.72
65. Real image, magnified 2.74 times
67. 3.33 diopters, converging lens
69. $20.8' \approx \dfrac{1}{3}°$
71. 72 cm
83. 0.257 mm
85. 45 mm

Chapter 32

11. 1.65 cm
13. 420 nm
15. 4
17. (a) 4.8°, 9.7°   (b) 2.9°, 6.8°
19. (a) 2nd (b) 1st
21. 103 nm
23. 424 nm, 594 nm
25. the top 4.5 cm
27. 29.3°
29. 0.0162
31. 36 cm
33. 1.3 m
35. 94.1°
37. (a) $m_{max} = 38$ (b) $m_{max} = 3$

39. The fourth bright 550-nm fringe ($m = 4$) coincides with the sixth dark 400-nm fringe ($m = 5$).
41. 5.33°
43. (a) 500 nm, 560 nm   (b) same as (a)
    (c) $m = 2$, no overlap
45. 0.01 nm
47. 77.2 nm
49. Echelle grating has 60% greater resolving power
51. 3.28 Å
53. 200 nm
55. 2.16 cm
57. Next integer greater than
    $nN + (1/2)(n - 1)$
59. 545 nm
61. 83.8 cm
63. (a, b) objects separated by 22.8 cm can be resolved; no to newspaper headlines, yes to billboards   (c) objects separated by 4.6 cm (1.8 in) can be resolved; yes to banner headlines and billboards
65. 6.9 km
67. 484 nm
69. $n - 1 = m\lambda_{vac}/2L$
73. $y_1 = 17.7$ cm, $y_2 = 44.7$ cm
77. No, $\Delta\lambda = 6.33$ pm.
79. 210 nm

Chapter 33

13. (a) 4.50 h   (b) 4.56 h   (c) 4.62 h
15. 33.1 ly
17. 40.4 m
19. $0.14c$
21. (a) 2   (b) 2.53
23. 0.141
25. (a) 2.1 MeV   (b) 1.6 MeV
29. (a) $0.857c$ (b) 9.69 min
31. $c/\sqrt{2}$
33. Twin A: 83 years; twin B: 40 years
35. $0.96c$
39. B before A by 3.47×10^5 y
41. In Problem 39, the events are simultaneous in a frame moving from A to B at $v/c = 0.5$; in Problem 40, there is no frame in which they are simultaneous.
45. 0.936
47. (a)   12.5 y   (b) 7.50 y
51. (a) 4.2 ly   (b) B − 2.4 y (i.e., B occurs earlier)
53. (a) 0.758   (b) 5.82×10^{-19} kg·m/s
55. 25 h
57. (a) 0.256 eV   (b) 128 keV   (c) 3.11 MeV
63. $0.866c$
65. $0.95c$

Chapter 34

15. 16
17. $\lambda_{peak} = 10\ \mu$m; $\lambda_{medium} = 14.3\ \mu$m
19. (a) 500 nm, yello-green   (b) 710 nm, deep red near the limit of the visible spectrum
21. 191 nm; ultraviolet
23. $P_{blue} = 1.44 P_{red}$

25. $\lambda = 122$ nm, 103 nm, 97.2nm
27. 91.2 nm; ultraviolet
29. (a) 3.7×10^{-63} m   (b) 73 nm
31. $v_p = 5.46 \times 10^{-4} v_e$
33. 6.3×10^7 m/s
35. 126 nm
37. 23 keV
39. $R_{200\,nm} = 0.056 R_{500\,nm}$
41. (a) 4.39×10^3 K   (b) 0.474
43. (a) 1.7×10^{28} s^{-1} (b) 3.2×10^{15} s^{-1}
    (c) 1.3×10^{18} s^{-1}
45. (a) 1.12×10^{15} Hz   (b) 2.80 eV
47. Copper, nickel, silicon
49. 443 nm
51. (a) 154 pm   (b) 222 eV
53. No
55. (a) 26.4 cm   (b) 4.70 μeV
57. $n_1 = 229$
61. (a) 27 pm   (b) 40.8 eV
63. 1.62 km/s
65. 2.90 km/s
67. 1.16 ps
71. $\frac{1}{2} m_e c^2 \left[(\gamma - 1) + \sqrt{(\gamma - 1)(\gamma + 3)} \right]$
79. 5.49×10^{-2} eV

Chapter 35

11. (a) At $x = 0$ (b) At $x = \pm 0.589 a$
13. 0.0625
15. $n = 5$
17. 6.03×10^{-22} J
19. $h = 312$ J·s, 4.71×10^{-35} times the actual
    value of Planck's constant
21. 6.79×10^{26}
23. 0.23 MeV
25. 2.66×10^{14} Hz
27. 3×10^{-32} J
29. 9.87×10^{-11} J $= 616$ MeV
31. $0.968 b^{-2.5}$
35. (a) 3.48×10^{-19} J $= 2.18$ eV
    (b) 571 nm
37. 21.3 μm
39. 2.20 nm, 2.75 nm, 4.71 nm, 4.12 nm,
    6.59 nm, 11.0 nm
41. (a) $\Psi_{n\,odd}(x') = \sqrt{\dfrac{2}{L}} \cos\left(\dfrac{n\pi x'}{L} \right)$;

$\Psi_{n\,even}(x') = \sqrt{\dfrac{2}{L}} \sin\left(\dfrac{n\pi x'}{L} \right)$

(b) $E_n = n^2 h^2 / 8mL^2$
43. 0.759 nm
45. 8.36×10^{31}
47. $n > 99$
49. (a) 0.48   (b) 0.091   (c) 0.20   (d) 0.26
    (e) 0.25
53. $n = 3$
55. (a) 4   (b) 2
59. (a) no bound states   (b) one bound state,
    at $\mathcal{E} = 6.44$ (c) two bound states, at
    $\mathcal{E} = 7.53$, $\mathcal{E} = 29.3$
63. $B/A = (k_1 - k_2)/(k_1 + k_2)$,
    $C/A = 2k_1/(k_1 + k_2)$

65. 2.38 nm, yes
67. 2×10^{13} K

Chapter 36

15. $\sqrt{13.6/1.5} = 3$
17. d.
19. $\ell = 5$
21. $3d$
23. $\sqrt{6} \hbar$
25. $\dfrac{3}{2}, \dfrac{5}{2}$
27. $\dfrac{21}{2} \hbar \omega$
29. $1s^2 2s^2 2p^6 3s^2 3p^6 3d^1 4s^2$
33. 1.14 meV
35. $n = 4$, $\ell = 3$
37. about 3×10^{68}
39. 90°, 65.9°, 114°, 35.3°, 145°
41. 0, ± 1, ± 2, ± 3
45. $\frac{1}{2}, \frac{1}{2}, \frac{3}{2}, \frac{3}{2}$, and $\frac{5}{2}$
47. (a) $E_4 = E_1/16$ (b) $L = \sqrt{12} \hbar$.
    (c) $J = \sqrt{35} \hbar / 2$ (d) The magnitude of L
    is greater than J.
49. (a) $16 \hbar \omega$ (b) $\hbar \omega / 2$
51. $1s^2 2s^2 2p^6 3s^2 3p^6 3d^{10} 4s^1$
53. 3.02×10^{17}
55. 0.11
57. $3.03 \hbar$
59. (a) $\ell \pm \dfrac{1}{2} = \dfrac{5}{2}$ or $\dfrac{7}{2}$ (b) $8m_j$-values
67. $r_{av} = 1.5 a_0$
71. (a) 18.7 eV   (b) 9.5%

Chapter 37

17. 3.48 mm
19. 9.40×10^{-46} kg·m^2
21. 7.07×10^{13} Hz
23. 181 kcal/mol
25. 548.7 nm
27. InAs, 3.54 μm
29. 1.25 meV
31. $\ell \hbar^2 / I$
33. 0.121 nm
35. (a) 0.179 eV   (b) 0.359 eV
37. (a) 4.72×10^{13} Hz
    (b) 1.95×10^{-46} kg·m^2
39. 35.8 μm
41. 10.2
43. -8.40 eV
47. 4.7 eV
49. 6.35×10^4 K, over 200 times room
    temperature
51. 709 nm, zinc selenide wouldn't make a
    good photovoltaic cell
53. 1.79 kA
55. (a) 0.368 nm   (b) 12.3
57. 35.4%
59. (a) $U_{min} = 0$ (b) $r_{min} = r_0$
61. $E_{av} = (3/5) E_F$
63. 2.5 meV
65. 2.78 eV. Yes, it works okay.

Chapter 38

13. $^{211}_{86}$Ra, $^{220}_{86}$Ra, $^{222}_{86}$Ra
15. (a) same   (b) $Q_K - Q_{Cl} + 2e$
17. 5.9 fm
19. $^{64}_{29}$Cu $\rightarrow ^{64}_{30}$Zn $+ \beta^- + \bar{\nu}$ (40%)
    $^{64}_{29}$Cu $\rightarrow ^{64}_{28}$Ni $+ \beta^+ + \nu$ (19%)
    $^{64}_{29}$Cu $+ e^- \rightarrow ^{64}_{28}$Ni $+ \nu$ (41%)
21. (a) $^{10}_{6}$C $\longrightarrow ^{10}_{5}$B $+ e^+ + \bar{\nu}_e$ (b) 5.56 MBq
23. (a) 193 y   (b) 289 y
25. 59.95 u
27. 5.61 MeV/nucleon
29. 2
31. 1.0×10^{20} fission/s
33. 2×10^{20} m^{-3}
35. 10 s
37. 5.3×10^{-12} eV
39. 8.80 meV/nucleaon
41. (a) 7.85×10^5, essentially 0   (b) 1.009,
    5.0×10^5 (c) 3.6, 8.2×10^4
43. (a) 0.834   (b) 0.281   (c) 0.432%
45. (a) $^{228}_{90}$Th
47. 8.92×10^3 y
49. 8.04 d for Poland, 16.2 d for Austria,
    10.0 d for Germany
51. 3.0×10^9 y
53. 23%
55. 1.68 mCi/L
57. 1.02 kg
59. 8.63 s
61. (a) 3.41 GW   (b) 35.2%
63. 454 kg
65. (a) 4.4×10^{10} J   (b) 340 gal
67. 4.3 MeV
69. $^{1}_{0}n + ^{198}_{80}$Hg $\longrightarrow ^{197}_{79}$Au $+ ^{A}_{Z}$X
71. (a) 1.1 GJ   (b) 2.8 pellets/s   (c) 450 kg
75. $\lambda_A N_A + \lambda_B N_B = \lambda_A N_0 [(\lambda_A - 2\lambda_B)$
    $\exp(-\lambda_A t) + \lambda_B \exp(-\lambda_B t)]/(\lambda_A - \lambda_B)$
77. 1568 Mev
79. 10 mg left, barely enough!

Chapter 39

17. 3×10^{-16} s
19. $\pi^+ \rightarrow \mu^+ + \nu_\mu$
21. $\eta \rightarrow \pi^+ \pi^- \pi^0$, conserves charge
    $(1 - 1 + 0 = 0)$
23. No, the conservation of baryon number
25. sss
27. 4.54×10^4 m^3, about 1.2×10^7 gallons
29. 10^{28} K
31. 1.33 Gly
33. 12 Gy
35. a, violates conservation of baryon
    number
37. (a) No   (b) yes
39. $c\bar{c}$
41. (a) 0.16 μJ   (b) 16.3 μm
43. 90 μs
45. 313 ly
47. (a) 256 fm   (b) -2.81 keV
49. (a) 5.74×10^3 km/s   (b) 383 Mly
51. 2.6×10^{-25} s
53. 18 Gy, 13.6 Gy
55. 5 km/s/Mly

Credits

Front Matter

Pages i and iii (Volume I): Tyler Boley, Workbook. Pages i and iii (Volume II): John Eisele, Colorado State University. Page vi (Volumes I and II): Tad Merrick Photography.

Chapter 1

Page 1: David Parker/Photo Researchers. Page 4, Figure 1.3: Digital Vision/Getty Images. Page 4: NASA. Page 5, Figure 1.4:(a) NASA (b) Andrew Syred /Photo Researchers.

Chapter 2

Page 13 T: Ap Wide World Photos. Page 13 B: Getty Images/ Aurora Creative. Page 14: Nik Wheeler/Corbis. Page 23, Figure 2.11: Richard Megna/Fundamental Photographs. Page 25: National Institute of Standards and Technology (NIST).

Chapter 3

Page 31: Martha Holmes/Nature Picture Library. Page 37, Figure 3.11: Richard Megna/Fundamental Photographs. Page 41: Preston Wilson/ Getty Images.

Chapter 4

Page 49: Superstock/AGEfotostock. Page 52, Figure 4.4: NASA. Page 54, Figure 4.7: Solstice Photography/Getty Images. Page 55, Figure 4.9: NASA. Page 58: Reuters/Corbis.

Chapter 5

Page 67: Peter/Georgina Bowater/Mira. Page 73, Figure 5.10: Nic Bothma/AP Wide World.

Chapter 6

Page 86: Paolo Cocco/AFP/Getty Images.

Chapter 7

Page 102: Kennan Ward/Corbis. Page 105: Courtesy Northfield Mountain Pumped Storage Project. Page 109 L, R: Erik Borg/Addison Wesley Longman.

Chapter 8

Page 118 L: istockphoto.com. Page 118 R: Courtesy Northrup Grumman. Page 119, Figure 8.2: Lowell Observatory. Page 123, Figure 8.6: David Hardy/Photo Researchers.

Chapter 9

Page 132: Patrick Reeves. Page 136, Figure 9.7: Wally MacNamee/ Corbis. Page 141: NASA Headquarters. Page 143, Figure 9.13: Richard Megna/Fundamental Photographs.

Chapter 10

Page 155: Wind Prospect Ltd. Page 166: Courtesy Mark Flynn/ University of Texas-Austin.

Chapter 11

Page 174: NASA/Goddard Space Flight Center. Page 181: California Institute of Technology Archives.

Chapter 12

Page 186: Robert Harding Picture Library Ltd./Alamy.

Chapter 13

Page 203: Damir Frkovic/Masterfile. Page 204: Courtesy Project Bandaloop. Page 209 L: Richard Chung. Page 209 R: Robert Holmes/Alamy. Page 216 L, Figure 13.25: University of Washington Libraries. Page 216 R, Figure 13.25: AP Wide World.

Chapter 14

Page 222: Pixtal/AGEfotostock. Page 231, Figure 14.21: Uri Haber-Schaim/Kendall/Hunt Publishing. Page 233, Figure 14.26: Nordic Photos/Getty Images. Page 233, Figure 14.27: Richard Megna/Fundamental Photographs. Page 235, Figure 14.30: Michael Freeman.

Chapter 15

Page 243: Corbis/Bettmann. Page 249: Courtesy Pagani Automobili s.p.a. Page 252, Figure 15.16: Courtesy Columbia University Physics Dept. Page 253: Lester Lefkowitz/Getty Images. Page 254, Figure 15.21: Anupam Pal.

Chapter 16

Page 261: Mark Antman/The Imageworks, Page 261 (inset): Steve Allen/Getty Images. Page 262: Ted Kinsman/Photo Researchers. Page 269, Figure 16.8: Hal Lott/The Stock Connection. Page 270, Figure 16.10: Manuel G. Velarde.

Chapter 17

Page 279: Digital Vision/Getty Images. Page 286, Figure 17.7: AP/Wide World. Page 287: Brian J. Skerry/National Geographic Image Collection.

Chapter 18

Page 292: Photodisc/Getty Images. Page 299, Figure 18.1: Jon Arnold/AGEfotostock.

Chapter 19

Page 309: James Hardy/AGEfotostock. Page 315: Peter Bowater/ Photo Researchers. Page 317: Courtesy Richard Wolfson.

Chapter 20

Page 327: NASA Earth Observing System. Page 328: Photodisc Red/Getty Images. Page 329, Figure 20.1: Richard Megna/Fundamental Photographs. Page 340: David Paul Morris/Getty Images. Page 340 (inset): Michael Davison/Photo Researchers

Chapter 21

Page 347: Peter Menzel. Page 361: Mark Antman/The Image Works.

Chapter 22

Page 367: AP/Wide World. Page 379: Courtesy of EPRI.

Chapter 23

Page 384: Yoav Levy/Phototake. Page 387, Figure 23.5: Loren Winters/Visuals Unlimited. Page 390: Cindy Charles/PhotoEdit.

Chapter 24
Page 398: Danial Sambraus/Photo Researchers. Page 410, Figure 24.16: Erik Borg/Addison Wesley Longman. Page 410, Figure 24.17: William Ferguson.

Chapter 25
Page 415: Tom Pantages. Page 419, Figure 25.7: Erik Borg/Addison Wesley Longman.

Chapter 26
Page 436: M. Aschwanden et al. (LMSAL), TRACE, NASA. Page 437, Figure 26.1: Stuart Field. Page 440: Ivan Massar. Page 446, Figure 26.22: NASA. Page 448: John Wilson White/AWL. Page 452, Figure 26.30: University Corporation for Atmospheric Research UCAR. Page 456, Figure 26.37: Richard Megna/Fundamental Photographs.

Chapter 27
Page 464: Charles E. Rotkin/Corbis. Page 479, Figure 27.28: NASA. Page 483, Figure 27.34: Milt and Joan Mann/Cameramann International, Ltd.

Chapter 28
Page 489: Lester Lefkowitz/Corbis. Page 497, Figure 28.13: Erik Borg/Addison Wesley. Page 501, Figure 28.21 TL: Norbert Schaefer/Corbis. Page 501, Figure 28.21 TM: Taxi/Getty Images. Page 501, Figure 28.21 TR: Jonathan Nourak/PhotoEdit. Page 501, Figure 28.21 ML: Jonathan Nourak/PhotoEdit. Page 501, Figure 28.21 MR: D2 Productions/Omni-Photo Communications. Page 501, Figure 28.21 BL: Jonathan Nourak/PhotoEdit. Page 501, Figure 28.21 BR: Michael Mathers/PeterArnold.

Chapter 29
Page 508 T: PhotoDisc/Getty Images. Page 508 BL: PhotoDisc Red/Getty Images. Page 508 BR: Michael Keller/Corbis. Page 518, Figure 29.8: Richard Megna/Fundamental Photographs. Page 523, 29.14: Illustration: Babakin Space Center, courtesy of The Planetary Society.

Chapter 30
Page 529: Craig Tuttle/Corbis. Page 530: Courtesy of Schott Corporation. Page 535, 30.10: Richard Megna/Fundamental Photographs. Page 536: Spencer Grant/Photo Researchers. Page 537, Figure 30.16: Courtesy of R. Giovanelli and H. R. Gillett/CSIRO National Measurement Laboratory, Australia. Page 539, Figure 30.18: Library of Congress.

Chapter 31
Page 542: Yoav Levy/Phototake. Page 543, Figure 31.2: Martin Bough/Fundamental Photographs. Page 544, Figure 31.4: Courtesy of the Space Telescope Institute/NASA. Page 545, Figure 31.7: Erik Borg/Addison Wesley Longman. Page 545, Figure 31.9: Stewart Weir/Corbis. Page 547, Figure 31.11: Perkin Elmer. Page 551, Figure 31.21: Erik Borg/Addison Wesley Longman. Page 555: Chris Barry/Phototake. Page 558, Figure 31.34: Giant Magellan Telescope, Carnegie Observatories.

Chapter 32
Page 563: Chris Collins/Corbis. Page 565: Andrew Syred/Photo Researchers. Page 565, Figure 32.3: M. Cagnet et al. Atlas of Optical Phenomena Springer-Cerlag, 1962. Page 568, Figure 32.8: Chris Jones. Page 571, Figure 32.13: Richard Megna/Fundamental Photographs. Page 571, Figure 32.14: Courtesy of Jay M. Pasachoff. Page 572, 32.16: Erik Borg/Addison Wesley. Page 573, Figure 32.17: NASA/Ames Research Center/Ligo Project. Page 576, Figure 32.23: Courtesy of Cagnet Atlas. Page 576, Figure 32.24: Courtesy of Cagnet Atlas. Page 577, 32.27: Chris Jones. Page 579: Courtesy of the Blu-Ray Disc Association. Page 580: Chris Jones.

Chapter 33
Page 585: Delft University of Technology/Photo Researchers, Inc. Page 586: AP/Wide World. Page 589, Figure 33.4: Einstein Archives, Hebrew University, Jerusalem. Page 604, Figure 33.20: Space Telescope Science Institute/Photo Researchers.

Chapter 34
Page 609: Courtesy Wolfgang Ketterle, MIT. Page 620: Andrew Syred/Photo Researchers. Page 621, Figure 34.13: Millie LeBlanc/Courtesy of The Educational Development Center.

Chapter 35
Page 629: IBM Corporation. Page 630, Figure 35.2: Courtesy of Elisha Huggins, Dartmouth. Page 639: Courtesy of IBM.

Chapter 36
Page 646: Chip Clark. Page 654, Figure 36.14: Courtesy Wolfgang Ketterle, MIT.

Chapter 37
Page 666: Alfred Pasieka/Photo Researchers. Page 679: Koichi Kamoshida/Liaison/Getty Images.

Chapter 38
Page 684: WDCN/Univ. College London/Photo Researchers. Page 688: David Job/Getty Images. Page 703, Figure 38.18: Lawrence Livermore National Laboratory/Photo Researchers.

Chapter 39
Page 709: Brookhaven National Laboratory. Page 717, Figure 39.7: Patrice Loiez, CERN/Photo Researchers. Page 718, Figure 39.8: ICRR Institute for Cosmic Ray Research. Page 719: CERN. Page 720, Figure 39.11: NASA Headquarters. Page 721, Figure 39.13: NASA Headquarters.

Index